BIOPHYSICAL CHEMISTRY OF BIOINTERFACES

BIOPHYSICAL CHEMISTRY OF BIOINTERFACES

Hiroyuki Ohshima

WILEY

A JOHN WILEY & SONS, INC., PUBLICATION

Published by John Wiley & Sons, Inc., Hoboken, New Jersey
Published simultaneously in Canada

Library of Congress Cataloging-in-Publication Data:

Ohshima, Hiroyuki, 1944-
Biophysical chemistry of biointerfaces / Hiroyuki Ohshima.
p. cm.
Includes bibliographical references and index.
ISBN 978-0-470-16935-3 (cloth)
1. Biological interfaces. 2. Physical biochemistry. 3. Surface chemistry.
I. Title.
QP517.S87O36 2010
612'.01583–dc22

2010013122

Printed in the United States of America

10 9 8 7 6 5 4 3 2 1

CONTENTS

Preface xiii

List of Symbols xv

PART I POTENTIAL AND CHARGE AT INTERFACES 1

1 Potential and Charge of a Hard Particle 3

1.1 Introduction, 3
1.2 The Poisson-Boltzmann Equation, 3
1.3 Plate, 6
1.3.1 Low Potential, 8
1.3.2 Arbitrary Potential: Symmetrical Electrolyte, 8
1.3.3 Arbitrary Potential: Asymmetrical Electrolyte, 13
1.3.4 Arbitrary Potential: General Electrolyte, 14
1.4 Sphere, 16
1.4.1 Low Potential, 17
1.4.2 Surface Charge Density-Surface Potential Relationship: Symmetrical Electrolyte, 18
1.4.3 Surface Charge Density-Surface Potential Relationship: Asymmetrical Electrolyte, 21
1.4.4 Surface Charge Density-Surface Potential Relationship: General Electrolyte, 22
1.4.5 Potential Distribution Around a Sphere with Arbitrary Potential, 25
1.5 Cylinder, 31
1.5.1 Low Potential, 32
1.5.2 Arbitrary Potential: Symmetrical Electrolyte, 33
1.5.3 Arbitrary Potential: General Electrolytes, 34
1.6 Asymptotic Behavior of Potential and Effective Surface Potential, 37
1.6.1 Plate, 38
1.6.2 Sphere, 41
1.6.3 Cylinder, 42
1.7 Nearly Spherical Particle, 43
References, 45

2 Potential Distribution Around a Nonuniformly Charged Surface and Discrete Charge Effects **47**

2.1 Introduction, 47
2.2 The Poisson-Boltzmann Equation for a Surface with an Arbitrary Fixed Surface Charge Distribution, 47
2.3 Discrete Charge Effect, 56
References, 62

3 Modified Poisson-Boltzmann Equation **63**

3.1 Introduction, 63
3.2 Electrolyte Solution Containing Rod-like Divalent Cations, 63
3.3 Electrolyte Solution Containing Rod-like Zwitterions, 70
3.4 Self-atmosphere Potential of Ions, 77
References, 82

4 Potential and Charge of a Soft Particle **83**

4.1 Introduction, 83
4.2 Planar Soft Surface, 83
4.2.1 Poisson–Boltzmann Equation, 83
4.2.2 Potential Distribution Across a Surface Charge Layer, 87
4.2.3 Thick Surface Charge Layer and Donnan Potential, 90
4.2.4 Transition Between Donnan Potential and Surface Potential, 91
4.2.5 Donnan Potential in a General Electrolyte, 92
4.3 Spherical Soft Particle, 93
4.3.1 Low Charge Density Case, 93
4.3.2 Surface Potential–Donnan Potential Relationship, 95
4.4 Cylindrical Soft Particle, 100
4.4.1 Low Charge Density Case, 100
4.4.2 Surface Potential–Donnan Potential Relationship, 101
4.5 Asymptotic Behavior of Potential and Effective Surface Potential of a Soft Particle, 102
4.5.1 Plate, 102
4.5.2 Sphere, 103
4.5.3 Cylinder, 104
4.6 Nonuniformly Charged Surface Layer: Isoelectric Point, 104
References, 110

5 Free Energy of a Charged Surface **111**

5.1 Introduction, 111
5.2 Helmholtz Free Energy and Tension of a Hard Surface, 111
5.2.1 Charged Surface with Ion Adsorption, 111
5.2.2 Charged Surface with Dissociable Groups, 116

5.3 Calculation of the Free Energy of the Electrical Double Layer, 118
5.3.1 Plate, 119
5.3.2 Sphere, 120
5.3.3 Cylinder, 121
5.4 Alternative Expression for F_{el}, 122
5.5 Free Energy of a Soft Surface, 123
5.5.1 General Expression, 123
5.5.2 Expressions for the Double-Layer Free Energy for a Planar Soft Surface, 127
5.5.3 Soft Surface with Dissociable Groups, 128
References, 130

6 Potential Distribution Around a Charged Particle in a Salt-Free Medium **132**

6.1 Introduction, 132
6.2 Spherical Particle, 133
6.3 Cylindrical Particle, 143
6.4 Effects of a Small Amount of Added Salts, 146
6.5 Spherical Soft Particle, 152
References, 162

PART II INTERACTION BETWEEN SURFACES **163**

7 Electrostatic Interaction of Point Charges in an Inhomogeneous Medium **165**

7.1 Introduction, 165
7.2 Planar Geometry, 166
7.3 Cylindrical Geometry, 180
References, 185

8 Force and Potential Energy of the Double-Layer Interaction Between Two Charged Colloidal Particles **186**

8.1 Introduction, 186
8.2 Osmotic Pressure and Maxwell Stress, 186
8.3 Direct Calculation of Interaction Force, 188
8.4 Free Energy of Double-Layer Interaction, 198
8.4.1 Interaction at Constant Surface Charge Density, 199
8.4.2 Interaction at Constants Surface Potential, 200
8.5 Alternative Expression for the Electric Part of the Free Energy of Double-Layer Interaction, 201
8.6 Charge Regulation Model, 201
References, 202

9 Double-Layer Interaction Between Two Parallel Similar Plates 203

9.1 Introduction, 203
9.2 Interaction Between Two Parallel Similar Plates, 203
9.3 Low Potential Case, 207
9.3.1 Interaction at Constant Surface Charge Density, 208
9.3.2 Interaction at Constant Surface Potential, 211
9.4 Arbitrary Potential Case, 214
9.4.1 Interaction at Constant Surface Charge Density, 214
9.4.2 Interaction at Constant Surface Potential, 224
9.5 Comparison Between the Theory of Derjaguin and Landau and the Theory of Verwey and Overbeek, 226
9.6 Approximate Analytic Expressions for Moderate Potentials, 227
9.7 Alternative Method of Linearization of the Poisson–Boltzmann Equation, 231
9.7.1 Interaction at Constant Surface Potential, 231
9.7.2 Interaction at Constant Surface Charge Density, 234
References, 240

10 Electrostatic Interaction Between Two Parallel Dissimilar Plates 241

10.1 Introduction, 241
10.2 Interaction Between Two Parallel Dissimilar Plates, 241
10.3 Low Potential Case, 244
10.3.1 Interaction at Constant Surface Charge Density, 244
10.3.2 Interaction at Constant Surface Potential, 251
10.3.3 Mixed Case, 252
10.4 Arbitrary Potential: Interaction at Constant Surface Charge Density, 252
10.4.1 Isodynamic Curves, 252
10.4.2 Interaction Energy, 258
10.5 Approximate Analytic Expressions for Moderate Potentials, 262
References, 263

11 Linear Superposition Approximation for the Double-Layer Interaction of Particles at Large Separations 265

11.1 Introduction, 265
11.2 Two Parallel Plates, 265
11.2.1 Similar Plates, 265
11.2.2 Dissimilar Plates, 270
11.2.3 Hypothetical Charge, 276
11.3 Two Spheres, 278
11.4 Two Cylinders, 279
References, 281

12 Derjaguin's Approximation at Small Separations 283

12.1 Introduction, 283
12.2 Two Spheres, 283
12.2.1 Low Potentials, 285
12.2.2 Moderate Potentials, 286
12.2.3 Arbitrary Potentials: Derjaguin's Approximation Combined with the Linear Superposition Approximation, 288
12.2.4 Curvature Correction to Derjaguin' Approximation, 290
12.3 Two Parallel Cylinders, 292
12.4 Two Crossed Cylinders, 294
References, 297

13 Donnan Potential-Regulated Interaction Between Porous Particles 298

13.1 Introduction, 298
13.2 Two Parallel Semi-infinite Ion-penetrable Membranes (Porous Plates), 298
13.3 Two Porous Spheres, 306
13.4 Two Parallel Porous Cylinders, 310
13.5 Two Parallel Membranes with Arbitrary Potentials, 311
13.5.1 Interaction Force and Isodynamic Curves, 311
13.5.2 Interaction Energy, 317
13.6 pH Dependence of Electrostatic Interaction Between Ion-penetrable Membranes, 320
References, 322

14 Series Expansion Representations for the Double-Layer Interaction Between Two Particles 323

14.1 Introduction, 323
14.2 Schwartz's Method, 323
14.3 Two Spheres, 327
14.4 Plate and Sphere, 342
14.5 Two Parallel Cylinders, 348
14.6 Plate and Cylinder, 353
References, 356

15 Electrostatic Interaction Between Soft Particles 357

15.1 Introduction, 357
15.2 Interaction Between Two Parallel Dissimilar Soft Plates, 357
15.3 Interaction Between Two Dissimilar Soft Spheres, 363
15.4 Interaction Between Two Dissimilar Soft Cylinders, 369
References, 374

16 Electrostatic Interaction Between Nonuniformly Charged Membranes **375**

16.1 Introduction, 375
16.2 Basic Equations, 375
16.3 Interaction Force, 376
16.4 Isoelectric Points with Respect To Electrolyte Concentration, 378
Reference, 380

17 Electrostatic Repulsion Between Two Parallel Soft Plates After Their Contact **381**

17.1 Introduction, 381
17.2 Repulsion Between Intact Brushes, 381
17.3 Repulsion Between Compressed Brushes, 382
References, 387

18 Electrostatic Interaction Between Ion-Penetrable Membranes In a Salt-free Medium **388**

18.1 Introduction, 388
18.2 Two Parallel Hard Plates, 388
18.3 Two Parallel Ion-Penetrable Membranes, 391
References, 398

19 van der Waals Interaction Between Two Particles **399**

19.1 Introduction, 399
19.2 Two Molecules, 399
19.3 A Molecule and a Plate, 401
19.4 Two Parallel Plates, 402
19.5 A Molecule and a Sphere, 404
19.6 Two Spheres, 405
19.7 A Molecule and a Rod, 407
19.8 Two Parallel Rods, 408
19.9 A Molecule and a Cylinder , 408
19.10 Two Parallel Cylinders, 410
19.11 Two Crossed Cylinders, 412
19.12 Two Parallel Rings, 412
19.13 Two Parallel Torus-Shaped Particles, 413
19.14 Two Particles Immersed In a Medium, 417
19.15 Two Parallel Plates Covered with Surface Layers, 418
References, 419

20 DLVO Theory of Colloid Stability **420**

20.1 Introduction, 420

20.2 Interaction Between Lipid Bilayers, 420
20.3 Interaction Between Soft Spheres, 425
References, 429

PART III ELECTROKINETIC PHENOMENA AT INTERFACES 431

21 Electrophoretic Mobility of Soft Particles 433

21.1 Introduction, 433
21.2 Brief Summary of Electrophoresis of Hard Particles, 433
21.3 General Theory of Electrophoretic Mobility of Soft Particles, 435
21.4 Analytic Approximations for the Electrophoretic Mobility of Spherical Soft Particles, 440
21.4.1 Large Spherical Soft Particles, 440
21.4.2 Weakly Charged Spherical Soft Particles, 444
21.4.3 Cylindrical Soft Particles, 447
21.5 Electrokinetic Flow Between Two Parallel Soft Plates, 449
21.6 Soft Particle Analysis of the Electrophoretic Mobility of Biological Cells and Their Model Particles, 454
21.6.1 RAW117 Lymphosarcoma Cells and Their Variant Cells, 454
21.6.2 Poly(*N*-isopropylacrylamide) Hydrogel-Coated Latex, 455
21.7 Electrophoresis of Nonuniformly Charged Soft Particles, 457
21.8 Other Topics of Electrophoresis of Soft Particles, 463
References, 464

22 Electrophoretic Mobility of Concentrated Soft Particles 468

22.1 Introduction, 468
22.2 Electrophoretic Mobility of Concentrated Soft Particles, 468
22.3 Electroosmotic Velocity in an Array of Soft Cylinders, 475
References, 479

23 Electrical Conductivity of a Suspension of Soft Particles 480

23.1 Introduction, 480
23.2 Basic Equations, 480
23.3 Electrical Conductivity, 481
References, 484

24 Sedimentation Potential and Velocity in a Suspension of Soft Particles 485

24.1 Introduction, 485
24.2 Basic Equations, 485
24.3 Sedimentation Velocity of a Soft Particle, 490

24.4 Average Electric Current and Potential, 490
24.5 Sedimentation Potential, 491
24.6 Onsager's Reciprocal Relation, 494
24.7 Diffusion Coefficient of a Soft Particle, 495
References, 495

25 Dynamic Electrophoretic Mobility of a Soft Particle **497**

25.1 Introduction, 497
25.2 Basic Equations, 497
25.3 Linearized Equations, 499
25.4 Equation of Motion of a Soft Particle, 501
25.5 General Mobility Expression, 501
25.6 Approximate Mobility Formula, 503
References, 506

26 Colloid Vibration Potential in a Suspension of Soft Particles **508**

26.1 Introduction, 508
26.2 Colloid Vibration Potential and Ion Vibration Potential, 508
References, 513

27 Effective Viscosity of a Suspension of Soft Particles **515**

27.1 Introduction, 515
27.2 Basic Equations, 516
27.3 Linearized Equations, 518
27.4 Electroviscous Coefficient, 520
27.5 Approximation for Low Fixed-Charge Densities, 523
27.6 Effective Viscosity of a Concentrated Suspension of Uncharged Porous Spheres, 527
Appendix 27A, 530
References, 531

PART IV OTHER TOPICS **533**

28 Membrane Potential and Donnan Potential **535**

28.1 Introduction, 535
28.2 Membrane Potential and Donnan Potential, 535
References, 541

Index **543**

PREFACE

The principal aim of this book is to provide a tool for discussing various phenomena at biointerfaces such as the surface of cells on the basis of biophysical chemistry. For nonbiological interfaces, colloid and interface science, one of the major branches of physical chemistry, forms a powerful basis for understanding various interfacial phenomena. The Derjaguin–Landau–Verwey–Overbeek (DLVO) theory explains well the stability of colloidal suspensions in terms of the electrostatic and van der Waals interactions between the particles. The behavior of colloidal particles in an applied electric filed is analyzed by electrophoresis theories of Smoluchowski, Hückel, and Henry. The charge or potential of the particle surface plays an essential role in the above-mentioned phenomena. It must be noted here that the particle-fixed charges are assumed to be located only at the particle surface (of zero thickness). This model, however, is by no means a good approximation for biocolloids such as biological cells. For such particles, fixed charges are distributed over some depth on the particle surface, or the particle surface is covered with a polyelectrolyte layer. We call polyelectrolyte-coated particles soft particles. In this book, we discuss various phenomena at biointerfaces, that is, potential and charge at interfaces, electrokinetic phenomena at interfaces, and interactions between surfaces, on the basis of the soft particle model. We will see that the Donnan potential as well as the surface potential is an important factor controlling electric properties of soft particles or soft surfaces.

I would like to express my sincere thanks to Professor Tom Healy and Professor Lee White, who introduced me into the field of electrokinetic phenomena when I stayed as a postdoctoral fellow at the University of Melbourne in 1981–1983. I would like to thank Professor Shinpei Ohki at the State University of New York at Buffalo, where I stayed as a postdoctoral fellow. He pointed out to me the important role of the Donnan potential in electric phenomena of soft particles. I am happy to thank my sons Manabu and Nozomu and their wives Yumi and Michiyo for their understanding and help during the writing of this book.

Finally, I would like to gratefully acknowledge the assistance provided by Ms. Anita Lekhwani, Senior Acquisitions Editor, and Ms. Rebekah Amos, Editorial Program Coordinator.

HIROYUKI OHSHIMA

LIST OF SYMBOLS

a	particle radius
d	thickness of the surface charge layer
e	elementary electric charge
g	gravity
k	Bolzmann's constant
K^{∞}	electrical conductivity of an electrolyte solution in the absence of particles
K^{*}	complex conductivity of an electrolyte solution
n_i^{∞}	bulk concentration (number density) of the ith ionic species
N	number density of ionized groups in the surface charge layer
N_A	Avogadro's constant
p	pressure
T	absolute temperature
u	liquid velocity
U	electrophoretic velocity
U_{SED}	sedimentation velocity
y	scaled electric potential
z_i	valence of the ith ionic species
Z	valence of ionized groups in the surface charge layer
ε_o	permittivity of a vacuum
ε_r	relative permittivity of an electrolyte solution
ϕ	particle volume fraction
γ	frictional coefficient of the forces exerted by the polymer segments on the liquid flow
η	viscosity
η_s	effective viscosity of a suspension of particles
κ	Debye–Hückel parameter
κ_m	Debye–Hückel parameter in the surface charge layer
λ	$(\gamma/\eta)^{1/2}$
λ_i	Drag coefficient of the ith ionic species
$1/\lambda$	softness parameter
μ	electrophoretic mobility
ρ_{el}	volume charge density resulting from electrolyte ions

ρ_{fix}	volume density of fixed charges distributed in the surface charge layer
ρ_{o}	mass density of a medium
σ	surface charge density
ω	angular frequency
ψ	electric potential
$\psi^{(0)}$	equilibrium electric potential
ψ_{o}	surface potential
ψ_{DON}	Donnan potential
ζ	zeta potential

PART I
Potential and Charge at Interfaces

1 Potential and Charge of a Hard Particle

1.1 INTRODUCTION

The potential and charge of colloidal particles play a fundamental role in their interfacial electric phenomena such as electrostatic interaction between them and their motion in an electric field [1–4]. When a charged colloidal particle is immersed in an electrolyte solution, mobile electrolyte ions with charges of the sign opposite to that of the particle surface charges, which are called counterions, tend to approach the particle surface and neutralize the particle surface charges, but thermal motion of these ions prevents accumulation of the ions so that an ionic cloud is formed around the particle. In the ionic cloud, the concentration of counterions becomes very high while that of coions (electrolyte ions with charges of the same sign as the particle surface charges) is very low, as schematically shown in Fig. 1.1, which shows the distribution of ions around a charged spherical particle of radius a. The ionic cloud together with the particle surface charge forms an electrical double layer. Such an electrical double layer is often called an electrical diffuse double layer, since the distribution of electrolyte ions in the ionic cloud takes a diffusive structure due to thermal motion of ions. The electric properties of charged colloidal particles in an electrolyte solution strongly depend on the distributions of electrolyte ions and of the electric potential across the electrical double layer around the particle surface. The potential distribution is usually described by the Poisson–Boltzmann equation [1–4].

1.2 THE POISSON–BOLTZMANN EQUATION

Consider a uniformly charged particle immersed in a liquid containing N ionic species with valence z_i and bulk concentration (number density) n_i^∞ ($i = 1, 2, \ldots, N$) (in units of m^{-3}). From the electroneutrality condition, we have

$$\sum_{i=1}^{N} z_i n_i^\infty = 0 \tag{1.1}$$

Usually we need to consider only electrolyte ions as charged species. The electric potential $\psi(\boldsymbol{r})$ at position $\boldsymbol{r}$ outside the particle, measured relative to the bulk

Biophysical Chemistry of Biointerfaces By Hiroyuki Ohshima

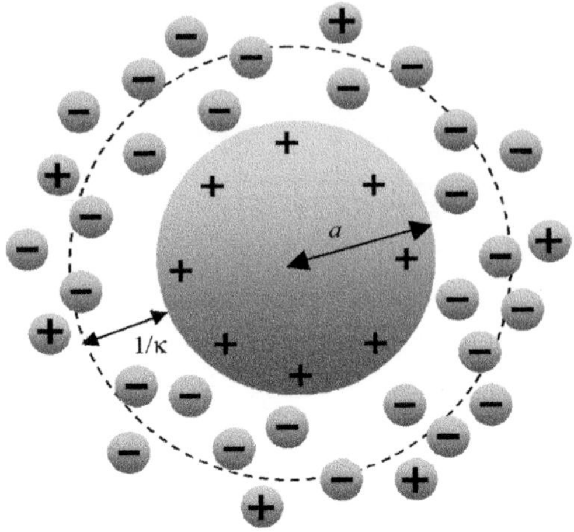

FIGURE 1.1 Electrical double layer around a positively charged colloidal particle. The particle is surrounded by an ionic cloud, forming the electrical double layer of thickness $1/\kappa$, in which the concentration of counterions is greater than that of coions.

solution phase, where ψ is set equal to zero, is related to the charge density $\rho_{\rm el}(\boldsymbol{r})$ at the same point by the Poisson equation, namely,

$$\Delta\psi(\boldsymbol{r}) = -\frac{\rho_{\rm el}(\boldsymbol{r})}{\varepsilon_{\rm r}\varepsilon_{\rm o}} \tag{1.2}$$

where Δ is the Laplacian, $\varepsilon_{\rm r}$ is the relative permittivity of the electrolyte solution, and $\varepsilon_{\rm o}$ is the permittivity of the vacuum. We assume that the distribution of the electrolyte ions $n_i(\boldsymbol{r})$ obeys Boltzmann's law, namely,

$$n_i(\boldsymbol{r}) = n_i^{\infty} \exp\left(-\frac{z_i e\psi(\boldsymbol{r})}{kT}\right) \tag{1.3}$$

where $n_i(\boldsymbol{r})$ is the concentration (number density) of the ith ionic species at position $\boldsymbol{r}$, e is the elementary electric charge, k is Boltzmann's constant, and T is the absolute temperature. The charge density $\rho_{\rm el}(\boldsymbol{r})$ at position $\boldsymbol{r}$ is thus given by

$$\rho_{\rm el}(\boldsymbol{r}) = \sum_{i=1}^{N} z_i e n_i(\boldsymbol{r}) = \sum_{i=1}^{N} z_i e n_i^{\infty} \exp\left(-\frac{z_i e\psi(\boldsymbol{r})}{kT}\right) \tag{1.4}$$

which is the required relation between $\psi(\boldsymbol{r})$ and $\rho_{\rm el}(\boldsymbol{r})$.

Combining Eqs. (1.2) and (1.4) gives

$$\Delta\psi(\boldsymbol{r}) = -\frac{1}{\varepsilon_{\rm r}\varepsilon_{\rm o}} \sum_{i=1}^{N} z_i e n_i^{\infty} \exp\left(-\frac{z_i e\psi(\boldsymbol{r})}{kT}\right) \tag{1.5}$$

This is the Poisson–Boltzmann equation for the potential distribution $\psi(\boldsymbol{r})$. The surface charge density σ of the particle is related to the potential derivative normal to the particle surface as

$$\varepsilon_p \frac{\partial \psi}{\partial n} - \varepsilon_r \frac{\partial \psi}{\partial n} = \frac{\sigma}{\varepsilon_o} \tag{1.6}$$

where ε_p is the relative permittivity of the particle and n is the outward normal at the particle surface. If the internal electric fields inside the particle can be neglected, then the boundary condition (1.6) reduces to

$$\frac{\partial \psi}{\partial n} = -\frac{\sigma}{\varepsilon_r \varepsilon_o} \tag{1.7}$$

If the potential ψ is low, namely,

$$\left| \frac{z_i e \psi}{kT} \right| \ll 1 \tag{1.8}$$

then Eq. (1.5) reduces to

$$\Delta \psi = \kappa^2 \psi \tag{1.9}$$

with

$$\kappa = \left(\frac{1}{\varepsilon_r \varepsilon_o kT} \sum_{i=1}^{N} z_i^2 e^2 n_i^\infty \right)^{1/2} \tag{1.10}$$

Equation (1.9) is the linearized Poisson–Boltzmann equation and κ in Eq. (1.10) is the Debye–Hückel parameter. This linearization is called the Debye–Hückel approximation and Eq. (1.9) is called the Debye–Hückel equation. The reciprocal of κ (i.e., $1/\kappa$), which is called the Debye length, corresponds to the thickness of the double layer. Note that n_i^∞ in Eqs. (1.5) and (1.10) is given in units of m^{-3}. If one uses the units of M (mol/L), then n_i^∞ must be replaced by $1000N_A n_i^\infty$, N_A being Avogadro's number.

Expressions for κ for various types of electrolytes are explicitly given below.

(i) For a symmetrical electrolyte of valence z and bulk concentration n,

$$\kappa = \left(\frac{2z^2 e^2 n}{\varepsilon_r \varepsilon_o kT} \right)^{1/2} \tag{1.11}$$

(ii) For a 1-1 symmetrical electrolyte of bulk concentration n,

$$\kappa = \left(\frac{2e^2 n}{\varepsilon_r \varepsilon_o kT} \right)^{1/2} \tag{1.12}$$

(iii) For a 2-1 electrolyte of bulk concentration n,

$$\kappa = \left(\frac{6e^2 n}{\varepsilon_r \varepsilon_o kT} \right)^{1/2} \tag{1.13}$$

(iv) For a mixed solution of 1-1 electrolyte of bulk concentration n_1 and 2-1 electrolyte of bulk concentration n_2,

$$\kappa = \left(\frac{2(n_1 + 3n_2)e^2}{\varepsilon_r \varepsilon_o kT} \right)^{1/2} \tag{1.14}$$

(v) For a 3-1 electrolyte of bulk concentration n,

$$\kappa = \left(\frac{12e^2 n}{\varepsilon_r \varepsilon_o kT} \right)^{1/2} \tag{1.15}$$

(vi) For a mixed solution of 1-1 electrolyte of concentration n_1 and 3-1 electrolyte of concentration n_2,

$$\kappa = \left(\frac{2(n_1 + 6n_2)e^2}{\varepsilon_r \varepsilon_o kT} \right)^{1/2} \tag{1.16}$$

1.3 PLATE

Consider the potential distribution around a uniformly charged plate-like particle in a general electrolyte solution composed of N ionic species with valence z_i and bulk concentration (number density) n_i^∞ ($i = 1, 2, \ldots, N$) (in units of m^{-3}). We take an x-axis perpendicular to the plate surface with its origin $x = 0$ at the plate surface so that the region $x < 0$ corresponds to the internal region of the plate while the region $x > 0$ corresponds to the solution phase (Fig. 1.2). The electric potential $\psi(x)$ at position x outside the plate, measured relative to the bulk solution phase, where ψ is set equal to zero, is related to the charge density $\rho_{el}(x)$ of free mobile charged species by the Poisson equation (Eq. (1.2)), namely,

$$\frac{d^2\psi(x)}{dx^2} = -\frac{\rho_{el}(x)}{\varepsilon_r \varepsilon_o} \tag{1.17}$$

We assume that the distribution of the electrolyte ions $n_i(x)$ obeys Boltzmann's law, namely,

$$n_i(x) = n_i^\infty \exp\left(-\frac{z_i e \psi(x)}{kT} \right) \tag{1.18}$$

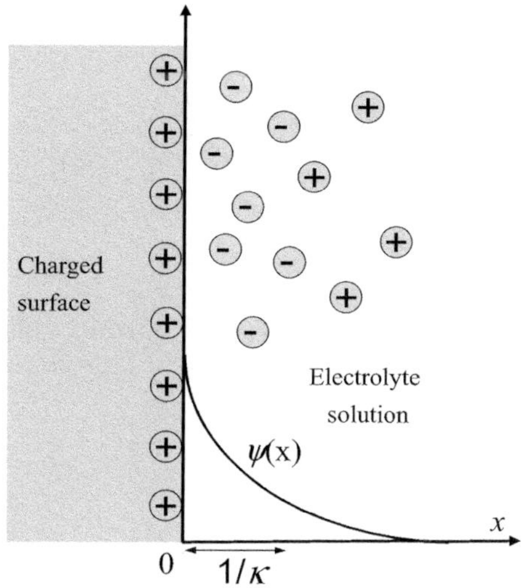

FIGURE 1.2 Schematic representation of potential distribution $\psi(x)$ near the positively charged plate.

where $n_i(x)$ is the concentration (number density) of the ith ionic species at position x. The charge density $\rho_{el}(x)$ at position x is thus given by

$$\rho_{el}(x) = \sum_{i=1}^{N} z_i e n_i(x) = \sum_{i=1}^{N} z_i e n_i^{\infty} \exp\left(-\frac{z_i e\psi(x)}{kT}\right) \tag{1.19}$$

Combining Eqs. (1.17) and (1.19) gives the following Poisson–Boltzmann equation for the potential distribution $\psi(x)$:

$$\frac{d^2\psi(x)}{dx^2} = -\frac{1}{\varepsilon_r\varepsilon_o}\sum_{i=1}^{N} z_i e n_i^{\infty} \exp\left(-\frac{z_i e\psi(x)}{kT}\right) \tag{1.20}$$

We solve the planar Poisson–Boltzmann equation (1.20) subject to the boundary conditions:

$$\psi = \psi_o \quad \text{at} \quad x = 0 \tag{1.21}$$

$$\psi \to 0, \ \frac{d\psi}{dx} \to 0 \quad \text{as} \quad x \to \infty \tag{1.22}$$

where ψ_o is the potential at the plate surface $x = 0$, which we call the surface potential.

If the internal electric fields inside the particle can be neglected, then the surface charge density σ of the particle is related to the potential derivative normal to the particle surface by (see Eq. (1.7))

$$\left.\frac{d\psi}{dx}\right|_{x=0^+} = -\frac{\sigma}{\varepsilon_r\varepsilon_o} \tag{1.23}$$

1.3.1 Low Potential

If the potential ψ is low (Eq. (1.8)), then Eq. (1.20) reduces to the following linearized Poisson–Boltzmann equation (Eq. (1.9)):

$$\frac{d^2\psi}{dx^2} = \kappa^2\psi \tag{1.24}$$

The solution to Eq. (1.24) subject to Eqs. (1.21) and (1.22) can be easily obtained:

$$\psi(x) = \psi_o e^{-\kappa x} \tag{1.25}$$

Equations (1.23) and (1.25) give the following surface charge density–surface potential (σ–ψ_o) relationship:

$$\psi_0 = \frac{\sigma}{\varepsilon_r\varepsilon_o\kappa} \tag{1.26}$$

Equation (1.26) has the following simple physical meaning. Since ψ decays from ψ_o to zero over a distance of the order of κ^{-1} (Eq. (1.25)), the electric field at the particle surface is approximately given by ψ_o/κ^{-1}. This field, which is generated by σ, is equal to $\sigma/\varepsilon_r\varepsilon_o$. Thus, we have $\psi_o/\kappa^{-1} = \sigma/\varepsilon_r\varepsilon_o$, resulting in Eq. (1.26).

1.3.2 Arbitrary Potential: Symmetrical Electrolyte

Now we solve the original nonlinear Poisson–Boltzmann equation (1.20). If the plate is immersed in a symmetrical electrolyte of valence z and bulk concentration n, then Eq. (1.20) becomes

$$\begin{aligned}\frac{d^2\psi(x)}{dx^2} &= -\frac{zen}{\varepsilon_r\varepsilon_o}\left[\exp\left(-\frac{ze\psi(x)}{kT}\right) - \exp\left(\frac{ze\psi(x)}{kT}\right)\right] \\ &= \frac{2zen}{\varepsilon_r\varepsilon_o}\sinh\left(\frac{ze\psi(x)}{kT}\right)\end{aligned} \tag{1.27}$$

We introduce the dimensionless potential $y(x)$

$$y = \frac{ze\psi}{kT} \tag{1.28}$$

then Eq. (1.27) becomes

$$\frac{d^2y}{dx^2} = \kappa^2 \sinh y \tag{1.29}$$

where the Debye–Hückel parameter κ is given by Eq. (1.11). Note that $y(x)$ is scaled by kT/ze, which is the thermal energy measured in units of volts. At room temperatures, kT/ze (with $z=1$) amounts to ca. 25 mV. Equation (1.29) can be solved by multiplying dy/dx on its both sides to give

$$\frac{dy}{dx}\frac{d^2y}{dx^2} = \kappa^2 \sinh y \frac{dy}{dx} \tag{1.30}$$

which is transformed into

$$\frac{1}{2}\frac{d}{dx}\left\{\left(\frac{dy}{dx}\right)^2\right\} = \kappa^2 \frac{d}{dx}\cosh y \tag{1.31}$$

Integration of Eq. (1.31) gives

$$\left(\frac{dy}{dx}\right)^2 = 2\kappa^2 \cosh y + \text{constant} \tag{1.32}$$

By taking into account Eq. (1.22), we find that constant $= -2\kappa^2$ so that Eq. (1.32) becomes

$$\left(\frac{dy}{dx}\right)^2 = 2\kappa^2(\cosh y - 1) = 4\kappa^2 \sinh^2\left(\frac{y}{2}\right) \tag{1.33}$$

Since y and dy/dx are of opposite sign, we obtain from Eq. (1.33)

$$\frac{dy}{dx} = -2\kappa \sinh(y/2) \tag{1.34}$$

Equation (1.34) can be further integrated to give

$$\int_y^{y_o} \frac{dy}{2\sinh(y/2)} = \kappa \int_0^x dx \tag{1.35}$$

where

$$y_o = \frac{ze\psi_o}{kT} \tag{1.36}$$

is the scaled surface potential. Thus, we obtain

$$y(x) = 4\,\mathrm{arctanh}(\gamma e^{-\kappa x}) = 2\ln\left(\frac{1+\gamma e^{-\kappa x}}{1-\gamma e^{-\kappa x}}\right) \tag{1.37}$$

or

$$\psi(x) = \frac{2kT}{ze}\ln\left(\frac{1+\gamma e^{-\kappa x}}{1-\gamma e^{-\kappa x}}\right) \tag{1.38}$$

with

$$\gamma = \tanh\left(\frac{ze\psi_o}{4kT}\right) = \frac{\exp(ze\psi_o/2kT)-1}{\exp(ze\psi_o/2kT)+1} = \frac{\exp(y_o/2)-1}{\exp(y_o/2)+1} \tag{1.39}$$

Figure 1.3 exhibits γ as a function of y_o, showing that γ is a linearly increasing function of y_o for low y_o, namely,

$$\gamma \approx \frac{y_o}{4} = \frac{ze\psi_o}{4kT} \tag{1.40}$$

but reaches a plateau value at 1 for $|y_o| \geq 8$.

Figure 1.4 shows $y(x)$ for several values of y_o calculated from Eq. (1.37) in comparison with the Debye–Hückel linearized solution (Eq. (1.25)). It is seen that the Debye–Hückel approximation is good for low potentials ($|y_o| \leq 1$). As seen from Eqs. (1.25) and (1.37), the potential $\psi(x)$ across the electrical double layer varies nearly

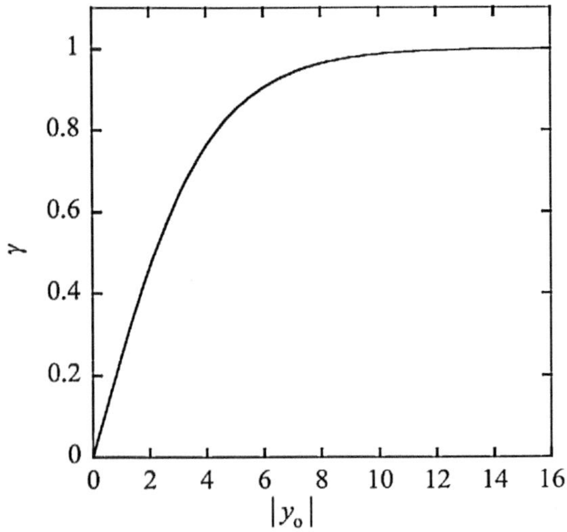

FIGURE 1.3 γ as a function of $|y_o|$ (Eq. (1.39)).

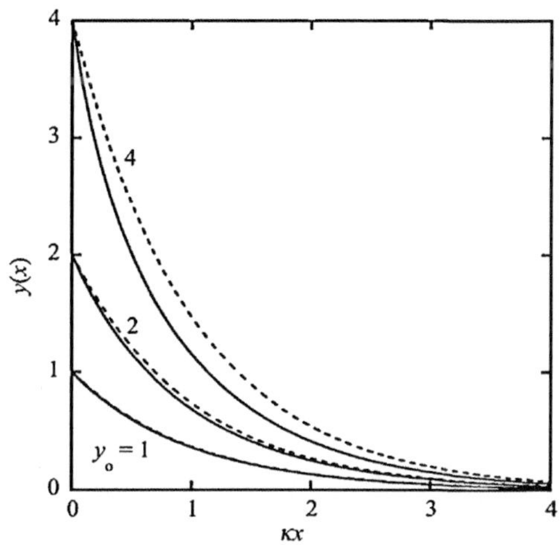

FIGURE 1.4 Potential distribution $y(x) \equiv ze\psi(x)/kT$ around a positively charged plate with scaled surface potential $y_o \equiv ze\psi_o/kT$. Calculated for $y_o = 1$, 2, and 4. Solid lines, exact solution (Eq. (1.37)); dashed lines, the Debye–Hückel linearized solution (Eq. (1.25)).

exponentially (Eqs. (1.37)) or exactly exponentially (Eq. (1.25)) with the distance x from the plate surface, as shown in Fig. 1.4. Equation (1.25) shows that the potential $\psi(x)$ decays from ψ_o at $x = 0$ to ψ_o/e ($\psi_o/3$) at $x = 1/\kappa$. Thus, the reciprocal of the Debye–Hückel parameter κ (the Debye length), which has the dimension of length, serves as a measure for the thickness of the electrical double layer. Figure 1.5 plots the

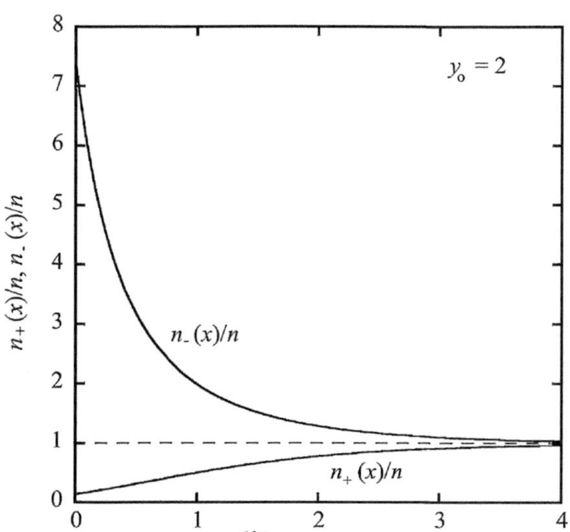

FIGURE 1.5 Concentrations of counterions (anions) $n_-(x)$ and coions (cations) $n_+(x)$ around a positively charged planar surface (arbitrary scale). Calculated from Eqs. (1.3) and (1.26) for $y_o = 2$.

concentrations of counterion ($n_-(x) = n\exp(y(x))$) and coions ($n_+(x) = n\exp(-y(x))$) around a positively charged plate as a function of the distance x from the plate surface, showing that these quantities decay almost exponentially over the distance of the Debye length $1/\kappa$ just like the potential distribution $\psi(x)$ in Fig. 1.4.

By substituting Eq. (1.38) into Eq. (1.23), we obtain the following relationship connecting ψ_o and σ:

$$\sigma = \frac{2\varepsilon_r\varepsilon_o\kappa kT}{ze}\sinh\left(\frac{ze\psi_o}{2kT}\right) = (8n\varepsilon_r\varepsilon_o kT)^{1/2}\sinh\left(\frac{ze\psi_o}{2kT}\right) \tag{1.41}$$

or inversely,

$$\begin{aligned}\psi_o &= \frac{2kT}{ze}\text{arcsinh}\left(\frac{\sigma}{\sqrt{8n\varepsilon_r\varepsilon_o kT}}\right) = \frac{2kT}{ze}\text{arcsinh}\left(\frac{ze\sigma}{2\varepsilon_r\varepsilon_o\kappa kT}\right) \\ &= \frac{2kT}{ze}\ln\left[\frac{ze\sigma}{2\varepsilon_r\varepsilon_o\kappa kT} + \left\{\left(\frac{ze\sigma}{2\varepsilon_r\varepsilon_o\kappa kT}\right)^2 + 1\right\}\right]\end{aligned} \tag{1.42}$$

If σ is small and thus ψ_o is low, that is, the condition (Eq. (1.8)) is fulfilled, then Eq. (1.38) reduces to Eq. (1.25) with the surface potential given by Eq. (1.26). Figure 1.6 shows the σ–ψ_o relationship calculated from Eq. (1.42) in comparison with the approximate results (Eq. (1.26)). The deviation of Eq. (1.26) from Eq. (1.42) becomes significant as the charge density σ increases.

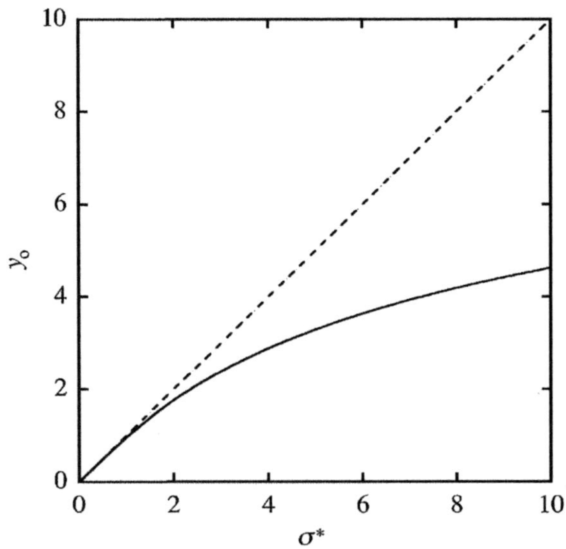

FIGURE 1.6 Scaled surface potential $y_o = ze\psi_o/kT$ as a function of the scaled surface charge density $\sigma^* = ze\sigma/\varepsilon_r\varepsilon_o\kappa kT$ for a positively charged planar plate in a symmetrical electrolyte solution of valence z. Solid line, exact solution (Eq. (1.41)); dashed line, Debye–Hückel linearized solution (Eq. (1.26)).

1.3.3 Arbitrary Potential: Asymmetrical Electrolyte

When a charged plate is immersed in a 2-1 electrolyte (e.g., $CaCl_2$) of bulk concentration n, the Poisson–Boltzmann equation (1.20) becomes

$$\frac{d^2\psi(x)}{dx^2} = -\frac{2en}{\varepsilon_r\varepsilon_o}\left[\exp\left(-\frac{2e\psi(x)}{kT}\right) - \exp\left(\frac{e\psi(x)}{kT}\right)\right] \tag{1.43}$$

where the first term on the right-hand side of Eq. (1.43) corresponds to divalent cations while the second to monovalent anions. Equation (1.43) subject to the boundary conditions (1.21) and (1.22) can be integrated by multiplying dy/dx on both sides of Eq. (1.43) to give

$$\psi(x) = \frac{kT}{e}\ln\left[\frac{3}{2}\left(\frac{1+\frac{2}{3}\gamma' e^{-\kappa x}}{1-\frac{2}{3}\gamma' e^{-\kappa x}}\right)^2 - \frac{1}{2}\right] \tag{1.44}$$

with

$$\gamma' = \frac{3}{2}\left\{\frac{\left(\frac{2}{3}e^{y_o}+\frac{1}{3}\right)^{1/2}-1}{\left(\frac{2}{3}e^{y_o}+\frac{1}{3}\right)^{1/2}+1}\right\} \tag{1.45}$$

$$y_o = \frac{e\psi_o}{kT} \tag{1.46}$$

where y_o is the scaled surface potential and κ is given by Eq. (1.13). By substituting Eq. (1.44) into Eq. (1.23), we obtain the following relationship between the surface potential ψ_o and the surface charge density σ:

$$\sigma = \frac{\varepsilon_r\varepsilon_o\kappa kT}{e}\left\{1-\exp\left(-\frac{e\psi_o}{kT}\right)\right\}\left\{\frac{2}{3}\exp\left(\frac{e\psi_o}{kT}\right)+\frac{2}{3}\right\}^{1/2} \tag{1.47}$$

If σ is small and thus ψ_o is low, then the potential distribution $\psi(x)$ (Eq. (1.44)) and the σ–ψ_o relationship (Eq. (1.47)) are given by Eqs. (1.25) and (1.26), respectively, which also hold for general electrolytes.

Next consider the case of a mixed solution of 1-1 electrolyte of bulk concentration n_1 and 2-1 electrolyte of bulk concentration n_2. The Poisson–Boltzmann equation (1.5) becomes

$$\frac{d^2\psi(x)}{dx^2} = -\frac{e}{\varepsilon_r\varepsilon_o}\left[n_1\exp\left(-\frac{e\psi(x)}{kT}\right) + 2n_2\exp\left(-\frac{2e\psi(x)}{kT}\right) - (n_1+2n_2)\exp\left(\frac{e\psi(x)}{kT}\right)\right] \tag{1.48}$$

Equation (1.48) subject to the boundary conditions (1.21) and (1.22) can be easily integrated to give

$$\psi(x) = \frac{kT}{e}\ln\left[\left(\frac{1}{1-\eta/3}\right)\left\{\frac{1+(1-\eta/3)\gamma'' e^{-\kappa x}}{1-(1-\eta/3)\gamma'' e^{-\kappa x}}\right\}^2 - \left(\frac{\eta/3}{1-\eta/3}\right)\right] \tag{1.49}$$

with

$$\eta = \frac{3n_2}{n_1 + 3n_2} \tag{1.50}$$

$$\gamma'' = \left(\frac{1}{1-\eta/3}\right)\left[\frac{\{(1-\eta/3)e^{y_o} + \eta/3\}^{1/2} - 1}{\{(1-\eta/3)e^{y_o} + \eta/3\}^{1/2} + 1}\right] \tag{1.51}$$

where y_o is the scaled surface potential defined by Eq. (1.46) and κ is given by Eq. (1.14).

The relationship between σ and ψ_o, which is derived from Eqs. (1.23) and (1.49), is given by

$$\sigma = \frac{\varepsilon_r \varepsilon_o \kappa kT}{e}(1 - e^{-y_o})\left[\left(1 - \frac{\eta}{3}\right)e^{y_o} + \frac{\eta}{3}\right]^{1/2} \tag{1.52}$$

1.3.4 Arbitrary Potential: General Electrolyte

In the case of a charged plate immersed in a general electrolyte, the Poisson–Boltzmann equation (1.20), in terms of the scaled potential $y(x) = e\psi/kT$, is rewritten as

$$\frac{d^2y}{dx^2} = -\frac{\kappa^2 \sum_{i=1}^{N} z_i n_i e^{-z_i y}}{\sum_{i=1}^{N} z_i^2 n_i} \tag{1.53}$$

where the Debye–Hückel parameter κ is given by Eq. (1.10). The boundary conditions are given by Eqs. (1.21) and (1.22). Since y_o and dy/dx are of opposite sign, integration of Eq. (1.53) gives

$$\frac{dy}{dx} = -\mathrm{sgn}(y_o)\kappa\left[\frac{2\sum_{i=1}^{N} n_i(e^{-z_i y} - 1)}{\sum_{i=1}^{N} z_i^2 n_i}\right]^{1/2} \tag{1.54}$$

where $\mathrm{sgn}(y_o) = +1$ for $y_o > 0$ and -1 for $y_o < 0$. Note that the sign of $1 - \exp(-y)$ equals that of y_o. Equation (1.54) is thus rewritten as

$$\frac{dy}{dx} = -\kappa f(y) \tag{1.55}$$

with

$$f(y) = \mathrm{sgn}(y_o)\left[\frac{2\sum_{i=1}^{N} n_i(e^{-z_i y} - 1)}{\sum_{i=1}^{N} z_i^2 n_i}\right]^{1/2} = (1 - e^{-y})\left[\frac{2\sum_{i=1}^{N} n_i(e^{-z_i y} - 1)}{(1 - e^{-y})^2 \sum_{i=1}^{N} z_i^2 n_i}\right]^{1/2} \tag{1.56}$$

Note that as $y \to 0$, $f(y) \to y$. Explicit expressions for $f(y)$ for some simple cases are given below.

(i) For a symmetrical electrolyte of valence z,

$$f(y) = 2\sinh(zy/2) \tag{1.57}$$

(ii) For a monovalent electrolyte,

$$f(y) = 2\sinh(y/2) \tag{1.58}$$

(iii) For a 2-1 electrolyte,

$$f(y) = (1 - e^{-y})\left(\frac{2}{3}e^{y} + \frac{1}{3}\right)^{1/2} \tag{1.59}$$

(iv) For a mixed solution of 2-1 electrolyte of concentration n_2 and 1-1 electrolyte of concentration n_1,

$$f(y) = (1 - e^{-y})\left[\left(1 - \frac{\eta}{3}\right)e^{y} + \frac{\eta}{3}\right]^{1/2} \tag{1.60}$$

with

$$\eta = \frac{3n_2}{n_1 + 3n_2} \tag{1.61}$$

(v) For a 3-1 electrolyte,

$$f(y) = (1 - e^{-y})\left(\frac{1}{2}e^{y} + \frac{1}{3} + \frac{1}{6}e^{-y}\right)^{1/2} \tag{1.62}$$

(vi) For a mixed solution of 3-1 electrolyte of concentration n_2 and 1-1 electrolyte of concentration n_1,

$$f(y) = (1 - e^{-y})\left[\left(1 - \frac{\eta'}{2}\right)e^{y} + \frac{\eta'}{3} + \frac{\eta'}{6}e^{-y}\right]^{1/2} \tag{1.63}$$

with

$$\eta' = \frac{6n_2}{n_1 + 6n_2} \tag{1.64}$$

By integrating Eq. (1.55) between $x = 0$ ($y = y_o$) and $x = x$ ($y = y$), we obtain

$$\kappa x = \int_y^{y_o} \frac{dy}{f(y)} \tag{1.65}$$

which gives the relationship between y and x. For a symmetrical electrolyte of valence z, 2-1 electrolytes, and a mixed solution of 2-1 electrolyte of concentration n_2 and 1-1 electrolyte of concentration n_1, Eq. (1.65) reproduces Eqs. (1.38), (1.44), and (1.49), respectively.

The surface charge density–surface potential (σ–y_o) relationship is obtained from Eqs. (1.23) and (1.55) and given in terms of $f(y_o)$ as

$$\sigma = -\varepsilon_r\varepsilon_o \left.\frac{d\psi}{dx}\right|_{x=0^+} = \frac{\varepsilon_r\varepsilon_o\kappa kT}{e} f(y_o) \tag{1.66}$$

or

$$\begin{aligned}\sigma &= \frac{\varepsilon_r\varepsilon_0\kappa kT}{e}\,\mathrm{sgn}(y_o)\left[\frac{2\sum_{i=1}^{N} n_i(e^{-z_i y_o}-1)}{\sum_{i=1}^{N} z_i^3 n_i}\right]^{1/2}\\ &= \frac{\varepsilon_r\varepsilon_0\kappa kT}{e}\,\mathrm{sgn}(y_o)(1-e^{-y_o})\left[\frac{2\sum_{i=1}^{N} n_i(e^{-z_i y_o}-1)}{(1-e^{-y_o})^2\sum_{i=1}^{N} z_i^2 n_i}\right]^{1/2}\end{aligned} \tag{1.67}$$

Note that as $y_o \to 0$, $f(y_o) \to y_o$ so that for low y_o Eq. (1.67) reduces to Eq. (1.26). For a symmetrical electrolyte of valence z, 2-1 electrolytes, and a mixed solution of 2-1 electrolyte of concentration n_2 and 1-1 electrolyte of concentration n_1, Eq. (1.67) combined with Eqs. (1.57), (1.59), and (1.60) reproduces Eqs. (1.41), (1.47), and (1.52), respectively.

1.4 SPHERE

Consider a spherical particle of radius a in a general electrolyte solution. The electric potential $\psi(r)$ at position r obeys the following spherical Poisson–Boltzmann equation [3]:

$$\frac{d^2\psi}{dr^2} + \frac{2}{r}\frac{d\psi}{dr} = -\frac{1}{\varepsilon_r\varepsilon_o}\sum_{i=1}^{N} z_i e n_i^{\infty} \exp\left(-\frac{z_i e\psi}{kT}\right) \tag{1.68}$$

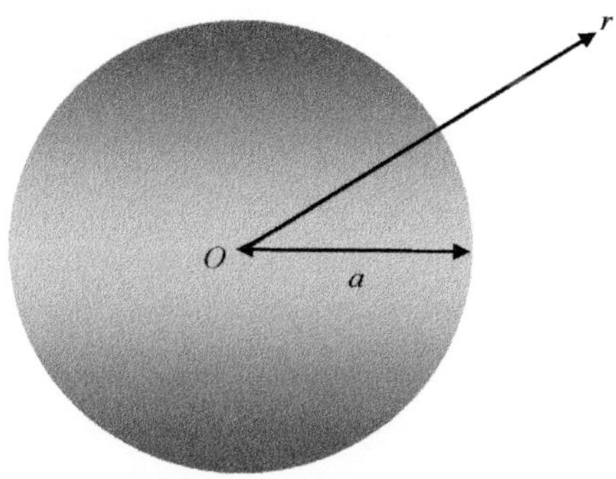

FIGURE 1.7 A sphere of radius a.

where we have taken the spherical coordinate system with its origin $r = 0$ placed at the center of the sphere and r is the distance from the center of the particle (Fig. 1.7). The boundary conditions for $\psi(r)$, which are similar to Eqs. (1.21) and (1.22) for a planar surface, are given by

$$\psi = \psi_o \quad \text{at} \quad r = a^+ \tag{1.69}$$

$$\psi \to 0, \quad \frac{d\psi}{dr} \to 0 \quad \text{as} \quad zr \to \infty \tag{1.70}$$

1.4.1 Low Potential

When the potential is low, Eq. (1.68) can be linearized to give

$$\frac{d^2\psi}{dr^2} + \frac{2}{r}\frac{d\psi}{dr} = \kappa^2\psi \tag{1.71}$$

The solution to Eq. (1.71) subject to Eqs. (1.69) and (1.70) is

$$\psi(r) = \psi_o \frac{a}{r} e^{-\kappa(r-a)} \tag{1.72}$$

If we introduce the distance $x = r - a$ measured from the sphere surface, then Eq. (1.72) becomes

$$\psi(x) = \psi_o \frac{a}{a+x} e^{-\kappa x} \tag{1.73}$$

For $x \ll a$, Eq. (1.73) reduces to the potential distribution around the planar surface given by Eq. (1.19). This result implies that in the region very near the particle surface, the surface curvature may be neglected so that the surface can be regarded as planar. In the limit of $\kappa \to 0$, Eq. (1.72) becomes

$$\psi(r) = \psi_o \frac{a}{r} \tag{1.74}$$

which is the Coulomb potential distribution around a sphere as if there were no electrolyte ions or electrical double layers. Note that Eq. (1.74) implies that the potential decays over distances of the particle radius a, instead of the double layer thickness $1/\kappa$.

The surface charge density σ of the particle is related to the particle surface potential ψ_o obtained from the boundary condition at the sphere surface,

$$\left.\frac{\partial \psi}{\partial r}\right|_{r=a^+} = -\frac{\sigma}{\varepsilon_r \varepsilon_o} \tag{1.75}$$

which corresponds to Eq. (1.23) for a planar surface. By substituting Eq. (1.72) into Eq. (1.75), we find the following ψ_o–σ relationship:

$$\psi_o = \frac{\sigma}{\varepsilon_r \varepsilon_o \kappa (1 + 1/\kappa a)} \tag{1.76}$$

For $\kappa a \gg 1$, Eq. (1.76) tends to Eq. (1.26) for the ψ_o–σ relationship for the plate case. That is, for $\kappa a \gg 1$, the curvature of the particle surface may be neglected so that the particle surface can be regarded as planar. In the opposite limit of $\kappa a \ll 1$, Eq. (1.76) tends to

$$\psi_o = \frac{\sigma a}{\varepsilon_r \varepsilon_o} \tag{1.77}$$

If we introduce the total charge $Q = 4\pi a^2$ on the particle surface, then Eq. (1.77) can be rewritten as

$$\psi_o = \frac{Q}{4\pi \varepsilon_r \varepsilon_o a} \tag{1.78}$$

which is the Coulomb potential. This implies that for $\kappa a \ll 1$, the existence of the electrical double layer may be ignored. Figures 1.8 and 1.9, respectively, show that Eq. (1.76) tends to Eq. (1.26) for a planar ψ_o–σ relationship for $\kappa a \gg 1$ and to the Coulomb potential given by Eq. (1.78) for $\kappa a \ll 1$. It is seen that the surface potential ψ_o of a spherical particle of radius a can be regarded as ψ_o of a plate for $\kappa a \geq 10^2$ (Fig. 1.8) and as the Coulomb potential for $\kappa a \leq 10^{-2}$ (Fig. 1.9). That is, for $\kappa a \leq 10^{-2}$, the presence of the electrical double layer can be neglected.

1.4.2 Surface Charge Density–Surface Potential Relationship: Symmetrical Electrolyte

When the magnitude of the surface potential is arbitrary so that the Debye–Hückel linearization cannot be allowed, we have to solve the original nonlinear spherical Poisson–Boltzmann equation (1.68). This equation has not been solved but its approximate analytic solutions have been derived [5–8]. Consider a sphere of radius a with a

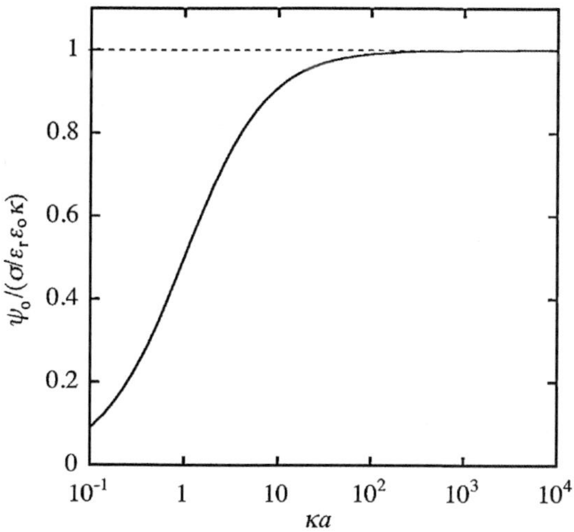

FIGURE 1.8 Surface potential ψ_o and surface charge density σ relationship of a spherical particle of radius a as a function of κa calculated with Eq. (1.76). For $\kappa a \geq 10^2$, the surface potential ψ_o can be regarded as $\psi_o = \sigma/\varepsilon_r\varepsilon_o\kappa$ for the surface potential of a plate.

surface charge density σ immersed in a symmetrical electrolyte solution of valence z and bulk concentration n. Equation (1.68) in the present case becomes

$$\frac{d^2\psi}{dr^2} + \frac{2}{r}\frac{d\psi}{dr} = \frac{2zen}{\varepsilon_r\varepsilon_o}\sinh\left(\frac{ze\psi}{kT}\right) \tag{1.79}$$

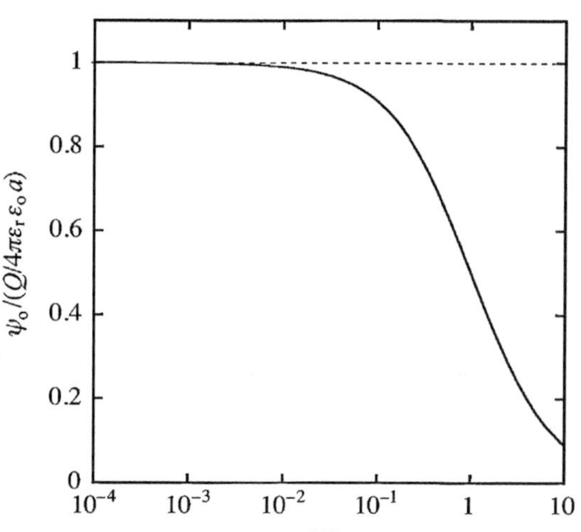

FIGURE 1.9 Surface potential ψ_o and surface charge density σ relationship of a spherical particle of radius a as a function of κa calculated with Eq. (1.76). For $\kappa a \leq 10^{-2}$, the surface potential ψ_o can be regarded as the Coulomb potential $\psi_o = Q/4\pi\varepsilon_r\varepsilon_o a$.

Loeb et al. tabulated numerical computer solutions to the nonlinear spherical Poisson–Boltzmann equation (1.63). On the basis of their numerical tables, they discovered the following empirical formula for the σ–ψ_o relationship:

$$\sigma = \frac{2\varepsilon_r\varepsilon_o\kappa kT}{ze}\left[\sinh\left(\frac{ze\psi_o}{2kT}\right) + \left(\frac{2}{\kappa a}\right)\tanh\left(\frac{ze\psi_o}{4kT}\right)\right] \tag{1.80}$$

where the Debye–Hückel parameter κ is given by Eq. (1.11). A mathematical basis of Eq. (1.80) was given by Ohshima et al. [7], who showed that if, on the left-hand side of Eq. (1.79), we replace $2/r$ with its large a limiting form $2/a$ and $d\psi/dr$ with that for a planar surface (the zeroth-order approximation given by Eq. (1.34)), namely,

$$\frac{2}{r}\frac{d\psi}{dr} \rightarrow \frac{2}{a}\frac{d\psi}{dr}\bigg|_{\text{zeroth-order}} = -\frac{4\kappa}{a}\sinh\left(\frac{ze\psi}{2kT}\right) \tag{1.81}$$

then Eq. (1.63) becomes

$$\frac{d^2y}{dr^2} = \kappa^2\left(\sinh y + \frac{4}{\kappa a}\sinh\left(\frac{y}{2}\right)\right) \tag{1.82}$$

where $y = ze\psi/kT$ is the scaled potential (Eq. (1.28)). Since the right-hand side of Eq. (1.82) involves only y (and does not involve r explicitly), Eq. (1.82) can be readily integrated by multiplying dy/dr on its both sides to yield

$$\frac{dy}{dr} = -2\kappa\sinh\left(\frac{y}{2}\right)\left[1 + \frac{2}{\kappa a\,\cosh^2(y/4)}\right]^{1/2} \tag{1.83}$$

By expanding Eq. (1.83) with respect to $1/\kappa a$ and retaining up to the first order of $1/\kappa a$, we obtain

$$\frac{dy}{dr} = -2\kappa\sinh\left(\frac{y}{2}\right)\left[1 + \frac{1}{\kappa a\cosh^2(y/4)}\right] \tag{1.84}$$

Substituting Eq. (1.84) into Eq. (1.75), we obtain Eq. (1.80), which is the first-order σ–ψ_o relationship.

A more accurate σ–ψ_o relationship can be obtained by using the first-order approximation given by Eq. (1.84) (not using the zeroth-order approximation for a planar surface given by Eq. (1.34)) in the replacement of Eq. (1.81), namely,

$$\frac{2}{r}\frac{dy}{dr} \rightarrow \frac{2}{a}\frac{dy}{dr}\bigg|_{\text{first-order}} = -\frac{4}{a}\kappa\sinh\left(\frac{y}{2}\right)\left[1 + \frac{1}{\kappa a\cosh^2(y/4)}\right] \tag{1.85}$$

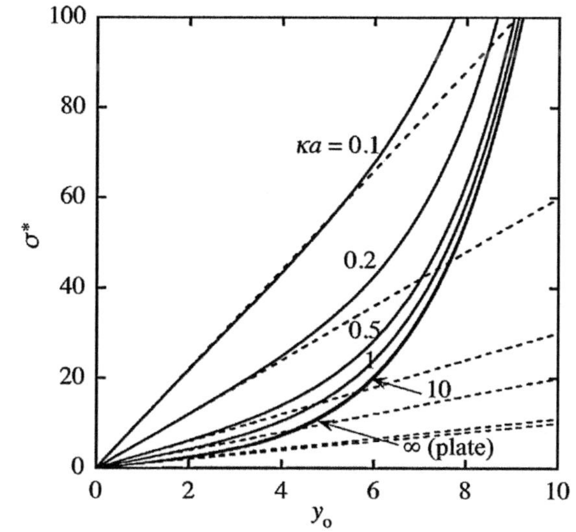

FIGURE 1.10 Scaled surface charge density $\sigma^* = ze\sigma/\varepsilon_r\varepsilon_o\kappa kT$ as a function of the scaled surface potential $y_o = ze\psi_o/kT$ for a positively charged sphere in a symmetrical electrolyte solution of valence z for various values of κa. Solid line, exact solution (Eq. (1.86)); dashed line, Debye–Hückel linearized solution (Eq. (1.76)).

The result is

$$\sigma = \frac{2\varepsilon_r\varepsilon_o\kappa kT}{ze}\sinh\left(\frac{ze\psi_o}{2kT}\right)\left[1 + \frac{1}{\kappa a}\frac{2}{\cosh^2(ze\psi_o/4kT)} + \frac{1}{(\kappa a)^2}\frac{8\ln[\cosh(ze\psi_o/4kT)]}{\sinh^2(ze\psi_o/2kT)}\right]^{1/2} \tag{1.86}$$

which is the second-order σ–ψ_o relationship [4]. The relative error of Eq. (1.80) is less than 1% for $\kappa a \geq 5$ and that of Eq. (1.86) is less than 1% for $\kappa a \geq 1$. Note that as y_o increases, Eqs. (1.80) and (1.86) approach Eq. (1.41). That is, as the surface potential y_o increases, the dependence of y_o on κa becomes smaller. Figure 1.10 gives the σ–ψ_o relationship for various values of κa calculated from Eq. (1.86) in comparison with the low-potential approximation (Eq. (1.76)).

1.4.3 Surface Charge Density–Surface Potential Relationship: Asymmetrical Electrolyte

The above approximation method can also be applied to the case of a sphere in a 2-1 symmetrical solution, yielding [4]

$$\sigma = \frac{\varepsilon_r\varepsilon_o\kappa kT}{e}\left[pq + \frac{2\{(3-p)q - 3\}}{\kappa apq}\right] \tag{1.87}$$

as the first-order σ–ψ_o relationship and

$$\sigma = \frac{\varepsilon_r \varepsilon_o \kappa kT}{e} pq \left[1 + \frac{4}{\kappa a} \frac{(3-p)q-3}{(pq)^2} + \frac{4}{(\kappa a)^2 (pq)^2} \left\{ 6 \ln\left(\frac{q+1}{2}\right) + \ln(1-p) \right\} \right]^{1/2} \tag{1.88}$$

as the second-order σ–ψ_o relationship, where

$$p = 1 - \exp(-e\psi_o/kT) \tag{1.89}$$

$$q = \left[\frac{2}{3} \exp\left(\frac{e\psi_o}{kT}\right) + \frac{1}{3} \right]^{1/2} \tag{1.90}$$

and κ is the Debye–Hückel parameter for a 2-1 electrolyte solution (Eq. (1.13)).

For the case of a mixed solution of 1-1 electrolyte of concentration n_1 and 2-1 electrolyte of concentration n_2, the first-order σ–ψ_o relationship is given by [7]

$$\sigma = \frac{\varepsilon_r \varepsilon_o \kappa kT}{e} \left[pt + \frac{2}{\kappa apt} \left\{ (3-p)t - 3 - \frac{3^{1/2}(1-\eta)}{2\eta^{1/2}} \ln\left(\frac{\{1+(\eta/3)^{1/2}\}\{t-(\eta/3)^{1/2}\}}{\{1-(\eta/3)^{1/2}\}\{t+(\eta/3)^{1/2}\}} \right) \right\} \right] \tag{1.91}$$

with

$$t = \left[\left(1 - \frac{\eta}{3}\right) \exp\left(\frac{e\psi_o}{kT}\right) + \frac{\eta}{3} \right]^{1/2} \tag{1.92}$$

$$\eta = \frac{3n_2}{n_1 + 3n_2} \tag{1.93}$$

where κ is the Debye–Hückel parameter for a mixed solution of 1-1 and 2-1 electrolytes (Eq. (1.14)) and p is given by Eq. (1.89).

1.4.4 Surface Charge Density–Surface Potential Relationship: General Electrolyte

The above method can be extended to the case of general electrolytes composed of N ionic species with valence z_i and bulk concentration (number density) n_i^∞ ($i = 1, 2, \ldots, N$) (in units of m^{-3}) [8]. The spherical Poisson–Boltzmann equation (1.68) can be rewritten in terms of the scaled potential $y \equiv e\psi/kT$ as

$$\frac{d^2 y}{dr^2} + \frac{2}{r}\frac{dy}{dr} = -\frac{\kappa^2 \sum_{i=1}^{N} z_i n_i e^{-z_i y}}{\sum_{i=1}^{N} z_i^2 n_i} \tag{1.94}$$

where κ is the Debye–Hückel parameter of the solution and defined by Eq. (1.10). In the limit of large κa, Eq. (1.94) reduces to the planar Poisson–Boltzmann equation (1.53), namely,

$$\frac{d^2y}{dr^2} = -\frac{\kappa^2 \sum_{i=1}^{N} z_i n_i e^{-z_i y}}{\sum_{i=1}^{N} z_i^2 n_i} \tag{1.95}$$

Integration of Eq. (1.95) gives

$$\frac{dy}{dr} = -\mathrm{sgn}(y_o)\kappa \left[\frac{2\sum_{i=1}^{N} n_i(e^{-z_i y} - 1)}{\sum_{i=1}^{N} z_i^2 n_i}\right]^{1/2} \tag{1.96}$$

Equation (1.96) is thus rewritten as

$$\frac{dy}{dr} = -\kappa f(y) \tag{1.97}$$

where $f(y)$ is defined by Eq. (1.56). Note that as $y \to 0$, $f(y)$ tends to y and the right--hand side of Eq. (1.94) is expressed as $\kappa^2 f(y)df/dy$. By combining Eqs. (1.75) and (1.97), we obtain

$$\sigma = -\varepsilon_r\varepsilon_o \left.\frac{d\psi}{dr}\right|_{r=a^+} = \frac{\varepsilon_r\varepsilon_o\kappa kT}{e} f(y_o) \tag{1.98}$$

where $y_o \equiv e\psi/kT$ is the scaled surface potential. Equation (1.98) is the zeroth-order σ–y_o relationship.

To obtain the first-order σ–y_o relationship, we replace the second term on the left-hand side of Eq. (1.94) by the corresponding quantity for the planar case (Eq. (1.88)), namely,

$$\frac{2}{r}\frac{dy}{dr} \to \frac{2}{a}\left.\frac{dy}{dr}\right|_{\text{zeroth-order}} = -\frac{2\kappa}{a} f(y) \tag{1.99}$$

Equation (1.94) thus becomes

$$\frac{d^2y}{dr^2} = \frac{2\kappa}{a} f(y) + \kappa^2 f(y)\frac{df}{dy} \tag{1.100}$$

which is readily integrated to give

$$\frac{dy}{dr} = -\kappa f(y)\left[1 + \frac{4}{\kappa a f^2(y)}\int_0^y f(u)du\right]^{1/2} \tag{1.101}$$

By expanding Eq. (1.101) with respect to $1/\kappa a$ and retaining up to the first order of $1/\kappa a$, we have

$$\frac{dy}{dr} = -\kappa f(y)\left[1 + \frac{2}{\kappa a f^2(y)}\int_0^y f(u)du\right] \tag{1.102}$$

From Eqs. (1.75) and (1.102) we obtain the first-order σ–y_o relationship, namely,

$$\sigma = \frac{\varepsilon_r \varepsilon_o \kappa kT}{e} f(y_o)\left[1 + \frac{2}{\kappa a f^2(y_o)}\int_0^{y_o} f(u)du\right] \tag{1.103}$$

Note that since $f(y_o) \to y_o$ as $y_o \to 0$, Eq. (1.103) tends to the correct form in the limit of small y_o (Eq. (1.76)), namely,

$$\sigma = \varepsilon_r \varepsilon_o \kappa \left(1 + \frac{1}{\kappa a}\right)\psi_o \tag{1.104}$$

We can further obtain the second-order σ–y_o relationship by replacing the second term on the left-hand side of Eq. (1.94) by the corresponding quantity for the first-order case (i.e., by using Eq. (1.101) instead of Eq. (1.97)), namely,

$$\frac{2}{r}\frac{dy}{dr} \to \frac{2}{a}\frac{dy}{dr}\bigg|_{\text{first-order}} = -\frac{2\kappa}{a} f(y)\left[1 + \frac{2}{\lambda \kappa a f^2(y)}\int_0^y f(u)du\right] \tag{1.105}$$

where we have introduced a fitting parameter λ. Then, Eq. (1.94) becomes

$$\frac{d^2y}{dr^2} = \frac{2\kappa}{a} f(y)\left[1 + \frac{2}{\lambda \kappa a f^2(y)}\int_0^y f(u)du\right] + \kappa^2 f(y)\frac{df}{dy} \tag{1.106}$$

which is integrated to give

$$\sigma = \frac{\varepsilon_r \varepsilon_o \kappa kT}{e} f(y_o)\left[1 + \frac{4}{\kappa a f^2(y_o)}\int_0^{y_o} f(y)dy + \frac{8}{\lambda(\kappa a)^2 f^2(y_o)}\int_0^{y_o}\frac{1}{f(y)}\left\{\int_0^y f(u)du\right\}dy\right]^{1/2} \tag{1.107}$$

In the limit of small y_o, Eq. (1.107) tends to

$$\sigma = \varepsilon_r \varepsilon_o \kappa \psi_o \left[1 + \frac{2}{\kappa a} + \frac{2}{\lambda(\kappa a)^2}\right]^{1/2} \tag{1.108}$$

We thus choose $\lambda = 2$ to obtain Eq. (1.104). Therefore, Eq. (1.107) becomes

$$\sigma = \frac{\varepsilon_r \varepsilon_o \kappa kT}{e} f(y_o) \left[1 + \frac{4}{\kappa a f^2(y_o)} \int_0^{y_o} f(y) dy + \frac{4}{(\kappa a)^2 f^2(y_o)} \int_0^{y_o} \frac{1}{f(y)} \left\{ \int_0^{y} f(u) du \right\} dy \right]^{1/2} \tag{1.109}$$

which is the required second-order σ–y_o relationship.

For some simple cases, from Eqs. (1.103) and (1.109) with Eq. (1.56) one can derive explicit expressions for the σ–y_o relationship. Indeed, for 1-1 and 2-1 electrolyte solutions and their mixed solution, Eqs. (1.103) and (1.109) with Eqs. (1.58)–(1.60) yield Eqs. (1.80) and (1.86)–(1.88). As another example, one can derive expressions for the σ–y_o relationship for the case of 3-1 electrolytes of concentration n. In this case, the Debye–Hückel parameter κ and $f(y)$ are given by Eqs. (1.15) and (1.62), respectively. By substituting Eq. (1.62) into Eqs. (1.103) and (1.109) and carrying out numerical integration, we can derive the first-order and second-order σ–y_o relationships, respectively. The relative error of Eq. (1.103) is less than 1% for $\kappa a \geq 5$ and that of Eq. (1.109) is less than 1% for $\kappa a \geq 1$.

1.4.5 Potential Distribution Around a Sphere with Arbitrary Potential

By using an approximation method similar to the above method and the method of White [6], one can derive an accurate analytic expression for the potential distribution around a spherical particle. Consider a sphere of radius a in a symmetrical electrolyte solution of valence z and bulk concentration n[7]. The spherical Poisson–Boltzmann equation (1.68) in this case becomes

$$\frac{d^2 y}{dr^2} + \frac{2}{r}\frac{dy}{dr} = \kappa^2 \sinh y \tag{1.110}$$

where $y = ze\psi/kT$ is the scaled potential. Making the change of variables

$$s = \frac{a}{r} \exp(-\kappa(r-a)) \tag{1.111}$$

we can rewrite Eq. (1.110) as

$$s^2 \frac{d^2 y}{ds^2} + s\frac{dy}{ds} = \sinh y - \frac{2\kappa a + 1}{(\kappa a + 1)^2} G(y) \tag{1.112}$$

where

$$G(y) = \left(\frac{2\kappa r + 1}{2\kappa a + 1} \right) \left(\frac{\kappa a + 1}{\kappa r + 1} \right)^2 \left(\sinh y - s\frac{dy}{ds} \right) \tag{1.113}$$

and the boundary conditions (1.69) and (1.70) as

$$y = y_o \text{ for } s = 1 \tag{1.114}$$

$$y = \frac{dy}{ds} = 0 \text{ for } s = 0 \tag{1.115}$$

When $\kappa a \gg 1$, Eq. (1.112) reduces to

$$s^2 \frac{d^2y}{ds^2} + s\frac{dy}{ds} = \sinh y \tag{1.116}$$

with solution

$$y(r) = 2\ln\left[\frac{1 + \tanh(y_o/4)s}{1 - \tanh(y_o/4)s}\right] \tag{1.117}$$

an approximate expression obtained by White [6].

To obtain a better approximation, we replace $G(y)$ in Eq. (1.112) by its large κa limiting value

$$\begin{aligned}\lim_{\kappa a \to \infty} G(y) &= \sinh y - \lim_{\kappa a \to \infty}\left(s\frac{dy}{ds}\right) \\ &= \sinh y - 2\sinh(y/2)\end{aligned} \tag{1.118}$$

to obtain

$$s^2 \frac{d^2y}{ds^2} + s\frac{dy}{ds} = \sinh y - \frac{2\kappa a}{\kappa a + 1}\left\{y - 2\sinh\left(\frac{y}{2}\right)\right\} \tag{1.119}$$

This equation can be integrated once to give

$$s\frac{dy}{ds} = \frac{2\kappa a + 1}{(\kappa a + 1)^2}\sinh\left(\frac{y}{2}\right)\left[1 + \frac{2\kappa a + 1}{(\kappa a)^2\cosh^2(y/4)}\right]^{1/2} \tag{1.120}$$

which is further integrated to give

$$y(r) = 2\ln\left[\frac{(1 + Bs)(1 + Bs/(2\kappa a + 1))}{(1 - Bs)(1 - Bs/(2\kappa a + 1))}\right] \tag{1.121}$$

or

$$\psi(r) = \frac{2kT}{ze}\ln\left[\frac{(1 + Bs)(1 + Bs/(2\kappa a + 1))}{(1 - Bs)(1 - Bs/(2\kappa a + 1))}\right] \tag{1.122}$$

where

$$B = \frac{((2\kappa a + 1)/(\kappa a + 1))\tanh(y_o/4)}{1 + \left\{1 - ((2\kappa a + 1)/(\kappa a + 1)^2)\tanh^2(y_o/4)\right\}^{1/2}} \tag{1.123}$$

The relative error of Eq. (1.122) is less than 1% for $\kappa a \geq 1$. Note also that Eq. (1.121) or (1.122) exhibits the correct asymptotic form, namely,

$$y(r) = \text{costant} \times s \tag{1.124}$$

Figure 1.11 gives the scaled potential distribution $y(r)$ around a positively charged spherical particle of radius a with $y_o = 2$ in a symmetrical electrolyte solution of valence z for several values of κa. Solid lines are the exact solutions to Eq. (1.110) and dashed lines are the Debye–Hückel linearized results (Eq. (1.72)). Note that Eq. (1.122) is in excellent agreement with the exact results. Figure 1.12 shows the plot of the equipotential lines around a sphere with $y_o = 2$ at $\kappa a = 1$ calculated from Eq. (1.121). Figures 1.13 and 1.14, respectively, are the density plots of counterions (anions) ($n_-(r) = n\exp(+y(r))$) and coions (cations) ($n_+(r) = n\exp(-y(r))$) around the sphere calculated from Eq. (1.121).

Note that one can obtain the σ–y_o relationship from Eq. (1.120),

$$\begin{aligned}\sigma &= -\varepsilon_r\varepsilon_o \left.\frac{dy}{dr}\right|_{r=a^+} = \frac{\varepsilon_r\varepsilon_o\kappa kT}{e}\frac{\kappa a+1}{\kappa a} s\left.\frac{dy}{ds}\right|_{s=1} \\ &= \frac{2\varepsilon_r\varepsilon_o\kappa kT}{e}\sinh\left(\frac{y_o}{2}\right)\left[1+\frac{2\kappa a+1}{(\kappa a)^2\cosh^2(y_o/4)}\right]^{1/2}\end{aligned} \tag{1.125}$$

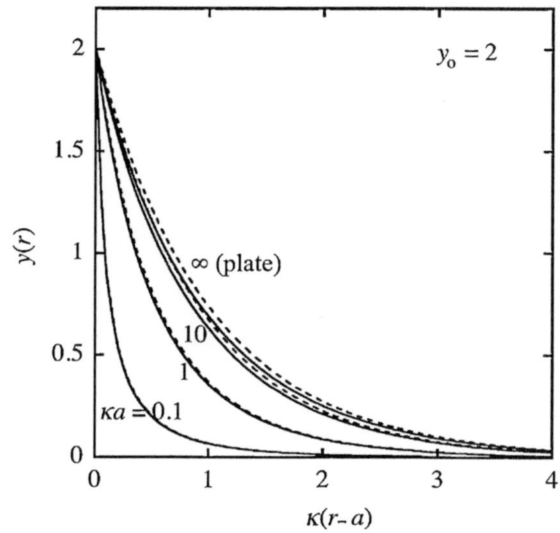

FIGURE 1.11 Scaled potential distribution $y(r)$ around a positively charged spherical particle of radius a with $y_o = 2$ in a symmetrical electrolyte solution of valence z for several values of κa. Solid lines, exact solution to Eq. (1.110); dashed lines, Debye–Hückel linearized solution (Eq. (1.72)). Note that the results obtained from Eq. (1.122) agree with the exact results within the linewidth.

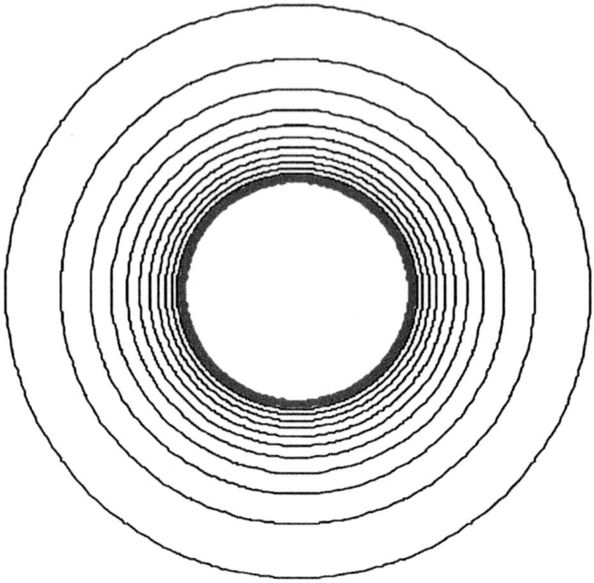

FIGURE 1.12 Contour lines (isopotential lines) for $\psi(r)$ around a positively charged sphere with $y_o = 2$ at $\kappa a = 1$. Arbitrary scale.

FIGURE 1.13 Density plots of counterions (anions) around a positively charged spherical particle with $y_o = 2$ at $\kappa a = 1$. Calculated from $n_{-}(r) = n\exp(+y(r))$ with the help of Eq. (1.121). The darker region indicates the higher density and $n_{-}(r)$ tends to its bulk value n far from the particle. Arbitrary scale.

FIGURE 1.14 Density plots of coions (cations) around a positively charged spherical particle with $y_o = 2$ at $\kappa a = 1$. Calculated from $n_+(r) = n\exp(-y(r))$ with the help of Eq. (1.121). The darker region indicates the higher density and $n_+(r)$ tends to its bulk value n far from the particle.

which corresponds to Eq. (1.80), but one cannot derive the second-order σ–y_o relationship by this method. The advantage in transforming the spherical Poisson–Boltzmann equation (1.79) into Eq. (1.112) lies in its ability to yield the potential distribution $y(r)$ that shows the correct asymptotic form (Eq. (1.124)).

We obtain the potential distribution around a sphere of radius a having a surface potential ψ_o immersed in a solution of general electrolytes [9]. The Poisson–Boltzmann equation for the electric potential $\psi(r)$ is given by Eq. (1.94), which, in terms of $f(r)$, is rewritten as

$$\frac{d^2y}{dr^2} + \frac{2}{r}\frac{dy}{dr} = \kappa^2 f(y)\frac{dy}{dr} \tag{1.126}$$

with $f(r)$ given by Eq. (1.56). We make the change of variables (Eq. (1.111)) and rewrite Eq. (1.126) as

$$s^2\frac{d^2y}{ds^2} + s\frac{dy}{ds} = f(y)\frac{df}{dy} - \frac{2\kappa r + 1}{(\kappa r + 1)^2}\left\{f(y)\frac{df}{dy} - s\frac{dy}{ds}\right\} \tag{1.127}$$

which is subject to the boundary conditions: $y = y_o$ at $s = 1$ and $y = dy/ds = 0$ at $s = 0$ (see Eqs. (1.114) and (1.115)). When $\kappa a \gg 1$, Eq. (1.127) reduces to

$$s^2\frac{d^2y}{ds^2} + s\frac{dy}{ds} = f(y)\frac{df}{dy} \tag{1.128}$$

which is integrated once to give

$$s\frac{dy}{ds} = f(y) \tag{1.129}$$

We then replace the second term on the right-hand side of Eq. (1.127) with its large κa limiting form, that is, $\kappa r \rightarrow \kappa a$ and $s\,dy/ds \rightarrow f(y)$ (Eq. (1.120)),

$$\frac{2\kappa r + 1}{(\kappa r + 1)^2}\left\{f(y)\frac{df}{dy} - s\frac{dy}{ds}\right\} \rightarrow \frac{2\kappa a + 1}{(\kappa a + 1)^2}\left\{f(y)\frac{df}{dy} - f(y)\right\} \tag{1.130}$$

Equation (1.127) then becomes

$$s^2\frac{d^2y}{ds^2} + s\frac{dy}{ds} = +f(y)\frac{df}{dy} - \frac{2\kappa a + 1}{(\kappa a + 1)^2}\left\{f(y)\frac{df}{dy} - s\frac{dy}{ds}\right\} \tag{1.131}$$

and integrating the result once gives

$$s\frac{dy}{ds} = F(y) \tag{1.132}$$

with

$$F(y) = \frac{\kappa a}{\kappa a + 1}f(y)\left[1 + \frac{2(2\kappa a + 1)}{(\kappa a)^2}\frac{1}{f^2(y)}\int_0^y f(u)du\right]^{1/2} \tag{1.133}$$

Note that $F(y) \rightarrow y$ as $y \rightarrow 0$ and that $F(y) \rightarrow f(y)$ as $\kappa a \rightarrow \infty$. Expressions for $F(y)$ for several cases are given below.

(i) For a 1-1 electrolyte solution,

$$F(y) = \frac{2\kappa a}{\kappa a + 1}\sinh\left(\frac{y}{2}\right)\left[1 + \frac{2(2\kappa a + 1)}{(\kappa a)^2}\frac{1}{\cosh^2(y/4)}\right]^{1/2} \tag{1.134}$$

(ii) For the case of 2-1 electrolytes,

$$\begin{aligned} F(y) &= \frac{\kappa a}{\kappa a + 1}(1 - e^{-y})\left(\frac{2}{3}e^y + \frac{1}{3}\right)^{1/2} \\ &\left[\frac{1 + 2(2\kappa a + 1)(2 + e^{-y})(\frac{2}{3}e^y + \frac{1}{3})^{1/2} - 3}{(\kappa a)^2(1 - e^{-y})^2(\frac{2}{3}e^y + \frac{1}{3})}\right]^{1/2} \end{aligned} \tag{1.135}$$

(iii) For the case of a mixed solution of 1-1 electrolyte of concentration n_1 and 2-1 electrolyte of concentration n_2,

$$\begin{aligned} F(y) = {} & \frac{\kappa a}{\kappa a+1}(1-e^{-y})\left[\left(1-\frac{\eta}{3}\right)e^{y}+\frac{\eta}{3}\right]^{1/2} \\ & \times\left[1+\frac{2(2\kappa a+1)}{(\kappa a)^2(1-e^{-y})^2\{(1-\eta/3)e^{y}+\eta/3\}}\right. \\ & \times\left\{(2+e^{-y})\sqrt{\left(1-\frac{\eta}{3}\right)e^{y}+\frac{\eta}{3}}-3-\frac{\sqrt{3}(1-\eta)}{2\sqrt{\eta}}\right. \\ & \left.\left.\times\ln\left(\frac{\{\sqrt{(1-\eta/3)e^{y}+\eta/3}-\sqrt{\eta/3}\}(1+\sqrt{\eta/3})}{\{\sqrt{(1-\eta/3)e^{y}+\eta/3}+\sqrt{\eta/3}\}(1-\sqrt{\eta/3})}\right)\right\}\right]^{1/2} \end{aligned} \tag{1.136}$$

where η is defined by Eq. (1.61).

Equation (1.132) is integrated again to give

$$-\ln s = \int_{y}^{y_o}\frac{dy}{F(y)} \tag{1.137}$$

Substituting Eq. (1.134) into Eq. (1.137), we obtain Eq. (1.122) with $z = 1$. For a 2-1 electrolyte and a mixture of 1-1 and 2-1 electrolytes, one can numerically calculate $y(r)$ from Eq. (1.137) with the help of Eqs. (1.135) and (1.136) for $F(y)$. For a 3-1 electrolyte and a mixture of 1-1 and 3-1 electrolytes, one can numerically calculate $F(y)$ from $f(y)$ (Eqs. (1.62) and (1.63)) with the help of Eq. (1.133) and then calculate $y(r)$ from Eq. (1.137).

1.5 CYLINDER

A similar approximation method can be applied for the case of infinitely long cylindrical particles of radius a in a general electrolyte composed of N ionic species with valence z_i and bulk concentration n_i ($i = 1, 2, \ldots, N$). The cylindrical Poisson–Boltzmann equation is

$$\frac{d^2\psi}{dr^2}+\frac{1}{r}\frac{d\psi}{dr} = -\frac{1}{\varepsilon_r\varepsilon_o}\sum_{i=1}^{N} z_i e n_i^{\infty}\exp\left(-\frac{z_i e\psi}{kT}\right) \tag{1.138}$$

where r is the radial distance measured from the center of the cylinder (Fig. 1.15). The conditions (1.69), (1.70), and (1.75) for a spherical particle of radius a are also

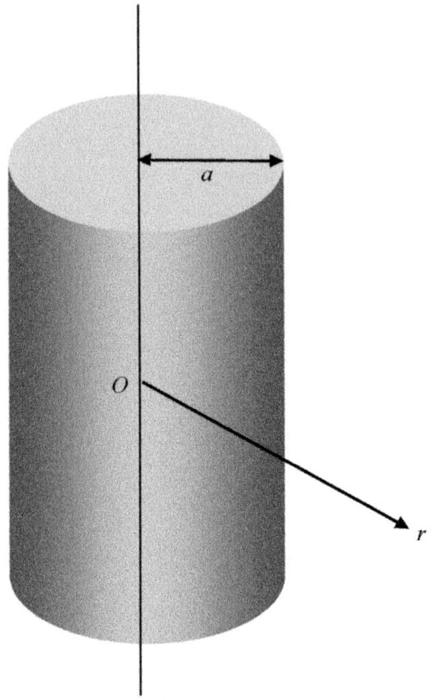

FIGURE 1.15 A cylinder of radius a.

applied for a cylindrical particle of radius a, namely,

$$\psi = \psi_o \text{ at } r = a^+ \tag{1.139}$$

$$\psi \to 0, \; \frac{d\psi}{dr} \to 0 \text{ as } r \to \infty \tag{1.140}$$

$$\left. \frac{\partial \psi}{\partial r} \right|_{r=a^+} = -\frac{\sigma}{\varepsilon_r \varepsilon_o} \tag{1.141}$$

where σ is the surface charge density of the cylinder.

1.5.1 Low Potential

For low potentials, Eq. (1.128) reduces to

$$\frac{d^2\psi}{dr^2} + \frac{1}{r}\frac{d\psi}{dr} = \kappa^2 \psi \tag{1.142}$$

where κ is given by Eq. (1.10). The solution is

$$\psi(r) = \psi_o \frac{K_0(\kappa r)}{K_0(\kappa a)} \tag{1.143}$$

where ψ_o is the surface potential of the particle and $K_n(z)$ is the modified Bessel function of the second kind of order n. The surface charge density σ of the particle is obtained from Eq. (1.141) as

$$\sigma = \varepsilon_r \varepsilon_o \kappa \psi_o \frac{K_1(\kappa a)}{K_0(\kappa a)} \tag{1.144}$$

or

$$\psi_o = \frac{\sigma}{\varepsilon_r \varepsilon_o \kappa} \frac{K_0(\kappa a)}{K_1(\kappa a)} \tag{1.145}$$

In the limit of $\kappa a \to \infty$, Eq. (1.145) approaches Eq. (1.26) for the plate case.

1.5.2 Arbitrary Potential: Symmetrical Electrolyte

For arbitrary ψ_o, accurate approximate analytic formulas have been derived [7,10], as will be shown below. Consider a cylinder of radius a with a surface charge density σ immersed in a symmetrical electrolyte solution of valence z and bulk concentration n. Equation (1.138) in this case becomes

$$\frac{d^2y}{dr^2} + \frac{1}{r}\frac{dy}{dr} = \kappa^2 \sinh y \tag{1.146}$$

where $y = ze\psi/kT$ is the scaled potential. By making the change of variables [6]

$$c = \frac{K_0(\kappa r)}{K_0(\kappa a)} \tag{1.147}$$

we can rewrite Eq. (1.146) as

$$c^2 \frac{d^2y}{dc^2} + c\frac{dy}{dc} = \sinh y - (1 - \beta^2)H(y) \tag{1.148}$$

where

$$H(y) = \left[\frac{1 - \{K_0(\kappa r)/K_1(\kappa r)\}^2}{1 - \beta^2}\right]\left(\sinh y - c\frac{dy}{dc}\right) \tag{1.149}$$

$$\beta = \frac{K_0(\kappa a)}{K_1(\kappa a)} \tag{1.150}$$

and the boundary conditions (1.139) and (1.140) as

$$y = y_o \quad \text{for } c = 1 \tag{1.151}$$

$$y = \frac{dy}{ds} = 0 \quad \text{for } c = 0 \tag{1.152}$$

In the limit $\kappa a \gg 1$, Eq. (1.148) reduces to

$$c^2 \frac{d^2y}{dc^2} + c\frac{dy}{dc} = \sinh y \tag{1.153}$$

with solution

$$y(c) = 2\ln\left[\frac{1 + \tanh(y_o/4)c}{1 - \tanh(y_o/4)c}\right] \tag{1.154}$$

an expression obtained by White [6]. We note that from Eq. (1.149)

$$\underset{\kappa a \to \infty}{H(y)} = \sinh y - 2\sinh(y/2) \tag{1.155}$$

and replacing $H(y)$ in Eq. (148) by its large κa limiting form (Eq. (1.155)) we obtain

$$c^2 \frac{d^2y}{dc^2} + c\frac{dy}{dc} = \sinh y - (1 - \beta^2)\{\sinh y - 2\sinh(y/2)\} \tag{1.156}$$

This equation can be integrated to give

$$y(r) = 2\ln\left[\frac{(1 + Dc)\{1 + ((1 - \beta)/(1 + \beta))Dc\}}{(1 - Dc)\{1 - ((1 - \beta)/(1 + \beta))Dc\}}\right] \tag{1.157}$$

and

$$\sigma = \frac{2\varepsilon_r\varepsilon_o\kappa kT}{ze}\sinh\left(\frac{y_o}{2}\right)\left[1 + \left(\frac{1}{\beta^2} - 1\right)\frac{1}{\cosh^2(y_o/4)}\right]^{1/2} \tag{1.158}$$

with

$$D = \frac{(1 + \beta)\tanh(y_o/4)}{1 + \{1 - (1 - \beta^2)\tanh^2(y_o/4)\}^{1/2}} \tag{1.159}$$

where $y_o = ze\psi_o/kT$ is the scaled surface potential of the cylinder. For low potentials, Eq. (1.158) reduces to Eq. (1.144).

1.5.3 Arbitrary Potential: General Electrolytes

We start with Eq. (1.138), which can be rewritten as

$$\frac{d^2y}{dr^2} + \frac{1}{r}\frac{dy}{dr} = -\frac{\kappa^2 \sum_{i=1}^{N} z_i n_i e^{-z_i y}}{\sum_{i=1}^{N} z_i^2 n_i} \tag{1.160}$$

where $y = e\psi/kT$

Equation (1.160) may further be rewritten as

$$\frac{d^2y}{dr^2} + \frac{1}{r}\frac{dy}{dr} = \kappa^2 f(y)\frac{df}{dy} \tag{1.161}$$

where $f(y)$ is defined by Eq. (1.56). Making the change of variables (Eq. (1.147)), we can rewrite Eq. (1.161) as

$$c^2\frac{d^2y}{dc^2} + c\frac{dy}{dc} = f(y)\frac{df}{dy} - \left[1 - \left\{\frac{K_0(\kappa r)}{K_1(\kappa r)}\right\}^2\right]\left\{f(y)\frac{df}{dy} - c\frac{dy}{dc}\right\} \tag{1.162}$$

When $\kappa a \gg 1$, Eq. (1.162) reduces to

$$c^2\frac{d^2y}{dc^2} + c\frac{dy}{dc} = f(y)\frac{df}{dy} \tag{1.163}$$

Equation (1.163) is integrated once to give

$$c\frac{dy}{dc} = f(y) \tag{1.164}$$

We then replace the second term on the right-hand side of Eq. (1.162) with its large κa limiting form, namely,

$$\left[1 - \left\{\frac{K_0(\kappa r)}{K_1(\kappa r)}\right\}^2\right]\left\{f(y)\frac{df}{dy} - c\frac{dy}{dc}\right\} \to (1-\beta^2)\left\{f(y)\frac{df}{dy} - f(y)\right\} \tag{1.165}$$

Equation (1.162) then becomes

$$c^2\frac{d^2y}{dc^2} + c\frac{dy}{dc} = f(y)\frac{df}{dy} - (1-\beta^2)\left\{f(y)\frac{df}{dy} - f(y)\right\} \tag{1.166}$$

and integrating the result once gives

$$c\frac{dy}{dc} = F(y) \tag{1.167}$$

with

$$F(y) = \beta f(y)\left[1 + 2\left(\frac{1}{\beta^2} - 1\right)\frac{1}{f^2(y)}\int_0^y f(u)du\right]^{1/2} \tag{1.168}$$

Note that $F(y) \to y$ as $y \to 0$ and that $F(y) \to f(y)$ as $\kappa a \to \infty$. Expressions for $F(y)$ for several cases are given below.

(i) For a 1-1 electrolyte solution,

$$F(y) = 2\beta \sinh\left(\frac{y}{2}\right)\left[1 + \left(\frac{1}{\beta^2} - 1\right)\frac{1}{\cosh^2(y/4)}\right]^{1/2} \tag{1.169}$$

(ii) For the case of 2-1 electrolytes,

$$F(y) = \beta(1 - e^{-y})\left(\frac{2}{3}e^y + \frac{1}{3}\right)^{1/2}\left[1 + 2\left(\frac{1}{\beta^2} - 1\right)\frac{(2 + e^{-y})(\frac{2}{3}e^y + \frac{1}{3})^{1/2} - 3}{(1 - 2^{-y})^2(\frac{2}{3}e^y + \frac{1}{3})}\right]^{1/2} \tag{1.170}$$

(iii) For the case of a mixed solution of 1-1 electrolyte of concentration n_1 and 2-1 electrolyte of concentration n_2,

$$\begin{aligned} F(y) &= \beta(1 - e^{-y})\left[\left(1 - \frac{\eta}{3}\right)e^y + \frac{\eta}{3}\right]^{1/2} \\ &\times\left[1 + \frac{2(\beta^{-2} - 1)}{(1 - e^{-y})^2\{(1 - \eta/3)e^y + \eta/3\}}\right. \\ &\times\left\{(2 + e^{-y})\sqrt{\left(1 - \frac{\eta}{3}\right)e^y + \frac{\eta}{3}} - 3 - \frac{\sqrt{3}(1 - \eta)}{2\sqrt{\eta}}\right. \\ &\left.\left.\times\ln\left(\frac{\{\sqrt{(1 - \eta/3)e^y + \eta/3} - \sqrt{\eta/3}\}(1 + \sqrt{\eta/3})}{\{\sqrt{(1 - \eta/3)e^y + \eta/3} + \sqrt{\eta/3}\}(1 - \sqrt{\eta/3})}\right)\right]\right\}^{1/2} \end{aligned} \tag{1.171}$$

where η is defined by Eq. (1.61).

The relationship between the reduced surface charge density σ and the reduced surface potential $y_o = e\psi_o/KT$ follows immediately from Eq. (1.141), namely [10],

$$\begin{aligned} \sigma &= -\varepsilon_r\varepsilon_o\frac{\partial\psi}{\partial r}\bigg|_{r=a^+} = \frac{\varepsilon_r\varepsilon_o\kappa kT}{e}\frac{K_1(\kappa r)}{K_0(\kappa a)}\frac{dy}{dc}\bigg|_{c=1} \\ &= \frac{\varepsilon_r\varepsilon_o\kappa kT}{e}\frac{1}{\beta}c\frac{dy}{dc}\bigg|_{c=1} = \frac{\varepsilon_r\varepsilon_o\kappa kT}{e}\frac{1}{\beta}F(y_o)| \end{aligned} \tag{1.172}$$

Note that when $|y_o| \ll 1$, Eq. (1.172) gives the correct limiting form (Eq. (1.144)), since $F(y) \to y$ as $y \to 0$.

For the case of a cylinder having a surface potential y_o in a 1-1 electrolyte solution, $F(y)$ is given by Eq. (1.169). Substitution of Eq. (1.169) into Eq. (1.172) yields Eq. (1.158) with $z = 1$. For the case of 2-1 electrolytes, $F(y)$ is given by Eq. (1.170). The σ–y_o relationship is thus given by

$$\sigma = \frac{\varepsilon_r \varepsilon_o \kappa kT}{e} pq \left[1 + 2\left(\frac{1}{\beta^2} - 1 \right) \frac{(3-p)q - 3}{(pq)^2} \right]^{1/2} \tag{1.173}$$

For a mixed solution of 1-1 electrolyte of concentration n_1 and 2-1 electrolyte of concentration n_2, $F(y)$ is given by Eq. (1.171). The σ–y_o relationship is thus given by

$$\sigma = \frac{\varepsilon_r \varepsilon_o \kappa kT}{e} pt \left[1 + \frac{2(\beta^{-2} - 1)}{(pt)^2} \left\{ (3-p)t - 3 - \frac{\sqrt{3}(1-\eta)}{2\sqrt{\eta}} \ln\left(\frac{(t - \sqrt{\eta/3})(1 + \sqrt{\eta/3})}{(t + \sqrt{\eta/3})(1 - \sqrt{\eta/3})} \right) \right\} \right]^{1/2} \tag{1.174}$$

with

$$t = \sqrt{\left(1 - \frac{\eta}{3}\right) e^{y_o} + \frac{\eta}{3}} \tag{1.175}$$

Similarly, for the case of 3-1 electrolytes, $F(y)$ is calculated from $f(y)$ (Eq. (1.62)). For a mixed solution of 3-1 electrolyte of concentration n_2 and 1-1 electrolyte of concentration n_1, $F(y)$ is calculated from $f(y)$ (Eq. (1.63)). By substituting the obtained expressions for $F(y)$ into Eq. (1.172) and carrying out numerical integration, we can derive the σ–y_o relationship.

Equation (1.167) is integrated again to give

$$-\ln c = \int_y^{y_o} \frac{dy}{F(y)} \tag{1.176}$$

Equation (1.176) gives the general expression for the potential distribution around a cylinder. For the special case of a cylinder in a 1-1 electrolyte, in which case $F(y)$ is given by Eq. (1.134), we obtain Eq. (1.157) with $z = 1$. For other types of electrolytes, one can calculate by using Eq. (1.176) with the help of the corresponding expression for $F(y)$.

1.6 ASYMPTOTIC BEHAVIOR OF POTENTIAL AND EFFECTIVE SURFACE POTENTIAL

Consider here the asymptotic behavior of the potential distribution around a particle (plate, sphere, or cylinder) at large distances, which will also be used for calculating the electrostatic interaction between two particles.

1.6.1 Plate

Consider a charged plate with arbitrary surface potential ψ_o in a symmetrical electrolyte solution of valence z and Debye–Hückel parameter κ. We take an x-axis perpendicular to the plate surface with its origin at the plate surface so that the region $x > 0$ corresponds to the solution phase while the region $x < 0$ to the plate interior. Equation (1.37) (or Eq. (1.38)) for the potential distribution $\psi(x)$ around the surface in the region far from the surface, that is, at large κx, takes the form

$$\psi(x) = \frac{4kT}{ze}\gamma e^{-\kappa x} = \frac{4kT}{ze}\tanh\left(\frac{ze\psi_o}{4kT}\right)e^{-\kappa x} \tag{1.177}$$

or

$$y(x) \equiv \frac{ze\psi(x)}{kT} = 4\gamma e^{-\kappa x} = 4\tanh\left(\frac{ze\psi_o}{4kT}\right)e^{-\kappa x} \tag{1.178}$$

Comparing Eqs. (1.25) and (1.177), we find that the effective surface potential ψ_{eff} of the plate is given by

$$\psi_{\text{eff}} = \frac{4kT}{ze}\gamma = \frac{kT}{ze}\cdot 4\tanh\left(\frac{ze\psi_o}{4kT}\right) \tag{1.179}$$

and the scaled effective surface potential $Y = zey_{\text{eff}}/kT$ is given by

$$Y = 4\gamma = 4\tanh\left(\frac{ze\psi_o}{4kT}\right) \tag{1.180}$$

where γ is defined by Eq. (1.39). For small potentials, the effective surface potential ψ_{eff} tends to the real surface potential ψ_o. Figure 1.16 shows the asymptotic solution (Eq. (1.178)) in comparison with the exact solution (Eq. (1.37)).

For a sphere in a 2-1 electrolyte solution, it follows from Eq. (1.44) that $\psi(x)$ asymptotes

$$\psi(x) = \frac{4kT}{e}\gamma' e^{-\kappa x} = \frac{kT}{e}\cdot 6\left\{\frac{\left(\frac{2}{3}e^{y_o}+\frac{1}{3}\right)^{1/2}-1}{\left(\frac{2}{3}e^{y_o}+\frac{1}{3}\right)^{1/2}+1}\right\}e^{-\kappa x} \tag{1.181}$$

where γ' is defined by Eq. (1.45). The effective surface potential ψ_{eff} is thus given by

$$\psi_{\text{eff}} = \frac{4kT}{e}\gamma' = \frac{kT}{e}\cdot 6\left\{\frac{\left(\frac{2}{3}e^{y_o}+\frac{1}{3}\right)^{1/2}-1}{\left(\frac{2}{3}e^{y_o}+\frac{1}{3}\right)^{1/2}+1}\right\} \tag{1.182}$$

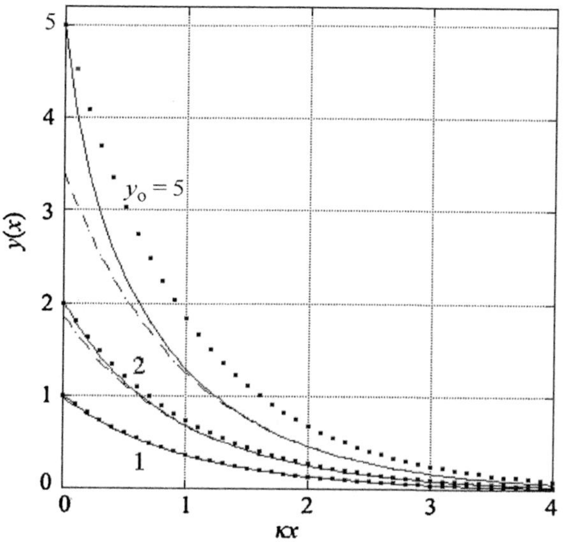

FIGURE 1.16 Potential distribution $y(x)$ around a positively charged plate with scaled surface potential $y_o \equiv ze\psi_o/kT$ in a symmetrical electrolyte solution for $y_o = 1$, 2, and 5. Solid lines are exact results (Eq. (1.37)), dashed lines are asymptotic results (Eq. (1.178)), and dotted lines are the Debye–Hückel approximation (Eq. (1.25)).

and the scaled effective surface potential $Y = zey_{\text{eff}}/kT$ is given by

$$Y = 4\gamma' = 6\left\{\frac{\left(\frac{2}{3}e^{y_o} + \frac{1}{3}\right)^{1/2} - 1}{\left(\frac{2}{3}e^{y_o} + \frac{1}{3}\right)^{1/2} + 1}\right\} \tag{1.183}$$

For the case of a mixed solution of 1-1 electrolyte of bulk concentration n_1 and 2-1 electrolyte of bulk concentration n_2, Eq. (1.48) in the region far from the surface, that is, at large κx, takes the form

$$\psi(x) = \frac{4kT}{e}\gamma'' e^{-\kappa x} = \frac{4kT}{e}\left(\frac{1}{1-\eta/3}\right)\left[\frac{\{(1-\eta/3)e^{y_o} + \eta/3\}^{1/2} - 1}{\{(1-\eta/3)e^{y_o} + \eta/3\}^{1/2} + 1}\right] e^{-\kappa x} \tag{1.184}$$

where η and γ'' are defined by Eqs. (1.50) and (1.51), respectively. The effective surface potential ψ_{eff} is thus given by

$$\psi_{\text{eff}} = \frac{4kT}{e}\gamma'' = \frac{4kT}{e}\left(\frac{1}{1-\eta/3}\right)\left[\frac{\{(1-\eta/3)e^{y_o} + \eta/3\}^{1/2} - 1}{\{(1-\eta/3)e^{y_o} + \eta/3\}^{1/2} + 1}\right] \tag{1.185}$$

and the scaled effective surface potential $Y = zey_{\text{eff}}/kT$ is given by

$$Y = 4\gamma'' = 4\left(\frac{1}{1-\eta/3}\right)\left[\frac{\{(1-\eta/3)e^{y_o} + \eta/3\}^{1/2} - 1}{\{(1-\eta/3)e^{y_o} + \eta/3\}^{1/2} + 1}\right] \tag{1.186}$$

We obtain the scaled effective surface potential Y for a plate having a surface potential ψ_o (or scaled surface potential y_o) immersed in a solution of general electrolytes [9]. Integration of the Poisson–Boltzmann equation for the electric potential $\psi(x)$ is given by Eq. (1.65), namely,

$$\kappa x = \int_y^{y_o} \frac{dy}{f(y)} \tag{1.187}$$

where $y = e\psi/kT$ and $f(x)$ is defined by Eq. (1.56). Equation (1.187) can be rewritten as

$$\begin{aligned} \kappa x &= \int_y^{y_o} \left\{ \frac{1}{f(y)} - \frac{1}{y} \right\} dy + \int_y^{y_o} \frac{1}{y} dy \\ &= \int_y^{y_o} \left\{ \frac{1}{f(y)} - \frac{1}{y} \right\} dy + \ln|y_o| - \ln|y| \end{aligned} \tag{1.188}$$

Note here that the asymptotic form $y(x)$ must be

$$y(x) = \text{constant} \times \exp(-\kappa x) \tag{1.189}$$

or

$$\kappa x = \text{constant} - \ln|y| \tag{1.190}$$

Therefore, the integral term of (1.188) must become independent of y at large x. Since y tends to zero in the limit of large x, the lower limit of the integration may be replaced by zero. We thus find that the asymptotic form $y(x)$ satisfies

$$\kappa x = \int_0^{y_o} \left\{ \frac{1}{f(y)} - \frac{1}{y} \right\} dy + \ln|y_o| - \ln|y| \tag{1.191}$$

It can be shown that the asymptotic form of $y(x)$ satisfies

$$y(x) = Ye^{-\kappa x} \tag{1.192}$$

with

$$Y = y_o \exp\left[\int_0^{y_o} \left\{ \frac{1}{f(y)} - \frac{1}{y} \right\} dy \right] \tag{1.193}$$

Equation (1.193) is the required expression for the scaled effective surface potential (or the asymptotic constant) Y. Wilemski [11] has derived an expression for Y (Eq. (15) with Eq. (12) in his paper [11]), which can be shown to be equivalent to Eq. (1.193). For a planar surface having scaled surface potentials y_o in a z-z symmetrical electrolyte solution, Eq. (1.193) reproduces Eq. (1.180). Similarly, for the case of a planar surface having a scaled surface potential y_o in a 2-1 electrolyte solution, Eq. (1.193) reproduces Eq. (1.183), while for the case of a mixed solution of 1-1 electrolyte of concentration n_1 and 2-1 electrolyte of concentration n_2, it gives Eq. (1.186).

1.6.2 Sphere

The asymptotic expression for the potential of a spherical particle of radius a in a symmetrical electrolyte solution of valence z and Debye–Hückel parameter κ at a large distance r from the center of the sphere may be expressed as

$$\psi(r) = \psi_{\text{eff}} \frac{a}{r} e^{-\kappa(r-a)} \tag{1.194}$$

$$y(r) = \frac{ze}{kT}\psi(r) = Y\frac{a}{r}e^{-\kappa(r-a)} \tag{1.195}$$

where r is the radial distance measured from the sphere center, ψ_{eff} is the effective surface potential, and $Y = ze\psi_{\text{eff}}/kT$ is the scaled effective surface potential of a sphere. From Eq. (1.122) we obtain

$$\psi_{\text{eff}} = \frac{kT}{ze} \cdot \frac{8\tanh(y_o/4)}{1 + \{1 - ((2\kappa a + 1)/(\kappa a + 1)^2)\tanh^2(y_o/4)\}^{1/2}} \tag{1.196}$$

or

$$Y = \frac{8\tanh(y_o/4)}{1 + \{1 - ((2\kappa a + 1)/(\kappa a + 1)^2)\tanh^2(y_o/4)\}^{1/2}} \tag{1.197}$$

It can be shown that ψ_{eff} reduces to the real surface potential ψ_o in the low-potential limit.

We obtain an approximate expression for the scaled effective surface potential Y for a sphere of radius a having a surface potential ψ_o (or scaled surface potential $y_o = ey_o/kT$) immersed in a solution of general electrolytes [9]. The Poisson–Boltzmann equation for the scaled electric potential $y(r) = e\psi/kT$ is approximately given by Eq. (1.137), namely,

$$-\ln s = \int_y^{y_o} \frac{dy}{F(y)} \tag{1.198}$$

with

$$s = \frac{a}{r}\exp(-\kappa(r-a)) \tag{1.199}$$

where $F(y)$ is defined by Eq. (1.133). It can be shown that the scaled effective surface potential $Y = e\psi_{\text{eff}}/kT$ is given by

$$Y = y_o \exp\left[\int_0^{y_o} \left\{\frac{1}{F(y)} - \frac{1}{y}\right\} dy\right] \tag{1.200}$$

Equation (1.200) is the required expression for the scaled effective surface potential (or the asymptotic constant) Y and reproduces Eq. (1.197) for a sphere of radius a having a surface potential ψ_o in a symmetrical electrolyte solution of valence z. The relative error of Eq. (1.200) is less than 1% for $\kappa a \geq 1$.

1.6.3 Cylinder

The effective surface potential ψ_{eff} or scaled effective surface potential $Y = ze\psi_{\text{eff}}/kT$ of a cylinder in a symmetrical electrolyte solution of valence z can be obtained from the asymptotic form of the potential around the cylinder, which in turn is derived from Eq. (1.157) as [7]

$$y(r) = Yc \tag{1.201}$$

with

$$c = \frac{K_0(\kappa r)}{K_0(\kappa a)} \tag{1.202}$$

and

$$Y = \frac{8\tanh(y_o/4)}{1 + \{1 - (1-\beta^2)\tanh^2(y_o/4)\}^{1/2}} \tag{1.203}$$

where r is the distance from the axis of the cylinder and

$$\beta = \frac{K_0(\kappa a)}{K_1(\kappa a)} \tag{1.204}$$

We obtain an approximate expression for the scaled effective surface potential Y for a cylinder of radius a having a surface potential ψ_{eff} (or scaled surface potential $Y = e\psi_{\text{eff}}/kT$) immersed in a solution of general electrolytes . The Poisson–Boltzmann equation for the scaled electric potential $y(r) = e\psi/kT$ is approximately given by Eq. (1.176), namely,

$$-\ln c = \int_y^{y_o} \frac{dy}{F(y)} \tag{1.205}$$

where $F(y)$ is defined by Eq. (1.168). It can be shown that the scaled effective surface potential $Y = e\psi_{\text{eff}}/kT$ is given by

$$Y = y_o \exp\left[\int_0^{y_o} \left\{\frac{1}{F(y)} - \frac{1}{y}\right\} dy\right] \tag{1.206}$$

Equation (1.206) reproduces Eq. (1.203) for a cylinder in a symmetrical electrolyte solution of valence z. The relative error of Eq. (1.206) is less than 1% for $\kappa a \geq 1$.

1.7 NEARLY SPHERICAL PARTICLE

So far we have treated uniformly charged planar, spherical, or cylindrical particles. For general cases other than the above examples, it is not easy to solve analytically the Poisson–Boltzmann equation (1.5). In the following, we give an example in which one can derive approximate solutions.

We give below a simple method to derive an approximate solution to the linearized Poisson–Boltzmann equation (1.9) for the potential distribution $\psi(\boldsymbol{r})$ around a nearly spherical spheroidal particle immersed in an electrolyte solution [12]. This method is based on Maxwell's method [13] to derive an approximate solution to the Laplace equation for the potential distribution around a nearly spherical particle.

Consider first a prolate spheroid with a constant uniform surface potential ψ_o in an electrolyte solution (Fig. 1.17a). The potential ψ is assumed to be low enough to obey the linearized Poisson–Boltzmann equation (1.9). We choose the z-axis as the axis of symmetry and the center of the prolate as the origin. Let a and b be the major and minor axes of the prolate, respectively. The equation for the surface of the prolate is then given by

$$\frac{x^2+y^2}{b^2}+\frac{z^2}{a^2}=1 \tag{1.207}$$

We introduce the spherical polar coordinate (r, θ, ϕ), that is, $r^2=x^2+y^2+z^2$ and $z=r\cos\theta$, and the eccentricity of the prolate

$$e_p=\sqrt{1-(b/a)^2} \tag{1.208}$$

Then, when the spheroid is nearly spherical (i.e., for low e_p), Eq. (1.207) becomes

$$r=a\left(1-\frac{e_p^2}{2}\sin^2\theta\right)=a\left[1+\frac{e_p^2}{3}\left\{\frac{1}{2}(3\cos^2\theta-1)-1\right\}\right] \tag{1.209}$$

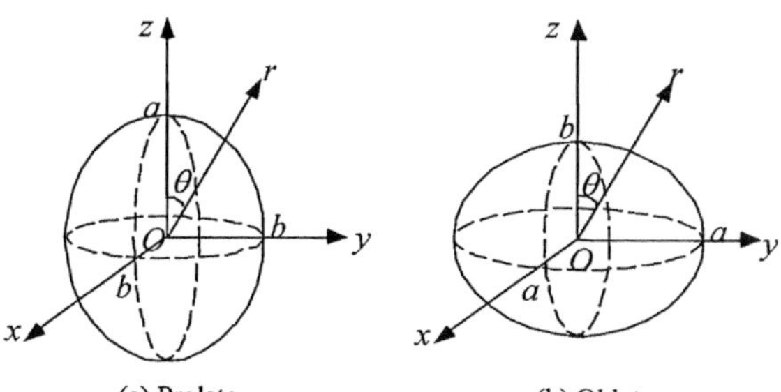

FIGURE 1.17 Prolate spheroid (a) and oblate spheroid (b). a and b are the major and minor semiaxes, respectively. The z-axis is the axis of symmetry.

which is an approximate equation for the surface of the prolate with low eccentricity e_p (which is correct to order e_p^2).

The solution to Eq. (1.9) must satisfy the boundary conditions that ψ tends to zero as $r \to \infty$ and $\psi = \psi_o$ at the prolate surface (given by Eq. (1.209)). We thus obtain

$$\psi(r,\theta) = \psi_o \frac{a}{r} e^{-\kappa(r-a)} - \psi_o \frac{(1+\kappa a)}{3} e_p^2 \left\{ \frac{a}{r} e^{-\kappa(r-a)} - \frac{k_2(\kappa r)}{4k_2(\kappa a)}(3\cos^2\theta - 1) \right\} \tag{1.210}$$

where $k_n(z)$ is the modified spherical Bessel function of the second kind of order n.

We can also obtain the surface charge density $\sigma(\theta)$ from Eq. (1.210), namely,

$$\sigma(\theta) = -\varepsilon_r \varepsilon_o \frac{\partial \psi}{\partial n} = -\varepsilon_r \varepsilon_o \cos\alpha \frac{\partial \psi}{\partial r} \text{ at } r = a\left[1 + \frac{e_p^2}{3}\left\{\frac{1}{2}(3\cos^2\theta - 1) - 1\right\}\right] \tag{1.211}$$

where α is the angle between n and r. It can be shown that $\cos\alpha = 1 + O(e_p^4)$. Then we find from Eqs. (1.210) and (1.211) that

$$\begin{aligned}\frac{\sigma(\theta)}{\varepsilon_r \varepsilon_o \kappa \psi_o} = {} & 1 + \frac{1}{\kappa a} + \frac{e_p^2}{3\kappa a}\Bigg[1 - \Bigg\{2 + 2\kappa a + (\kappa a)^2 \\ & -(1+\kappa a)\frac{\{9 + 9\kappa a + 4(\kappa a)^2 + (\kappa a)^3\}}{2\{3 + 3\kappa a + (\kappa a)^2\}}(3\cos^2\theta - 1)\Bigg\}\Bigg]\end{aligned} \tag{1.212}$$

Figure 1.18 shows equipotential lines (contours) around a prolate spheroid on the z–x plane at $y = 0$, calculated from Eq. (1.210) at $\kappa a = 1.5$ and $\kappa b = 1$.

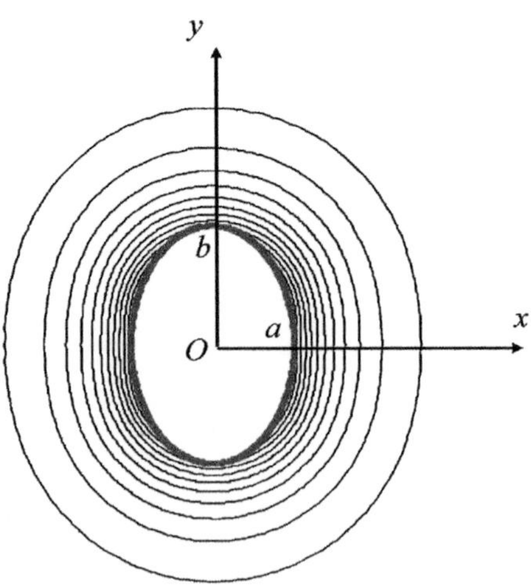

FIGURE 1.18 Equipotential lines (contours) around a prolate spheroid on the z–x plane at $y = 0$. Calculated from Eq. (1.210) at $\kappa a = 1.5$ and $\kappa b = 1$ (arbitrary size).

We next consider the case of an oblate spheroid with constant surface potential ψ_o (Fig. 1.17b). The surface of the oblate is given by

$$\frac{x^2+y^2}{a^2}+\frac{z^2}{b^2}=1 \tag{1.213}$$

where the z-axis is again the axis of symmetry, a and b are the major and minor semiaxes, respectively. Equation (1.213) can be approximated by

$$r=a\left(1+\frac{e_o^2}{2}\sin^2\theta\right)=a\left[1-\frac{e_o^2}{3}\left\{\frac{1}{2}(3\cos^2\theta-1)-1\right\}\right] \tag{1.214}$$

where the eccentricity e_o of the oblate is given by

$$e_o=\sqrt{(a/b)^2-1} \tag{1.215}$$

After carrying out the same procedure as employed for the case of the prolate spheroid, we find that $\psi(r,\theta)$ and the σ–ψ_o relationship, both correct to order e_o^2, are given by

$$\psi(r,\theta)=\psi_o\frac{b}{r}e^{-\kappa(r-b)}-\psi_o\frac{(1+\kappa b)}{3}e_o^2\left\{\frac{b}{r}e^{-\kappa(r-b)}-\frac{k_2(\kappa r)}{4k_2(\kappa b)}(3\cos^2\theta-1)\right\} \tag{1.216}$$

$$\begin{aligned}\frac{\sigma(\theta)}{\varepsilon_r\varepsilon_o\kappa\psi_o} = {} & 1+\frac{1}{\kappa b}-\frac{e_o^2}{3\kappa b}\Big[1-\Big\{2+2\kappa b+(\kappa b)^2 \\ & -(1+\kappa b)\frac{\{9+9\kappa b+4(\kappa b)^2+(\kappa b)^3\}}{2\{3+3\kappa b+(\kappa b)^2\}}(3\cos^2\theta-1)\Big\}\Big]\end{aligned} \tag{1.217}$$

which can also be obtained directly from Eqs. (1.210) and (1.212) by interchanging $a \leftrightarrow b$ and replacing e_p^2 by $-e_o^2$.

The last term on the right-hand side of Eqs. (1.210), (1.212), (1.216), and (1.217) corresponds to the deviation of the particle shape from a sphere.

REFERENCES

1. B. V. Derjaguin and L. Landau, *Acta Physicochim.* 14 (1941) 633.
2. E. J. W. Verwey and J. Th. G. Overbeek, *Theory of the Stability of Lyophobic Colloids*, Elsevier, Amsterdam, 1948.
3. H. Ohshima and K. Furusawa (Eds.), *Electrical Phenomena at Interfaces: Fundamentals, Measurements, and Applications*, 2nd edition, revised and expanded, Dekker, New York, 1998.
4. H. Ohshima, *Theory of Colloid and Interfacial Electric Phenomena*, Elsevier/Academic Press, Amsterdam, 2006.

5. A. L. Loeb, J. Th. G. Overbeek, and P. H. Wiersema, *The Electrical Double Layer Around a Spherical Colloid Particle*, MIT Press, Cambridge, MA, 1961.
6. L. R. White, *J. Chem. Soc., Faraday Trans.* 2 73 (1977) 577.
7. H. Ohshima, T. W. Healy, and L. R. White, *J. Colloid Interface Sci.* 90 (1982) 17.
8. H. Ohshima, *J. Colloid Interface Sci.* 171 (1995) 525,
9. H. Ohshima, *J. Colloid Interface Sci.* 174 (1995) 45.
10. H. Ohshima, *J. Colloid Interface Sci.* 200 (1998) 291.
11. G. Wilemski, *J. Colloid Interface Sci.* 88 (1982) 111.
12. H. Ohshima, *Colloids Surf. A: Physicochem. Eng. Aspects* 169 (2000) 13.
13. J. C. Maxwell, *A Treatise on Electricity and Magnetism*, Vol. 1, Dover, New York, 1954 p. 220.

2 Potential Distribution Around a Nonuniformly Charged Surface and Discrete Charge Effects

2.1 INTRODUCTION

In the previous chapter we assumed that particles have uniformly charged hard surfaces. This is called the smeared charge model. Electric phenomena on the particle surface are usually discussed on the basis of this model [1–4]. In this chapter we present a general method of solving the Poisson–Boltzmann equation for the electric potential around a nonuniformly charged hard surface. This method enables us to calculate the potential distribution around surfaces with arbitrary fixed surface charge distributions. With this method, we can discuss the discrete nature of charges on a particle surface (the discrete charge effects).

2.2 THE POISSON–BOLTZMANN EQUATION FOR A SURFACE WITH AN ARBITRARY FIXED SURFACE CHARGE DISTRIBUTION

Here we treat a planar plate surface immersed in an electrolyte solution of relative permittivity ε_r and Debye–Huckel parameter κ. We take x- and y-axes parallel to the plate surface and the z-axis perpendicular to the plate surface with its origin at the plate surface so that the region $z > 0$ corresponds to the solution phase (Fig. 2.1). First we assume that the surface charge density σ varies in the x-direction so that σ is a function of x, that is, $\sigma = \sigma(x)$. The electric potential ψ is thus a function of x and z. We assume that the potential $\psi(x, z)$ satisfies the following two-dimensional linearized Poison–Boltzmann equation, namely,

$$\left(\frac{\partial^2}{\partial x^2} + \frac{\partial^2}{\partial z^2}\right)\psi(x, z) = \kappa^2\psi(x, z) \tag{2.1}$$

Equation (2.1) is subject to the following boundary conditions at the plate surface and far from the surface:

$$\left.\frac{\partial\psi(x, z)}{\partial z}\right|_{z=0} = -\frac{\sigma(x)}{\varepsilon_r\varepsilon_o} \tag{2.2}$$

Biophysical Chemistry of Biointerfaces By Hiroyuki Ohshima

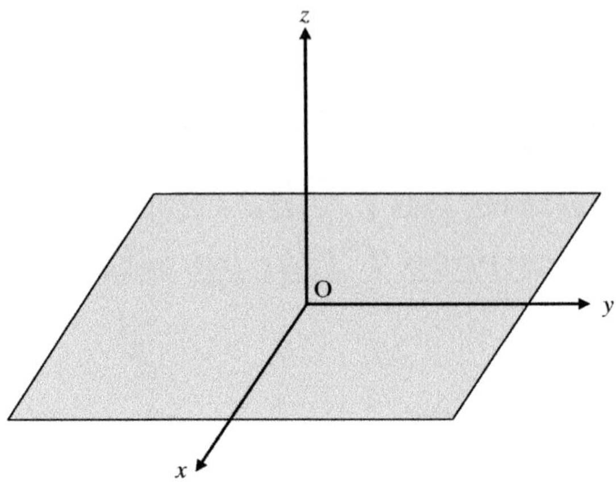

FIGURE 2.1 A charged plate in an electrolyte solution. The x- and y-axes are taken to be parallel to the plate surface and the z-axis perpendicular to the plate.

and

$$\psi(x, z) \to 0, \ \frac{\partial \psi}{\partial z} \to 0 \text{ as } z \to \infty \tag{2.3}$$

To solve Eq. (2.1) subject to Eq. (2.3), we write $\psi(x, z)$ and $\sigma(x)$ by their Fourier transforms,

$$\psi(x, z) = \frac{1}{2\pi} \int \hat{\psi}(k, z) e^{ikx} \, dk \tag{2.4}$$

$$\sigma(x) = \frac{1}{2\pi} \int \hat{\sigma}(k) e^{iks} \, dk \tag{2.5}$$

where $\hat{\psi}(x, z)$ and $\hat{\sigma}(k)$ are the Fourier coefficients. We thus obtain

$$\hat{\psi}(k, z) = \int \psi(x, z) e^{-iks} \, dx \tag{2.6}$$

$$\hat{\sigma}(k) = \int \sigma(x) e^{-ikx} \, dx \tag{2.7}$$

Substituting Eq. (2.4) into Eq. (2.1) subject to Eq. (2.3), we have

$$\frac{\partial^2 \hat{\psi}(k, z)}{\partial z^2} = (\kappa^2 + k^2) \hat{\psi}(k, z) \tag{2.8}$$

which is solved to give

$$\hat{\psi}(k, z) = C(k) \exp(-\sqrt{\kappa^2 + k^2} z) \tag{2.9}$$

where $C(k)$ is the Fourier coefficient independent of z. Equation (2.4) thus becomes

$$\psi(x,z) = \int C(k)\exp[ikx - z\sqrt{k^2+\kappa^2}]dk \tag{2.10}$$

The unknown coefficient $C(k)$ can be determined to satisfy the boundary condition (Eq. (2.2)) to give

$$C(k) = \frac{\hat{\sigma}(k)}{\varepsilon_r\varepsilon_o\sqrt{\kappa^2+k^2}} \tag{2.11}$$

Substituting Eq. (2.11) into Eq. (2.10), we have

$$\psi(x,z) = \frac{1}{2\pi\varepsilon_r\varepsilon_o}\int \frac{\hat{\sigma}(k)}{\sqrt{\kappa^2+k^2}}\exp\left[ikx - \sqrt{k^2+\kappa^2}z\right]dk \tag{2.12}$$

which is the general expression for $\psi(x, z)$.

Consider several cases of charge distributions.

(i) *Uniform smeared charge density* Consider a plate with a uniform surface charge density σ

$$\sigma(x) = \sigma \tag{2.13}$$

From Eq. (2.7), we have

$$\hat{\sigma}(k) = \sigma\int \exp(-ikx)dx = 2\pi\sigma\delta(k) \tag{2.14}$$

where $\delta(k)$ is Dirac's delta function and we have used the following relation:

$$\delta(k) = \frac{1}{2\pi}\int \exp(ikx)dx \tag{2.15}$$

Substituting this result into Eq. (2.12), we have

$$\psi(x,z) = \frac{\sigma}{\varepsilon_r\varepsilon_o}\int \frac{\delta(k)}{\sqrt{\kappa^2+k^2}}\exp\left[ikx - \sqrt{k^2+\kappa^2}z\right]dk \tag{2.16}$$

Carrying out the integration, we obtain

$$\psi(z) = \frac{\sigma}{\varepsilon_r\varepsilon_o\kappa}e^{-\kappa z} \tag{2.17}$$

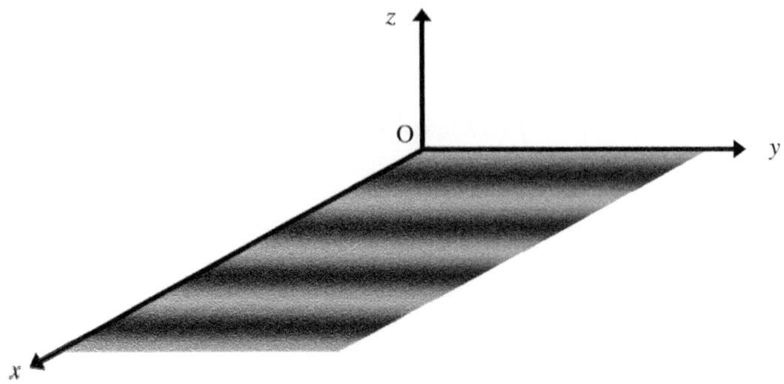

FIGURE 2.2 A plate surface with a sinusoidal charge distribution.

which is independent of x and agrees with Eq. (1.25) combined with Eq. (1.26), as expected.

(ii) *Sinusoidal charge distribution* Consider the case where the surface charge density $\sigma(x)$ varies sinusoidally (Fig. 2.2), namely,

$$\sigma(x) = \cos(qx) \tag{2.18}$$

From Eq. (2.7), we have

$$\begin{aligned}\hat{\sigma}(k) &= \int \cos(qx)e^{-ikx}\,dx \\ &= \frac{1}{2}\int (e^{iqx} + e^{-iqx})e^{-ikx}\,dx \\ &= \frac{\sigma}{2}(2\pi)\{\delta(k+q) + \delta(k-q)\}\end{aligned} \tag{2.19}$$

Substituting this result into Eq. (2.12), we have

$$\begin{aligned}\psi(x,z) &= \frac{\sigma}{2\varepsilon_r\varepsilon_o}\int \frac{\{\delta(k+q)+\delta(k-q)\}}{\sqrt{\kappa^2+k^2}}\exp\left[ikx - \sqrt{k^2+\kappa^2}z\right]dk \\ &= \frac{\sigma}{2\varepsilon_r\varepsilon_o}\frac{1}{\sqrt{\kappa^2+q^2}}\exp\left[iqx - \sqrt{q^2+\kappa^2}z\right] + \exp\left[-iqx - \sqrt{q^2+\kappa^2}z\right] \\ &= \frac{\sigma}{\varepsilon_r\varepsilon_o}\frac{\cos(qx)}{\sqrt{\kappa^2+q^2}}\exp\left[-\sqrt{q^2+\kappa^2}z\right]\end{aligned} \tag{2.20}$$

When $q = 0$, Eq. (2.20) reduces to Eq. (2.17) for the uniform charge distribution case. Figure 2.3 shows an example of the potential distribution $\psi(x, z)$ calculated from Eq. (2.20). It is to be noted that comparison of Eqs. (2.20) and (2.18) shows

$$\psi(x,z) = \frac{\sigma(x)}{\varepsilon_r\varepsilon_o}\frac{\exp[-\sqrt{q^2+\kappa^2}z]}{\sqrt{\kappa^2+q^2}} \tag{2.21}$$

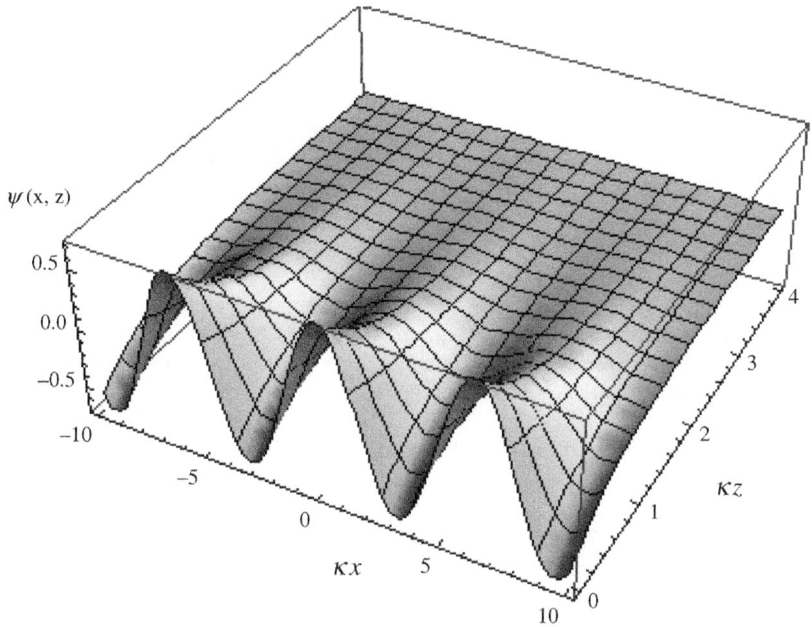

FIGURE 2.3 Scaled potential distribution $\psi^*(x, z) = \psi(x, z)/(\sigma/\varepsilon_r\varepsilon_o k)$ around a surface with a sinusoidal charge density distribution $\sigma(x) = \sigma\cos(qx)$ as a function of κx and κz, where κx and κz are the scaled distances in the x and y directions, respectively. Calculated for $q/\kappa = 1$.

That is, in this case the surface potential $\psi(x, z)$ is proportional to the surface charge density $\sigma(x)$.

The surface potential $\psi(x, 0)$ calculated from Eq. (2.20) is given by

$$\psi(x, 0) = \frac{\sigma}{\varepsilon_r\varepsilon_o}\frac{\cos(qx)}{\sqrt{\kappa^2 + q^2}} \tag{2.22}$$

which is a function of x.

(iii) *Sigmoidal distribution* We treat a planar charged plate with a gradient in surface charge density, which is σ_1 at one end and σ_2 at the other, varying sigmoidally between σ_1 and σ_2 along the plate surface [5] (Fig. 2.4). We may thus assume that the surface charge density σ varies in the x-direction according to the following form:

$$\begin{aligned}\sigma(x) &= \sigma_1 + \frac{\sigma_2 - \sigma_1}{1 + \exp(-\beta x)} \\ &= \frac{1}{2}(\sigma_1 + \sigma_2) + \frac{1}{2}(\sigma_2 - \sigma_1)\tanh\left(\frac{\beta x}{2}\right)\end{aligned} \tag{2.23}$$

where $\beta(\geq 0)$ is a parameter proportional to the slope of $\sigma(x)$ at $x = 0$. An example of charge distribution calculated for several values of β/κ at

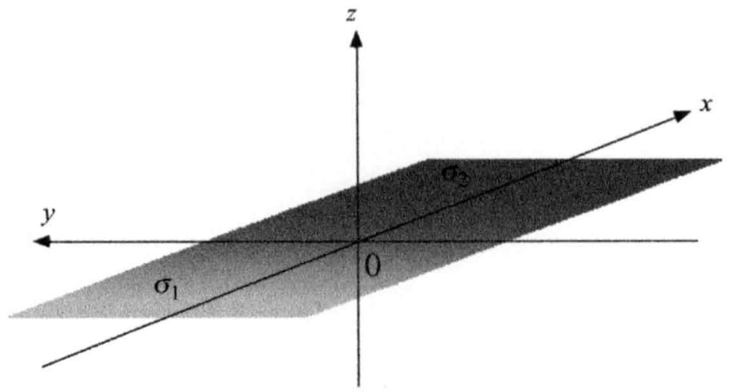

FIGURE 2.4 A plate surface with a sigmoidal gradient in surface charge density. The surface charge density σ varies in the x-direction. $\sigma(x)$ tends to σ_1 as $x \to -\infty$ and to σ_2 as $x \to +\infty$, varying sigmoidally around $x=0$. From Ref. 5.

$\sigma_2/\sigma_1 = 3$ is given in Fig. 2.5. From Eq. (2.7), we have

$$\hat{\sigma}(k) = \frac{1}{2}\left[(\sigma_1 + \sigma_2)\delta(k) - \frac{i(\sigma_2 - \sigma_1)}{\beta \sinh(\pi k/\beta)}\right] \tag{2.24}$$

Then by substituting Eq. (2.23) into Eq. (2.12), we obtain the solution to Eq. (2.1) as

$$\psi(x, z) = \frac{(\sigma_1 + \sigma_2)}{2\varepsilon_r\varepsilon_o\kappa} e^{-\kappa z} + \frac{(\sigma_2 - \sigma_1)}{\beta\varepsilon_r\varepsilon_o} \int_0^\infty \frac{\exp(-\kappa z\sqrt{t^2+1})\sin(\kappa x t)}{\sqrt{t^2+1}\,\sinh(\pi\kappa t/\beta)} dt \tag{2.25}$$

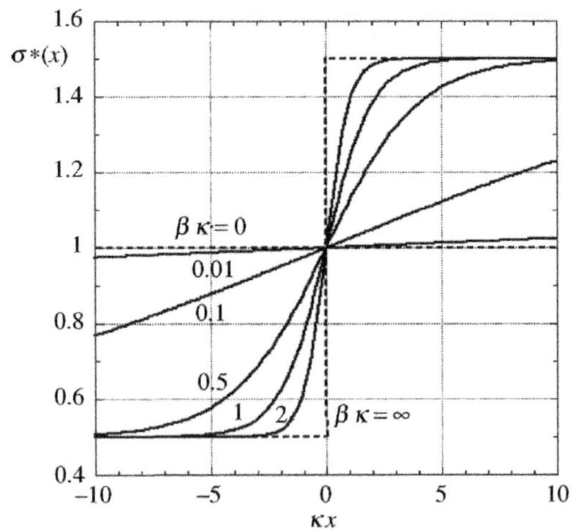

FIGURE 2.5 Sigmoidal distribution of reduced surface charge density $\sigma^*(x)$, defined by $\sigma^*(x) = 2\sigma(x)/(\sigma_1 + \sigma_2)$, as a function of κx for several values of β/κ at $\sigma_2/\sigma_1 = 3$. The limiting case $\beta/\kappa = \infty$ corresponds to a sharp boundary between two regions of σ_1 and σ_2. The opposite limiting case $\beta/\kappa = 0$ corresponds to a uniform charge distribution with a density $(\sigma_1 + \sigma_2)/2$. From Ref. 5.

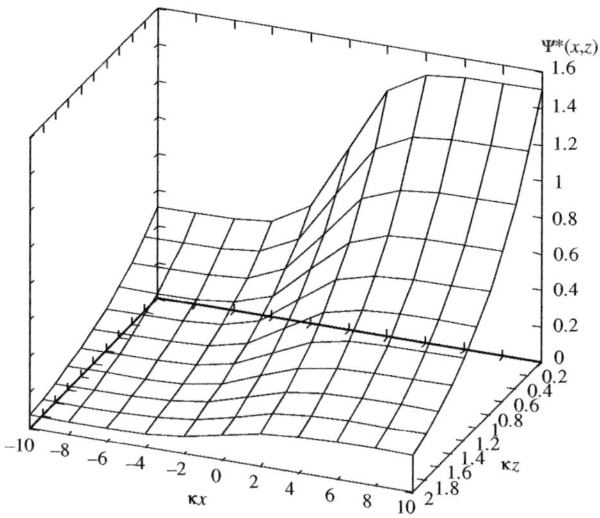

FIGURE 2.6 Reduced potential distribution $\psi(x, z) = 2\varepsilon_r\varepsilon_o\kappa\psi(x, z)/(\sigma_1 + \sigma_2)$ around a surface with a sigmoidal gradient in surface charge density as a function of κx and κz calculated from Eq. (2.25) for $\beta/\kappa = 1$ at $\sigma_2/\sigma_1 = 3$. From Ref. 5.

which gives the potential distribution ψ (x, z) around a charged plate with a sigmoidal gradient in surface charge density given by Eq. (2.23) as a function of x and z. The first term on the right-hand side of Eq. (2.25) corresponds to the potential for a plate carrying the average charge of σ_1 and σ_2. Equation (2.25) corresponds to the situation in which there is no flow of electrolyte ions along the plate surface, although a potential gradient (i.e., electric field) is formed along the plate surface. The potential gradient is counterbalanced by the concentration gradient of electrolyte ions. The electrochemical potential of electrolyte ions thus takes the same value everywhere in the solution phase so that there is no ionic flow.

To calculate the potential distribution via Eq. (2.25), one needs numerical integration. Figure 2.6 shows an example of the numerical calculation of $\psi(x, z)$ for $\beta/\kappa = 1$ at $\sigma_2/\sigma_1 = 3$. As will be shown later, the potential $\psi(x, z)$ is not always proportional to the surface charge density $\sigma(x)$ unlike the case where the plate is uniformly charged (σ = constant).

We now consider several limiting cases of Eq. (2.25).

(a) When $\beta/\kappa \gg 1$, Eq. (2.25) tends to the following limiting value:

$$\psi(x, z) = \frac{\sigma_1 + \sigma_2}{2\varepsilon_r\varepsilon_o\kappa}\left[\exp(-\kappa z) + \frac{2(\sigma_2 - \sigma_1)}{\pi(\sigma_1 + \sigma_2)}\int_0^{\infty}\frac{\exp(-\kappa z\sqrt{t^2+1})\sin(\kappa x t)}{t\sqrt{t^2+1}}dt\right] \tag{2.26}$$

which is independent of β/κ. Note that even in the limit of $\beta/\kappa = \infty$, which corresponds to a sharp boundary between two regions of σ_1 and σ_2 (Eq. (2.23)), the potential varies sigmoidally in the x-direction unlike $\sigma(x)$. This will be seen later again in Fig. 2.7, which shows the surface potential $\psi(x, 0)$.

(b) In the opposite limit $\beta/\kappa \ll 1$ (the charge gradient becomes small), Eq. (2.25) becomes

$$\psi(x,z) = \frac{\sigma_1 + \sigma_2}{2\varepsilon_r\varepsilon_o\kappa}\exp(-\kappa z) \tag{2.27}$$

Equation (2.27), which is independent of β/κ and x, coincides with the potential distribution produced by a plate with a uniform charge density $(\sigma_1 + \sigma_2)/2$ (the mean value of σ_1 and σ_2), as expected.

(c) When $\kappa z \gg 1$, Eq. (2.25) becomes

$$\begin{aligned}\psi(x,z) &= \frac{\sigma_1 + \sigma_2}{2\varepsilon_r\varepsilon_o\kappa}\left[\exp(-\kappa z) + \left(\frac{\sigma_2 - \sigma_1}{\sigma_1 + \sigma_2}\right)\tanh\left(\frac{\beta x}{2}\right)\exp(-\kappa z)\right] \\ &= \frac{\sigma(x)}{\varepsilon_r\varepsilon_o\kappa}\end{aligned} \tag{2.28}$$

That is, the potential $\psi(x, z)$ becomes directly proportional to the surface charge density $\sigma(x)$ in the region far from the plate surface, as in the case of uniform charge distribution ($\sigma =$ constant). The x-dependence of $\psi(x, z)$ thus becomes identical to that of $\sigma(x)$. It must be emphasized here that this is the case only for $\kappa z \gg 1$ and in general $\psi(x, z)$ is not proportional to $\sigma(x)$.

(d) As $\kappa x \to -\infty$ or $\kappa x \to +\infty$, Eq. (2.25) tends to

$$\psi(x,z) = \frac{\sigma_1}{\varepsilon_r\varepsilon_o\kappa}\exp(-\kappa z) \quad \text{as } x \to -\infty \tag{2.29}$$

$$\psi(x,z) = \frac{\sigma_2}{\varepsilon_r\varepsilon_o\kappa}\exp(-\kappa z) \quad \text{as } x \to +\infty \tag{2.30}$$

The potential distribution given by Eq. (2.29) (or Eq. (2.30)) agrees with that produced by a plate with surface charge density σ_1 (or σ_2), as expected.

The surface potential of the plate, $\psi(x, 0)$, is given by

$$\psi(x,0) = \frac{\sigma_1 + \sigma_2}{2\varepsilon_r\varepsilon_o\kappa}\left[1 + \frac{2\kappa(\sigma_2 - \sigma_1)}{\beta(\sigma_1 + \sigma_2)}\int_0^\infty \frac{\sin(\kappa x t)}{\sqrt{t^2 + 1}\,\sinh(\pi\kappa t/\beta)}dt\right] \tag{2.31}$$

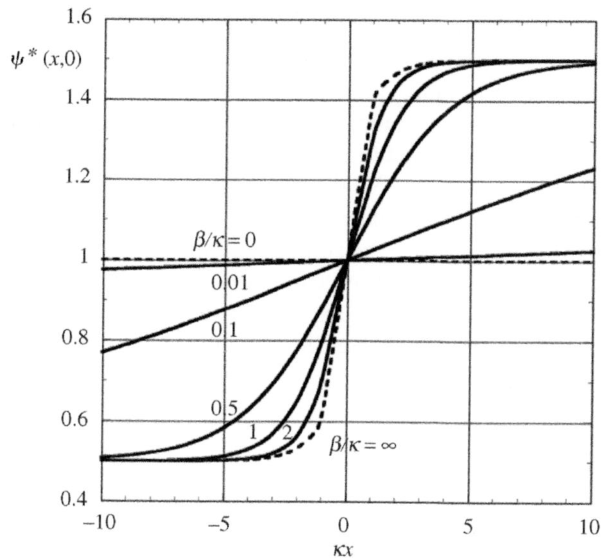

FIGURE 2.7 Distribution of the reduced surface potential $\psi^*(x, 0)$, defined by $\psi^*(x, 0) = 2\varepsilon_r\varepsilon_o\kappa\psi(x, 0)/(\sigma_1 + \sigma_2)$, along a surface with a sigmoidal gradient in surface charge density as a function of κx for several values of β/κ at $\sigma_2/\sigma_1 = 3$. From Ref. 5.

Figure 2.7 shows the distribution of the surface potential $\psi(x, 0)$ along the plate surface as a function of κx for several values of β/κ at $\sigma_2/\sigma_1 = 3$. Each curve for $\psi(x, 0)$ in Fig. 2.7 corresponds to a curve for $\sigma(x)$ with the same value of β/κ in Fig. 2.5. It is interesting to note that even for the limiting case of $\beta/\kappa = \infty$, which corresponds to a sharp boundary between two regions of σ_1 and σ_2, the potential does not show a step function type distribution but varies sigmoidally unlike $\sigma(x)$ in this limit.

We now derive expressions for the components of the electric field $\boldsymbol{E}$ (E_x, 0, E_z), which are obtained by differentiating $\psi(x, z)$ with respect to x or z, as potential gradient (i.e., electric field) is formed along the plate surface.

$$E_x = -\frac{\partial\psi}{\partial x} = -\frac{\sigma_2 - \sigma_1}{\varepsilon_r\varepsilon_o\beta}\kappa \int_0^\infty \frac{t\exp(-\kappa z\sqrt{t^2+1})\cos(\kappa x t)}{\sqrt{t^2+1}\,\sinh(\pi\kappa t/\beta)}dt \tag{2.32}$$

and

$$E_z = -\frac{\partial\psi}{\partial z} = \frac{\sigma_1 + \sigma_2}{2\varepsilon_r\varepsilon_o}\left[\exp(-\kappa z) + \frac{2\kappa(\sigma_2 - \sigma_1)}{\beta(\sigma_1 + \sigma_2)}\int_0^\infty \frac{\exp(-\kappa z\sqrt{t^2+1})\sin(\kappa x t)}{\sinh(\pi\kappa t/\beta)}dt\right] \tag{2.33}$$

2.3 DISCRETE CHARGE EFFECT

In some biological surfaces, especially the low charge density case, the discrete nature of the surface charge may become important. Here we treat a planar plate surface carrying discrete charges immersed in an electrolyte solution of relative permittivity ε_r and Debye–Huckel parameter κ. We take x- and y-axes parallel to the plate surface and the z-axis perpendicular to the plate surface with its origin at the plate surface so that the region $z > 0$ corresponds to the solution phase (Fig. 2.8). We treat the case in which the surface charge density σ is a function of x and y, that is, $\sigma = \sigma(x, y)$. The electric potential ψ is thus a function of x, y, and z. We assume that the potential $\psi(x, y, z)$ satisfies the following three-dimensional linearized Poison–Boltzmann equation:

$$\left(\frac{\partial^2}{\partial x^2} + \frac{\partial^2}{\partial y^2} + \frac{\partial^2}{\partial z^2}\right)\psi(x, y, z) = \kappa^2 \psi(x, y, z) \tag{2.34}$$

which is an extension of the two-dimensional linearized Poisson–Boltzmann equation (2.1) to the three-dimensional case. Equation (2.34) is subject to the following boundary conditions at the plate surface and far from the surface:

$$\left.\frac{\partial \psi(x, y, z)}{\partial z}\right|_{z=0} = -\frac{\sigma(x, y)}{\varepsilon_r \varepsilon_o} \tag{2.35}$$

and

$$\psi(x, y, z) \to 0, \quad \left.\frac{\partial \psi(x, y, z)}{\partial z}\right|_{z=0} \to 0 \quad \text{as} \quad z \to \infty \tag{2.36}$$

In Eq. (2.35) we have assumed that the electric filed within the plate can be neglected.

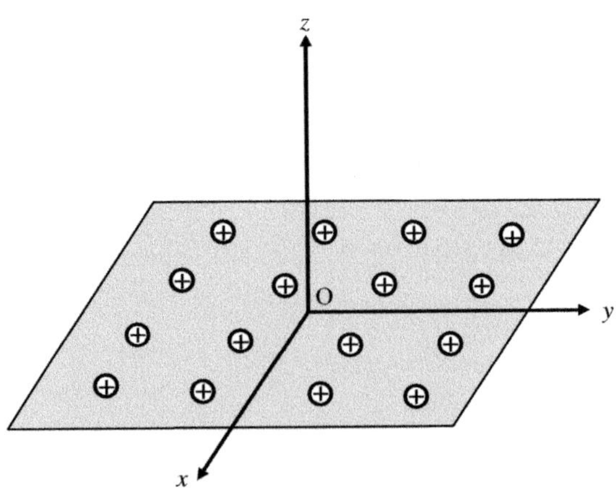

FIGURE 2.8 Plate surface carrying discrete charges in an electrolyte solution.

To solve Eq. (2.34), we write $\psi(x, y, z)$ and $\sigma(x, y)$ by their Fourier transforms,

$$\psi(\boldsymbol{r}) = \frac{1}{(2\pi)^2}\int \hat{\psi}(\boldsymbol{k}, z)\exp(i\boldsymbol{k}\cdot\boldsymbol{s})d\boldsymbol{k} \tag{2.37}$$

$$\sigma(\boldsymbol{s}) = \frac{1}{(2\pi)^2}\int \hat{\sigma}(\boldsymbol{k})\exp(i\boldsymbol{k}\cdot\boldsymbol{s})d\boldsymbol{k} \tag{2.38}$$

with

$$\boldsymbol{r} = (x, y, z),\ \boldsymbol{s} = (x, y),\ \text{and}\ \boldsymbol{k} = (k_x, k_y) \tag{2.39}$$

where $\hat{\psi}(\boldsymbol{k}, z)$ and $\hat{\sigma}(\boldsymbol{k})$ are the Fourier coefficients. We thus obtain

$$\hat{\psi}(\boldsymbol{k}, z) = \int \psi(\boldsymbol{r})\exp(-i\boldsymbol{k}\cdot\boldsymbol{s})d\boldsymbol{s} \tag{2.40}$$

$$\hat{\sigma}(\boldsymbol{k}) = \int \sigma(\boldsymbol{s})\exp(-i\boldsymbol{k}\cdot\boldsymbol{s})d\boldsymbol{s} \tag{2.41}$$

Substituting Eq. (2.40) into Eq. (2.34), we have

$$\frac{\partial^2\hat{\psi}(\boldsymbol{k}, z)}{\partial z^2} = (\kappa^2 + k^2)\hat{\psi}(\boldsymbol{k}, z) \tag{2.42}$$

where

$$k = \sqrt{k_x^2 + k_y^2} \tag{2.43}$$

Equation (2.42) is solved to give

$$\hat{\psi}(\boldsymbol{k}, z) = A(\boldsymbol{k})\exp(-\sqrt{\kappa^2 + k^2}z) \tag{2.44}$$

The unknown coefficient $A(\boldsymbol{k})$ can be determined to satisfy the boundary condition (Eq. (2.35)) to give

$$A(\boldsymbol{k}) = \frac{\hat{\sigma}(\boldsymbol{k})}{\varepsilon_r\varepsilon_o\sqrt{\kappa^2 + k^2}} \tag{2.45}$$

We thus obtain from Eq. (2.37)

$$\psi(\boldsymbol{r}) = \frac{1}{(2\pi)^2\varepsilon_r\varepsilon_o}\int \frac{\hat{\sigma}(\boldsymbol{k})}{\sqrt{\kappa^2 + k^2}}\exp(i\boldsymbol{k}\cdot\boldsymbol{s})\exp(-\sqrt{\kappa^2 + k^2}z)d\boldsymbol{k} \tag{2.46}$$

This is the general expression for the electric potential $\psi(x, y, z)$ around a plate surface carrying a discrete charge $\sigma(x, y)$.

Consider several cases of charge distributions.

(i) *Uniform smeared charge density* Consider a plate with a uniform surface charge density σ

$$\sigma(\boldsymbol{s}) = \sigma \tag{2.47}$$

From Eq. (2.41), we have

$$\begin{aligned}\hat{\sigma}(\boldsymbol{k}) &= \sigma \int \exp(-i\boldsymbol{k}\cdot\boldsymbol{s})d\boldsymbol{s} \\ &= (2\pi)^2 \sigma \delta(\boldsymbol{k})\end{aligned} \tag{2.48}$$

where we have used the following relation:

$$\delta(\boldsymbol{k}) = \frac{1}{(2\pi)^2} \int \exp(i\boldsymbol{k}\cdot\boldsymbol{s})d\boldsymbol{s} \tag{2.49}$$

Substituting this result into Eq. (2.46), we have

$$\psi(\boldsymbol{r}) = \frac{\sigma}{\varepsilon_r \varepsilon_o} \int \frac{\delta(\boldsymbol{k})}{\sqrt{\kappa^2 + k^2}} \exp(i\boldsymbol{k}\cdot\boldsymbol{s}) \exp(-\sqrt{\kappa^2 + k^2} z) d\boldsymbol{k} \tag{2.50}$$

Carrying out the integration, we obtain

$$\psi(\boldsymbol{r}) = \frac{\sigma}{\varepsilon_r \varepsilon_o \kappa} e^{-\kappa z} \tag{2.51}$$

which agrees with Eq. (1.25) combined with Eq. (1.26), as expected.

(ii) *A point charge* Consider a plate carrying only one point charge q situated at $(x, y) = (0, 0)$ (Fig. 2.9).

$$\sigma(\boldsymbol{s}) = q\delta(\boldsymbol{s}) \tag{2.52}$$

From Eq. (2.41), we have

$$\hat{\sigma}(\boldsymbol{k}) = q \int \delta(\boldsymbol{s}) \exp(-i\boldsymbol{k}\cdot\boldsymbol{s}) d\boldsymbol{s} = q \tag{2.53}$$

Substituting Eq. (2.53) into Eq. (2.46), we have

$$\psi(\boldsymbol{r}) = \frac{q}{(2\pi)^2 \varepsilon_r \varepsilon_o} \int \frac{1}{\sqrt{\kappa^2 + k^2}} \exp(i\boldsymbol{k}\cdot\boldsymbol{s}) \exp(-\sqrt{\kappa^2 + k^2} z) d\boldsymbol{k} \tag{2.54}$$

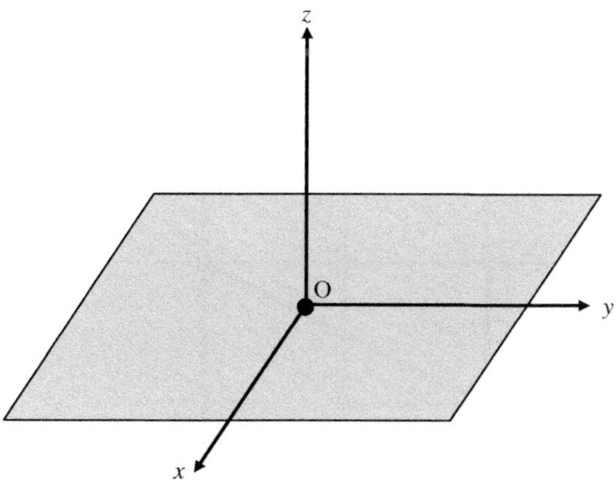

FIGURE 2.9 A point charge q on a plate surface.

We rewrite Eq. (2.54) by putting $\boldsymbol{k}\cdot\boldsymbol{s} = ks\cos\theta$:

$$\psi(\boldsymbol{r}) = \frac{q}{(2\pi)^2\varepsilon_r\varepsilon_o}\int_{k=0}^{2\pi}\int_{k=0}^{\infty}\frac{1}{\sqrt{\kappa^2+k^2}}\exp(iks\cos\theta - \sqrt{\kappa^2+k^2}z)k\,dk\,d\theta \tag{2.55}$$

With the help of the following relation:

$$\int_0^{2\pi}\exp(iks\cos\theta)d\theta = 2\pi J_0(ks) \tag{2.56}$$

where $J_0(z)$ is the zeroth-order Bessel function of the first kind, we can integrate Eq. (2.55) to obtain

$$\begin{aligned}\psi(\boldsymbol{r}) &= \frac{q}{2\pi\varepsilon_r\varepsilon_o}\int_{k=0}^{\infty}\frac{1}{\sqrt{\kappa^2+k^2}}\exp(-\sqrt{\kappa^2+k^2}z)J_0(ks)k\,dk \\ &= \frac{q}{2\pi\varepsilon_r\varepsilon_o}\frac{1}{\sqrt{s^2+z^2}}\exp(-\kappa\sqrt{s^2+z^2})\end{aligned} \tag{2.57}$$

which is rewritten as

$$\psi(r) = \frac{q}{2\pi\varepsilon_r\varepsilon_o r}e^{-\kappa r} \tag{2.58}$$

Note that Eq. (2.58) is twice the Coulomb potential $\psi(r) = qe^{-\kappa r}/4\pi_r\varepsilon_o r$ produced by a point charge q at distances r from the charge in an electrolyte solution of relative permittivity ε_r and Debye–Hückel parameter κ. This is because we have assumed that there is no electric field within the plate $x < 0$.

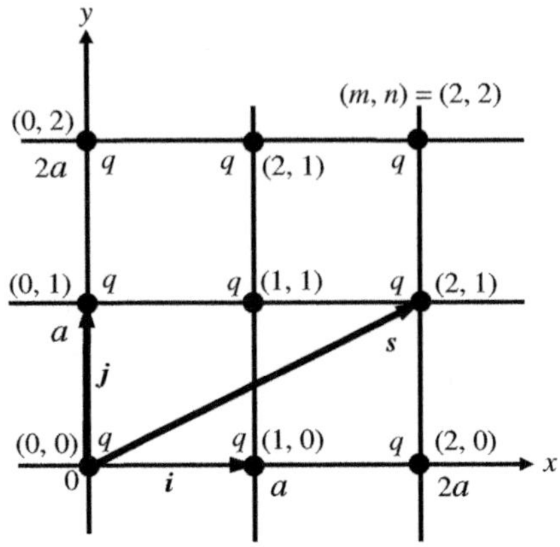

FIGURE 2.10 A squared lattice of point charges q with spacing a on a plate surface.

(iii) *Squared lattice of point charges* Consider a squared lattice of point charges q with spacing a (Fig. 2.10). The surface charge density $\boldsymbol{\sigma}(x, y)$ in this case is expressed as

$$\sigma(\boldsymbol{s}) = q\sum_{m,n}\delta(\boldsymbol{s} - m a\boldsymbol{i} - n a\boldsymbol{j}) \tag{2.59}$$

where $\boldsymbol{i}$ and $\boldsymbol{j}$ are unit vectors in the x and y directions, respectively, and m and n are integers. From Eq. (2.41), we have

$$\begin{aligned}\hat{\sigma}(\boldsymbol{k}) &= q\sum_{m,n}\int\delta(\boldsymbol{s} - m a\boldsymbol{i} - n a\boldsymbol{j})\exp(-i\boldsymbol{k}\cdot\boldsymbol{s})d\boldsymbol{s}\\ &= q\sum_{m,n}\exp[-i\boldsymbol{k}\cdot(m a\boldsymbol{i} + n a\boldsymbol{j})]\end{aligned} \tag{2.60}$$

Substituting this result into Eq. (2.46), we have

$$\psi(\boldsymbol{r}) = \frac{1}{(2\pi)^2}\frac{q}{\varepsilon_r\varepsilon_o}\sum_{m,n}\int\frac{\exp[i\boldsymbol{k}\cdot(\boldsymbol{s} - m a\boldsymbol{i} - n a\boldsymbol{j})]}{\sqrt{\kappa^2 + k^2}}\exp(-\sqrt{\kappa^2 + k^2}z)d\boldsymbol{k} \tag{2.61}$$

Carrying out the integration, we obtain

$$\psi(x, y, z) = \frac{q}{2\pi\varepsilon_r\varepsilon_o}\frac{\exp[-\kappa\sqrt{(x - ma)^2 + (x - na)^2 + z^2}]}{\sqrt{(x - ma)^2 + (x - na)^2 + z^2}} \tag{2.62}$$

which agrees with the results obtained by Nelson and McQuarrie [6].

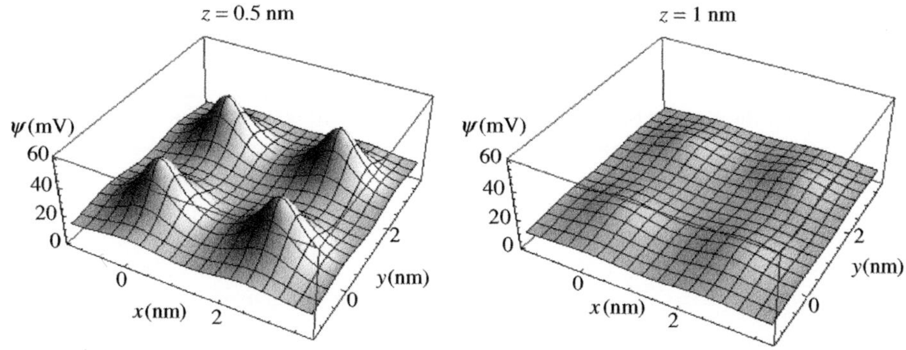

FIGURE 2.11 Potential distribution ψ near a plate surface carrying a squared lattice of point charges $q = e = 1.6 \times 10^{-19}$ C with spacing $a = 2.5$ nm in an aqueous 1-1 electrolyte solution of concentration 0.1 M at 25°C ($\varepsilon_r = 78.55$) calculated via Eq. (2.62) at distances $z = 0.5$ nm and $z = 1$ nm.

Figure 2.11 shows the potential distribution ψ near a plate surface carrying a squared lattice of point charges $q = e = 1.6 \times 10^{-19}$ C with spacing $a = 2.5$ nm in an aqueous 1-1 electrolyte solution of concentration 0.1 M at 25°C ($\varepsilon_r = 78.55$) calculated via Eq. (2.62) at distances $z = 0.5$ nm and $z = 1$ nm. Figure 2.12 shows the potential distribution $\psi(x, y, z)$ as a function of

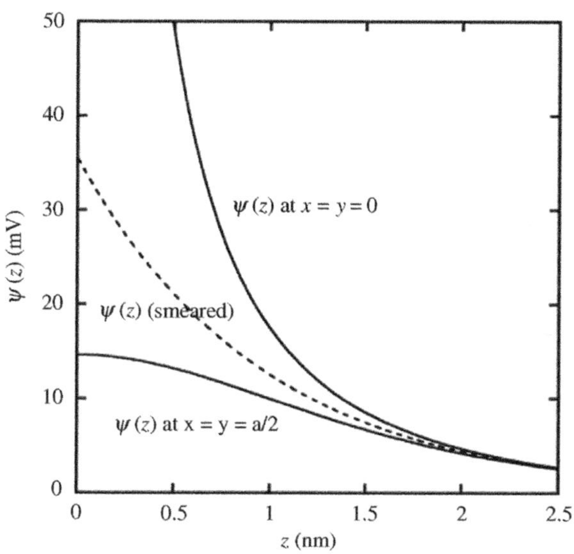

FIGURE 2.12 Potential distribution $\psi(x, y, z)$ near a plate surface carrying a squared lattice of point charges $q = e = 1.6 \times 10^{-19}$ C with spacing $a = 2.5$ nm in an aqueous 1-1 electrolyte solution of concentration 0.1 M at 25°C ($\varepsilon_r = 78.55$) as a function of the distance z from the plate surface at $x = y = 0$ and at $x = y = a/2 = 1.25$ nm, in comparison with the result for the smeared charge model (dashed line).

the distance z from the plate surface at $x = y = 0$ and at $x = y = a/2 = 1.25$ nm, in comparison with the result for the smeared charge model. We see that for distances away from the plate surface greater than the Debye length $1/\kappa$ (1 nm in this case), the electric potential due to a lattice of point charges is essentially the same as that for the smeared charge model. For small distances less than $1/\kappa$, however, the potential due to the array of point charges differs significantly from that for the smeared charge model.

REFERENCES

1. B. V. Derjaguin and L. Landau, *Acta Physicochim.* 14 (1941) 633.
2. E. J. W. Verwey and J. Th. G. Overbeek, *Theory of the Stability of Lyophobic Colloids*, Elsevier, Amsterdam, 1948.
3. H. Ohshima and K. Furusawa (Eds.), *Electrical Phenomena at Interfaces: Fundamentals, Measurements, and Applications*, 2nd edition, revised and expanded, Dekker, New York, 1998.
4. H. Ohshima, *Theory of Colloid and Interfacial Electric Phenomena*, Elsevier/Academic Press, Amsterdam, 2006.
5. H. Ohshima, *Colloids Surf. B: Biointerfaces* 15 (1999) 31.
6. A. P. Nelson and D. A. McQuarrie, *J. Theor. Biol.* 55 (1975) 13.

3 Modified Poisson–Boltzmann Equation

3.1 INTRODUCTION

The Poisson–Boltzmann equation for the electric potential in the vicinity of the surface of colloidal particles in an electrolyte solution, which is usually used to describe the electric properties of the particle surface [1–4], has some deficiencies. One is that electrolyte ions are treated as point charges: in other words, the size effects of ions are neglected. Carnie and McLaughlin [5] proposed a modified Poisson–Boltzmann equation taking into account the size effects of large divalent cations. In this chapter, we consider the potential distribution around a charged surface in an electrolyte solution containing rod-like ions. The Poisson–Boltzmann equation also neglects the self-atmosphere potential of ions [8]. In this chapter, we consider the effects of the self-atmosphere (or fluctuation) potential.

3.2 ELECTROLYTE SOLUTION CONTAINING ROD-LIKE DIVALENT CATIONS

Consider a negatively charged plate in a mixed solution of 1-1 electrolyte of concentration n_1 and 2-1 electrolyte of concentration n_2. We assume a divalent cation to be two positive point charges connected by a rod of length a (Fig. 3.1), while monovalent cations as well as anions are all assumed to be point charges. We take an x-axis in the direction normal to the plate so that the plane $x = 0$ coincides with the plate surface and the region $x > 0$ is the solution phase (Fig. 3.2). Let $\psi(x)$ be the potential at x relative to the bulk solution phase where $\psi(x)$ is set equal to zero. Consider one end of a rod. The probability of finding this end at position x is inevitably influenced by the other end. That is, when one end is located at position x, the other end can exist only in the region $(x - a, x + a)$ if $x > a$, or in the region $(0, x - a)$ if $x < a$ (these two regions can be combined in the form $(\max[0, x - a], x + a)$, as is seen in Fig. 3.3. Thus, the probability of finding one end of a rod at position x is the product of probabilities P_1 and P_2. Here P_1 is the probability that one end would be found at position x if it existed alone (i.e., in the absence of the other end) and P_2 is the probability of finding the other end at an arbitrary point in the region

Biophysical Chemistry of Biointerfaces By Hiroyuki Ohshima

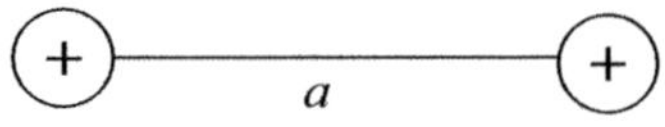

FIGURE 3.1 Rod-like divalent cation of length a.

(max[0, $x-a$], $x+a$) with the location of the first end kept at position x. The probability P_1 is proportional to

$$\exp\left[-\frac{e\psi(x)}{kT}\right] \tag{3.1}$$

and P_2 is proportional to

$$\frac{1}{2a}\int_{\max[0,x-a]}^{x+a} \exp\left[-\frac{e\psi(x')}{kT}\right]dx' \tag{3.2}$$

Here we have assumed a Boltzmann distribution for charged species. Since a rod has two ends, the concentration $n_{2+}(x)$ of one end of the rods at position x is given by twice the product of Eqs. (3.1), (3.2), and the bulk concentration n_2, namely,

$$n_{2+}(x) = 2n_2\exp\left[-\frac{e\psi(x)}{kT}\right]\frac{1}{2a}\int_{\max[0,x-a]}^{x+a} \exp\left[-\frac{e\psi(x')}{kT}\right]dx' \tag{3.3}$$

The concentration $n_+(x)$ of monovalent cations is

$$n_+(x) = n_1\exp\left[-\frac{e\psi(x)}{kT}\right] \tag{3.4}$$

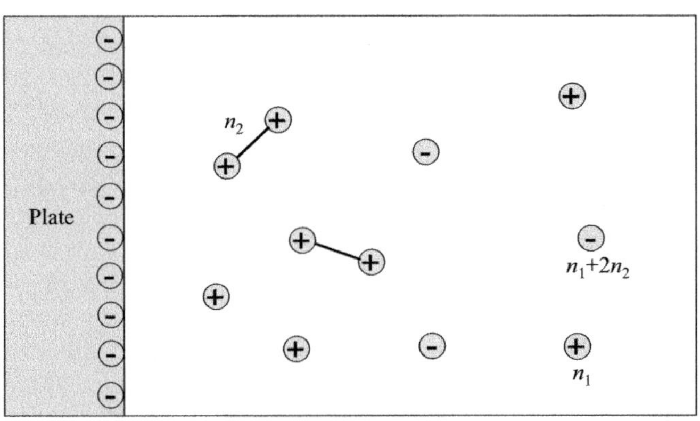

FIGURE 3.2 Negatively charged plate in contact with an aqueous solution of rod-like divalent cations of length and concentration n_2 and monovalent electrolytes of concentration n_1.

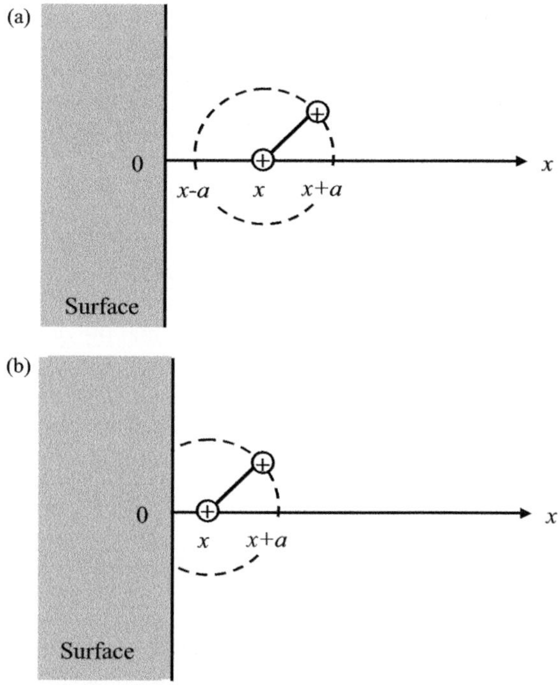

FIGURE 3.3 Rod-like divalent cation of length a near the charged surface: (a) $x \geq a$ and (b) $0 < x < a$.

while the concentration $n_-(x)$ of monovalent anions is

$$n_-(x) = (n_1 + 2n_2)\exp\left[\frac{e\psi(x)}{kT}\right] \tag{3.5}$$

The modified Poisson–Boltzmann equation derived by Carnie and McLaughlin [5] for the above system then reads

$$\frac{d^2\psi}{dx^2} = -\frac{e}{\varepsilon_r\varepsilon_o}\left[n_{2+}(x) + n_+(x) - n_-(x)\right]$$

or

$$\frac{d^2\psi}{dx^2} = -\frac{e}{\varepsilon_r\varepsilon_o}\left[n_1\exp\left(-\frac{e\psi(x)}{kT}\right) + n_2\exp\left(-\frac{e\psi(x)}{kT}\right)\right.$$
$$\left.\times\left\{\frac{1}{a}\int_{\max[0,x-a]}^{x+a}\exp\left(-\frac{e\psi(z)}{kT}\right)dz\right\} - (n_1+2n_2)\exp\left(\frac{ze\psi(x)}{kT}\right)\right] \tag{3.6}$$

which is subject to the following boundary conditions:

$$\psi = \psi_o \text{ at } x = 0 \tag{3.7}$$

$$\psi \to 0, \ \frac{d\psi}{dx} \to 0 \text{ as } x \to \infty \tag{3.8}$$

Here ε_r is the relative permittivity of the electrolyte solution.

We derive an approximate solution to Eq. (3.6) accurate for $x \ll a$ (i.e., very near the surface) [6]. For this region, $\max[0, x - a] = 0$ so that the quantity in the brackets on the right-hand side of Eq. (3.6) (which is the probability that either end of a rod-like cation is found at position x) becomes

$$\begin{aligned} \frac{1}{a}\int_0^{x+a} \exp\left(-\frac{e\psi(z)}{kT}\right)dz &\approx \frac{1}{a}\int_0^{a} \exp\left(-\frac{e\psi(z)}{kT}\right)dz \\ &= \frac{1}{a}\left[\int_0^{a} \left\{\exp\left(-\frac{e\psi(z)}{kT}\right) - 1\right\}dz + a\right] \end{aligned} \tag{3.9}$$

In the last expression, $\exp(-e\psi(z)/kT) - 1$ tends to zero rapidly as z increases. Thus, the upper limit of the z integration can be replaced by infinity so that Eq. (3.9) is approximated by the following quantity:

$$A = \frac{1}{a}\left[\int_0^{\infty} \left\{\exp\left(-\frac{e\psi(z)}{kT}\right) - 1\right\}dz + a\right] \tag{3.10}$$

which is a constant independent of x. Substituting Eq. (3.10) into Eq. (3.6) gives

$$\frac{d^2\psi}{dx^2} = -\frac{e}{\varepsilon_r\varepsilon_o}\left[(n_1 + An_2)\exp\left(-\frac{e\psi(x)}{kT}\right) + (n_1 + 2n_2)\exp\left(\frac{e\psi(x)}{kT}\right)\right] \tag{3.11}$$

Equation (3.11) may be a good approximation for $x \ll a$ but becomes invalid at large x. Indeed, Eq. (3.11) does not satisfy the electroneutrality condition at $x \to +\infty$. Therefore, we must replace the anion concentration $n_1 + 2n_2$ by $n_1 + An_2$. This replacement affects little the potential $\psi(x)$ for $x \ll a$, since contributions from anions are very small near the negatively charged surface. Hence, Eq. (3.11) becomes

$$\frac{d^2\psi}{dx^2} = -\frac{e}{\varepsilon_r\varepsilon_o}(n_1 + An_2)\left[\exp\left(-\frac{e\psi(x)}{kT}\right) - \exp\left(\frac{e\psi(x)}{kT}\right)\right] \tag{3.12}$$

It is of interest to note that rod-like divalent cations behave formally like monovalent cations of effective concentration An_2. Equation (3.12) under Eqs. (3.7) and (3.8) is easily integrated to give (see Eq. (1.38))

$$\psi(x) = \frac{2kT}{e}\ln\left[\frac{1 + \tanh(e\psi_o/4kT)\exp(-\kappa_{eff}x)}{1 - \tanh(e\psi_o/4kT)\exp(-\kappa_{eff}x)}\right] \tag{3.13}$$

where $\psi_o = \psi(0)$ is the surface potential of the plate and κ_{eff} is the effective Debye–Hückel parameter defined by

$$\kappa_{eff} = \left[\frac{2e^2(n_1 + An_2)}{\varepsilon_r\varepsilon_o kT}\right]^{1/2} \tag{3.14}$$

For the case when $n_1 \gg n_2$, κ_{eff} differs little from κ. Substituting Eq. (3.13) into Eq. (3.10), we obtain

$$A = 1 + \frac{2}{\kappa_{eff}a}\left[\exp\left(-\frac{e\psi_o}{2kT}\right) - 1\right] \tag{3.15}$$

The potential $\psi(x)$ must satisfy the following boundary condition at $x = +0$:

$$\left.\frac{d\psi}{dx}\right|_{x=+0} = -\frac{\sigma}{\varepsilon_r\varepsilon_o} \tag{3.16}$$

Here σ is the surface charge density of the plate and the electric fields inside the plate have been neglected. We now assume that negatively charged groups are distributed on the plate surface at a uniform density $1/S$, S being the area per negative charge, and that each charged group serves as a site capable of binding one monovalent cation or either end of a rod-like divalent cation. We introduce the equilibrium binding constants K_1 and K_2 for monovalent and divalent cations, respectively. Then, from the mass action law for cation binding, we find

$$\sigma = -\frac{e}{S}\cdot\frac{1}{1 + K_1 n_1^s + K_2 n_2^s} \tag{3.17}$$

where

$$n_1^s = n_1\exp\left(-\frac{e\psi_o}{kT}\right) \tag{3.18}$$

is the surface concentration of monovalent cations and

$$\begin{aligned} n_2^s &= n_2\exp\left(-\frac{e\psi_o}{kT}\right)\left[\frac{1}{a}\int_0^a \exp\left(-\frac{e\psi(z)}{kT}\right)dz\right] \\ &\approx An_2\exp\left(-\frac{e\psi_o}{kT}\right) \end{aligned} \tag{3.19}$$

is the surface concentration of the ends of divalent cations. Equation (3.19) is consistent with the observation that rod-like divalent cations behave formally like monovalent cations of concentration An_2. Estimating $d\psi/dx$ at $x = +0$ through Eq. (3.13) and combining Eqs. (3.16)–(3.19), we obtain

$$\exp\left(\frac{e\psi_o}{2kT}\right) - \exp\left(-\frac{e\psi_o}{2kT}\right) + \frac{e^2}{\varepsilon_r\varepsilon_o kT\kappa_{\text{eff}}S} \cdot \frac{1}{1 + (K_1 n_1 + K_2 A n_2)\exp(-e\psi_o/kT)} = 0 \tag{3.20}$$

Equations (3.14), (3.15), and (3.20) form coupled equations for ψ_o and κ_{eff}. To solve these equations, we obtain κ_{eff} as a function of ψ_o from Eq. (3.20) and substitute the result into Eq. (3.14). Then we have a single transcendental equation for ψ_o, the solution of which can be obtained numerically. Using results obtained for ψ_o and κ_{eff}, we can calculate $\psi(x)$ at any position x via Eq. (3.13).

In Figs 3.4 and 3.5, we give some results obtained by the present approximation method. Figure 3.4 shows the result for $a = 2$ nm, $S = 0.7$ nm^2, $n_1 = n_2 = 0.1$ M, $K_1 = K_2 = 0$ (non-ion adsorbing plate), $\varepsilon_r = 78.5$, and $T = 298$ K. For comparison, Fig. 3.4 shows also the results for $a \to 0$ and $a \to \infty$. As $a \to 0$, rod-like divalent cations to point-like divalent cations so that the electrolyte solution surrounding the plate becomes a mixed solution of 1-1 electrolyte of bulk concentration n_1 and 2-1 electrolyte of bulk concentration n_2. In this case, the potential distribution is given by Eqs. (1.49)–(1.51) and (1.14), namely,

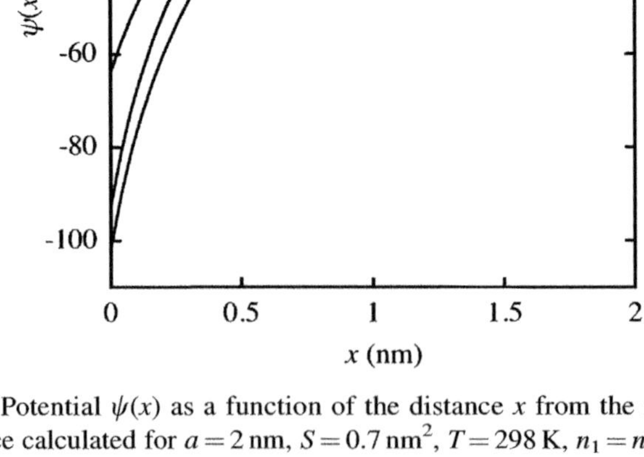

FIGURE 3.4 Potential $\psi(x)$ as a function of the distance x from the non-ion adsorbing membrane surface calculated for $a = 2$ nm, $S = 0.7$ nm^2, $T = 298$ K, $n_1 = n_2 = 0.1$ M, in comparison with the results for $a \to 0$ and $a \to \infty$.

$$\psi(x) = \frac{kT}{e}\ln\left[\left(\frac{1}{1-\eta/3}\right)\left\{\frac{1+(1-\eta/3)\gamma'' e^{-\kappa x}}{1-(1-\eta/3)\gamma'' e^{-\kappa x}}\right\}^2 - \left(\frac{\eta/3}{1-\eta/3}\right)\right] \tag{3.21}$$

with

$$\eta = \frac{3n_2}{n_1 + 3n_2} \tag{3.22}$$

$$\gamma'' = \left(\frac{1}{1-\eta/3}\right)\left[\frac{\{(1-\eta/3)e^{y_o} + \eta/3\}^{1/2} - 1}{\{(1-\eta/3)e^{y_o} + \eta/3\}^{1/2} + 1}\right] \tag{3.23}$$

$$\kappa = \left(\frac{2(n_1 + 3n_2)e^2}{\varepsilon_r \varepsilon_o kT}\right)^{1/2} \tag{3.24}$$

On the other hand, we regard the case of $a \to \infty$ as the situation in which rod-like divalent cations become two independent monovalent cations. The electrolyte solution surrounding the plate now becomes a monovalent electrolyte solution of concentration $n_1 + 2n_2$. In this case, the potential distribution is given by Eqs. (1.38) and (1.12), namely,

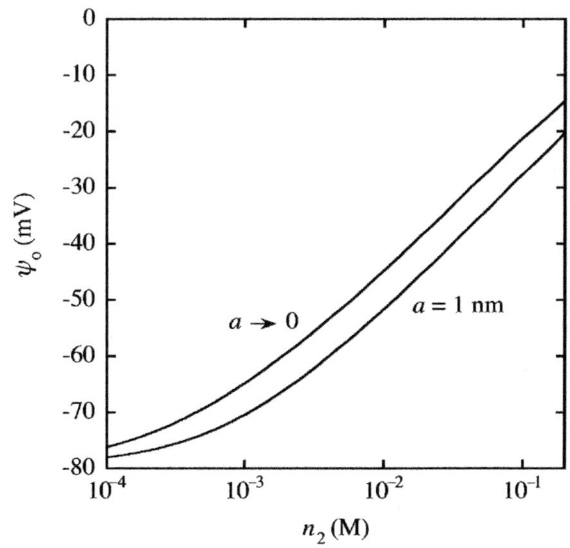

FIGURE 3.5 Surface potential ψ_o of the ion adsorbing plate as a function of divalent cation concentration n_2. Comparison of rod model (rod length $a = 1$ nm) and point charge model ($a \to 0$). $S = 0.7\ \mathrm{nm}^2$, $T = 298$ K, $n_1 = 0.1$ M, $K_1 = 0.8\ \mathrm{M}^{-1}$, $K_2 = 10\ \mathrm{M}^{-1}$.

$$\psi(x) = \frac{2kT}{e}\ln\left[\frac{1 + \tanh(e\psi_o/4kT)\exp(-\kappa x)}{1 - \tanh(e\psi_o/4kT)\exp(-\kappa x)}\right] \tag{3.25}$$

with

$$\kappa = \left(\frac{2(n_1 + 2n_2)e^2}{\varepsilon_r\varepsilon_o kT}\right)^{1/2} \tag{3.26}$$

It is seen that the potential distribution and the plate surface potential depend strongly on the size of rod-like ions.

Figure 3.5 shows the result for the ion adsorbing plate. We plot the plate surface potential ψ_o as a function of rod-like divalent cation (of length a) concentration n_2 in comparison with that for point-like divalent cations ($a \to 0$). We put $a = 1$ nm, $n_1 = 0.1$ M, $K_1 = 0.8\,\mathrm{M}^{-1}$, and $K_2 = 10\,\mathrm{M}^{-1}$. The values of the other parameters are the same as in Fig. 3.4. For the point-like divalent cation model, instead of Eq. (3.17), we have used the following equation for cation binding:

$$\sigma = -\frac{e}{S}\cdot\frac{1}{1 + K_1 n_1 \exp(-e\psi_o/kT) + K_2 n_2 \exp(-2e\psi_o/kT)} \tag{3.27}$$

We see that the magnitude of ψ_o in the rod model is larger than that predicted from the point charge model.

3.3 ELECTROLYTE SOLUTION CONTAINING ROD-LIKE ZWITTERIONS

Consider a plate-like charged colloidal particle immersed in an electrolyte solution composed of monovalent point-like electrolytes of concentration n_1 and zwitterions of concentrations n_z (Fig. 3.6) [7]. The zwitterion is regarded as a rod of length a, carrying two point charges $+e$ and $-e$ at its ends (Fig. 3.7). We take an x-axis perpendicular to the particle surface with its origin at the surface so that the region $x > 0$ corresponds to the electrolyte solution and $x < 0$ to the particle interior, as shown in Fig. 3.8. Since zwitterions carry no net charges, any effects of zwitterions on the particle surface potential cannot be expected from the original Poisson–Boltzmann approach in which ions are all treated as point charges. This is the case only in the limit of small zwitterions. Obviously, however, this approach would not be a good approximation for systems involving large zwitterions such as neutral amino acids. In the limit of large zwitterions their behavior becomes similar to that of two separate charges.

On the basis of the theory of Carnie and McLaughlin [5], the Poisson–Boltzmann equation for $\psi(x)$ can be modified in the present case as follows. Consider the end of a rod that carries the $+e$ charge. As in the problem of rod-like divalent cations,

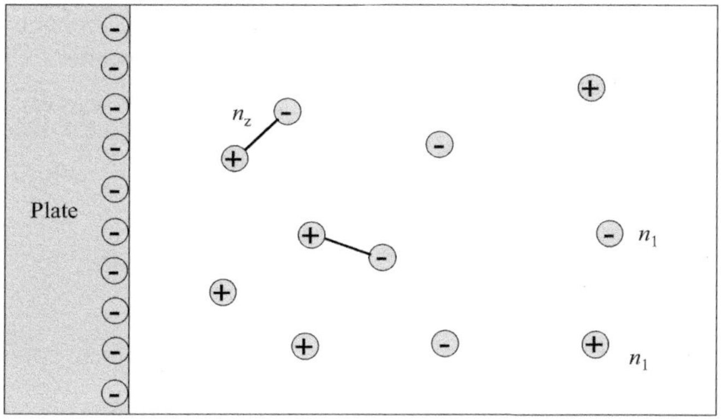

FIGURE 3.6 Positively charged plate in contact with an aqueous solution of rod-like divalent cations of length and concentration n_z and monovalent electrolytes of concentration n_1.

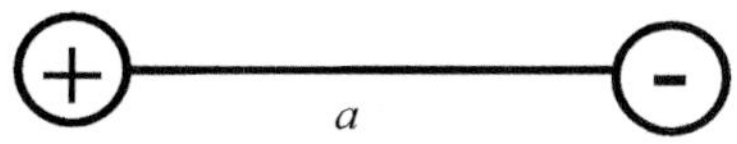

FIGURE 3.7 Rod-like zwitterions of length a carrying $+e$ and $-e$ at its ends.

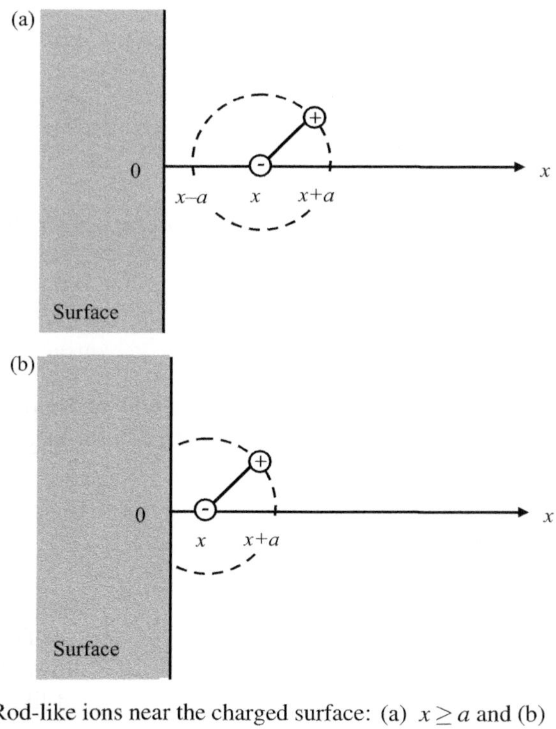

FIGURE 3.8 Rod-like ions near the charged surface: (a) $x \geq a$ and (b) $0 < x < a$.

the probability of finding this end at position x is the product of probabilities P_1 and P_2. Here P_1 is the probability that $+e$ end would be found at position x if it existed alone (i.e., in the absence of the $-e$ end) and P_2 is the probability of finding the $-e$ end at an arbitrary point in the region (max[0,$x-a$], $x+a$) with the location of the $+e$ end kept at position x. The probability P_1 is proportional to

$$\exp\left[-\frac{e\psi(x)}{kT}\right] \tag{3.28}$$

and P_2 is proportional to

$$\frac{1}{2a}\int_{\max[0,x-a]}^{x+a} \exp\left[+\frac{e\psi(x')}{kT}\right]dx' \tag{3.29}$$

Here we have assumed a Boltzmann distribution for charged species. Thus, the concentration $C_{z+}(x)$ of the $+e$ end of the rods at position x is given by the product of Eqs. (3.28), (3.29) and the bulk concentration n_z, namely,

$$n_{z+}(x) = n_z \exp\left[-\frac{e\psi(x)}{kT}\right]\left\{\frac{1}{2a}\int_{\max[0,x-a]}^{x+a} \exp\left[+\frac{e\psi(x')}{kT}\right]dx'\right\} \tag{3.30}$$

Similarly, the concentration $n_+(x)$ of the $-e$ end at position x, $C_{z-}(x)$ is given by

$$n_{z-}(x) = n_z \exp\left[+\frac{e\psi(x)}{kT}\right]\left\{\frac{1}{2a}\int_{\max[0,x-a]}^{x+a} \exp\left[-\frac{e\psi(x')}{kT}\right]dx'\right\} \tag{3.31}$$

The modified Poisson–Boltzmann equation derived for this system then reads

$$\frac{d^2\psi}{dx^2} = -\frac{e}{\varepsilon_r\varepsilon_o}[n_{z+}(x) - n_{z-}(x) + n_+(x) - n_-(x)] \tag{3.32}$$

where $n_+(x)$ and $n_-(x)$ are, respectively, the concentrations of coexisting monovalent cations and anions. If we assume a Boltzmann distribution for these ions, then $n_+(x)$ and $n_-(x)$ are given by

$$n_+(x) = n_1 \exp\left[-\frac{e\psi(x)}{kT}\right] \tag{3.33}$$

$$n_-(x) = n_1 \exp\left[+\frac{e\psi(x)}{kT}\right] \tag{3.34}$$

Combining Eqs. (3.31)–(3.35), we obtain

$$\frac{d^2\psi}{dx^2} = -\frac{e}{\varepsilon_r\varepsilon_o}\left\{n_z \exp\left[-\frac{e\psi(x)}{kT}\right]\frac{1}{2a}\int_{\max[0,x-a]}^{x+a}\exp\left[\frac{e\psi(x')}{kT}\right]dx'\right.$$
$$-n_z \exp\left[\frac{e\psi(x)}{kT}\right]\frac{1}{2a}\int_{\max[0,x-a]}^{x+a}\exp\left[-\frac{e\psi(x')}{kT}\right]dx' \tag{3.35}$$
$$\left.+n_1 \exp\left[-\frac{e\psi(x)}{kT}\right] - n_1 \exp\left[\frac{e\psi(x)}{kT}\right]\right\}$$

The boundary conditions at the particle surface $x = 0$ and at $x = a$ are given by

$$\left.\frac{d\psi}{dx}\right|_{x=0^+} = -\frac{\sigma}{\varepsilon_r\varepsilon_o} \tag{3.36}$$

$$\psi(x) \text{ and } d\psi/dx \text{ are, respectively, continuous at } x = a \tag{3.37}$$

Equation (3.35) subject to the boundary conditions (3.36) and (3.37) can be solved numerically to provide the potential distribution $\psi(x)$. The relationship between the surface charge density σ and the surface potential $\psi_o = \psi(0)$ derived as follows. Multiplying $d\psi/dx$ on both sides of Eq. (3.35) for both regions $0 < x < a$ and $x > a$, integrating the result once, we have

$$\left(\frac{d\psi}{dx}\right)^2 = \kappa^2\left[e^{y(x)} + e^{-y(x)} - 2 + \frac{r}{2a}\left\{(e^{-y(x)} - 1)\int_0^{x+a} e^{y(x')}dx'\right.\right.$$
$$+\int_x^a (e^{-y(x')} - 1)e^{y(x'+a)}dx' + \int_a^\infty (e^{-y(x')} - 1)(e^{y(x'+a)} - e^{y(x'-a)})dx'$$
$$+(e^{y(x)} - 1)\int_0^{x+a} e^{-y(x')}dx' + \int_x^a (e^{y(x')} - 1)e^{-y(x'+a)}dx'$$
$$\left.\left.+\int_x^\infty (e^{y(x')} - 1)(e^{-y(x'+a)} - e^{-y(x'-a)})dx'\right\}\right], 0 < x \leq a \tag{3.38}$$

$$\left(\frac{d\psi}{dx}\right)^2 = \kappa^2\left[e^{y(x)} + e^{-y(x)} - 2 + \frac{r}{2a}\left\{(e^{-y(x)} - 1)\int_{x-a}^{x+a} e^{y(x')}dx'\right.\right.$$
$$+\int_x^\infty (e^{-y(x')} - 1)(e^{y(x'+a)} - e^{y(x'-a)})dx' + (e^{y(x)} - 1)\int_{x-a}^{x+a} e^{-y(x')}dx'$$
$$\left.\left.+\int_x^\infty (e^{y(x')} - 1)(e^{-y(x'+a)} - e^{-y(x'-a)})dx'\right\}\right], \quad x \geq a \tag{3.39}$$

with

$$y(x) = e\psi(x)/kT \tag{3.40}$$

$$\kappa = \left[\frac{2e^2 n_1}{\varepsilon_r \varepsilon_o kT}\right]^{1/2} \tag{3.41}$$

$$r = n_z/n_1 \tag{3.42}$$

where $y(x)$ is the scaled potential, κ is the Debye–Hückel parameter of the electrolyte solution in the absence of rod-like zwitterions (i.e., $n_z = 0$), and r is the concentration ratio of rod-like zwitterions and univalent point-like electrolytes.
Evaluating Eq. (3.38) at $x = +0$

$$\left(\left.\frac{dy}{dx}\right|_{x=0^+}\right)^2 = \kappa^2\left[e^{y_o} + e^{-y_o} - 2 + \frac{r}{2a}\int_0^a \{e^{-y_o+y(x')} + e^{y_o-y(x')} - 2\}dx'\right] \tag{3.43}$$

where

$$y_0 = e\psi_0/kT \tag{3.44}$$

Combining this result with Eq. (3.36), we obtain

$$\left(\frac{e\sigma}{\varepsilon_r \varepsilon_o kT}\right)^2 = \kappa^2\left[e^{y_o} + e^{-y_o} - 2 + \frac{r}{2a}\int_0^a \{e^{-y_o+y(x')} + e^{y_o-y(x')} - 2\}dx'\right] \tag{3.45}$$

or

$$\frac{e\sigma \quad 2}{\varepsilon_r \varepsilon_o kT} = sgn(y_0)\kappa\left[e^{y_o} + e^{-y_o} - 2 + \frac{r}{2a}\int_0^a \{e^{-y_o+y(x')} + e^{y_o-y(x')} - 2\}dx'\right]^{1/2} \tag{3.46}$$

Equation (3.46) is the required relationship between y_o and σ. The second term in the brackets on the right-hand side of Eq. (3.46) represents the contribution from rod-like zwitterions. Calculation of y_o as a function of n_1 and n_z for given σ via Eq. (3.46) requires $y(x)$, which is obtained numerically from the solution to Eq. (3.35). Figure 3.9 shows some results of the calculation of the surface potential y_o as a function of the scaled rod length κa for several values of $r = n_z/n_1$ (Fig. 3.9).

We derive approximate analytic expressions for the relationship between y_o and σ. Consider first the limiting case of small a. In this case one may expand $\exp(\pm y(x))$ around $x = 0$ in the integrand of Eq. (3.46)

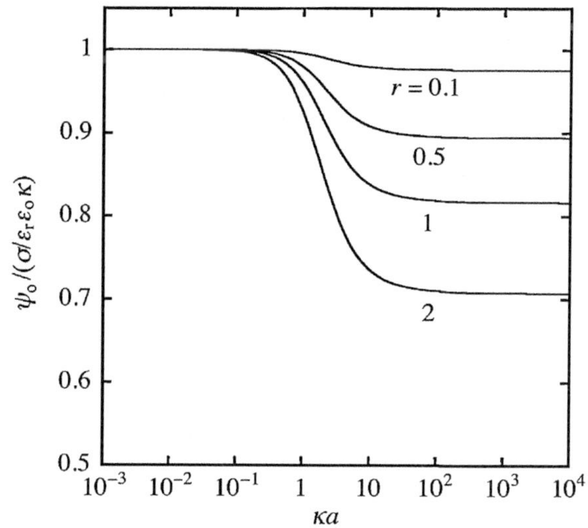

FIGURE 3.9 Scaled surface potential ψ_o of a plate immersed in a mixed solution of a monovalent electrolyte of concentration n_1 and rod-like zwitterions of length a and concentration n_z as a function of reduced rod length κa for several values of $r = n_z/n_1$.

$$e^{\pm y(x)} = e^{\pm y_o}\left[1 \pm \left.\frac{dy}{dx}\right|_{x=0^+} x + \frac{1}{2}\left\{\left(\left.\frac{dy}{dx}\right|_{x=0^+}\right)^2 \pm \left.\frac{d^2y}{dx^2}\right|_{x=0^+}\right\}x^2 + \cdots\right] \tag{3.47}$$

Substituting Eq. (3.47) into Eq. (3.43) yields

$$\left(\left.\frac{dy}{dx}\right|_{x=0^+}\right)^2 = \kappa^2\left[e^{y_o} + e^{-y_o} - 2 + \frac{r}{6}\left(\left.\frac{dy}{dx}\right|_{x=0^+}\right)^2 a^2\right] \tag{3.48}$$

or

$$\begin{aligned}\left(\left.\frac{dy}{dx}\right|_{x=0^+}\right)^2 &= \frac{\kappa^2}{1 - r(\kappa a)^2/6}(e^{y_o} + e^{-y_o} - 2) \\ &\approx \kappa^2\left[1 + \frac{r}{6}(\kappa a)^2\right](e^{y_o} + e^{-y_o} - 2)\end{aligned} \tag{3.49}$$

By combining with Eq. (3.36), we thus obtain

$$\left(\frac{e\sigma}{\varepsilon_r\varepsilon_o kT}\right)^2 = \kappa^2\left[1 + \frac{r}{6}(\kappa a)^2\right](e^{y_o/2} - e^{-y_o/2})^2 \tag{3.50}$$

For $\kappa a \rightarrow 0$, Eq. (3.50) becomes

$$\left(\frac{e\sigma}{\varepsilon_r \varepsilon_o kT}\right)^2 = \kappa^2 (e^{y_o/2} - e^{-y_o/2})^2 \tag{3.51}$$

which is the σ/y_o relationship for the electrolyte solution containing only monovalent electrolytes (Eq. (1.41)). This is physically obvious because for $a \rightarrow 0$, zwitterions become neutral species so that they exert no electrostatic effects on the potential distribution. Also it follows from comparison between Eqs. (3.50) and (3.51) that in the case of $\kappa a \ll 1$, the effective Debye–Hückel parameter is given by

$$\kappa_{\mathrm{eff}} = \kappa \sqrt{1 + \frac{r}{6} (\kappa a)^2} \tag{3.52}$$

Consider next the opposite limiting case of $a \rightarrow \infty$. Note that $y(x)$ should rapidly tend to zero beyond the electrical double layer (or beyond a distance of the order of $1/\kappa$ from the particle surface), that is, $y(x) \rightarrow 0$ as $x \rightarrow \infty$, so that

$$\frac{1}{2a} \int_0^a e^{\pm y(x')} dx' \rightarrow 1 \text{ as } a \rightarrow \infty \tag{3.53}$$

Equation (3.45) thus becomes

$$\left(\frac{e\sigma}{\varepsilon_r \varepsilon_o kT}\right)^2 = \kappa^2 \left(1 + \frac{r}{2}\right) (e^{y_o/2} - e^{-y_o/2})^2 \tag{3.54}$$

which is the relationship between y_o and σ at $\kappa a \gg 1$. Equation (3.54) implies that the effective Debye–Hückel parameter is given by

$$\kappa_{\mathrm{eff}} = \kappa \left(1 + \frac{r}{2}\right)^{1/2} = \left[\frac{2e^2 (n_1 + n_z/2)}{\varepsilon_r \varepsilon_o kT}\right]^{1/2} \tag{3.55}$$

This means that for $\kappa a \gg 1$ zwitterions behave like univalent electrolytes of concentration $n_z/2$ instead of n_z. That is, the probability of finding of one of the rod ends near the particle surface is half that for the case of point charges due to the presence of the other end.

We give below a simple method to obtain an approximate relationship between ψ_o and σ. In the iteration procedure, the zeroth-order solution is the one for the case of no divalent cations. The next-order solution is obtained by replacing $y(x)$ in Eq. (3.46) by the potential distribution $y_1(x)$ for the case of $n_z = 0$, which is given by Eq. (1.37), namely,

$$y_1(x) = 2 \ln \left(\frac{1 + \gamma e^{-\kappa x}}{1 - \gamma e^{-\kappa x}}\right) \tag{3.56}$$

with

$$\gamma = \tanh(y_o/4) \tag{3.57}$$

Substituting Eq. (3.56) into Eq. (3.45) and integrating, we obtain

$$\left(\frac{e\sigma}{\varepsilon_r\varepsilon_o kT}\right)^2 = \kappa^2\left[\left(1+\frac{r}{2}\right)(e^{y_o/2}-e^{-y_o/2})^2+\frac{2r}{\kappa a}\left\{e^{y_o}\left(\frac{1}{1+\gamma}-\frac{1}{1+\gamma e^{-\kappa a}}\right)\right.\right.$$
$$\left.\left.+e^{-y_o}\left(\frac{1}{1-\gamma}-\frac{1}{1-\gamma e^{-\kappa a}}\right)\right\}\right] \tag{3.58}$$

Equation (3.58) is a transcendental equation for y_o. The solution of Eq. (3.58) can easily be obtained numerically for given values of σ, n_1, and n_z. Equation (3.58) is found to provide a quite accurate relationship between y_o and σ for practical purposes. The reason for this is that the solution to Eq. (3.58) yields correct limiting forms both at $\kappa a \gg 1$and at $\kappa a \ll 1$, that is, Eqs. (3.51) and (3.55).

Finally, we give below an approximate explicit expression for ψ_o applicable for small ψ_o obtained by linearizing Eq. (3.58) with respect to ψ_o, namely,

$$\psi_o = \frac{\sigma}{\varepsilon_r\varepsilon_o\kappa}\frac{1}{\left[1+\frac{r}{2}-\frac{r}{\kappa a}\left\{1-e^{-\kappa a}-\frac{1}{4}(1-e^{-2\kappa a})\right\}\right]^{1/2}} \tag{3.59}$$

In the limit of small κa, we have

$$\psi_o = \frac{\sigma}{\varepsilon_r\varepsilon_o\kappa}\frac{1}{\left[1+\frac{r}{6}(\kappa a)^2\right]^{1/2}} \tag{3.60}$$

In the opposite limit of large κa, Eq. (3.59) reduces

$$\psi_o = \frac{\sigma}{\varepsilon_r\varepsilon_o\kappa}\frac{1}{(1+r/2)^{1/2}} \tag{3.61}$$

We again see that in the limit of small κa, the effective Debye–Hückel parameter is given by Eq. (3.52) and that in the limit of large κa, rod-like zwitterions behave like monovalent electrolytes of concentration $n_z/2$ and the effective Debye–Hückel parameter is given by Eq. (3.55).

3.4 SELF-ATMOSPHERE POTENTIAL OF IONS

Another deficiency of the Poisson–Boltzmann equation is that as Kirkwood showed [8], it neglects the effects of the self-atmosphere (or fluctuation) potential. The existence of the self-atmosphere potential at ions in bulk electrolyte was first discussed by the Debye–Hückel theory of strong electrolytes [9]. This theory showed that the electric potential energy of an ion is lower than the mean potential in the electrolyte solution, since an ionic atmosphere consisting of a local space charge of opposite sign is formed around this ion, the difference being larger for higher electrolyte concentrations. The self-atmosphere potential plays a particularly important role in determining the surface tension of an aqueous electrolyte solution.

We calculate the self-atmosphere potential of an ion near the planar surface. Imagine a planar uncharged plate in contact with a solution of general electrolytes composed of N ionic mobile species of valence z_i and bulk concentration (number density) n_i^{∞} ($i = 1, 2, \ldots, N$) (Fig. 3.10). Let ε_r and ε_p, respectively, be the relative permittivities of the electrolyte solution and the plate. Consider the potential distribution around an ion. By symmetry, we use a cylindrical coordinate system $\boldsymbol{r}(s, x)$ with its origin 0 at the plate surface and the x-axis perpendicular to the surface.

Following Loeb [10] and Williams [11], we express the potential $\Psi_i(\boldsymbol{r}, \boldsymbol{r}_c)$ at a field point $\boldsymbol{r}(s, x)$ around an ion of ith species at position $\boldsymbol{r}_c(0, x_c)$ in the solution phase ($x_c > 0$) as the sum of two potentials [4, 12]

$$\Psi(\boldsymbol{r}, \boldsymbol{r}_c) = \psi(x) + \phi_i(\boldsymbol{r}, \boldsymbol{r}_c) \tag{3.62}$$

where $\psi(x)$ is the mean potential, which depends only on x, and $\phi_i(\boldsymbol{r}, \boldsymbol{r}_c)$ is the self--atmosphere potential. We assume that $\phi_i(\boldsymbol{r}, \boldsymbol{r}_c)$ is small.

The concentration (number density) $n_i(x)$ of the ith ions of charge q_i (where $q_i = z_i e$, z_i being the ionic valence) at position $\boldsymbol{r}(s, x)$ may be given by

$$n_i(x) = n_i^{\infty} \exp\left[-\frac{q_i \psi(x)}{kT} + W_i(x)\right] \tag{3.63}$$

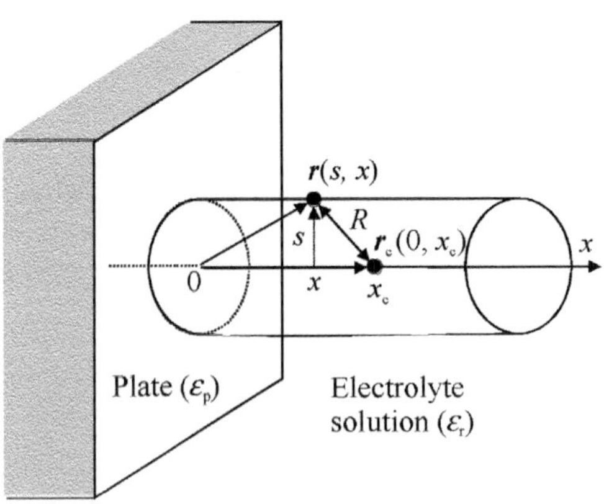

FIGURE 3.10 Field point at $\boldsymbol{r}(s, x)$ around an ion located at $\boldsymbol{r}_p(0, x_c)$

with

$$W_i(x) = \int_0^{q_i} \lim_{R \to 0} \left[\phi_i(\boldsymbol{r}, \boldsymbol{r}_c) - \frac{q'_i}{4\pi\varepsilon_r\varepsilon_o R} \right] dq'_i \tag{3.64}$$

where $R = \sqrt{(x - x_c)^2 + s^2}$ is the distance between the field position $\boldsymbol{r}(x, s)$ and the position of the central ion at $\boldsymbol{r}_c(0, x_c)$. The quantity in the bracket on the right-hand side of Eq. (3.64) is the self-atmosphere potential minus the self-potential (i.e., Coulomb potential) of the central ion, that is, the potential felt by this ion and $W_i(x)$ is the work of bringing an ion of the ith species from the bulk phase to the position x against its ionic atmosphere around this ion. With the help of Eq. (3.63), the Poisson equation for the mean potential $\psi(x)$,

$$\frac{d^2\psi(x)}{dx^2} = -\frac{1}{\varepsilon_r\varepsilon_o} \sum_{i=1}^{N} q_i n_i(x), \ x > 0 \tag{3.65}$$

becomes the following modified Poisson–Boltzmann equation:

$$\frac{d^2\psi(x)}{dx^2} = -\frac{1}{\varepsilon_r\varepsilon_o} \sum_{i=1}^{N} q_i n_i^\infty \exp\left[-\frac{q_i\psi(x)}{kT} + W_i(x) \right], \ x > 0 \tag{3.66}$$

Similarly, the Poisson–Boltzmann for the total potential $\Psi(\boldsymbol{r}, \boldsymbol{r}_p)$ is given by

$$\begin{aligned} \Delta\Psi(\boldsymbol{r}, \boldsymbol{r}_c) &= \left(\frac{\partial^2}{\partial s^2} + \frac{1}{s}\frac{\partial}{\partial s} + \frac{\partial^2}{\partial x^2} \right) \Psi(\boldsymbol{r}, \boldsymbol{r}_c) \\ &= -\frac{1}{\varepsilon_r\varepsilon_o} \sum_{i=1}^{N} q_i n_i^\infty \exp\left[-\frac{q_i\Psi(\boldsymbol{r}, \boldsymbol{r}_c)}{kT} + W_i(x) \right] + \frac{q_i}{4\pi\varepsilon_r\varepsilon_o} \delta(\boldsymbol{r} - \boldsymbol{r}_c) \end{aligned} \tag{3.67}$$

Subtracting Eq. (3.66) from Eq. (3.67) and linearizing the result with respect to the self-atmosphere potential $\phi_i(\boldsymbol{r}, \boldsymbol{r}_c)$, we obtain

$$\begin{aligned} &\frac{\partial^2\phi_i}{\partial s^2} + \frac{1}{s}\frac{\partial\phi_i}{\partial s} + \frac{\partial^2\phi_i}{\partial x^2} \\ &= \frac{1}{\varepsilon\varepsilon_o k_B T} \sum_{i=1}^{N} q_i^2 n_i \exp\left[-\frac{q_i\psi(x)}{kT} + W_i(x) \right] \cdot \phi_i(\boldsymbol{r}, \boldsymbol{r}_c) \\ &+ \frac{q_i}{4\pi\varepsilon_r\varepsilon_o} \delta(\boldsymbol{r} - \boldsymbol{r}_p) \end{aligned} \tag{3.68}$$

The above two equations for $\psi(x)$ and $\phi_i(\boldsymbol{r}, \boldsymbol{r}_c)$ may be solved in the following way [10, 11]. (i) As a first approximation, an expression for $\psi(x)$ is obtained by neglecting $\phi_i(\boldsymbol{r}, \boldsymbol{r}_c)$. (ii) The obtained approximate expression for $\psi(x)$ is introduced into the equation for $\phi_i(\boldsymbol{r}, \boldsymbol{r}_c)$, which is then solved. (iii) The obtained solution for $\phi_i(\boldsymbol{r}, \boldsymbol{r}_c)$ is used to set up an improved equation for $\psi(x)$, which is then solved numerically.

As a first approximation, we put $W_i(x) = 0$ on the right-hand side of Eq. (3.66), obtaining

$$\frac{d^2\psi(x)}{dx^2} = -\frac{1}{\varepsilon_r\varepsilon_o}\sum_{i=1}^{N} q_i n_i^{\infty} \exp\left[-\frac{q_i\psi(x)}{kT}\right], \; x > 0 \tag{3.69}$$

which agrees with the usual planar Poisson–Boltzmann equation (1.20). Similarly, as a first approximation, we put $W_i(x) = 0$ on the right-hand side of Eq. (3.67), obtaining

$$\frac{\partial^2\phi_i}{\partial s^2} + \frac{1}{s}\frac{\partial\phi_i}{\partial s} + \frac{\partial^2\phi_i}{\partial x^2} = \kappa^2(x)\phi_i(\boldsymbol{r}, \boldsymbol{r}_c) + \frac{q_i}{4\pi\varepsilon_r\varepsilon_o}\delta(\boldsymbol{r} - \boldsymbol{r}_c), \; x > 0 \tag{3.70}$$

where $\kappa(x)$ is the local Debye–Hückel parameter, given by

$$\kappa(x) = \left(\frac{1}{\varepsilon_r\varepsilon_o kT}\sum_{i=1}^{N} q_i^2 n_i^{\infty} \exp\left[-\frac{q_i\psi(x)}{kT}\right]\right)^{1/2} \tag{3.71}$$

For the region inside the plate $x < 0$, ϕ_i satisfies the Laplace equation, that is,

$$\frac{\partial^2\phi_i}{\partial s^2} + \frac{1}{s}\frac{\partial\phi_i}{\partial s} + \frac{\partial^2\phi_i}{\partial x^2} = 0, \; x < 0 \tag{3.72}$$

As a first approximation, we may replace $\kappa(x)$ by its bulk value $\kappa = (\Sigma q_i^2 n_i^{\infty}/\varepsilon_r\varepsilon_o kT)^{1/2}$ so that Eq. (3.70) for the self-atmosphere potential $\phi_i(\boldsymbol{r}, \boldsymbol{r}_c) = \phi_i(s, x)$ becomes

$$\frac{\partial^2\phi_i}{\partial s^2} + \frac{1}{s}\frac{\partial\phi_i}{\partial s} + \frac{\partial^2\phi_i}{\partial x^2} = \kappa^2\phi_i(s, x) + \frac{q_i}{4\pi\varepsilon_r\varepsilon_o}\delta(\boldsymbol{r} - \boldsymbol{r}_c), \; x > 0 \tag{3.73}$$

The boundary condition for ϕ_i is given by

$$\phi_i(s, 0^-) = \phi_i(s, 0^+) \tag{3.74}$$

$$\varepsilon_p \left.\frac{\partial\phi_i}{\partial x}\right|_{x=0^-} = \varepsilon_r \left.\frac{\partial\phi_i}{\partial x}\right|_{x=0^+} \tag{3.75}$$

In order to solve Eqs. (3.72) and (3.73) subject to Eqs. (3.74) and (3.75), we express $\phi_i(x)$ as

$$\phi_i(s,x) = \int_0^\infty e^{kx} A(k) J_0(ks) k\, dk, \ x < 0 \tag{3.76}$$

$$\phi_i(s,x) = \int_0^\infty e^{-px} B(k) J_0(ks) k\, dk + \frac{q_i \exp\left(-\sqrt{(x-x_c)^2+s^2}\right)}{4\pi\varepsilon_r\varepsilon_o\sqrt{(x-x_c)^2+s^2}}, \ x > 0 \tag{3.77}$$

with

$$p = \sqrt{k^2+\kappa^2} \tag{3.78}$$

where $J_0(ks)$ is the zeroth order Bessel function and the last term on the right-hand side of Eq. (3.77) can be expressed in terms of $J_0(z)$ by using the following relation:

$$\frac{\exp\left(-\sqrt{(x-x_c)^2+s^2}\right)}{\sqrt{(x-x_c)^2+s^2}} = \int_0^\infty \frac{\exp(-p|x-x_c|)}{p} J_0(ks) k\, dk \tag{3.79}$$

The functions $A(x)$ and $B(x)$ are to be determined so as to satisfy Eqs. (3.74) and (3.75). In this way we obtain

$$\phi_i(s,x) = \frac{q_i}{4\pi\varepsilon_r\varepsilon_o}\int_0^\infty e^{-p(x+x_c)}\frac{1}{p}\left(\frac{\varepsilon_r p - \varepsilon_p k}{\varepsilon_r p + \varepsilon_p k}\right) J_0(ks) k\, dk + \frac{q_i}{4\pi\varepsilon_r\varepsilon_o R}\exp(-\kappa R) \tag{3.80}$$

Substituting Eq. (3.80) into Eq. (3.64) gives

$$W_i(x) = \frac{q_i^2}{8\pi\varepsilon_r\varepsilon_o}\int_0^\infty e^{-2px}\frac{1}{p}\left(\frac{\varepsilon_r p - \varepsilon_p k}{\varepsilon_r p + \varepsilon_p k}\right) k\, dk \tag{3.81}$$

For the special cases where $\varepsilon_p = 0$ (an insulating plate) or $\varepsilon_p = \infty$ (a metallic plate), Eq. (3.81) reduces to

$$\phi_i(s,x) = \frac{q_i}{4\pi\varepsilon_r\varepsilon_o}\left\{\frac{1}{R}\exp(-\kappa R) \pm \frac{1}{R'}\exp(-\kappa R')\right\} \tag{3.82}$$

where $R = \sqrt{(x-x_c)^2+s^2}$, $R' = \sqrt{(x+x_c)^2+s^2}$, the plus and minus signs refer to insulating and metallic plates, respectively, and R' denotes the distance between the field point $\boldsymbol{r}$ and the image of the central ion. For such cases, Eq. (3.81) gives

$$W_i(x) = \pm\frac{q_i^2 \exp(-2\kappa x)}{16\pi\varepsilon_r\varepsilon_o x} \tag{3.83}$$

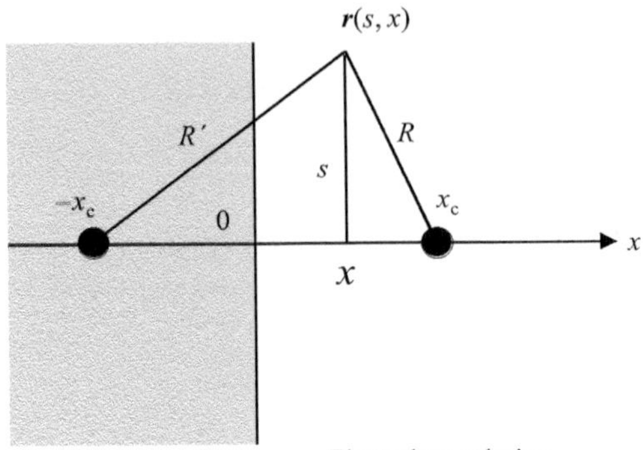

FIGURE 3.11 Interaction between an ion at $x = x_c$ and its image at $x = -x_c$, which is repulsion for an insulating plate ($\varepsilon_p = 0$) and attraction for a metallic plate ($\varepsilon_p = \infty$).

where the plus and minus signs refer to insulating and metallic walls, respectively. That is, the image interaction between an ion and an insulating plate is repulsive while that between an ion and a metallic plate is attractive (Fig. 3.11).

REFERENCES

1. B. V. Derjaguin and L. Landau, *Acta Physicochim.* 14 (1941) 633.
2. E. J. W. Verwey and J. Th. G. Overbeek, *Theory of the Stability of Lyophobic Colloids*, Elsevier, Amsterdam, 1948.
3. H. Ohshima and K. Furusawa (Eds.), *Electrical Phenomena at Interfaces, Fundamentals, Measurements, and Applications, 2nd edition*, revised and expanded, Dekker, New York, 1998.
4. H. Ohshima, *Theory of Colloid and Interfacial Electric Phenomena*, Elsevier/Academic Press, Amsterdam, 2006.
5. S. Carnie and S. McLaughlin, *Biophys. J.* 44 (1983) 325.
6. H. Ohshima and S. Ohki, *J. Colloid Interface Sci.* 142 (1991) 596.
7. H. Ohshima, K. Nakamori, and T. Kondo, *Biophys. Chem.* 53 (1993) 59.
8. J. G. Kirkwood, *J. Chem. Phys.* 2 (1934) 767.
9. P. Debye and E. Hückel, *Phyz. Z.* 24 (1923) 305.
10. A. Loeb, *J. Colloid Sci.* 6 (1951) 75.
11. W. E. Williams, *Proc. Phys. Soc. A* 66 (1953) 372.
12. H. Ohshima, in: J. Luetzenkirchen (Ed.), *Surface Complexation Modeling*, Elsevier, 2006.

4 Potential and Charge of a Soft Particle

4.1 INTRODUCTION

In Chapter 1, we have discussed the potential and charge of hard particles, which colloidal particles play a fundamental role in their interfacial electric phenomena such as electrostatic interaction between them and their motion in an electric field [1–4]. In this chapter, we focus on the case where the particle core is covered by an ion-penetrable surface layer of polyelectrolytes, which we term a surface charge layer (or, simply, a surface layer). Polyelectrolyte-coated particles are often called soft particles [3–16]. It is shown that the Donnan potential plays an important role in determining the potential distribution across a surface charge layer. Soft particles serve as a model for biocolloids such as cells. In such cases, the electrical double layer is formed not only outside but also inside the surface charge layer Figure 4.1 shows schematic representation of ion and potential distributions around a hard surface (Fig. 4.1a) and a soft surface (Fig. 4.1b).

4.2 PLANAR SOFT SURFACE

4.2.1 Poisson–Boltzmann Equation

Consider a surface charge layer of thickness d coating a planar hard surface in a general electrolyte solution containing M ionic species with valence z_i and bulk concentration (number density) n_i^∞ ($i = 1, 2, \ldots, M$). We treat the case where fully ionized groups of valence Z are distributed at a uniform density of N in the surface charge layer and the particle core is uncharged. We take an x-axis perpendicular to the surface charge layer with its origin $x = 0$ at the boundary between the surface charge layer and the surrounding electrolyte solution so that the surface charge layer corresponds to the region $-d < x < 0$ and the electrolyte solution to $x > 0$ (Fig. 4.1b).

The charge density $\rho_{el}(x)$ resulting from the mobile charged ionic species is related to the electric potential $\psi(x)$ by the Poisson equation

Biophysical Chemistry of Biointerfaces By Hiroyuki Ohshima

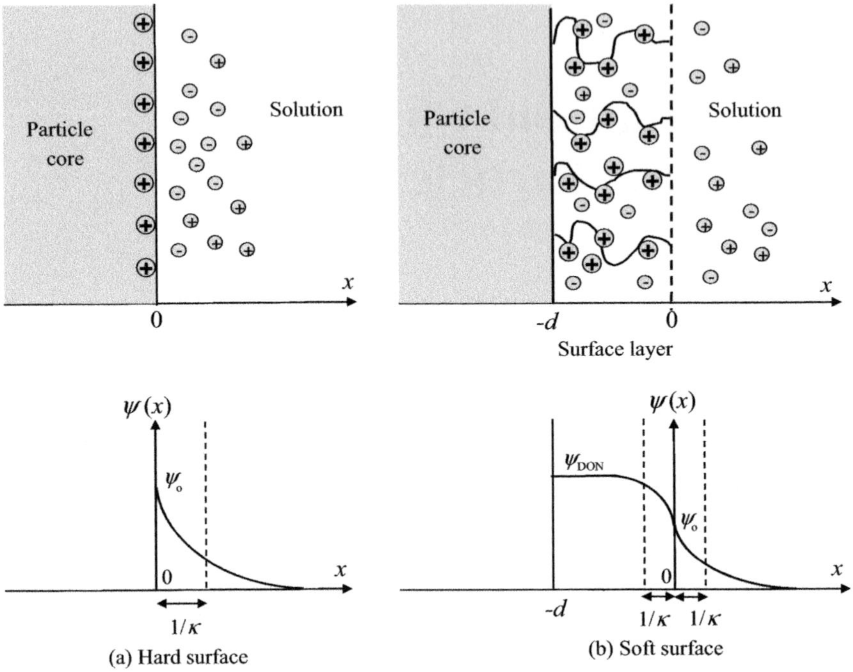

FIGURE 4.1 Ion and potential distribution around a hard surface (a) and a soft surface (b). When the surface layer is thick, the potential deep inside the surface layer becomes the Donnan potential.

$$\frac{d^2\psi}{dx^2} = -\frac{\rho_{\rm el}(x)}{\varepsilon_r\varepsilon_o}, \; 0 < x < +\infty \tag{4.1}$$

$$\frac{d^2\psi}{dx^2} = -\frac{\rho_{\rm el}(x) + ZeN}{\varepsilon_r\varepsilon_o}, \; -d < x < 0 \tag{4.2}$$

where ε_r is the relative permittivity of the solution, ε_o is the permittivity of a vacuum, and e is the elementary electric charge. We have here assumed that the relative permittivity takes the same value in the regions inside and outside the surface charge layer. Note that the right-hand side of Eq. (4.2) contains the contribution of the fixed charges of density $\rho_{\rm fix} = ZeN$ in the polyelectrolyte layer. We also assume that the distribution of the electrolyte ions $n_i(x)$ obeys Boltzmann's law. Then, we have

$$n_i(x) = n_i^{\infty}\exp\left(-\frac{z_i e\psi(x)}{kT}\right) \tag{4.3}$$

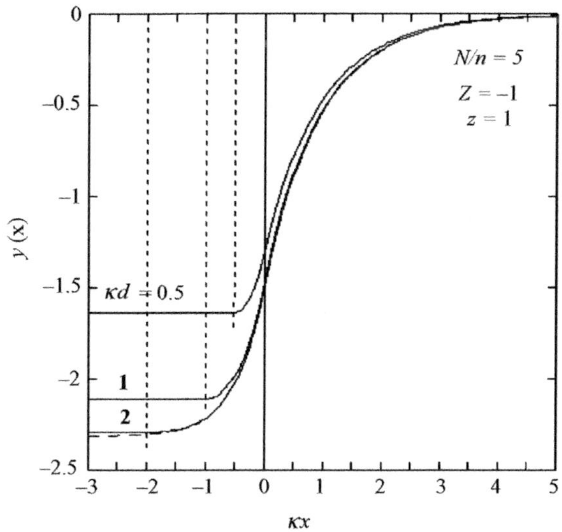

FIGURE 4.2 The scaled potential $y(x)$ as a function of the scaled distance κx across an ion-penetrable surface charge layer of the scaled thickness κd for several values of κd ($\kappa d = 0.5$, 1, and 2) at $N/n = 5$, $Z = -1$, and $z = 1$. The vertical dotted line stands for the position of the surface of the particle core for the respective cases. The dashed curve corresponds to the limiting case of $\kappa d \to \infty$. From Ref. [4].

and the charge density $\rho_{\mathrm{el}}(x)$ at x is given by

$$\rho_{\mathrm{el}}(x) = \sum_{i=1}^{M} z_i e n_i^{\infty} \exp\left(-\frac{z_i e \psi(x)}{kT}\right) \tag{4.4}$$

The potential $\psi(x)$ at position x in the regions $x > 0$ and $-d < x < 0$ thus satisfies the following Poisson–Boltzmann equations:

$$\frac{d^2y}{dx^2} = -\frac{\kappa^2 \sum\limits_{i=1}^{M} z_i n_i^{\infty} \exp(-z_i y)}{\sum\limits_{i=1}^{M} z_i^2 n_i^{\infty}}, \quad 0 < x < +\infty \tag{4.5}$$

$$\frac{d^2y}{dx^2} = -\frac{\kappa^2 \sum\limits_{i=1}^{M} z_i n_i^{\infty} \exp(-z_i y) + ZN}{\sum\limits_{i=1}^{M} z_i^2 n_i^{\infty}}, \quad -d < x < 0 \tag{4.6}$$

with

$$y = \frac{e\psi}{kT} \tag{4.7}$$

$$\kappa = \left(\frac{1}{\varepsilon_r \varepsilon_o kT} \sum_{i=1}^{M} z_i^2 e^2 n_i^\infty \right)^{1/2} \tag{4.8}$$

where y is the scaled potential and κ is the Debye–Hückel parameters of the solution. The boundary conditions are

$$\left. \frac{d\psi}{dx} \right|_{x=-d^+} = 0 \tag{4.9}$$

$$\psi(-0^-) = \psi(-0^+) \tag{4.10}$$

$$\left. \frac{d\psi}{dx} \right|_{x=-0^-} = \left. \frac{d\psi}{dx} \right|_{x=-0^+} \tag{4.11}$$

$$\psi(x) \to 0 \ \text{as}\ x \to \infty \tag{4.12}$$

$$\frac{d\psi}{dx} \to 0 \ \text{as}\ x \to \infty \tag{4.13}$$

Equation (4.9) corresponds to the situation in which the particle core is uncharged.

For the special case where the electrolyte is symmetrical with valence z and bulk concentration n, we have

$$\frac{d^2\psi}{dx^2} = \frac{2zen}{\varepsilon_r \varepsilon_o} \sinh\left(\frac{ze\psi}{kT} \right), \ x > 0 \tag{4.14}$$

$$\frac{d^2\psi}{dx^2} = \frac{2zen}{\varepsilon_r \varepsilon_o} \sinh\left(\frac{ze\psi}{kT} \right) - \frac{ZeN}{\varepsilon_r \varepsilon_o}, \quad -d < x < 0 \tag{4.15}$$

or

$$\frac{d^2y}{dx^2} = \kappa^2 \sinh y, \ x > 0 \tag{4.14a}$$

$$\frac{d^2y}{dx^2} = \kappa^2 \left(\sinh y - \frac{ZN}{2zn} \right), \quad -d < x < 0 \tag{4.15a}$$

with

$$y = \frac{ze\psi}{kT} \tag{4.16}$$

$$\kappa = \left(\frac{z^2 e^2 n}{\varepsilon_r \varepsilon_o kT} \right)^{1/2} \tag{4.17}$$

where y and κ are, respectively, the scaled potential and the Debye–Hückel parameter of a symmetrical electrolyte solution. Note that the definition of y for symmetrical electrolytes contains z.

4.2.2 Potential Distribution Across a Surface Charge Layer

If the thickness of the surface layer d is much greater than the Debye length $1/\kappa$, then the potential deep inside the surface layer becomes the Donnan potential ψ_{DON}, which is obtained by setting the right-hand side of Eq. (4.15) to zero, namely,

$$\psi_{\text{DON}} = \left(\frac{kT}{ze} \right) \text{arcsinh} \left(\frac{ZN}{2zn} \right) = \left(\frac{kT}{ze} \right) \ln \left[\frac{ZN}{2zn} + \left\{ \left(\frac{ZN}{2zn} \right)^2 + 1 \right\}^{1/2} \right] \tag{4.18}$$

Equation (4.16) may be rewritten as

$$\frac{d^2\psi}{dx^2} = \frac{2zen}{\varepsilon_r \varepsilon_o} \left\{ \sinh \left(\frac{ze\psi}{kT} \right) - \sinh \left(\frac{ze\psi_{\text{DON}}}{kT} \right) \right\}, \quad -d < x < 0 \tag{4.19}$$

When $|ZN/zn| \ll 1$, Eq. (4.18) gives the following linearized Donnan potential:

$$\psi_{\text{DON}} = \frac{ZNkT}{2z^2 ne} = \frac{ZeN}{\varepsilon_r \varepsilon_o \kappa^2} \tag{4.20}$$

Also we term $\psi_o \equiv \psi(0)$ (which is the potential at the boundary between the surface layer and the surrounding electrolyte solution), the surface potential of the polyelectrolyte layer.

For the simple case where $\psi(x)$ is low, Eqs. (4.14) and (4.15) can be linearized to become

$$\frac{d^2\psi}{dx^2} = \kappa^2 \psi, \ x > 0 \tag{4.21}$$

$$\frac{d^2\psi}{dx^2} = \kappa^2 \left(\psi - \frac{ZNkT}{2z^2 ne} \right), \quad -d < x < 0 \tag{4.22}$$

The solution to Eqs. (4.21) and (4.22) subject to Eqs. (4.9)–(4.13) is given by

$$\psi(x) = \frac{ZNkT}{4z^2ne}(1 - e^{-2\kappa d})e^{-\kappa x}, \ x > 0 \tag{4.23}$$

$$\psi(x) = \frac{ZNkT}{2z^2ne}\left\{1 - \frac{e^{\kappa x} + e^{-\kappa(x+2d)}}{2}\right\}, \quad -d < x < 0 \tag{4.24}$$

and the surface potential $\psi_o \equiv \psi(0)$ is given by

$$\psi_o = \frac{ZNkT}{4z^2ne}(1 - e^{-2\kappa d}) = \frac{ZeN}{2\varepsilon_r\varepsilon_o\kappa^2}(1 - e^{-2\kappa d}) \tag{4.25}$$

If we take the limit $d \rightarrow 0$ in Eq. (4.25), keeping the product Nd constant, that is, keeping the total amount of fixed charges $\sigma = ZeNd$ constant, then Eq. (4.25) becomes

$$\psi_o = \frac{\sigma}{\varepsilon_r\varepsilon_o\kappa} \tag{4.26}$$

where we have defined σ as

$$\sigma = Ze \lim_{\substack{d \rightarrow 0 \\ Nd = \text{constant}}} (Nd) \tag{4.27}$$

Equation (4.26) is the surface potential of a hard surface carrying a surface charge density σ [2–4].

It is to be noted that when $\kappa d \ll 1$, the potential deep inside the surface layer tends to the linearized Donnan potential (Eq. (4.20)), that is,

$$\psi(x) \approx \psi(-d) \approx \psi_{\text{DON}} = \frac{ZNkT}{2z^2ne} = \frac{ZeN}{\varepsilon_r\varepsilon_o\kappa^2} \tag{4.28}$$

and that the surface potential ψ_o (Eq. (1.25)) becomes half the Donna potential

$$\psi_o = \frac{\psi_{\text{DON}}}{2} = \frac{ZNkT}{4z^2ne} = \frac{ZeN}{2\varepsilon_r\varepsilon_o\kappa^2} \tag{4.29}$$

The solution to the nonlinear differential equations (4.14) and (4.15) can be obtained as follows. Equation (4.14) subject to Eqs. (4.12) and (4.13) can be integrated to give

$$\frac{dy}{dx} = -2\kappa\sinh\left(\frac{y}{2}\right), \; x > 0 \tag{4.30}$$

which is further integrated to give

$$y = 4 \text{ arctanh}\left[\tanh\left(\frac{y_o}{4}\right)e^{-\kappa x}\right], \; x > 0 \tag{4.31}$$

where $y_o \equiv y(0) = ze\psi(0)/kT$ is the scaled surface potential. Equation (4.31) coincides with the potential distribution around a hard planar surface (Eq. (1.38)). That is, the potential distribution outside the surface charge layer is the same as that for a hard surface.

Integration of Eq. (4.15) subject to Eq. (4.9) yields

$$\frac{dy}{dx} = -\text{sgn}(y_o)\kappa\left[\cosh y - \cosh y(-d) - \frac{ZN}{zn}\{y - y(-d)\}\right]^{1/2}, \; -d < x < 0 \tag{4.32}$$

Equation (4.32) can further be integrated to give

$$\kappa x = \text{sgn}(y_o)\int_y^{y_o} \frac{dy}{[2\{\cosh y - \cosh y(-d)\} - (ZN/zn)\{y - y(-d)\}]^{1/2}} \tag{4.33}$$

which determines y as a function of x. Evaluation Eq. (4.33) at $x = -d$, we obtain

$$\kappa d = -\text{sgn}(y_o)\int_{y(-d)}^{y_o} \frac{dy}{[2\{\cosh y - \cosh y(-d)\} - (ZN/zn)\{y - y(-d)\}]^{1/2}} \tag{4.34}$$

On the other hand, by evaluating Eqs. (4.31) and (4.33) at $x = 0$ and substituting the results into Eq. (4.11), we obtain

$$4\sinh^2\left(\frac{y_o}{2}\right) = 2\{\cosh y_o - \cosh y(-d)\} - \frac{ZN}{zn}\{y_o - y(-d)\} \tag{4.35}$$

which can be rewritten as

$$2\sinh^2\left(\frac{y_o}{2}\right) = \{\cosh y_o - \cosh y(-d)\} - \sinh y_{DON}\{y_o - y(-d)\} \tag{4.35a}$$

where Eq. (4.18) (i.e., $\sinh y_{DON} = zN/2zn$) has been used and $y_{DON} \equiv ze\psi_{DON}/kT$ is the scaled Donnan potential. Equations (4.34) and (4.35) form coupled transcendental and integral equations for y_o and $y(-d)$. By use of obtained values for y_o and $y(-d)$, the potential y at an arbitrary point x can be calculated from Eqs. (4.31) and (4.33).

Figure 4.2 shows the scaled potential $y(x)$ across a negatively charged surface layer as a function of the scaled distance κx for three values of the scaled thickness of the surface layer κd at $N/n = 5$, $Z = -1$, and $z = 1$. We see that the potential varies exponentially in the both regions inside and outside the surface layer. In Fig. 4.2, $y(x)$ for the limiting case of $\kappa d \to \infty$ is also given. It is seen that $y(x)$ for $\kappa d \geq 2$ almost coincides with that for $\kappa d \to \infty$, implying that the potential $\psi(x)$ deep inside the surface layer for $\kappa d \geq 2$ is practically equal to the Donnan potential ψ_{DON}.

4.2.3 Thick Surface Charge Layer and Donnan Potential

Consider the case where the thickness of the surface layer d is much larger than the Debye length $1/\kappa$. In this case, the electric field $(d\psi/dx)$ and its derivative $(d^2\psi/dx^2)$ deep inside the surface layer become zero so that the potential deep inside the surface layer becomes the Donnan potential ψ_{DON}, given by Eq. (4.18). It must be noted here that this is the case only for $d \gg 1/\kappa$. If the condition $d \gg 1/\kappa$ does not hold, there is no region where the potential reaches the Donnan potential. When d $1/\kappa$, by replacing $y(-d)$ by y_{DON} in Eq. (4.35), we obtain

$$\begin{aligned}\psi_o &= \psi_{DON} - \left(\frac{kT}{ze}\right)\tanh\left(\frac{ze\psi_{DON}}{2kT}\right)\\ &= \left(\frac{kT}{ze}\right)\left(\ln\left[\frac{ZN}{2zn}+\left\{\left(\frac{ZN}{2zn}\right)^2+1\right\}^{1/2}\right]+\frac{2zn}{ZN}\left[1-\left\{\left(\frac{ZN}{2zn}\right)^2+1\right\}^{1/2}\right]\right)\end{aligned} \tag{4.36}$$

or

$$\begin{aligned}y_o &= y_{DON} - \tanh(y_{DON}/2)\\ &= \ln\left[\frac{ZN}{2zn}+\left\{\left(\frac{ZN}{2zn}\right)^2+1\right\}^{1/2}\right]+\frac{2zn}{ZN}\left[1-\left\{\left(\frac{ZN}{2zn}\right)^2+1\right\}^{1/2}\right]\end{aligned} \tag{4.36a}$$

An approximate expression for the potential $\psi(x)$ at an arbitrary point in the surface layer can be obtained by expanding $y(x)$ in Eqs. (4.33) around y_{DON}. Alternatively, it can be obtained in the following way. If we put $\psi = \psi_{DON} + \Delta\psi$ in Eq. (4.19) and linearize it with respect to $\Delta\psi$, then we obtain

$$\frac{d^2\psi}{dx^2} = \kappa_m^2(\psi - \psi_{DON}) \tag{4.37}$$

with

$$\kappa_m = \kappa \cos h^{1/2} y_{DON} = \kappa\left[1+\left(\frac{ZN}{2zn}\right)^2\right]^{1/4} \tag{4.38}$$

where κ_m may be interpreted as the effective Debye–Hückel parameter of the surface charge layer that involves the contribution of the fixed charges ZeN. Equation (4.37) is solved to give

$$\psi(x) = \psi_{DON} + (\psi_o - \psi_{DON})e^{-\kappa_m |x|} \tag{4.39}$$

We, thus, see that the potential distribution within the surface layer is characterized by ψ_{DON} and ψ_o, as schematically shown in Fig. 4.1b.

4.2.4 Transition Between Donnan Potential and Surface Potential

Consider the opposite limiting case of $\kappa d \ll 1$. In this case we may put $y = y(-d) + \Delta y$ in the integrand of Eq. (4.34) and linearize it with respect to Δy. Then we find

$$\kappa d = 2\left[\frac{y_o - y(-d)}{2\sinh y(-d) - ZN/zn}\right]^{1/2} \tag{4.40}$$

Note that in this case $y(-d)$ is no longer equal to the Donnan potential. Similarly, we obtain from Eq. (4.35)

$$4\sinh^2\left(\frac{y_o}{2}\right) = 2\left\{\sinh y(-d) - \frac{ZN}{zn}\right\}\{y_o - y(-d)\} \tag{4.41}$$

Eliminating $y_o - y(-d)$ from Eqs. (4.40) and (4.41), and using Eq. (4.17), we find

$$2\sinh\left(\frac{y_o}{2}\right) = \frac{ZeNd}{2n\varepsilon_r\varepsilon_o kT} - \kappa d \sinh y(-d) \tag{4.42}$$

If we take the limit $d \to 0$ in Eq. (4.42), keeping the product Nd constant, that is, keeping the total amount of fixed charges $ZeNd$ constant, then the second term on the right-hand side of Eq. (4.42) becomes negligible and we obtain the following result:

$$y_o = y(-d) = 2\mathrm{arcsinh}\left[\frac{\sigma}{(8n\varepsilon_r\varepsilon_o kT)^{1/2}}\right] \tag{4.43}$$

where we have defined σ as

$$\sigma = Ze \lim_{\substack{d\to 0 \\ Nd=\text{constant}}} (Nd) \tag{4.44}$$

which can be interpreted as the surface charge density of a rigid surface.

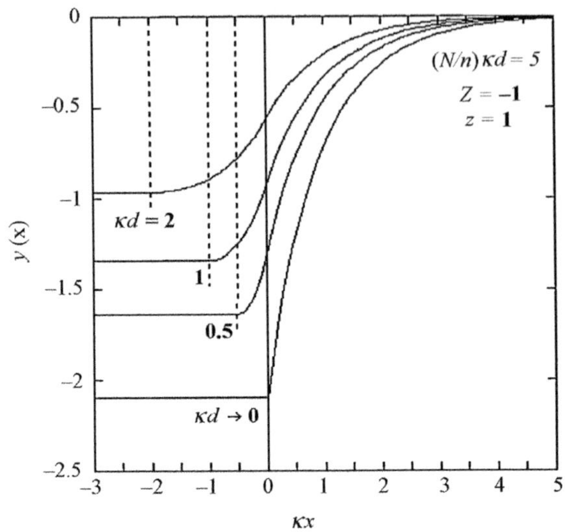

FIGURE 4.3 Scaled potential $y(x)$ as a function of the scaled distance κx across an ion-penetrable surface charge layer of the scaled thickness κd for several values of κd ($\kappa d = 0.5$, 1, and 2). The scaled charge amount contained in the surface layer is kept constant at $(N/n)\kappa d = 5$. The vertical dotted line stands for the position of the surface of the particle core for the respective cases. The curve with $\kappa d \to 0$ corresponds to the limiting case for the charged rigid surface with the scaled surface charge density equal to $(N/n)\kappa d = 5$. From Ref. [4].

Figure 4.3 shows numerical results for $y(x)$ when $\kappa d = 0$, 0.5, 1, and 2. Here we have varied d, keeping the total amount of the fixed charges $ZeNd$ per unit of the surface charge layer constant. Figure 4.3 indicates a continuous transition from the Donnan potential to the surface potential as d decreases to zero. Figure 4.3 also shows a strong dependence of $y(x)$ on the thickness d of the surface layer; that is, the magnitude of $y(x)$ decreases considerably as d increases. The reason for this lowering of $y(x)$ is that the enlarged area inside the surface layer into which electrolyte ions are allowed to penetrate increases the shielding effect of the ions upon the fixed charges in the surface layer.

4.2.5 Donnan Potential in a General Electrolyte

When a thick soft plate is immersed in a general electrolyte solution, the Donnan potential $y_{\rm DON}$ is given by setting the right-hand side of Eq. (4.6) (in which $-d$ is replaced by $-\infty$) to zero , that is, $y_{\rm DON}$ is given as a root of the following transcendental equation:

$$\sum_{i=1}^{M} z_i n_i^{\infty} \exp(-z_i y_{\rm DON}) + \frac{ZN}{\kappa^2} = 0 \tag{4.45}$$

The relationship between $y_{\rm DON}$ and $y_{\rm o}$ is obtained in the following way. Integrating Eqs. (4.5) and (4.6) gives

$$\left(\frac{dy}{dx}\right)^2 = \frac{2\kappa^2 \sum_{i=1}^{M} n_i^\infty (e^{-z_i y} - 1)}{\sum_{i=1}^{M} z_i^2 n_i^\infty}, \quad 0 < x < +\infty \tag{4.46}$$

$$\left(\frac{dy}{dx}\right)^2 = \frac{2\kappa^2 \sum_{i=1}^{M} n_i^\infty (e^{-z_i y} - e^{-z_i y_{\mathrm{DON}}}) - 2ZN(y - y_{\mathrm{DON}})}{\sum_{i=1}^{M} z_i^2 n_i^\infty}, \quad -\infty < x < 0 \tag{4.47}$$

By evaluating Eqs. (4.46) and (4.47) at $x=0$ and equating the results, we obtain

$$y_{\mathrm{o}} = y_{\mathrm{DON}} - \frac{\sum_{i=1}^{M} n_i^\infty (1 - e^{-z_i y_{\mathrm{DON}}})}{\sum_{i=1}^{M} z_i n_i^\infty e^{-z_i y_{\mathrm{DON}}}} \tag{4.48}$$

where Eq. (4.45) has been used. For a mixed solution of 1-1 electrolyte of bulk concentration n_1 and 2-1electrolyte of bulk concentration n_2, Eq. (4.48) becomes

$$y_{\mathrm{o}} = y_{\mathrm{DON}} - \left[\frac{(e^{y_{\mathrm{DON}}} - 1)\{(3-\eta)e^{y_{\mathrm{DON}}} + \eta\}}{(3-\eta)e^{y_{\mathrm{DON}}}(e^{y_{\mathrm{DON}}} + 1) + 2\eta}\right] \tag{4.49}$$

with $\eta = 3n_2/(n_1 + 3n_2)$.

4.3 SPHERICAL SOFT PARTICLE

4.3.1 Low Charge Density Case

Consider a spherical soft particle consisting the particle core of radius a covered by an ion-penetrable layer of polyelectrolytes of thickness d. The outer radius of the particle is thus given by $b = a + d$ (Fig. 4.4). Within the surface layer, ionized groups of valence Z are distributed at a constant density N. The Poisson–Boltzmann equations (4.1) and (4.2) are replaced by the following spherical Poisson–Boltzmann equations for electric potential $\psi(r)$, r being the distance from the center of the particle:

$$\frac{d^2\psi}{dr^2} + \frac{2}{r}\frac{d\psi}{dr} = -\frac{\rho_{\mathrm{el}}(r) + ZeN}{\varepsilon_{\mathrm{r}}\varepsilon_{\mathrm{o}}}, \quad a < r < b \tag{4.50}$$

$$\frac{d^2\psi}{dr^2} + \frac{2}{r}\frac{d\psi}{dr} = -\frac{\rho_{el}(r)}{\varepsilon_r\varepsilon_o}, \ r > b \tag{4.51}$$

We first treat the case in which the fixed-charge density ZeN is low. Then Eqs. (4.50) and (4.51) are linearized to give

$$\frac{d^2\psi}{dr^2} + \frac{2}{r}\frac{d\psi}{dr} = \kappa^2\psi - \frac{ZeN}{\varepsilon_r\varepsilon_o}, \ a < r < b \tag{4.52}$$

$$\frac{d^2\psi}{dr^2} + \frac{2}{r}\frac{d\psi}{dr} = \kappa^2\psi, \ r > b \tag{4.53}$$

The solution to Eqs. (4.52) and (4.53) subject to appropriate boundary conditions similar to Eqs. (4.9)–(4.13) is given by

$$\psi(r) = \frac{ZeN}{\varepsilon_r\varepsilon_o\kappa^2}\left[1 - \left(\frac{1+\kappa b}{1+\kappa a}\right)e^{-\kappa(b-a)} \times \left\{\frac{\sinh[\kappa(r-a)]}{\kappa r} + \frac{a\cosh[\kappa(r-a)]}{r}\right\}\right], \ a < r < b \tag{4.54}$$

$$\psi(r) = \psi_o \frac{b}{r}\exp[-\kappa(r-b)], \ r > b \tag{4.55}$$

with

$$\psi_o = \frac{ZeN}{2\varepsilon_r\varepsilon_o\kappa^2}\left\{1 - \frac{1}{\kappa b} + \frac{(1-\kappa a)(1+\kappa b)}{(1+\kappa a)\kappa b}e^{-2\kappa(b-a)}\right\} \tag{4.56}$$

where $\psi_o \equiv \psi(b)$ is the potential at $r = b$ (which we call the surface potential of a soft sphere). When $\kappa(b-a) = \kappa d \gg 1$, it follows from Eq. (4.54) that the potential deep inside the surface layer approaches the Donnan potential (Eq. (4.20)). With the help of Eq. (4.20), Eq. (4.56) gives the relationship between surface potential ψ_o and Donnan potential ψ_{DON} when the charge density ZeN is small, namely,

$$\psi_o = \frac{\psi_{DON}}{2}\left\{1 - \frac{1}{\kappa b} + \frac{(1-\kappa a)(1+\kappa b)}{(1+\kappa a)\kappa b}e^{-2\kappa(b-a)}\right\} \tag{4.57}$$

When κa 1, Eq. (4.57) tends to Eq. (4.29).

If we take the limit $b - a \to 0$ in Eq. (4.57), keeping the product Nd constant, that is, keeping the total amount of fixed charges $\sigma = ZeNd$ constant, then Eq. (4.57) becomes

$$\psi_o = \frac{\sigma}{\varepsilon_r\varepsilon_o\kappa(1 + 1/\kappa a)} \tag{4.58}$$

which is the surface potential ψ_o of a hard sphere with a surface charge density σ (Eq. (1.76)).

4.3.2 Surface Potential–Donnan Potential Relationship

Consider the case where the magnitude of the charge density ZeN is arbitrary. We rewrite Eqs. (4.50) and (4.51) as

$$\frac{d^2y}{dr^2}+\frac{2}{r}\frac{dy}{dr}=\kappa^2(\sinh y-\sinh y_{\mathrm{DON}}),\ a<r<b \tag{4.59}$$

$$\frac{d^2y}{dr^2}+\frac{2}{r}\frac{dy}{dr}=\kappa^2\sinh y,\ r>b \tag{4.60}$$

which cannot be solved analytically. In the following, we give a method of obtaining approximate relationship between surface potential and Donnan potential [11]. Note that the solution to Eq. (4.60) is the same as that for a hard sphere of radius b and its accurate analytic expressions are given in Refs [12] and [13]. We, thus, need to consider only the region inside the polyelectrolyte layer. In the following, we present two simple methods (i) and (ii) for solving Eq. (4.59). Method (i) is applied for the case of large $\kappa d \equiv \kappa(b-a)$ while method (ii) for small κd.

(i) *Large κd case.* In this case, the potential $y(r)$ deep inside the surface layer is practically equal to the Donnan potential y_{DON}, as is seen in Figs 4.4a and 4.5. Figure 4.5 shows the solution to Eqs. (4.59) and (4.65) for $y_o=1$, $\kappa a=10$, $\kappa b=20$, and $\kappa d=10$. We may thus expand $y(r)$ around y_{DON}

$$y(r)=y_{\mathrm{DON}}+\Delta y(r),\ a<r<b \tag{4.61}$$

By substituting Eq. (4.66) into Eq. (4.59), we obtain

$$\frac{d^2\Delta y}{dr^2}+\frac{2}{r}\frac{d\Delta y}{dr}=\kappa_{\mathrm{m}}^2\Delta y \tag{4.62}$$

with

$$\kappa_{\mathrm{m}}=\kappa\sqrt{\cosh y_{\mathrm{DON}}} \tag{4.63}$$

Equation (4.67) can easily be solved and we obtain

$$y(r)=y_{\mathrm{DON}}+\left(\frac{b}{r}\right)\left\{\frac{\kappa_{\mathrm{m}}a\cosh[\kappa_{\mathrm{m}}(r-a)]+\sinh[\kappa_{\mathrm{m}}(r-a)]}{\kappa_{\mathrm{m}}a\cosh[\kappa_{\mathrm{m}}(b-a)]+\sinh[\kappa_{\mathrm{m}}(b-a)]}\right\}\times(y_o-y_{\mathrm{DON}}),\ a<r<b \tag{4.64}$$

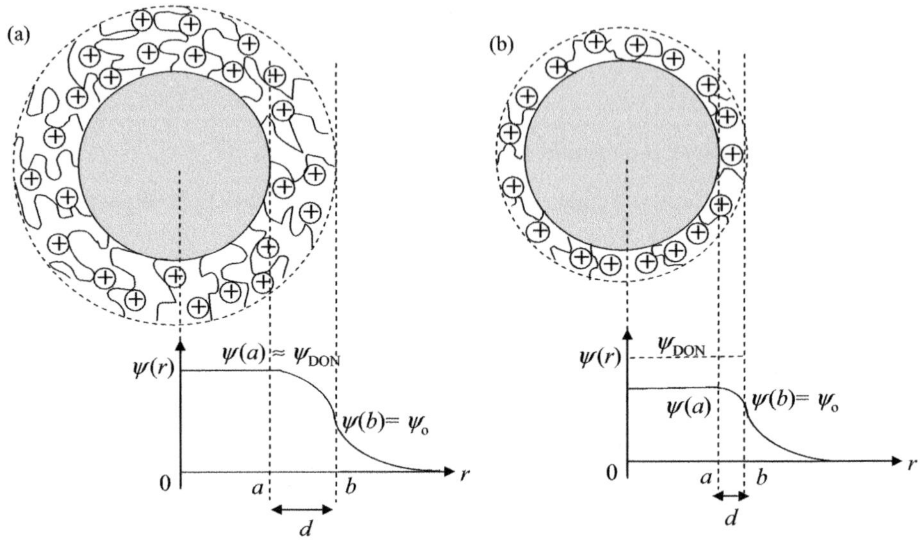

FIGURE 4.4 A spherical soft particle covered with a thick surface layer (a) or a thin surface layer (b) and the potential distribution. a = radius of the particle core. d = thickness of the polyelectrolyte layer covering the particle core. $b = a + d$. ψ_o is the surface potential. When the surface layer is thick, the potential deep inside the surface layer becomes the Donnan potential (a), while for thin surface layer, the potential within the surface layer cannot reach the Donnan potential (b).

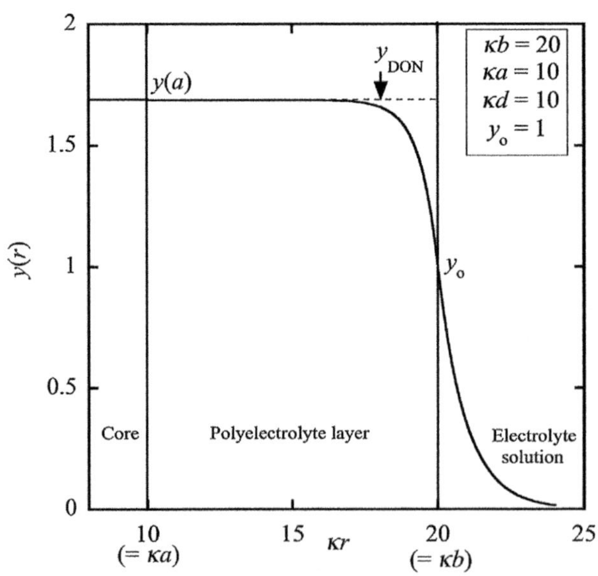

FIGURE 4.5 Potential distribution around a spherical particle for $y_o = 1$, $\kappa b = 20$, $\kappa a = 10$, and thus $\kappa d = 10$. From Ref. [11].

which gives

$$\left.\frac{dy}{dr}\right|_{r=b^-} = -\left(\frac{y_{\rm DON} - y_{\rm o}}{b}\right) \times \left\{\frac{\kappa_{\rm m}(b-a)\cosh[\kappa_{\rm m}(r-a)] + (\kappa_{\rm m}^2 ab - 1)\sinh[\kappa_{\rm m}(r-a)]}{\kappa_{\rm m}a\cosh[\kappa_{\rm m}(b-a)] + \sinh[\kappa_{\rm m}(b-a)]}\right\} \tag{4.65}$$

As for dy/dr at $r = b^+$ for a sphere of radius b carrying surface potential $y_{\rm o}$, the following accurate approximate expression has been derived [12,13]:

$$\left.\frac{dy}{dr}\right|_{r=b^+} = -2\kappa\sinh\left(\frac{y_{\rm o}}{2}\right)\left[1 + \frac{2}{\kappa b\cosh^2(y_{\rm o}/4)} + \frac{8\ln[\cosh(y_{\rm o}/4)]}{(\kappa b)^2\sinh^2(y_{\rm o}/2)}\right]^{1/2} \tag{4.66}$$

By equating Eqs. (4.65) and (4.66), we obtain

$$(y_{\rm DON} - y_{\rm o})\left\{\frac{\kappa_{\rm m}(b-a)\cosh[\kappa_{\rm m}(b-a)] + (\kappa_{\rm m}^2 ab - 1)\sinh[\kappa_{\rm m}(b-a)]}{\kappa_{\rm m}a\cosh[\kappa_{\rm m}(b-a)] + \sinh[\kappa_{\rm m}(b-a)]}\right\}$$
$$= 2\kappa b\sinh\left(\frac{y_{\rm o}}{2}\right)\left[1 + \frac{2}{\kappa b\cosh^2(y_{\rm o}/4)} + \frac{8\ln[\cosh(y_{\rm o}/4)]}{(\kappa b)^2\sinh^2(y_{\rm o}/2)}\right]^{1/2} \tag{4.67}$$

Equation (4.67) is the required transcendental equation determining $y_{\rm o}$ for given values of $y_{\rm DON}$, $\kappa d \equiv \kappa(b-a)$ and κb and is a good approximation for large κd. That is, the value of $y_{\rm o}$ can be obtained as a root of Eq. (4.67) with the help of computer programs such as Mathematica, by using the FindRoot function. By substituting the obtained value of $y_{\rm o}$ into Eq. (4.64), the potential distribution $y(r)$ inside the polyelectrolyte layer can be calculated. For the case where $\kappa a \gg 1$, $\kappa b \gg 1$, and $\kappa d \gg 1$, Eq. (4.64) is well approximated by

$$y(r) = y_{\rm DON} + (y_{\rm o} - y_{\rm DON})\exp[-\kappa_{\rm m}(r-a)], \quad a < r < b \tag{4.68}$$

(ii) *Small κd case.* In this case the potential never reaches the Donnan potential, as is seen in Figs 4.4b and 4.6. Figure 4.6 shows the solution to Eqs. (4.59) and (4.60) for $y_{\rm o} = 1$, $\kappa a = 9$, $\kappa b = 10$, and $\kappa d = 1$. It is thus better to expand $y(r)$ around $y(a)$ than $y_{\rm DON}$, that is,

$$y(r) = y(a) + \Delta y(r), \quad a < r < b \tag{4.69}$$

in Eq. (4.59) and linearize it with respect to $\Delta y(r)$, then we obtain

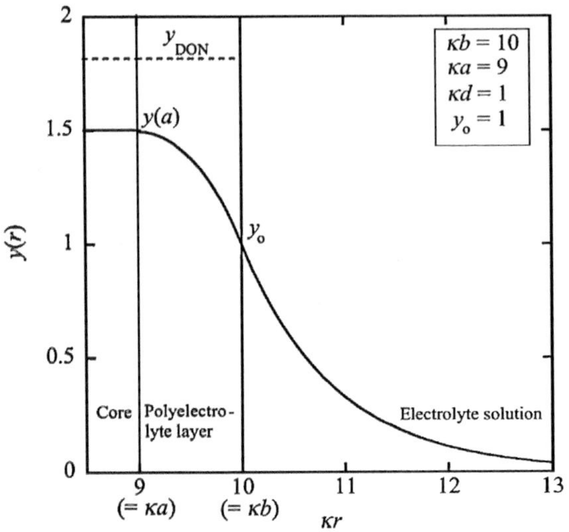

FIGURE 4.6 Potential distribution around a spherical particle for $y_o = 1$, $\kappa b = 10$, $\kappa a = 9$, and thus $\kappa d = 1$. From Ref. [11].

$$\frac{d^2\Delta y}{dr^2} + \frac{2}{r}\frac{d\Delta y}{dr} = \kappa_a^2\left[\Delta y - \left\{\frac{\sinh y_{DON} - \sinh[y(a)]}{\cosh[y(a)]}\right\}\right] \tag{4.70}$$

with

$$\kappa_a = \kappa\sqrt{\cosh[y(a)]} \tag{4.71}$$

Equation (4.70) can easily be solved to give

$$y(r) = y(a) + \left(\frac{\sinh y_{DON} - \sinh[y(a)]}{\cosh[y(a)]}\right) \times \left\{1 - \frac{a\cosh[\kappa_a(r-a)]}{r} - \frac{\sinh[\kappa_a(r-a)]}{\kappa_a r}\right\}, \quad a < r < b \tag{4.72}$$

from which we obtain

$$\left.\frac{dy}{dr}\right|_{r=b^-} = -\frac{1}{b}\left(\frac{\sinh y_{DON} - \sinh[y(a)]}{\cosh[y(a)]}\right) \times\left\{\left(1 - \frac{a}{b}\right)\cosh[\kappa_a(b-a)] + \left(\kappa_a a - \frac{1}{\kappa_a b}\right)\sinh[\kappa_a(b-a)]\right\} \tag{4.73}$$

As for dy/dr at $r = b^+$, we again use Eq. (4.66). By equating Eqs. (4.66) and (4.73), we obtain

$$\left(\frac{\sinh y_{\rm DON} - \sinh[y(a)]}{\cosh[y(a)]}\right) \times \left\{ \left(1 - \frac{a}{b}\right)\cosh[\kappa_{\rm a}(b-a)] + \left(\kappa_{\rm a}a - \frac{1}{\kappa_{\rm a}b}\right)\sinh[\kappa_{\rm a}(b-a)] \right\} = 2\kappa b \sinh\left(\frac{y_{\rm o}}{2}\right)\left[1 + \frac{2}{\kappa b \cosh^2(y_{\rm o}/4)} + \frac{8\ln[\cosh(y_{\rm o}/4)]}{(\kappa b)^2 \sinh^2(y_{\rm o}/2)}\right]^{1/2} \tag{4.74}$$

where

$$y_{\rm o} = y(a) + \left(\frac{\sinh y_{\rm DON} - \sinh[y(a)]}{\cosh[y(a)]}\right) \times \left\{1 - \frac{a}{b}\cosh[\kappa_{\rm a}(b-a)] - \frac{1}{\kappa_{\rm a}b}\sinh[\kappa_{\rm a}(b-a)]\right\} \tag{4.75}$$

Equation (4.74) (as combined with Eq. (4.75)) is the required transcendental equation determining $y(a)$ for given values of $y_{\rm DON}$, $\kappa d \equiv \kappa(b-a)$ and κb, applicable for small κd.

In Table 4.1, which shows the relationship between $y_{\rm DON}$ and $y_{\rm o}$, we compare exact numerical results obtained by solving the nonlinear Poisson–Boltzmann equations (4.59) and (4.60) and approximate results obtained via Eqs. (4.67) and (4.74) for several values of $y_{\rm o}$, κa, and κb. Good agreement is seen between exact and approximate results.

TABLE 4.1 Scaled Surface Potential $y_{\rm o}$ and Scaled Donnan Potential $y_{\rm DON}$ at $\kappa d = 1$ and 10[a]

	$\kappa b = 10$, $\kappa a = 9$, $\kappa d = 1$		$\kappa b = 20$, $\kappa a = 10$, $\kappa d = 10$		
	$y_{\rm DON}$ (exact)	$y_{\rm DON}$ (Eq. (4.74) or Eq. (4.67)*)	$y_{\rm DON}$ (exact)	$y_{\rm DON}$ (Eq. (4.74))	$y_{\rm DON}$ (Eq. (4.36))
1	1.836	1.817 (−1.0)	1.734	1.676 (−3.3)	1.688 (−2.7)
2	2.981	2.911 (−2.3)	2.936	2.846 (−3.1)	2.895 (−1.4)
4	5.028	4.923* (−2.1)	5.007	4.906 (−2.0)	4.986

[a] $y_{\rm DON}$ (exact): exact numerical values obtained by solving Eqs. (4.59) and (4.60); $y_{\rm DON}$ (Eq. (4.74) or Eq. (4.67)): approximate values calculated with Eq. (4.74) or Eq. (4.67) (numerical values without and with asterisk are obtained via Eq. (4.67) and (4.74), respectively); $y_{\rm DON}$ (Eq. (4.36)): approximate values calculated with Eq. (4.36) for a planar soft plate. The percentage relative error is indicated in parentheses only if more than 1%.

4.4 CYLINDRICAL SOFT PARTICLE

Consider a cylindrical soft particle, that is, an infinitely long cylindrical hard particle of core radius a covered with an ion-penetrable layer of polyelectrolytes of thickness d in a symmetrical electrolyte solution of valence z and bulk concentration (number density) n. The polymer-coated particle has thus an inner radius a and an outer radius $b = a + d$. The origin of the cylindrical coordinate system (r, z, φ) is held fixed on the cylinder axis. We consider the case where dissociated groups of valence Z are distributed with a uniform density N in the polyelectrolyte layer so that the density of the fixed charges ρ_{fix} in the surface layer is given by $\rho_{\text{fix}} = ZeN$. We assume that the potential $\psi(r)$ satisfies the following cylindrical Poisson–Boltzmann equations:

$$\frac{d^2\psi}{dr^2} + \frac{1}{r}\frac{d\psi}{dr} = -\frac{zen}{\varepsilon_r\varepsilon_o}\left[\exp\left(-\frac{ze\psi}{kT}\right) - \exp\left(\frac{ze\psi}{kT}\right)\right] - \frac{ZeN}{\varepsilon_r\varepsilon_o}, \quad a < r < b \tag{4.76}$$

$$\frac{d^2\psi}{dr^2} + \frac{1}{r}\frac{d\psi}{dr} = -\frac{zen}{\varepsilon_r\varepsilon_o}\left[\exp\left(-\frac{ze\psi}{kT}\right) - \exp\left(\frac{ze\psi}{kT}\right)\right], \quad r > b \tag{4.77}$$

where r denotes the radial distance measured from the cylinder axis. If the thickness of the surface layer d is much greater than the Debye length $1/\kappa$, then the potential deep inside the surface layer becomes the Donnan potential ψ_{DON}.

4.4.1 Low Charge Density Case

The nonlinear equations (4.76) and (4.77) have not been solved analytically except in the low potential case (i.e., the low fixed-charge density case). In this case, the solution to Eqs. (4.76) and (4.77) is

$$\psi(r) = \frac{ZeN}{\varepsilon_r\varepsilon_o\kappa^2}\left[1 - \kappa b K_1(\kappa b)\left\{I_0(\kappa r) + \frac{I_1(\kappa a)}{K_1(\kappa a)}K_0(\kappa r)\right\}\right], \quad a < r < b \tag{4.78}$$

$$\psi(r) = \psi_o \frac{K_0(\kappa r)}{K_0(\kappa b)}, \quad r > b \tag{4.79}$$

with

$$\psi_o = \frac{ZeN}{\varepsilon_r\varepsilon_o\kappa^2}\kappa b K_0(\kappa b)\left\{I_1(\kappa b) - \frac{I_1(\kappa a)}{K_1(\kappa a)}K_1(\kappa b)\right\} \tag{4.80}$$

where $I_n(x)$ and $K_n(x)$ are, respectively, the nth order modified Bessel functions of the first and second kinds (note that Eq. (2.63) in Ref. 4 is not correct and Eq. (4.78) is the corrected expression). We obtain the following relationship between ψ_{DON}

and ψ_o from Eqs. (4.78) and (4.79)

$$\psi_o = \frac{\psi_{DON}}{2} \kappa b K_0(\kappa b) \left\{ I_1(\kappa b) - \frac{I_1(\kappa a)}{K_1(\kappa a)} K_1(\kappa b) \right\} \tag{4.81}$$

If we take the limit $b - a \to 0$ in Eq. (4.81), keeping the product Nd constant, that is, keeping the total amount of fixed charges $\sigma = ZeNd$ constant, then Eq. (4.81) becomes

$$\psi_o = \frac{\sigma}{\varepsilon_r \varepsilon_o \kappa} \frac{K_0(\kappa a)}{K_1(\kappa a)} \tag{4.82}$$

which agrees with the surface potential ψ_o of a hard cylinder with a surface charge density σ.

4.4.2 Surface Potential–Donnan Potential Relationship

Transcendental equations for determining the relationship between y_{DON} and y_o can be obtained by using methods similar to those used for spherical soft particles, as shown below [15].

(i) *Large κd case.* An approximate expression for $y(r)$ is

$$\kappa_a^2 a \left(\frac{\sinh y_{DON} - \sinh[y(a)]}{\cosh[y(a)]} \right) \{K_1(\kappa_a a) I_1(\kappa_a b) - I_1(\kappa_a a) K_1(\kappa_a b)\}, \; a < r < b \tag{4.83}$$

and a transcendental equation for determining the y_{DON}–y_o relationship is given by

$$\begin{aligned} &\kappa_m \left\{ \frac{K_1(\kappa_m a) I_1(\kappa_m b) - I_1(\kappa_m a) K_1(\kappa_m b)}{K_1(\kappa_m a) I_0(\kappa_m b) - I_1(\kappa_m a) K_0(\kappa_m b)} \right\} (y_{DON} - y_o) \\ &= 2\kappa \sinh\left(\frac{y_o}{2}\right) \left[1 + \frac{\{K_1(\kappa b)/K_0(\kappa b)\}^2 - 1}{\cosh^2(y_o/4)} \right]^{1/2} \end{aligned} \tag{4.84}$$

where κ_m is given by Eq. (4.63) and the following accurate expression:

$$\left. \frac{dy}{dr} \right|_{r=b^+} = -2\kappa \sinh\left(\frac{y_o}{2}\right) \left[1 + \frac{\{K_1(\kappa_a b)/K_0(\kappa_a b)\}^2 - 1}{\cosh^2(y_o/4)} \right]^{1/2} \tag{4.85}$$

has been used [15].

(ii) *Small κd case.* An approximate expression for $y(r)$ is

$$y(r) = y(a) + \left(\frac{\sinh y_{\text{DON}} - \sinh[y(a)]}{\cosh[y(a)]}\right) \times [1 - \kappa_a a\{K_1(\kappa_a a)I_0(\kappa_a r) + I_1(\kappa_a a)K_0(\kappa_a r)\}], \qquad a < r < b \tag{4.86}$$

and a transcendental equation for determining the y_{DON}–y_o relationship is given by

$$\kappa_a^2 a\left(\frac{\sinh y_{\text{DON}} - \sinh[y(a)]}{\cosh[y(a)]}\right)\{K_1(\kappa_a a)I_1(\kappa_a b) - I_1(\kappa_a a)K_1(\kappa_a b)\} = 2\kappa \sinh\left(\frac{y_o}{2}\right)\left[1 + \frac{\{K_1(\kappa b)/K_0(\kappa b)\}^2 - 1}{\cosh^2(y_o/4)}\right]^{1/2} \tag{4.87}$$

with

$$y_o = y(a) + \left(\frac{\sinh y_{\text{DON}} - \sinh[y(a)]}{\cosh[y(a)]}\right) \times [1 - \kappa_a a\{K_1(\kappa_a a)I_0(\kappa_a b) + I_1(\kappa_a a)K_0(\kappa_a b)]\tag{4.88}$$

where κ_a is given by Eq. (4.71).

4.5 ASYMPTOTIC BEHAVIOR OF POTENTIAL AND EFFECTIVE SURFACE POTENTIAL OF A SOFT PARTICLE

The potential distribution outside the surface charge layer of a soft particle with surface potential ψ_o is the same as the potential distribution around a hard particle with a surface potential ψ_o. The asymptotic behavior of the potential distribution around a soft particle and that for a hard particle are the same provided they have the same surface potential ψ_o. The effective surface potential is an important quantity that determines the asymptotic behaviors of the electrostatic interaction between soft particles (see Chapter 15).

4.5.1 Plate

For a plate with scaled surface potential ψ_o in a symmetrical electrolyte of valence z, the asymptotic form of the potential outside the surface charge layer is derived from Eq. (4.31), namely,

$$\psi(x) = \frac{4kT}{ze} \tanh\left(\frac{ze\psi_o}{4kT}\right) e^{-\kappa x} \tag{4.89}$$

or

$$y(x) \equiv \frac{ze\psi(x)}{kT} = 4 \tanh\left(\frac{ze\psi_o}{4kT}\right) e^{-\kappa x} \tag{4.90}$$

The effective surface potential ψ_{eff} of the plate is given by

$$\psi_{eff} = \frac{4kT}{ze}\gamma = \frac{kT}{ze} \cdot 4 \tanh\left(\frac{ze\psi_o}{4kT}\right) \tag{4.91}$$

and the scaled effective surface potential $Y = zey_{eff}/kT$ is given by

$$Y = 4\gamma = 4 \tanh\left(\frac{ze\psi_o}{4kT}\right) \tag{4.92}$$

where $\gamma = \tanh(ze\psi_o/kT)$. It must be stressed here that for a hard plate, ψ_o is related to the surface charge density σ (Chapter 1), while for a soft plate, ψ_o is related to the volume charge density ZeN. For other types of electrolytes, expressions for Y are given in Chapter 1.

4.5.2 Sphere

The asymptotic expression for the potential of a spherical particle of radius a in a symmetrical electrolyte solution of valence z and Debye–Hückel parameter κ, a large distance r from the center of the sphere may be expressed as

$$\psi(r) = \psi_{eff} \frac{a}{r} e^{-\kappa(r-a)} \tag{4.93}$$

$$y(r) = \frac{ze}{kT}\psi(r) = Y\frac{a}{r}e^{-\kappa(r-a)} \tag{4.94}$$

where r is the radial distance measured from the sphere center, ψ_{eff} is the effective surface potential, and $Y = ze\psi_{eff}/kT$ is the scaled effective surface potential of a sphere. From Eq. (1.196) we obtain

$$\psi_{eff} = \frac{kT}{ze} \cdot \frac{8\tanh(y_o/4)}{1 + \{1 - (2\kappa a + 1/(\kappa a + 1)^2)\tanh^2(y_o/4)\}^{1/2}} \tag{4.95}$$

or

$$Y = \frac{8\tanh(y_o/4)}{1 + \{1 - (2\kappa a + 1/(\kappa a + 1)^2)\tanh^2(y_o/4)\}^{1/2}} \tag{4.96}$$

It must be stressed here that for a hard sphere, ψ_o is related to the surface charge density σ (Chapter 1), while for a soft sphere, ψ_o is related to the volume charge density ZeN. For other types of electrolytes, expressions for Y are given in Chapter 1.

4.5.3 Cylinder

The effective surface potential ψ_{eff} or scaled effective surface potential $Y = ze\psi_{eff}/kT$ of a cylinder in a symmetrical electrolyte solution of valence z can be obtained from the asymptotic form of the potential around the cylinder, which in turn is derived from Eq. (1.157) as

$$y(r) = Yc \tag{4.97}$$

with

$$c = \frac{K_0(\kappa r)}{K_0(\kappa a)} \tag{4.98}$$

and

$$Y = \frac{8\tanh(y_o/4)}{1 + \{1 - (1 - \beta^2)\tanh^2(y_o/4)\}^{1/2}} \tag{4.99}$$

where r is the distance from the axis of the cylinder and

$$\beta = \frac{K_0(\kappa a)}{K_1(\kappa a)} \tag{4.100}$$

For other types of electrolytes, expressions for Y are given in Chapter 1.

4.6 NONUNIFORMLY CHARGED SURFACE LAYER: ISOELECTRIC POINT

Consider an ion-penetrable planar soft surface consisting of two sublayers 1 and 2 (Fig. 4.7) . The surface is in equilibrium with a monovalent electrolyte solution of bulk concentration n. Note here that n represents the total concentration of monovalent cations including H^+ ions and that of monovalent anions including OH^- ions. Let n_H be the H^+ concentration in the bulk solution phase. Sublayer 2 is

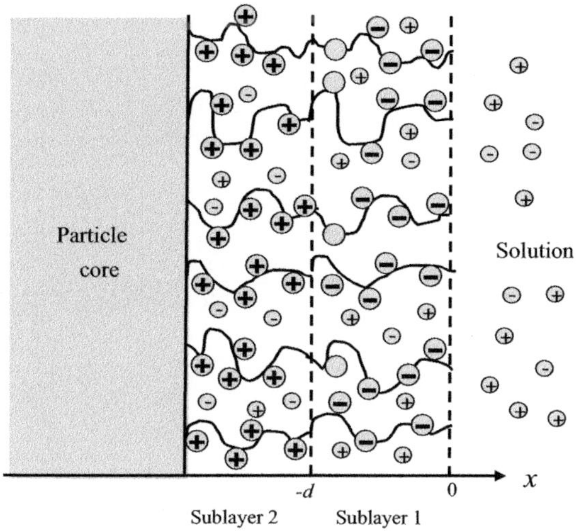

FIGURE 4.7 A soft surface consisting of two oppositely charged sublayers 1 and 2. The outer sublayer (layer 1) carries acidic groups and the inner sublayer (layer 2) basic groups. Layer 2 is assumed to be regarded as infinitely thick. The thickness of layer 1 is d, which is much thicker than $1/\kappa$. Fixed charges in the surface layer are represented by large circles with plus or minus signs, while electrolyte ions are by small circles with plus or minus signs.

assumed to be much thicker than $1/\kappa$ so that it can be regarded as infinitely thick. The thickness of sublayer 1 is denoted as d. We take an x-axis perpendicular to the surface with its origin at the outer edge of the surface layer so that the region $x > 0$ is the solution phase and $x < 0$ is the surface layer. In the outer layer (layer 1) monovalent acidic groups of dissociation constant K_a are distributed at a density N_1, while in the inner layer (layer 2) monovalent basic groups of dissociation constant K_b are distributed at a density N_2. The mass action law for the dissociation of acidic groups AH (AH $\Longleftrightarrow$ $A^- + H^+$) gives the number density of dissociated groups [10], namely,

$$\frac{N_1}{1 + (n_H/K_a)\exp(-e\psi(x)/kT)} \tag{4.101}$$

Here $\psi(x)$ is the electric potential at position x relative to the bulk solution phase, where $\psi(x)$ is set equal to zero, $n_H \exp(-e\psi(x)/kT)$ is the H^+ concentration at position x. The charge density resulting from dissociated acidic groups at position x in layer 1 $(-d < x < 0)$ is thus obtained by multiplying Eq. (4.101) by $-e$, namely,

$$-\frac{eN_1}{1 + (n_H/K_a)\exp(-e\psi(x)/kT)} \tag{4.102}$$

Similarly, the charge density resulting from the dissociated basic groups at x in layer 2 ($x < -d$) is given by

$$+\frac{eN_2}{1+(K_b/n_H)\exp(+e\psi(x)/kT)} \tag{4.103}$$

The Poisson–Boltzmann equation for $\psi(x)$ in layer 1 is then given by

$$\frac{d^2\psi}{dx^2} = \frac{2en}{\varepsilon_r\varepsilon_o}\sinh\left(\frac{e\psi}{kT}\right) + \frac{e}{\varepsilon_r\varepsilon_o}\frac{N_1}{1+(n_H/K_a)\exp(-e\psi(x)/kT)}, \quad -d < x < 0 \tag{4.104}$$

and for that in layer 2

$$\frac{d^2\psi}{dx^2} = \frac{2en}{\varepsilon_r\varepsilon_o}\sinh\left(\frac{e\psi}{kT}\right) - \frac{e}{\varepsilon_r\varepsilon_o}\frac{N_2}{1+(K_b/n_H)\exp(+e\psi(x)/kT)}, \quad x < -d \tag{4.105}$$

For the solution phase ($x > 0$), we have

$$\frac{d^2\psi}{dx^2} = \frac{2en}{\varepsilon_r\varepsilon_o}\sinh\left(\frac{e\psi}{kT}\right), \quad x > 0 \tag{4.106}$$

The boundary conditions are continuity conditions of $\psi(x)$ and $d\psi(x)/dx$, and $x = 0$ and $x = -d$.

Coupled equations for $\psi(x)$, Eqs. (4.104)–(4.106), subject to appropriate boundary conditions can be only numerically solved. We, thus, employ the following quasilinearization approximation method for obtaining the potential inside the surface layer. This method is based on the idea that the difference between the potential $\psi(x)$ at position x in sublayer 1 (or 2) and the Donnan potential of sublayer 1 (or 2) is not large so that the Poisson–Boltzmann equation for sublayer 1 (or 2) can be linearized with respect to this potential difference. The Donnan potentials of layers 1 and 2, which we denote by ψ_{DON1} and ψ_{DON2}, respectively, are obtained by setting the right-hand side of the Poisson–Boltzmann equations for the respective layers, Eqs. (4.104) and (4.105), equal to zero, that is, they are the solutions to the following transcendental equations (Fig. 4.8):

$$\sinh\left(\frac{e\psi_{\mathrm{DON1}}}{kT}\right) + \frac{N_1}{2n}\frac{1}{1+(n_H/K_a)\exp(-e\psi_{\mathrm{DON1}}/kT)} = 0 \tag{4.107}$$

and for that in layer 2

$$\sinh\left(\frac{e\psi_{\mathrm{DON2}}}{kT}\right) - \frac{N_2}{2n}\frac{1}{1+(K_b/n_H)\exp(+e\psi_{\mathrm{DON2}}/kT)} = 0 \tag{4.108}$$

Now we put

$$\psi(x) = \psi_{\mathrm{DON1}} + \Delta\psi, \quad -d < x < 0, \tag{4.109}$$

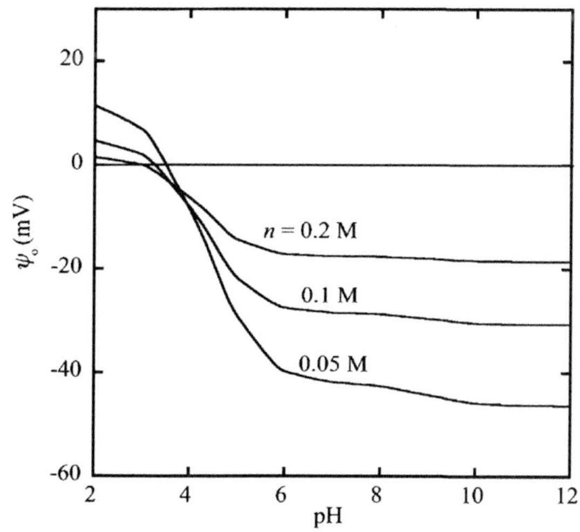

FIGURE 4.8 Surface potential ψ_o of a soft surface consisting of two oppositely charged sublayers 1 and 2 as a function of pH. Calculated with $N_1 = 0.725$ M, $N_2 = 0.235$ M, $K_a = 10^{-4}$ M (p$K_a = 4$), $K_b = 10^{-9}$ (p$K_b = 9$), $T = 298$ K, and $d = 1$ nm at several values of electrolyte concentration n. From Ref. [16].

$$\psi(x) = \psi_{\mathrm{DON2}} + \Delta\psi, \; x < -d \tag{4.110}$$

in Eqs. (4.104) and (4.105) and linearize them with respect to $\Delta\psi$. Then we have

$$\frac{d^2\Delta\psi}{dx^2} = \kappa^2 R_1^2 \Delta\psi, \quad -d < x < 0, \tag{4.111}$$

$$\frac{d^2\Delta\psi}{dx^2} = \kappa^2 R_2^2 \Delta\psi, \; x < -d \tag{4.112}$$

with

$$R_1 = \left[\cosh y_{\mathrm{DON1}} - \frac{\sinh y_{\mathrm{DON1}}}{1 + (K_a/n_H)\exp(+y_{\mathrm{DON1}})}\right]^{1/2} \tag{4.113}$$

$$R_2 = \left[\cosh y_{\mathrm{DON2}} - \frac{\sinh y_{\mathrm{DON2}}}{1 + (n_H/K_a)\exp(+y_{\mathrm{DON2}})}\right]^{1/2} \tag{4.114}$$

where

$$y_{\mathrm{DON1}} = \frac{e\psi_{\mathrm{DON1}}}{kT} \tag{4.115}$$

$$y_{\mathrm{DON1}} = \frac{e\psi_{\mathrm{DON2}}}{kT} \tag{4.116}$$

are the reduced Donnan potential and κ is the Debye–Hückel parameter. By combining the solutions to Eqs. (4.106), (4.111), and (4.112), we obtain

$$\begin{aligned}(y_{\mathrm{DON1}} - y_{\mathrm{DON2}}) + (y_o - y_{\mathrm{DON1}})\left[\cosh(R_2\kappa d) + \frac{R_1}{R_2}\sinh(R_2\kappa d)\right] \\ + 2\sinh\left(\frac{y_o}{2}\right)\left[\frac{1}{R_1}\sinh(R_1\kappa d) + \frac{1}{R_2}\cosh(R_1\kappa d)\right] = 0\end{aligned} \tag{4.117}$$

This is the required equation determining the reduced membrane surface potential y_o $(= e\psi_o/kT)$. Here y_{DON1} and y_{DON2} are the solutions to Eqs. (4.107) and (4.108).

When ψ_o is small, the transcendental equation, Eq. (4.117), can be simplified by approximating $2\sinh(y_o/2)$ by y_o in Eq. (4.117) so that one can obtain an explicit expression for y_o, namely,

$$y_o = \frac{y_{\mathrm{DON1}}[\cosh(R_1\kappa d) + (R_1/R_2)\sinh(R_1\kappa d) - 1] + y_{\mathrm{DON1}}}{(1 + 1/R_2)\cosh(R_1\kappa d) + (1/R_1 + R_1/R_2)\sinh(R_1\kappa d)} \tag{4.118}$$

The isoelectric point (IEP) is defined as the pH value at which y_o reverses its sign. As is seen from Eq. (4.117), IEP depends on the electrolyte concentration n.

For comparison, we give below equations determining the surface potential y_o for the case of uniform distribution of both acidic and basic groups. In this situation, layers 1 and 2 become identical so that the charge density at position x takes the same form throughout the surface layer:

$$-\frac{eN_1}{1 + (n_{\mathrm{H}}/K_a)\exp(-e\psi(x)/kT)} + \frac{eN_2}{1 + (K_b/n_{\mathrm{H}})\exp(+e\psi(x)/kT)} \tag{4.119}$$

and the reduced Donnan potential $y_{\mathrm{DON}} = ey_{\mathrm{DON}}/kT$ of the surface layer is given by the solution to the following transcendental equation:

$$\sinh y_{\mathrm{DON}} + \left(\frac{N_1}{2n}\right)\frac{1}{1 + (n_{\mathrm{H}}/K_a)e^{-y_{\mathrm{DON}}}} - \left(\frac{N_2}{2n}\right)\frac{1}{1 + (K_b/n_{\mathrm{H}})e^{+y_{\mathrm{DON}}}} = 0 \tag{4.120}$$

The same procedure we used in deriving Eq. (4.118) yields

$$y_o = \frac{R}{R+1}y_{\mathrm{DON}} \tag{4.121}$$

where

$$R = \left[\cos y_o + \left(\frac{N_1}{2n}\right)\frac{(n_{\mathrm{H}}/K_a)e^{-y_o}}{\{1 + (n_{\mathrm{H}}/K_a)e^{-y_o}\}^2} - \left(\frac{N_2}{2n}\right)\frac{(K_b/n_{\mathrm{H}})e^{y_o}}{\{1 + (n_{\mathrm{H}}/K_b)e^{-y_o}\}^2}\right]^{1/2} \tag{4.122}$$

In this case, the sign of the membrane surface potential y_o $(=e\psi_o/kT)$ coincides with that of the Donnan potential y_{DON} and becomes zero when y_o becomes zero. Namely, the H^+ concentration at which y_o changes its sign is given by setting y_{DON} equal to zero in Eq. (4.120), namely,

$$n_H = \frac{K_a}{2}\left[\left(\frac{N_1}{N_2}-1\right)+\left\{\left(\frac{N_1}{N_2}-1\right)^2+\frac{4K_bN_1}{K_aN_2}\right\}^{1/2}\right] \tag{4.123}$$

The IEP ($=-\log n_H$, n_H being given by Eq. (4.123)) is thus independent of the electrolyte concentration, in contrast to the case of nonuniform distribution of dissociated groups.

Figure 4.8 shows the surface potential ψ_o as a function of pH calculated for $N_1 = 0.725$ M, $N_2 = 0.235$ M, $K_a = 10^{-4}$ M (p$K_a = 4$), $K_b = 10^{-9}$ M (p$K_b = 9$), $T = 298$ K, and $d = 1$ nm at several values of the electrolyte concentration n. We see that the surface potential ψ_o and its IEP (i.e., the intersection of each curve with the $\psi_o = 0$ axis) vary with the electrolyte concentration n. For comparison, the corresponding results for a soft surface with uniform distribution of acidic and basic groups are given in Fig. 4.9, which shows that all intersections of curves calculated for different electrolyte concentrations with the $\psi_o = 0$ axis are the same. In other words, the IEP of a uniformly charged soft surface is independent of the electrolyte concentration. The dependence of the surface potential ψ_o on the electrolyte concentration corresponds to the following fact. The surface potential is determined by the

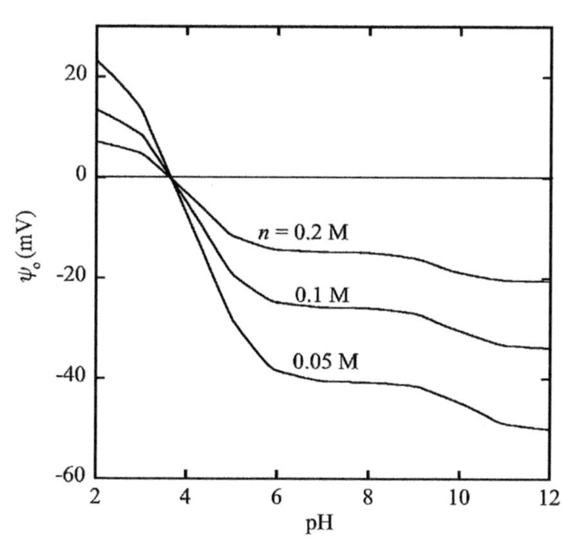

FIGURE 4.9 Surface potential ψ_o of a soft surface with a uniform distribution of both acidic and basic groups as a function of pH. The values of N_1, N_2, K_a, K_b, and T are the same as in Fig. 4.8. From Ref. [16].

membrane fixed charges located through the depth of $1/\kappa$ from the surface (i.e., the front edge of the surface layer). Thus, at high electrolyte concentrations (or small $1/\kappa$) the contribution of the inner layer (layer 2) is small, while at low electrolyte concentrations (or large $1/\kappa$) the contribution from layer 2 becomes larger. As a result, as the electrolyte concentration decreases, the IEP shifts to higher pH. It is interesting to note that the three curves in Fig. 4.8 intersect with each other at almost the same point, which lies below the $\psi_o = 0$ axis, while those curves in Fig. 4.9, on the other hand, intersect with each other at exactly the same point on the $\psi_o = 0$ axis.

REFERENCES

1. B. V. Derjaguin and L. Landau, *Acta Physicochim.* 14 (1941) 633.
2. E. J. W. Verwey and J. Th. G. Overbeek, *Theory of the Stability of Lyophobic Colloids*, Elsevier, Amsterdam, 1948.
3. H. Ohshima and K. Furusawa (Eds.), *Electrical Phenomena at Interfaces, Fundamentals, Measurements, and Applications, 2nd edition*, revised and expanded, Dekker, New York, 1998.
4. H. Ohshima, *Theory of Colloid and Interfacial Electric Phenomena*, Elsevier/Academic, 2006.
5. A. Mauro, *Biophys. J.* 2 (1962) 179.
6. H. Ohshima and S. Ohki, *Biophys. J.* 47 (1985) 673.
7. H. Ohshima and S. Ohki, *Bioelectrochem. Bioenerg.* 15 (1986) 173.
8. H. Ohshima and T. Kondo, *J. Colloid Interface Sci.* 116 (1987) 305.
9. H. Ohshima and T. Kondo, *Biophys. Chem.* 38 (1990) 117.
10. H. Ohshima and T. Kondo, *J. Theor. Biol.* 124 (1987) 191.
11. H. Ohshima, *J. Colloid Interface Sci.* 323 (2008) 92.
12. H. Ohshima, T. W. Healy, and L. R. White, *J. Colloid Interface Sci.* 90 (1982) 17.
13. H. Ohshima, *J. Colloid Interface Sci.* 171 (1995) 525.
14. H. Ohshima, *J. Colloid Interface Sci.* 323 (2008) 313.
15. H. Ohshima, *J. Colloid Interface Sci.* 200 (1998) 291.
16. T. Shinagawa, H. Ohshima, and T. Kondo, *Biophys. Chem.* 43 (1992) 149.

5 Free Energy of a Charged Surface

5.1 INTRODUCTION

In this chapter, we consider the Helmholtz free energy of a charged hard or soft surface in contact with an electrolyte solution. The free energy expressions will be used also for the calculation of the interaction energy between particles, as shown later in PART II.

5.2 HELMHOLTZ FREE ENERGY AND TENSION OF A HARD SURFACE

5.2.1 Charged Surface with Ion Adsorption

Consider a charged hard particle of surface area A carrying surface charges of uniform density σ immersed in an electrolyte solution on the basis of the smeared charge model. We first treat the case where the surface charges are due to the adsorption of N ions of valence Z and bulk concentration n onto the surface. These ions are called potential-determining ions. Then the surface charge density σ is given by

$$\sigma = \frac{ZeN}{A} \tag{5.1}$$

The Helmholtz free energy F_s of the particle surface, that is, the interface between the surface and the surrounding electrolyte solution, is given by [1–3]

$$F_s = F_s^o + F_{el} + N\mu_o^s - TS_c \tag{5.2}$$

with

$$F_s^o = \gamma_o A \tag{5.3}$$

where F_s^o and γ_o are, respectively, the Helmholtz free energy and tension of the uncharged surface in the absence of ion binding. The second term F_{el} is the

Biophysical Chemistry of Biointerfaces By Hiroyuki Ohshima

electrical work of charging the particle surface, that is, the electrical part of the double-layer free energy, which is given by

$$F_{\mathrm{el}} = A\int_0^{\sigma} \psi(\sigma')d\sigma' \tag{5.4}$$

where $\psi(\sigma')$ is the surface potential of the particle at a stage in the charging process at which the surface charge density is σ' $(0 \leq \sigma' \leq \sigma)$. The actual surface potential, which we denote by ψ_{o}, corresponds to the value at the final stage $\sigma' = \sigma$, that is, $\psi_{\mathrm{o}} = \psi(\sigma)$. The other two terms on the right-hand side of Eq. (5.2) are chemical components; $\mu_{\mathrm{o}}^{\mathrm{s}}$ is a constant term in the electrochemical potential μ of adsorbed ions on the particle surface, and S_{c} is the configurational entropy of the adsorbed ions on the surface. The electrochemical potential μ of adsorbed ions on the surface is given by

$$\mu = \left(\frac{\partial F_{\mathrm{s}}}{\partial N}\right)_A = \mu_{\mathrm{o}}^{\mathrm{s}} + Ze\psi_{\mathrm{o}} - T\frac{\partial S_{\mathrm{c}}}{\partial N} \tag{5.5}$$

$\mu_{\mathrm{o}}^{\mathrm{s}}$ being a constant. The electrochemical potential of the adsorbed ions must be equal to that of ions in the bulk solution phase, where the electric potential is zero, which is

$$\mu = \mu_{\mathrm{o}}^{\mathrm{b}} + kT\ln n \tag{5.6}$$

$\mu_{\mathrm{o}}^{\mathrm{b}}$ being a constant.

We now consider the following two cases [2,3]

.

(i) Case where S_{c} is independent of N and $\mu_{\mathrm{b}} = \mu_{\mathrm{s}}$

Consider the case where S_{c} is independent of N and $\mu_{\mathrm{b}} = \mu_{\mathrm{s}}$. In this case, the term TS_{c} in Eq. (5.2) can be dropped and Eqs. (5.2) and (5.5) become

$$F_{\mathrm{s}} = F_{\mathrm{s}}^{\mathrm{o}} + A\int_0^{\sigma} \psi(\sigma')d\sigma' + N\mu_{\mathrm{o}}^{\mathrm{s}} \tag{5.7}$$

$$\mu = \left(\frac{\partial F_{\mathrm{s}}}{\partial N}\right)_A = \mu_{\mathrm{o}}^{\mathrm{b}} + Ze\psi_{\mathrm{o}} \tag{5.8}$$

On multiplying N on both sides of Eq. (5.8), we obtain

$$N\mu = N\mu_{\mathrm{o}}^{\mathrm{s}} + A\sigma\psi_{\mathrm{o}} \tag{5.9}$$

where Eq. (5.1) has been used. By equating Eqs. (5.6) and (5.8), we obtain

$$\psi_o = \frac{kT}{Ze}\ln\left(\frac{n}{n_o}\right) \tag{5.10}$$

where n_o is the value of the concentration n of potential-determining ions at which ψ_o becomes zero. By using Eq. (5.9), Eq. (5.7) can be rewritten as

$$F_s = F_s^o + A\left\{\int_0^\sigma \psi(\sigma')d\sigma' - \sigma\psi_o\right\} + N\mu \tag{5.11}$$

Since the change in the Helmholtz free energy of the surrounding electrolyte solution caused by adsorption of N ions onto the particle surface is $-N\mu$, the total change F_t in the Helmholtz free energy of the particle and the surrounding solution is given by

$$F_t = F_s - N\mu = F_s^o + A\left\{\int_0^\sigma \psi(\sigma')d\sigma' - \sigma\psi_o\right\} \tag{5.12}$$

In Eq. (5.12), the second term on the right-hand side, which we denote by F, namely,

$$F = A\left\{\int_0^\sigma \psi(\sigma')d\sigma' - \sigma\psi_o\right\} \tag{5.13}$$

can be regarded as the free energy of the electrical double layer, which is equal to the amount of work carried out in building up the double layer around the particle. Equation (5.13) can further be rewritten as

$$F = -A\int_0^{\psi_o} \sigma(\psi_o')d\psi_o' \tag{5.14}$$

The tension of the interface between the particle and the electrolyte solution is obtained from Eq. (5.7), namely,

$$\gamma = \left(\frac{\partial F_s}{\partial A}\right)_N = \gamma_o + \int_0^\sigma \psi(\sigma')d\sigma' - \sigma\psi_o = \gamma_o - \int_0^{\psi_o} \sigma(\psi_o')d\psi_o' \tag{5.15}$$

where we have used

$$\left(\frac{\partial\sigma}{\partial A}\right)_N = -\frac{\sigma}{A} \tag{5.16}$$

which is obtained from Eq. (5.1). From Eqs. (5.11),(5.12) and (5.15), we find that

$$F_s = \gamma A + \mu N \tag{5.17}$$

and

$$F_t = \gamma A \tag{5.18}$$

(ii) Langmuir-type adsorption

If there are N_{max} binding sites on the particle surface and we assume 1:1 binding of the Langmuir type, then S_c is given by

$$S_c = k \ln \frac{N_{max}!}{N!(N_{max} - N)!} \tag{5.19}$$

which becomes, by using the Stirling formula,

$$S_c \approx k\{N_{max} \ln N_{max} - N \ln N - (N_{max} - N)\ln(N_{max} - N)\} \tag{5.20}$$

Equation (5.2) for the surface free energy F_s thus becomes

$$\begin{aligned} F_s &= F_s^o + A \int_0^\sigma \psi(\sigma')d\sigma' + N\mu_o^s \\ &-kT\{N_{max} \ln N_{max} - N \ln N - (N_{max} - N)\ln(N_{max} - N)\} \end{aligned} \tag{5.21}$$

The electrochemical potential μ of adsorbed ions is given by

$$\begin{aligned} \mu &= \left(\frac{\partial F_s}{\partial N}\right)_A = \mu_o^s + Ze\psi_o - T\frac{\partial S_c}{\partial N} \\ &= \mu_o^s + Ze\psi_o + kT \ln\left(\frac{N}{N_{max} - N}\right) \end{aligned} \tag{5.22}$$

By using Eqs. (5.9) and (5.22), Eq. (5.21) can be rewritten as

$$\begin{aligned} F_s &= F_s^o + A\{\textstyle\int_0^\sigma \psi(\sigma')d\sigma' - \sigma\psi_o\} \\ &-N_{max}kT \ln\left(\frac{N_{max}}{N_{max} - N}\right) + N\mu \end{aligned} \tag{5.23}$$

which gives the tension of the interface between the particle and the electrolyte solution, namely,

$$\begin{aligned}\gamma &= \left(\frac{\partial F_s}{\partial A}\right)_N \\ &= \gamma_o + \int_0^{\sigma} \psi(\sigma')d\sigma' - \sigma\psi_o - \frac{N_{max}kT}{A}\ln\left(\frac{N_{max}}{N_{max}-N}\right)\end{aligned} \tag{5.24}$$

Here we have used the fact that A is proportional to N_{max} so that

$$\frac{\partial N_{max}}{\partial A} = \frac{N_{max}}{A} \tag{5.25}$$

By equating Eqs. (5.6) and (5.22), we obtain

$$\frac{N}{N_{max}} = \frac{K_a n e^{-y_o}}{1 + K_a n e^{-y_o}} \tag{5.26}$$

with

$$K_a = \exp\left(\frac{\mu_o^b - \mu_o^s}{kT}\right) \tag{5.27}$$

$$y_o = \frac{Ze\psi_o}{kT} \tag{5.28}$$

where K_a is the adsorption constant and y_o is the scaled surface potential. By substituting Eq. (5.26) into Eqs. (5.23) and (5.24), we find that

$$\begin{aligned}F_s = F_s^o &+ A\left\{\int_0^{\sigma} \psi(\sigma')d\sigma' - \sigma\psi_o\right\} \\ &- N_{max}kT\ln(1 + K_a n e^{-y_o}) + N\mu\end{aligned} \tag{5.29}$$

and

$$\gamma = \gamma_o + \int_0^{\sigma} \psi(\sigma')d\sigma' - \sigma\psi_o - \frac{N_{max}kT}{A}\ln(1 + K_a n e^{-y_o}) \tag{5.30}$$

Also it follows from Eqs. (5.1) and (5.26) that

$$\sigma = \frac{ZeN_{max}}{A} \cdot \frac{K_a n e^{-y_o}}{1 + K_a n e^{-y_o}} \tag{5.31}$$

The total change F_t in the Helmholtz free energy of the particle and the surrounding solution is thus given by

$$
\begin{aligned}
F_t &= F_s - N\mu \\
&= F_s^o + A\{\textstyle\int_0^\sigma \psi(\sigma')d\sigma' - \sigma\psi_o\} - N_{max}\, kT \ln(1 + K_a n e^{-y_o})
\end{aligned}
\tag{5.32}
$$

We again find that Eqs. (5.17) and (5.18) hold among F_s, F_t, and γ. In this case, the free energy of the double layer $F = F_t - F_s^o$ is given by

$$
F = A\left\{\int_0^\sigma \psi(\sigma')d\sigma' - \sigma\psi_o\right\} - N_{max} kT \ln(1 + K_a n e^{-y_o}) \tag{5.33}
$$

which can be rewritten as

$$
F = -A\int_0^{\psi_o} \sigma(\psi_o')d\psi_o' - N_{max}\, kT \ln(1 + K_a\, n e^{-y_o}) \tag{5.34}
$$

5.2.2 Charged Surface with Dissociable Groups

Next we consider the case where the surface charge is due to dissociation of dissociable groups AH on the particle surface (of area A) in a solution containing H^+ ions of concentration n_H. We regard the dissociation reaction $AH = A^- + H^+$ as being desorption of H^+ from AH and adsorption of H^+ onto A^-. The potential-determining ions in this case are H^+ ions. Let the number of the dissociable groups be N_{max} and that of dissociated groups A^- be N so that the number of desorbed H^+ ions from AH is N and that of adsorbed H^+ ions onto A^- is $N_{max} - N$. Then, the configurational entropy S_c is given also by Eq. (5.19) and we find that the Helmholtz free energy F_s of the particle surface is

$$
\begin{aligned}
F_s &= F_s^o + A\int_0^\sigma \psi(\sigma')d\sigma' + (-N)\mu_o^s - TS_c \\
&= F_s^o + A\int_0^\sigma \psi(\sigma')d\sigma' + (-N)\mu_o^s \\
&\quad -kT\{N_{max}\ln N_{max} - N\ln N - (N_{max} - N)\ln(N_{max} - N)\}
\end{aligned}
\tag{5.35}
$$

where F_s^o is the Helmholtz free energy of the uncharged surface in the absence of the dissociation of dissociable groups. The surface charge density σ in this case is given by

$$
\sigma = -\frac{eN}{A} \tag{5.36}
$$

Since the number of adsorbed H^+ ions on the particle surface is $N_{max} - N$, the electrochemical potential μ of the H^+ ions on the surface is

$$\begin{aligned}\mu &= \left(\frac{\partial F}{\partial(N_{max} - N)}\right)_A = \mu_o^s + e\psi_o - T\frac{\partial S_c}{\partial(N_{max} - N)} \\ &= \mu_o^s + e\psi_o - kT\ln\left(\frac{N}{N_{max} - N}\right)\end{aligned} \tag{5.37}$$

which must be equal to that of H^+ ions in the bulk solution phase, that is,

$$\mu = \mu_o^b + kT\ln n_H \tag{5.38}$$

μ_o^b is a constant. By equating Eqs. (5.37) and (5.38), we obtain

$$\frac{N}{N_{max}} = \frac{1}{1 + n_H e^{-y_o}/K_d} \tag{5.39}$$

with

$$K_d = \exp\left(\frac{\mu_o^s - \mu_o^b}{kT}\right) \tag{5.40}$$

$$y_o = \frac{e\psi_o}{kT} \tag{5.41}$$

where K_d is the dissociation constant and $y_o = e\psi_o/kT$ is the scaled surface potential. By substituting Eq. (5.39) into Eq. (5.35), we obtain

$$\begin{aligned}F_s = {} & F_s^o + A\left\{\int_0^\sigma \psi(\sigma')d\sigma' - \sigma\psi_o\right\} \\ & - N_{max}kT\ln\left(1 + \frac{n_H e^{-y_o}}{K_d}\right) - N\mu\end{aligned} \tag{5.42}$$

where $n_H\exp(-y_o)$ is the surface concentration of H^+ ions.

Since the change in the Helmholtz free energy of the surrounding electrolyte solution caused by desorption of N ions from the particle surface is $N\mu$, the total change F_t in the Helmholtz free energy of the particle and the surrounding solution is thus given by

$$\begin{aligned}F_t &= F_s + N\mu \\ &= F_s^o + A\left\{\int_0^\sigma \psi(\sigma')d\sigma' - \sigma\psi_o\right\} - N_{max}\,kT\ln\left(1 + \frac{n_H e^{-y_o}}{K_d}\right)\end{aligned} \tag{5.43}$$

The free energy of the electrical double layer $F = F_t - F_s^o$ is thus given by

$$\begin{aligned} F &= A\left\{\int_0^{\sigma} \psi(\sigma')d\sigma' - \sigma\psi_o\right\} - N_{\max} kT \ln\left(1 + \frac{n_H e^{-y_o}}{K_d}\right) \\ &= -A\int_0^{\psi_o} \sigma(\psi_o')d\psi_o' - N_{\max} kT \ln\left(1 + \frac{n_H e^{-y_o}}{K_d}\right) \end{aligned} \tag{5.44}$$

From Eqs. (5.36) and (5.39) we obtain

$$\sigma = -\frac{eN_{\max}}{A}\cdot\frac{1}{1 + n_H e^{-y_o}/K_d} \tag{5.45}$$

If the dissociation of ionizable groups on the particle surface can be regarded as complete, then $N = N_{\max}$ and S_c can be dropped so that the surface free energy increase F_t is just equal to the electrical part of the double-layer free energy F_{el} (Eq. (5.4)), namely,

$$F_t = F_s = F_s^o + A\int_0^{\sigma} \psi(\sigma')d\sigma' \tag{5.46}$$

with

$$\sigma = -\frac{eN_{\max}}{A} \tag{5.47}$$

Note that Eqs. (5.17) and (5.18) hold among F_s, F, and γ, that is, $F = F_s = \gamma A$ and that in this case the electrical double-layer free energy F is equal to the electric work F_{el}, namely,

$$F = F_{el} = A\int_0^{\sigma} \psi(\sigma')d\sigma' \tag{5.48}$$

5.3 CALCULATION OF THE FREE ENERGY OF THE ELECTRICAL DOUBLE LAYER

All the expressions for the free energy of the electrical double layer F (Eqs. (5.13), (5.34) and (5.44)) contain the term

$$A\overline{F} \tag{5.49}$$

with

$$\overline{F} \equiv \int_0^{\sigma} \psi(\sigma')d\sigma' - \sigma\psi_o = -\int_0^{\psi_o} \sigma(\psi_o')d\psi_o' \tag{5.50}$$

For the low potential case where $\psi(\sigma')$ is proportional to σ' (or, $\sigma(\psi'_o)$ is proportional to ψ'_o), Eq. (5.50) then reduces to

$$\overline{F} = -\frac{1}{2}\sigma\psi_o \tag{5.51}$$

For the arbitrary potential case, we use the relationship between the surface charge density σ and the surface potential ψ_o derived in Chapter 1. In the following, we consider the cases of a plate-like particle, a spherical particle, and a cylindrical particle [3].

5.3.1 Plate

(i) For a plate-like particle in a z–z electrolyte, we have (Eq. (1.32))

$$\sigma = \frac{2\varepsilon_r\varepsilon_o\kappa kT}{ze}\sinh\left(\frac{ze\psi_o}{2kT}\right) = (8n\varepsilon_r\varepsilon_o kT)^{1/2}\sinh\left(\frac{ze\psi_o}{2kT}\right) \tag{5.52}$$

This relation holds at any stage in the charging process. By substituting Eq. (5.52) into Eq. (5.50), we obtain

$$\overline{F} = -8\varepsilon_r\varepsilon_o\kappa\left(\frac{kT}{ze}\right)^2\sinh^2\left(\frac{ze\psi_o}{4kT}\right) \tag{5.53}$$

where κ is the Debye–Hückel parameter for a z–z electrolyte (Eq. (1.11)).

(ii) For a plate-like particle in a 2-1 electrolytes, by substituting Eq. (1.43) into Eq. (5.50), we obtain

$$\overline{F} = -8\varepsilon_r\varepsilon_o\kappa\left(\frac{kT}{e}\right)^2\left\{(2+e^{-y_o})\left(\frac{2}{3}e^{y_o}+\frac{1}{3}\right)^{1/2} - 3\right\} \tag{5.54}$$

where $y_o = e\psi_o/kT$ is the scaled surface potential and κ is the Debye–Hückel parameter for a 2-1 electrolyte (Eq. (1.13)).

(iii) For a solution of a 1-1 electrolyte of bulk concentration n_1 and 2-1 electrolyte of bulk concentration n_2, we have

$$\begin{aligned}\overline{F} = {} & -8\varepsilon_r\varepsilon_o\kappa\left(\frac{kT}{e}\right)^2\left[(2+e^{-y_o})\left\{\left(1-\frac{\eta}{3}\right)e^{y_o}+\frac{\eta}{3}\right\}^{1/2} - 3 + \frac{1-\eta}{2}\left(\frac{3}{\eta}\right)^{1/2}\right. \\ & \left.\times \ln\frac{\{1-(\eta/3)^{1/2}\}\{((1-\eta/3)e^{y_o}+(\eta/3))^{1/2}+(\eta/3)\}^{1/2}}{\{1+(\eta/3)^{1/2}\}\{((1-\eta/3)e^{y_o}+(\eta/3))^{1/2}-(\eta/3)\}^{1/2}}\right]\end{aligned} \tag{5.55}$$

with

$$\eta = \frac{3n_2}{n_1+3n_2} \tag{5.56}$$

where κ is the Debye–Hückel parameter for the mixed solution of a 1-1 and 2-1electrolytes (Eq. (1.16)).

(iv) For a plate-like particle in a general electrolyte solution, the surface charge density s is given by Eq. (1.66). This result is substituted into Eq. (5.50) to yield

$$\overline{F} = -\frac{\varepsilon_r \varepsilon_o \kappa kT}{e} \int_0^{\psi_o} f(\psi_o') d\psi_o' = -\varepsilon_r \varepsilon_o \kappa \left(\frac{kT}{e}\right)^2 \int_0^{y_o} f(y_o') dy_o' \tag{5.57}$$

where $f(y)$ is defined by Eq. (1.56), namely,

$$f(y) = \mathrm{sgn}(y_o) \left[\frac{2\sum_{i=1}^{N} n_i (e^{-z_i y} - 1)}{\sum_{i=1}^{N} z_i^2 n_i} \right]^{1/2} = (1 - e^{-y}) \left[\frac{2\sum_{i=1}^{N} n_i (e^{-z_i y} - 1)}{(1 - e^{-y})^2 \sum_{i=1}^{N} z_i^2 n_i} \right]^{1/2} \tag{5.58}$$

and κ is the Debye–Hückel parameter for a general electrolyte solution (Eq. (1.10)).

For the low potential case, in which case ψ_o and σ are related to each other by Eq. (1.26), namely,

$$\psi_o = \frac{\sigma}{\varepsilon_r \varepsilon_o \kappa} \tag{5.59}$$

and Eqs.(5.53), (5.54), (5.55), and (5.57) all reduce to

$$\overline{F} = -\frac{1}{2} \varepsilon_r \varepsilon_o \kappa \psi_o^2 = -\frac{1}{2} \sigma \psi_o \tag{5.60}$$

5.3.2 Sphere

For the case of a spherical particle of radius a, the numerical values for the free energy F are given in a book by Loeb et al. . An approximate analytic expression for F for a spherical particle in a symmetrical electrolyte solution of valence z and bulk concentration n can be derived by using an approximate expression for the σ–ψ_o relationship (Eq. (1.80)), namely,

$$\sigma = \frac{2\varepsilon_r \varepsilon_o \kappa kT}{ze} \left[\sinh\left(\frac{ze\psi_o}{2kT}\right) + \left(\frac{2}{\kappa a}\right) \tanh\left(\frac{ze\psi_o}{4kT}\right) \right] \tag{5.61}$$

By substituting Eq. (5.61) into Eq. (5.50) we obtain

$$\overline{F} = -8\varepsilon_r\varepsilon_o\kappa\left(\frac{kT}{ze}\right)^2\left\{\sinh^2\left(\frac{ze\psi_o}{4kT}\right) + \left(\frac{2}{\kappa a}\right)\ln\left[\cosh\left(\frac{ze\psi_o}{4kT}\right)\right]\right\} \tag{5.62}$$

Equation (5.62) is in excellent agreement with the exact numerical results . For $\kappa a \geq 2$, the relative error is less than 1%. Even at $\kappa a = 1$, the maximum error is 1.7%. A better approximate expression for $\hat{F}$ can be derived with the help of the second-order surface charge density σ–surface potential y_o relationship (Eq. (1.86)) instead of the first-order σ–y_o relationship (Eq. (5.61)).

For a sphere in an general electrolyte solution, the first-order and second-order σ/y_o approximations are given by Eqs. (1.103) and (1.107). With the help of these results, we obtain the first-order approximation

$$\overline{F} = -\varepsilon_r\varepsilon_o\kappa\left(\frac{kT}{e}\right)^2\int_0^{y_o} f(y'_o)\left[1 + \frac{2}{\kappa a f^2(y'_o)}\int_0^{y'_o} f(u)du\right]dy'_o \tag{5.63}$$

and the second-order approximation is obtained with the help of Eq. (1.107).

For the low potential case, Eq. (5.62) reduces to

$$\overline{F} = -\frac{1}{2}\varepsilon_r\varepsilon_o\kappa\psi_o^2\left(1 + \frac{1}{\kappa a}\right) = -\frac{1}{2}\sigma\psi_o \tag{5.64}$$

where we have used Eq. (1.76) for the σ–ψ_o relationship at low potentials.

5.3.3 Cylinder

For a cylinder in a z–z symmetrical electrolyte solution, the σ–y_o approximation is given by Eq. (1.158) so that we obtain

$$\overline{F} = -2\varepsilon_r\varepsilon_o\kappa\left(\frac{kT}{ze}\right)^2\int_0^{y_o}\sinh\left(\frac{y'_o}{2}\right)\left[1 + \left(\frac{1}{\beta^2} - 1\right)\frac{1}{\cosh^2(y'_o/4)}\right]^{1/2}dy'_o \tag{5.65}$$

where β is defined by Eq. (1.150) (i.e., $\beta = K_0(\kappa a)/K_1(\kappa a)$). For a cylinder in a general electrolyte solution, the σ/y_o approximation is given by Eq. (1.172) and we have

$$\overline{F} = -\varepsilon_r\varepsilon_o\kappa\left(\frac{kT}{ze}\right)^2\frac{1}{\beta}\int_0^{y_o}F(y'_o)dy'_o \tag{5.66}$$

where $F(y_o)$ is defined by Eq. (1.168). For low potentials, Eqs. (5.65) and (5.66) reduce to

$$\overline{F} = -\frac{1}{2}\varepsilon_r\varepsilon_o\kappa\frac{1}{\beta}\psi_o^2 = -\frac{1}{2}\varepsilon_r\varepsilon_o\kappa\frac{K_0(\kappa a)}{K_1(\kappa a)}\psi_o^2 \tag{5.67}$$

5.4 ALTERNATIVE EXPRESSION FOR F_{EL}

Equation (5.4) for the electrical work F_{el} of charging the surface of a hard particle can be generalized for the case where σ is not uniform over the particles surface of area A and depends on the position $\boldsymbol{s}$ on the particle surface (Fig. 5.1). In this case Eq. (5.4) becomes [1]

$$F_{el} = \int_A \int_0^{\sigma} \psi(\sigma')d\sigma'\, dA \tag{5.68}$$

which can be rewritten as

$$F_{el} = \int_A \sigma(\boldsymbol{s}) \int_0^1 \psi(\boldsymbol{s},\nu)d\nu\, dA \tag{5.69}$$

where $\psi(\boldsymbol{s}, \nu)$ is the surface potential of the particle at position $\boldsymbol{s}$ on the particle surface at a stage ν in the charging process at which the surface charge density is $\nu\sigma$

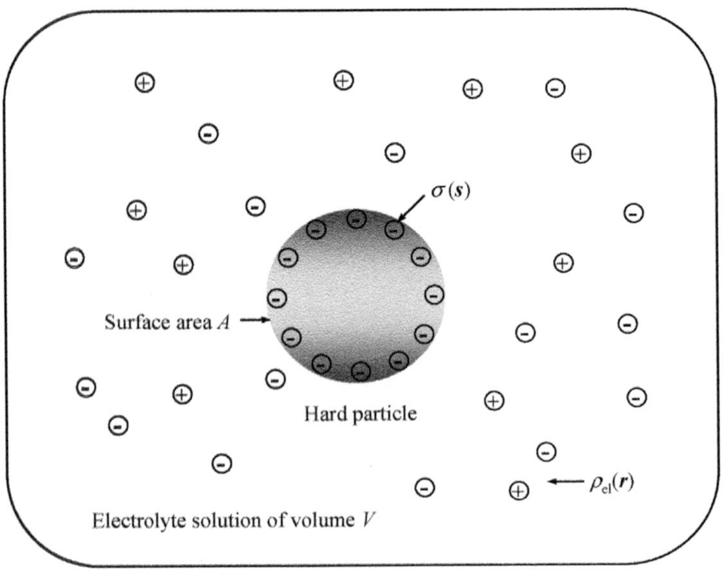

FIGURE 5.1 A hard particle of surface area A in an electrolyte solution of volume V. $\rho_{el}(\boldsymbol{r})$ is the volume charge density at position $\boldsymbol{r}$ resulting from electrolyte ions and $\sigma(\boldsymbol{s})$ is the charge density at position $\boldsymbol{s}$ on the particle surface.

($0 \leq v \leq 1$). The actual surface potential ψ_o corresponds to the value at the final stage $v = 1$.

An alternative expression for F_{el} (Eq. (5.69)) can be derived by considering a discharging process in which all the charges (both the surface charge density σ and the charge of electrolyte ions) in the system are decreased at the same rate. Consider a uniformly charged hard particle of surface area A carrying surface charges of uniform density σ immersed in an electrolyte solution of volume V. Let λ be a parameter that expresses a stage of the discharging process and varies from 1 to 0, and $\psi(\boldsymbol{s}, \lambda)$, $\psi(\boldsymbol{r}, \lambda)$, and $\rho_{el}(\boldsymbol{r}, \lambda)$ be, respectively, the particle surface potential at position $\boldsymbol{s}$ at stage λ, and the potential and volume charge density resulting from electrolyte ions at position $\boldsymbol{r}$ at stage λ. Then it can be shown that F_{el} is expressed by

$$F_{el} = \int_0^1 d\lambda \frac{2}{\lambda} E(\lambda) \tag{5.70}$$

where $E(\lambda)$ is the internal energy part of F_{el} given by

$$E(\lambda) = \frac{1}{2} \int_V \psi(\boldsymbol{r}, \lambda) \rho_{el}(\boldsymbol{r}, \lambda) dV + \frac{1}{2} \int_A \lambda \sigma(\boldsymbol{s}) \psi(\boldsymbol{s}, \lambda) dA \tag{5.71}$$

5.5 FREE ENERGY OF A SOFT SURFACE

5.5.1 General Expression

Now we consider the electrical work F_{el} of charging the surface of a soft particle consisting of the uncharged particle core of surface area A coated by an ion-penetrable surface layer of polyelectrolytes immersed in a liquid containing N ionic species with valence z_i and bulk concentration (number density) n_i^∞ ($i = 1, 2, \ldots, N$) of volume V (Fig. 5.2). We deal with the general case in which the density of particle fixed charges distributed in the polyelectrolyte layer is a function of position $\boldsymbol{r}$. Let $\rho_{fix}(\boldsymbol{r})$ be the volume density of particle fixed charges at position $\boldsymbol{r}$. Equation (5.69) for F_{el} in this case can be expressed in terms of the volume charge density $\rho_{fix}(\boldsymbol{r})$ instead of the surface charge density $\sigma(\boldsymbol{s})$ as

$$F_{el} = \int_V \rho_{fix}(\boldsymbol{r}) \int_0^1 \psi(\boldsymbol{r}, v) dv \, dV \tag{5.72}$$

where $\psi(\boldsymbol{r}, v)$ is the potential at position r at a stage v. For the low potential case, $\psi(\boldsymbol{r}, v)$ is proportional to v so that Eq. (5.72) reduces to

$$F_{el} = \frac{1}{2} \int_V \rho_{fix}(\boldsymbol{r}) \psi(\boldsymbol{r}) dV \tag{5.73}$$

where $\psi(\boldsymbol{r})$ is the actual potential at the final stage $v = 1$ in the charging process.

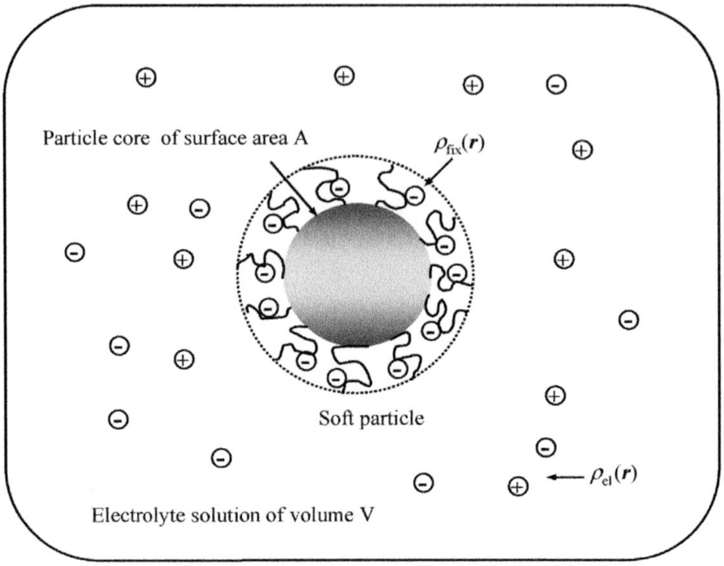

FIGURE 5.2 A soft particle composed of the particle core of surface area A coated by an ion-penetrable surface layer in an electrolyte solution of volume V. $\rho_{\rm el}(\boldsymbol{r})$ is the volume charge density at position $\boldsymbol{r}$ resulting from electrolyte ions and $\rho_{\rm fix}(\boldsymbol{r})$ is the volume charge density at position $\boldsymbol{r}$ distributed in the surface layer.

We derive an alternative expression for $F_{\rm el}$, which corresponds to Eq. (5.70) for a hard particle, by considering a discharging process in which all the charges (both the volume charge density $\rho_{\rm fix}(\boldsymbol{r})$ and the charge of electrolyte ions) in the system are decreased at the same rate. From Eq. (5.72), we obtain

$$\frac{\partial F_{\rm el}^{(\nu)}}{\partial \nu} = \int_V \rho_{\rm fix}(\boldsymbol{r})\psi(\boldsymbol{r},\nu)dV \tag{5.74}$$

where $F_{\rm el}^{(\nu)}$ is the value of $F_{\rm el}$ at stage ν. We consider the discharging process that starts from the stage ν. Let λ be a parameter that expresses a stage of the discharging process and varies from 1 to 0. The initial stage at $\lambda = 1$ corresponds to stage ν in the charging process. We rewrite Eq. (5.74) as

$$\frac{\partial F_{\rm el}^{(\nu)}}{\partial \nu} = \int_V \int_{\lambda=0}^{1} \frac{\partial}{\partial \lambda}\left[\lambda \rho_{\rm fix}(\boldsymbol{r})\psi(\boldsymbol{r},\nu,\lambda)\right] d\lambda\, dV \tag{5.75}$$

where $\psi(\boldsymbol{r}, \nu, \lambda)$ is the potential at position $\boldsymbol{r}$ at stage λ in the discharging process starting from the stage ν.

Equation (5.75) can be transformed into

$$\begin{aligned}\frac{\partial F_{\text{el}}^{(\nu)}}{\partial \nu} &= \int_V \int_{\lambda=0}^{1} \left[\rho_{\text{fix}}(\boldsymbol{r})\psi(\boldsymbol{r}, \nu, \lambda) + \lambda\rho_{\text{fix}}(\boldsymbol{r})\frac{\partial \psi(\boldsymbol{r}, \nu, \lambda)}{\partial \lambda}\right] d\lambda\, dV \\ &= \int_V \int_{\lambda=0}^{1} \left[\frac{\partial}{\partial \nu}\{\nu\rho_{\text{fix}}(\boldsymbol{r})\psi(\boldsymbol{r}, \nu, \lambda)\} - \nu\rho_{\text{fix}}(\boldsymbol{r})\frac{\partial \psi(\boldsymbol{r}, \nu, \lambda)}{\partial \nu}\right. \\ &\qquad \left. + \lambda\rho_{\text{fix}}(\boldsymbol{r})\frac{\partial \psi(\boldsymbol{r}, \nu, \lambda)}{\partial \lambda}\right] d\lambda\, dV \end{aligned} \tag{5.76}$$

Note that $\psi(\boldsymbol{r}, \nu, \lambda)$ is related to the volume charge density $\rho_{\text{el}}(\boldsymbol{r}, \nu, \lambda)$ resulting from electrolyte ions at position $\boldsymbol{r}$ at stage λ in the discharging process starting from the stage ν by the following Poisson–Boltzmann equation:

$$\Delta\psi(\boldsymbol{r}, \nu, \lambda) = -\frac{\rho_{\text{el}}(\boldsymbol{r}, \nu, \lambda)}{\varepsilon_r\varepsilon_o} - \frac{\nu\lambda\rho_{fix}({}^{\text{fix}}\boldsymbol{r})}{\varepsilon_r\varepsilon_o} \tag{5.77}$$

which yields

$$\Delta\frac{\partial \psi(\boldsymbol{r}, \nu, \lambda)}{\partial \nu} = -\frac{1}{\varepsilon_r\varepsilon_o}\frac{\partial \rho_{\text{el}}(\boldsymbol{r}, \nu, \lambda)}{\partial \nu} - \frac{\lambda\rho_{\text{fix}}(\boldsymbol{r})}{\varepsilon_r\varepsilon_o} \tag{5.78}$$

$$\Delta\frac{\partial \psi(\boldsymbol{r}, \nu, \lambda)}{\partial \lambda} = -\frac{1}{\varepsilon_r\varepsilon_o}\frac{\partial \rho_{\text{el}}(\boldsymbol{r}, \nu, \lambda)}{\partial \lambda} - \frac{\nu\rho_{\text{fix}}(\boldsymbol{r})}{\varepsilon_r\varepsilon_o} \tag{5.79}$$

By introducing Eqs. (5.78) and (5.79) into Eq. (5.76), we obtain

$$\begin{aligned}\frac{\partial F_{\text{el}}^{(\nu)}}{\partial \nu} = & \int_V \int_{\lambda=0}^{1} \left[\frac{\partial}{\partial \nu}\{\nu\rho_{\text{fix}}(\boldsymbol{r})\psi(\boldsymbol{r}, \nu, \lambda)\} + \frac{\partial \rho_{\text{el}}(\boldsymbol{r}, \nu, \lambda)}{\partial \lambda}\frac{\partial \psi(\boldsymbol{r}, \nu, \lambda)}{\partial \nu}\right. \\ & - \frac{\partial \rho_{\text{el}}(\boldsymbol{r}, \nu, \lambda)}{\partial \nu}\frac{\partial \psi(\boldsymbol{r}, \nu, \lambda)}{\partial \lambda} + \varepsilon_r\varepsilon_o\,\Delta\left(\frac{\partial \psi(\boldsymbol{r}, \nu, \lambda)}{\partial \lambda}\right)\frac{\partial \psi(\boldsymbol{r}, \nu, \lambda)}{\partial \nu} \\ & \left. - \varepsilon_r\varepsilon_o\,\Delta\left(\frac{\partial \psi(\boldsymbol{r}, \nu, \lambda)}{\partial \nu}\right)\frac{\partial \psi(\boldsymbol{r}, \nu, \lambda)}{\partial \lambda}\right] d\lambda\, dV \end{aligned} \tag{5.80}$$

Further, with the help of Green's theorem, we have

$$\begin{aligned} & \int_V \left[\Delta\left(\frac{\partial \psi(\boldsymbol{r}, \nu, \lambda)}{\partial \lambda}\right)\frac{\partial \psi(\boldsymbol{r}, \nu, \lambda)}{\partial \nu} - \varepsilon_r\varepsilon_o\,\Delta\left(\frac{\partial \psi(\boldsymbol{r}, \nu, \lambda)}{\partial \nu}\right)\frac{\partial \psi(\boldsymbol{r}, \nu, \lambda)}{\partial \lambda}\right] dV \\ & = \int_A \left[\frac{\partial^2 \psi(\boldsymbol{r}, \nu, \lambda)}{\partial n \partial \lambda}\frac{\partial \psi(\boldsymbol{r}, \nu, \lambda)}{\partial \nu} - \frac{\partial^2 \psi(\boldsymbol{r}, \nu, \lambda)}{\partial n \partial \nu}\frac{\partial \psi(\boldsymbol{r}, \nu, \lambda)}{\partial \lambda}\right] dA \end{aligned} \tag{5.81}$$

Here n is the outward normal from the particle core to the electrolyte solution and the integral taken on the external surface of V may be neglected. Sine we have assumed that the particle core is uncharged, that is, $\partial\psi/\partial n = 0$, the quantity given by Eq. (5.81) is zero. Also from the form of $\rho_{el}(\boldsymbol{r}, v, \lambda)$, namely,

$$\rho_{el}(\boldsymbol{r}, v, \lambda) = \sum_{i=1}^{N} \lambda z_i e n_i^{\infty} \exp\left(-\frac{\lambda z_i e \psi(\boldsymbol{r}, v, \lambda)}{kT}\right) \tag{5.82}$$

we find that the second and third terms in the brackets on the right-hand side of Eq. (5.80) becomes

$$\frac{\partial F_{el}^{(v)}}{\partial v} \frac{\partial \rho_{el}(\boldsymbol{r}, v, \lambda)}{\partial \lambda} \frac{\partial \psi(\boldsymbol{r}, v, \lambda)}{\partial v} - \frac{\partial \rho_{el}(\boldsymbol{r}, v, \lambda)}{\partial v} \frac{\partial \psi(\boldsymbol{r}, v, \lambda)}{\partial \lambda} = \frac{1}{\lambda} \frac{\partial}{\partial v} [\psi(\boldsymbol{r}, v, \lambda) \rho_{el}(\boldsymbol{r}, v, \lambda)] \tag{5.83}$$

Equation (5.80) thus becomes

$$\frac{\partial F_{el}^{(v)}}{\partial v} = \int_V \int_{\lambda=0}^{1} \left[\frac{1}{\lambda} \frac{\partial}{\partial v} \{\psi(\boldsymbol{r}, v, \lambda) \rho_{el}(\boldsymbol{r}, v, \lambda)\} + \frac{\partial}{\partial v} \{v \rho_{fix}(\boldsymbol{r}) \psi(\boldsymbol{r}, v, \lambda)\}\right] dV \tag{5.84}$$

By integrating Eq. (5.84) with respect to v from 0 to 1, we obtain

$$F_{el} = \int_V \int_{\lambda=0}^{1} \left[\frac{1}{\lambda} \{\psi(\boldsymbol{r}, v, \lambda) \rho_{el}(\boldsymbol{r}, v, \lambda)\} + \{v \rho_{fix}(\boldsymbol{r}) \psi(\boldsymbol{r}, v, \lambda)\}\right] dV \tag{5.85}$$

which can be rewritten as

$$F_{el} = \int_0^1 d\lambda \frac{2}{\lambda} E(\lambda) \tag{5.86}$$

where $E(\lambda)$ is the internal energy part of F_{el} given by

$$E(\lambda) = \frac{1}{2} \int_V \psi(\boldsymbol{r}, \lambda) \rho_{el}(\boldsymbol{r}, \lambda) dV + \frac{1}{2} \int_V \lambda \rho_{fix}(\boldsymbol{r}) \psi(\boldsymbol{r}, \lambda) dV \tag{5.87}$$

or

$$E(\lambda) = -\frac{1}{2} \varepsilon_r \varepsilon_o \int_V \psi(\boldsymbol{r}, \lambda) \Delta \psi(\boldsymbol{r}, \lambda) dV \tag{5.88}$$

Note that Eq. (5.85) becomes Eq. (5.71) by making the following replacement:

$$\rho_{fix}(\boldsymbol{r}) \rightarrow \sigma(\boldsymbol{s}) \tag{5.89}$$

5.5.2 Expressions for the Double-Layer Free Energy for a Planar Soft Surface

Consider the double-layer free energy per unit area of a planar soft surface, that is, an ion-penetrable polyelectrolyte layer of thickness d covering the plate core immersed in a solution containing a symmetrical electrolyte of valence z and bulk concentration n. We assume that charged groups are uniformly distributed in the surface layer. We denote by N and Z, respectively, the density and valence of charged groups in the surface layer so that the fixed-charge density ρ_{fix} in the polyelectrolyte layer is given by $\rho_{\text{fix}} = ZeN$. We take an x-axis perpendicular to the plate surface with its origin at the plate surface so that the region $x > 0$ corresponds to the solution phase and the region $-d < x < 0$ to the polyelectrolyte layer (Fig. 5.3). From Eq. (5.73), we find that the double-layer free energy per unit area F_{el} is given by [5]

$$F_{\text{el}} = ZeN \int_0^1 d\nu \int_{-d}^{0} \psi(x,\nu) dx \tag{5.90}$$

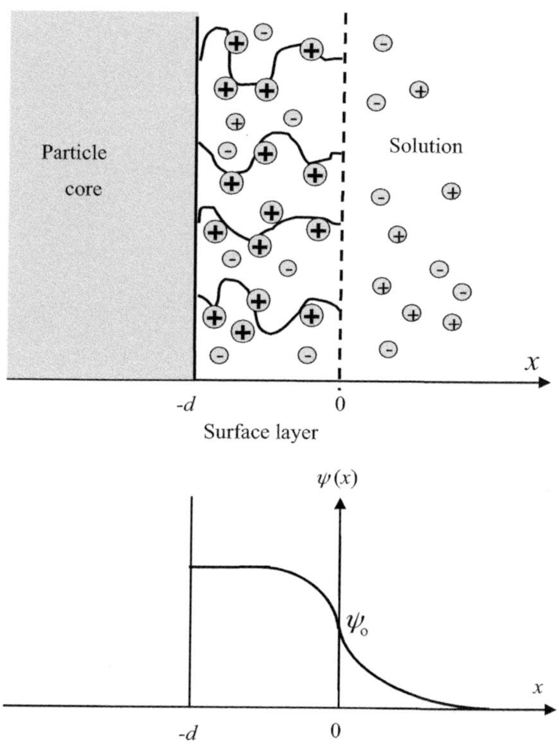

FIGURE 5.3 Planar soft surface of thickness d in an electrolyte solution.

For the low potential case, Eq. (5.90) reduces to

$$F_{\mathrm{el}} = \frac{1}{2} ZeN \int_{-d}^{0} \psi(x) dx \tag{5.91}$$

The potential distribution $\psi(x)$ in this case has been derived in Chapter 4 (Eq. (4.24)), namely,

$$\psi(x) = \frac{ZNkT}{2z^2 ne}\left\{1 - \frac{e^{\kappa x} + e^{-\kappa(x+2d)}}{2}\right\}, \quad -d < x < 0 \tag{5.92}$$

By substituting Eq. (4.24) into Eq. (5.91) and carrying out the integration, we obtain

$$F_{\mathrm{el}} = \frac{ZNkTd}{2z^2 ne}\left\{1 - \frac{1}{2\kappa d}\left(1 - e^{-2\kappa d}\right)\right\} \tag{5.93}$$

or

$$F_{\mathrm{el}} = \frac{\rho_{\mathrm{fix}}^2 d}{2\varepsilon_r \varepsilon_o \kappa^2}\left\{1 - \frac{1}{2\kappa d}\left(1 - e^{-2\kappa d}\right)\right\} \tag{5.94}$$

For a thick surface layer $\kappa d \gg 1$, Eq. (5.94) becomes

$$F_{\mathrm{el}} = \frac{\rho_{\mathrm{fix}}^2 d}{2\varepsilon_r \varepsilon_o \kappa^2} \tag{5.95}$$

Note that Eq. (5.95) is rewritten as

$$F_{\mathrm{el}} = \frac{1}{2}(\rho_{\mathrm{fix}}\, d)\psi_{\mathrm{DON}} \tag{5.96}$$

where

$$\psi_{\mathrm{DON}} = \frac{\rho_{\mathrm{fix}}}{\varepsilon_r \varepsilon_o \kappa^2} = \frac{ZeN}{\varepsilon_r \varepsilon_o \kappa^2} \tag{5.97}$$

is the Donnan potential in the surface charge layer (Eq. (4.20)).

5.5.3 Soft Surface with Dissociable Groups

Next we consider the case where the fixed charges in the surface charge layer are due to dissociation of dissociable groups AH in a solution containing H^+ ions of concentration $n_{H[6]}$. We regard the dissociation reaction $AH = A^- + H^+$ as being desorption of H^+ from AH and adsorption of H^+ onto A^-. As in Section 5.2.2, let the number of the dissociable groups per unit area be N_{max} and that of dissociated groups A^- at position x be $N(x)$ so that the number of desorbed H^+ ions from AH is $N(x)$ and that of adsorbed H^+ ions onto A^- is $N_{max} - N(x)$. Then, the configurational entropy $S_c(x)$ at position x is given also by Eq. (5.19) and we find that the Helmholtz free energy F_s of the surface charge layer per unit area is

$$F_s = \int_{-d}^{0} f_s(x)dx \tag{5.98}$$

where $f_s(x)$ is the corresponding free energy density at position x within the surface layer. We introduce the electric potential $\psi(x, \nu)$ at stage ν in the charging process at position x relative to the bulk solution phase. Then, $f_s(x)$ can be given by

$$f_s(x) = f_s^o(x) + \rho_{fix}(x)\int_0^1 \psi(x,\nu)d\nu - \mu_o N(x) - TS_c(x), \quad -d < x < 0 \tag{5.99}$$

with

$$\rho_{fix}(x) = -eN_H(x) \tag{5.100}$$

$$\begin{aligned} S_c(x) &= k\ln\frac{N_{max}!}{N(x)!(N_{max} - N(x))!} \\ &\approx k[N_{max}\ln N_{max} - N(x)\ln N(x) - \{N_{max} - N(x)\}\ln\{N_{max} - N(x)\}] \end{aligned} \tag{5.101}$$

where $f_s^o(x)$ is the free energy density in the absence of the dissociation of dissociable groups and $\rho_{fix}(x)$ is the density of fixed charges at position x. The electrochemical potential $\mu(x)$ of adsorbed H^+ at position x within the surface layer (whose number density equals that of undissociated groups AH, $N - N_H(x)$) is given by

$$\mu(x) = \frac{\partial f(x)}{\partial\{N_{max} - N(x)\}} = \mu_o + e\psi(x) - kT\ln\left[\frac{N(x)}{N_{max} - N(x)}\right] \tag{5.102}$$

At thermodynamic equilibrium, μ (x) must be equal to the chemical potential μ of H^+ in the bulk solution phase, which is

$$\mu = \mu_o^b + kT\ln n_H \tag{5.103}$$

μ_o^b being a constant. By equating Eqs. (5.102) and (5.103), we obtain

$$N(x) = \frac{N_{max}}{1 + (n_H/k)\exp(-e\psi(x)/kT)} \tag{5.104}$$

where

$$K_d = \exp\left(\frac{\mu_o^s - \mu_o^b}{kT}\right) \tag{5.105}$$

is the dissociation constant of the ionizable groups in the surface layer. Equation (5.104) represents the mass action law for the reaction $AH = A^- + H^+$ at position x.

The total change f_t in the free energy density in the surface charge layer at position x and the surrounding solution is thus given by

$$\begin{aligned} f_t(x) &= f_s(x) + N(x)\mu \\ &= f_s^o(x) + \rho_{fix}(x) \textstyle\int_0^1 \psi(x, v)dv - \rho_{fix}(x)\psi(x) \\ &- N(x)kT \ln\left[1 + \frac{n_H}{K}\exp(e\psi(x)]kT)\right] \end{aligned} \tag{5.106}$$

The free energy density of the electrical double layer $f(x) = f_t(x) - f_s^o(x)$ is thus given by

$$f(x) = \rho_{fix}(x) \int_0^1 \psi(x, v)dv - \rho_{fix}(x)\psi(x) - N(x)kT \ln\left[1 + \frac{n_H}{K}\exp(e\psi(x)]kT)\right] \tag{5.107}$$

and the free energy of the electrical double layer is given by

$$F = \int_{-d}^{0} f(x)dx \tag{5.108}$$

REFERENCES

1. E. J. W. Verwey and J. Th. G. Overbeek, *Theory of the Stability of Lyophobic Colloids*, Elsevier, Amsterdam, 1948.
2. H. Ohshima, in: H. Ohshima and K. Furusawa (Eds.), *Electrical Phenomena at Interfaces, Fundamentals, Measurements, and Applications*, 2nd edition, revised and expanded, Dekker, New York, 1998.

3. H. Ohshima, *Theory of Colloid and Interfacial Electric Phenomena*, Elsevier/Academic, Amsterdam, 2006.
4. A. L. Loeb, J. Th. G. Overbeek, and P. H. Wiersema, *The Electrical Double Layer Around a Spherical Colloid Particle*, MIT Press, Cambridge, MA, 1961.
5. H. Ohshima and T. Kondo, *J. Colloid Interface Sci.* 123 (1988) 136.
6. H. Ohshima and T. Kondo, *Biophys. Chem.*, 32 (1988) 161.

6 Potential Distribution Around a Charged Particle in a Salt-Free Medium

6.1 INTRODUCTION

The potential distribution around a charged colloidal particle in an electrolyte solution, which plays an essential role in various electric phenomena observed in colloidal suspensions, can be described by the Poisson–Boltzmann equation. Consider a particle carrying ionized groups on the particle surface in an electrolyte solution. In the suspension, in addition to electrolyte ions, there are counterions produced by dissociation of the particle surface groups. The following two assumptions are usually made: (i) the concentration of counterions from the surface groups can be neglected as compared with the electrolyte concentration and (ii) the suspension is assumed to be infinitely dilute so that any effects resulting from the finite particle volume fraction can be neglected. The assumption (i) becomes invalid where the electrolyte concentration is as low as or lower than that of counterions from the particle. The assumption (ii) does not hold when the concentration of all ions (electrolyte ions and counterions from the particle) is very low, since in this case the potential around each particle becomes quite long range and the effects of the finite particle volume fraction become appreciable.

The Poisson–Boltzmann equation for the potential distribution around a cylindrical particle without recourse to the above two assumptions for the limiting case of completely salt-free suspensions containing only particles and their counterions was solved analytically by Fuoss et al. [1] and Afrey et al. [2]. As for a spherical particle, although the exact analytic solution was not derived, Imai and Oosawa [3,4] studied the analytic properties of the Poisson–Boltzmann equation for dilute particle suspensions. The Poisson–Boltzmann equation for a salt-free suspension has recently been numerically solved [5–8].

The above theories, however, is based on an idealized model assuming that there are only counterions from the particle and no other ions. Actually, even salt-free media may contain other ions rather than counterions from the particle surface groups. Salt-free water at pH 7, for example, contains 10^{-7} M H^+ and OH^- ions. It is thus of interest to examine how the above behaviors of colloidal particles

Biophysical Chemistry of Biointerfaces By Hiroyuki Ohshima

in a salt-free medium are influenced if a small amount of salts is added to the medium. In their second paper [4], Imai and Oosawa studied the analytic properties of the Poisson–Boltzmann equation in the presence of added salts, which are found to exhibit similar behaviors to those of completely salt-free suspensions.

In this chapter, we first discuss the case of completely salt-free suspensions of spheres and cylinders. Then, we consider the Poisson–Boltzmann equation for the potential distribution around a spherical colloidal particle in a medium containing its counterions and a small amount of added salts [8]. We also deals with a soft particle in a salt-free medium [9].

6.2 SPHERICAL PARTICLE

Consider a dilute suspension of spherical colloidal particles of radius a with a surface charge density σ or the total surface charge $Q = 4\pi a^2 \sigma$ in a salt-free medium containing only counterions. We assume that each sphere is surrounded by a concentric spherical cell of radius R [5,7] (Fig. 6.1), within which counterions are distributed so that electrical neutrality as a whole is satisfied. The particle volume fraction ϕ is given by

$$\phi = (a/R)^3 \tag{6.1}$$

We treat the case of dilute suspensions, namely,

$$\phi \ll 1 \quad \text{or} \quad a/R \ll 1 \tag{6.2}$$

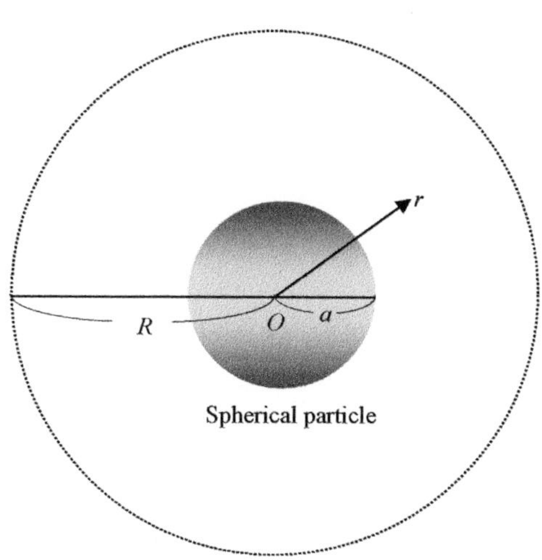

FIGURE 6.1 A spherical particle of radius a in a free volume of radius R. $(a/R)^3$ equals the particle volume fraction ϕ.

We denote the electric potential at a distance r from the center O of one particle by $\psi(r)$. Let the average number density and the valence of counterions be n and $-z$, respectively. Then we have

$$Q = 4\pi a^2 \sigma = \frac{4}{3}\pi(R^3 - a^3)zen \tag{6.3}$$

We set the electric potential $\psi(r)$ to zero at points where the volume charge density $\rho(r)$ resulting from counterions equals its average value ($-zen$). We assume that the distribution of counterions obeys a Boltzmann distribution, namely,

$$\rho(r) = -zen \exp\left[-\frac{-ze\psi(r)}{kT}\right] = -zen \exp\left[\frac{ze\psi(r)}{kT}\right] \tag{6.4}$$

The Poisson equation for the electric potential $\psi(r)$ around the particle is given by

$$\frac{d^2\psi(r)}{dr^2} + \frac{2}{r}\frac{d\psi(r)}{dr} = -\frac{\rho(r)}{\varepsilon_r\varepsilon_o} \tag{6.5}$$

where ε_r is the relative permittivity of the medium. By combining Eqs. (6.4) and (6.5), we obtain the following spherical Poisson–Boltzmann equation for a salt-free system:

$$\frac{d^2y}{dr^2} + \frac{2}{r}\frac{dy}{dr} = \kappa^2 e^y \tag{6.6}$$

where

$$y(r) = \frac{ze\psi(r)}{kT} \tag{6.7}$$

is the scaled potential and

$$\kappa = \left(\frac{z^2e^2n}{\varepsilon_r\varepsilon_o kT}\right)^{1/2} = \left[\frac{ze}{\varepsilon_r\varepsilon_o kT}\cdot\frac{3a^2\sigma}{(R^3 - a^3)}\right]^{1/2} = \left[\frac{3zeQ}{4\pi\varepsilon_r\varepsilon_o kT(R^3 - a^3)}\right]^{1/2} = \sqrt{\frac{3Q^*a}{R^3 - a^3}} \tag{6.8}$$

is the Debye–Hückel parameter in the present system. Here

$$Q^* = \frac{ze}{kT}\cdot\frac{Q}{4\pi\varepsilon_r\varepsilon_o a} \tag{6.9}$$

is the scaled particle charge. For the dilute case ($a/R \ll 1$), Eq. (6.8) becomes

$$\kappa = \left[\frac{3zea^2\sigma}{\varepsilon_r\varepsilon_o kTR^3}\right]^{1/2} = \left[\frac{3zeQ}{4\pi\varepsilon_r\varepsilon_o kTR^3}\right]^{1/2} = \sqrt{\frac{3Q^*a}{R^3}} \tag{6.10}$$

Note that the right-hand side of Eq. (6.6) contains only one term resulting from counterions, unlike the usual Poisson–Boltzmann equation. Since σ (or Q) and z are of the same sign, the product $z\sigma$ (or zQ) is always positive. Note that κ depends on σ (or Q), a, and R unlike the case of salt solutions, where κ is essentially independent of these parameters.

The boundary conditions are expressed as

$$\left.\frac{d\psi}{dr}\right|_{r=a} = -\frac{\sigma}{\varepsilon_r\varepsilon_o} = -\frac{Q}{4\pi\varepsilon_r\varepsilon_o a^2} = -\frac{kT}{ze}\cdot\frac{Q^*}{a} \tag{6.11}$$

$$\left.\frac{d\psi}{dr}\right|_{r=R} = 0 \tag{6.12}$$

which can be rewritten in terms of y and κ as

$$\left.\frac{dy}{dr}\right|_{r=a} = -\frac{Q^*}{a} = -\frac{\kappa^2(R^3 - a^3)}{3a^2} \tag{6.13}$$

$$\left.\frac{dy}{dr}\right|_{r=R} = 0 \tag{6.14}$$

For the dilute case, Eq. (6.13) becomes

$$\left.\frac{dy}{dr}\right|_{r=a} = -\frac{\kappa^2 R^3}{3a^2} \tag{6.15}$$

Also from Eqs. (6.6) and (6.13) we have

$$\int_a^R e^y r^2\, dr = \frac{1}{3}(R^3 - a^3) \tag{6.16}$$

Equation (6.6) cannot be solved analytically but its approximate solution for the case of dilute suspensions has been obtained by Imai and Oosawa [3,4]. They showed that there are two distinct cases separated by a certain critical value of the surface charge density σ or the total surface charge Q, that is, case 1: low surface charge density case and case 2: high surface charge density case, as schematically shown in Fig. 6.2. For case 1, there are two regions I ($R^* \le r \le R$) and II ($a \le r \le R^*$), while for case II, there are three regions I ($R^* \le r \le R$), II ($a^* \le r \le R^*$), and III

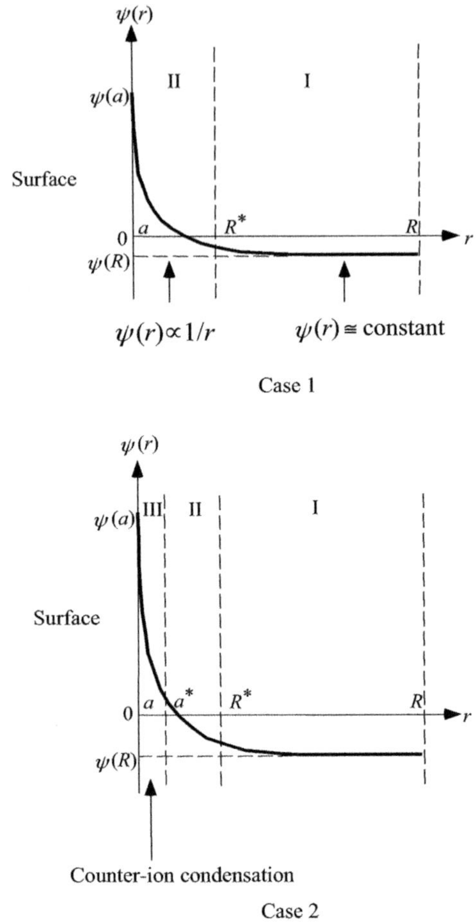

FIGURE 6.2 Three possible regions I, II, and III of r around a sphere of radius a. For case 1 (low surface charge densities), there are two regions I ($R^* \leq r \leq R$) and II ($a \leq r \leq R^*$), whereas for case 2 (high surface charge densities), there are three regions I ($R^* \leq r \leq R$), II ($a^* \leq r \leq R^*$), and III ($a \leq r \leq a^*$). Counterions are condensed in region III (counter-ion condensation). The particle surface potential ψ_o is defined by $\psi_o \equiv \psi(a) - \psi(R)$. From Ref. [5].

($a \leq r \leq a^*$). That is, for case 1, a^* reduces to a so that region III disappears. Note that counter-ion condensation occurs in the vicinity of the particle surface, that is, in region III because of very strong electric field there. This is a characteristic of a salt-free system.

Since in region I (far from the particle), the potential $y(r)$ is almost equal to $y(R)$, which is the potential at the outer boundary of the free volume of radius R, the Poisson–Boltzmann equation (6.6) can be approximated by

$$\frac{d^2y}{dr^2} + \frac{2}{r}\frac{dy}{dr} = \kappa^2 e^{y(R)} \tag{6.17}$$

which, subject to Eq. (6.12), can be solved to give

$$y(r) = Q^* \frac{a}{r} e^{y(R)} \left(1 + \frac{r^3}{2R^3} - \frac{3r}{2R}\right) + y(R), \quad (R^* \leq r \leq R) \tag{6.18}$$

In the dilute case ($a/R \ll 1$), as r decreases to R^*, approaching region II, the term proportional to $1/r$ in Eq. (6.18) becomes dominant. Since $y(r)$ must be continuous at $r = R^*$, $y(r)$ in region II ($a \leq r \leq R^*$ for case 1 and $a^* \leq r \leq R^*$ for case 2) should be of the form

$$y(r) = Q^* \frac{a}{r} e^{y(R)} + y(R) \tag{6.19}$$

which satisfies

$$\frac{d^2y}{dr^2} + \frac{2}{r}\frac{dy}{dr} = 0 \tag{6.20}$$

We now introduce

$$\xi(r) = -\frac{1}{\kappa^2 r} \cdot \frac{dy}{dr} e^{-y} \tag{6.21}$$

which is related to the ratio of the second term on the left-hand side to the right-hand side of Eq. (6.6). Since Eq. (6.6) tends to Eq. (6.20) when $\xi(r) \gg 1$, we see that the condition

$$\xi(r) \gg 1 \tag{6.22}$$

must hold in region II.

We evaluate $\xi(r)$ at $r = a^*$ from Eqs. (6.19) and (6.21) and obtain

$$\xi(a^*) = \frac{1}{3}\left(\frac{R}{a^*}\right)^3 \exp[-\{y(a^*) - y(R)\}] \tag{6.23}$$

which yields

$$y(a^*) - y(R) = 3 \ln\left(\frac{R}{a^*}\right) - \ln(3\xi(a^*)) \tag{6.24}$$

Since $\xi(a^*) \gg 1$ (Eq. (6.22)), we find that

$$y(a^*) - y(R) \leq 3 \ln\left(\frac{R}{a^*}\right) \tag{6.25}$$

which implies that the value of $y(a^*) - y(R)$ cannot exceed $3\ln(R/a^*)$ for the dilute case. In other words, the maximum value of $y(a^*) - y(R)$ is $3\ln(R/a^*)$. From Eq. (6.19), on the other hand, it follows that, for the dilute case,

$$y(a^*) - y(R) = -r\left.\frac{dy}{dr}\right|_{r=a*} \tag{6.26}$$

By combining Eqs. (6.25) and (6.26), we obtain

$$-\left.\frac{dy}{dr}\right|_{r=a^*} \leq \frac{3}{a^*}\ln\left(\frac{R}{a^*}\right) \tag{6.27}$$

We must thus consider two separate cases 1 and 2, depending on whether $-dy/dr|_{r=a}$ is larger or smaller than $(3/a)\ln(R/a)$.

(a) *Case 1 (low surface charge case).* Consider first case 1, in which case the following condition holds:

$$-\left.\frac{dy}{dr}\right|_{r=a} \leq \frac{3}{a}\ln\left(\frac{R}{a}\right) \tag{6.28}$$

By using Eq. (6.15) (for the dilute case), Eq. (6.28) can be rewritten as

$$(\kappa a)^2 \leq 3\phi\ln(1/\phi) \tag{6.29}$$

or by using Eq. (6.9)

$$Q^* \leq \ln(1/\phi) \tag{6.30}$$

In this case, region II can extend up to $r = a$, that is, a^* can reduce to a. Thus, the entire region of r can be covered only with regions I and II and one does not need region III. For the low surface charge case (small κ), Eq. (6.6) can be approximated by

$$\frac{d^2y}{dr^2} + \frac{2}{r}\frac{dy}{dr} = \kappa^2 = \frac{3Q^*}{R^3} \tag{6.31}$$

with the solution

$$y(r) = Q^*\frac{a}{r}\left(1 + \frac{r^3}{2R^3} - \frac{3r}{2R}\right) + y(R), \quad (a \leq r \leq R) \tag{6.32}$$

which is now applicable for the entire region of r. One can determine the value of $y(R)$ in Eq. (6.32) to satisfy Eq. (6.16) (in which e^y may be approximated by $1 + y$). The result for the dilute case is

$$y(R) = -\frac{\kappa^2 R^2}{10} = -\frac{Q^* a}{10R} \tag{6.33}$$

Thus, Eq. (6.32) becomes

$$y(r) = Q^* \frac{a}{r}\left(1 + \frac{r^3}{2R^3} - \frac{9r}{5R}\right), \quad (a \le r \le R) \tag{6.34}$$

and

$$y(R) \approx 0 \quad \text{for} \quad \phi \ll 1 \tag{6.35}$$

Note that Eq. (6.34) satisfies Eq. (6.12). From Eq. (6.34) we obtain

$$y(a) = Q^* \left(1 + \frac{a^3}{2R^3} - \frac{9a}{5R}\right) \approx Q^* \tag{6.36}$$

We define the particle surface potential ψ_o as the potential $\psi(a)$ at $r = a$ minus the potential $\psi(R)$ at $r = R$, that is, $\psi_o = \psi(a) - \psi(R)$. The scaled surface potential of the particle $y_o \equiv ze\psi_o/kT$ is thus given by $y_o = y(a) - y(R)$. From Eqs. (6.35) and (6.36), we obtain for the dilute case

$$y_o = Q^* \tag{6.37}$$

that is,

$$\psi_o = \frac{Q}{4\pi\varepsilon_r\varepsilon_o a} \tag{6.38}$$

Note that Eq. (6.37) (or Eq. (6.38)) does not depend on the particle volume fraction ϕ and coincides with the surface potential of a sphere of radius a carrying a total charge $Q = 4\pi a^2\sigma$ for the limiting case of $\kappa \to 0$, expressed by an unscreened Coulomb potential. That is, the surface potential in this case is the same as if the counterions were absent.

(b) *Case 2 (high surface charge case).* Consider next case 2, in which case the condition

$$-\left.\frac{dy}{dr}\right|_{r=a} > \frac{3}{a}\ln\left(\frac{R}{a}\right) \tag{6.39}$$

which can be rewritten as

$$(\kappa a)^2 > 3\phi \ln(1/\phi) \tag{6.40}$$

or

$$Q^* > \ln(1/\phi) \tag{6.41}$$

Thus, in this case, we find that

$$-\left.\frac{dy}{dr}\right|_{r=a} > -\left.\frac{dy}{dr}\right|_{r=a^*} \tag{6.42}$$

That is, region II cannot extend up to $r=a$. Therefore, the entire region of r cannot be covered only with regions I and II and thus one needs region III very near the particle surface between $r=a$ and $r=a^*$. In region III, the term $\kappa^2 e^y$ in Eq. (6.6) becomes very large so that $\xi(r) \ll 1$. The Poisson–Boltzmann equation in region III thus becomes

$$\frac{d^2y}{dr^2} = \kappa^2 e^y \tag{6.43}$$

which is of the same form as the Poisson–Boltzmann equation for a planar surface. Integration of Eq. (6.43) after multiplying dy/dr on both sides gives

$$\frac{1}{2}\left(\frac{dy}{dr}\right)^2 = \kappa^2 e^y + C \tag{6.44}$$

C being an integration constant. For high potentials, C may be ignored so that Eq. (6.44) yields

$$\frac{dy}{dr} = -\sqrt{2}\kappa e^{y/2} \tag{6.45}$$

By evaluating Eq. (6.45) at $r=a$ and using Eq. (6.15), we obtain for the dilute case

$$y(a) = \ln\left(\frac{\kappa^2 R^6}{18a^4}\right) = \ln\left[\frac{1}{6\phi}\left(\frac{ze}{kT}\right)\left(\frac{Q}{4\pi\varepsilon_r\varepsilon_o a}\right)\right] = \ln\left(\frac{Q^*}{6\phi}\right) \tag{6.46}$$

The value of $y(R)$ can be obtained as follows. In case 2, $y(a^*) - y(R)$ has already reached its maximum $3\ln(R/a^*)$. Thus from Eq. (6.19) we have

$$\frac{\kappa^2 R^3}{3a^*}\exp[y(R)] = 3\ln\left(\frac{R}{a^*}\right) \tag{6.47}$$

so that Eq. (6.19) becomes

$$y(r) = 3\ln\left(\frac{R}{a^*}\right)\frac{a^*}{r} + y(R), \quad (a^* \le r \le R) \tag{6.48}$$

The value of $y(R)$ in case 2 can be obtained from Eq. (6.46). For the purpose of determining $y(R)$ from Eq. (6.47), one may replace a^* by a in Eq. (6.47) so that

$$\begin{aligned} y(R) &= -\ln\left[\frac{(\kappa a)^2}{9\ln(R/a)}\left(\frac{R}{a}\right)^3\right] = -\ln\left[\frac{1}{\ln(1/\phi)}\left(\frac{ze}{kT}\right)\left(\frac{Q}{4\pi\varepsilon_r\varepsilon_o a}\right)\right] \\ &= -\ln\left[\frac{Q^*}{\ln(1/\phi)}\right] \end{aligned} \tag{6.49}$$

From Eqs. (6.46) and (6.49), we find that the scaled particle surface potential $y_s = y(a) - y(R)$ in case 2 is given by

$$\begin{aligned} y_o &= \ln\left[\frac{(\kappa a)^4}{162\ln(R/a)}\left(\frac{R}{a}\right)^9\right] = \ln\left[\frac{1}{6\phi\ln(1/\phi)}\left(\frac{ze}{kT}\right)^2\left(\frac{Q}{4\pi\varepsilon_r\varepsilon_o a}\right)^2\right] \\ &= \ln\left[\frac{(Q^*)^2}{6\phi\ln(1/\phi)}\right] \end{aligned} \tag{6.50}$$

that is, $\psi_o = \psi(a) - \psi(R)$ in case 2 is given by

$$\psi_o = \frac{kT}{ze}\ln\left[\frac{1}{6\phi\ln(1/\phi)}\left(\frac{ze}{kT}\right)^2\left(\frac{Q}{4\pi\varepsilon_r\varepsilon_o a}\right)^2\right] = \frac{kT}{ze}\ln\left[\frac{(Q^*)^2}{6\phi\ln(1/\phi)}\right] \tag{6.51}$$

Note that Eq. (6.51) implies that ψ_o depends less on Q than Eq. (6.38) and that ψ_o is a function of the particle volume fraction ϕ.

We introduce the effective surface potential ψ_{eff} and the effective surface charge Q_{eff} defined by

$$\psi_{eff} = \left(\frac{kT}{ze}\right)\ln(1/\phi) \tag{6.52}$$

$$Q_{eff} = 4\pi\varepsilon_r\varepsilon_o a\left(\frac{kT}{ze}\right)\ln(1/\phi) \tag{6.53}$$

and the corresponding dimensionless quantities y_{eff} and Q^*_{eff}

$$y_{eff} = \ln(1/\phi) \tag{6.54}$$

$$Q^*_{\mathrm{eff}} = \frac{Q_{\mathrm{eff}}}{4\pi\varepsilon_r\varepsilon_o a}\left(\frac{ze}{kT}\right) = \ln(1/\phi) \tag{6.55}$$

That is, $y_{\mathrm{eff}} = Q^*_{\mathrm{eff}}$. Then the potential distribution around the particle except the region very near the particle surface is approximately given by

$$\psi(r) = \psi_{\mathrm{eff}}\frac{a}{r} + y(R) = \frac{Q_{\mathrm{eff}}}{4\pi\varepsilon_r\varepsilon_o r} + y(R), \quad (a^* \le r \le R) \tag{6.56}$$

or

$$y(r) = y_{\mathrm{eff}}\frac{a}{r} + y(R) = Q^*_{\mathrm{eff}}\frac{a}{r} + y(R), \quad (a^* \le r \le R) \tag{6.57}$$

We thus see that for case 2 (high surface charge case), the particle behaves like a sphere showing a Coulomb potential with the effective surface charge Q_{eff} except the region very near the particle surface ($a^* \le r \le R$).

We compare the exact numerical solution to the Poisson–Boltzmann equation (6.6) and the approximate results, Eq. (6.37) for case 1 (low surface charge density case) and Eq. (6.50) for case 2 (high surface charge density case) in Fig. 6.3, in which the scaled surface potential $y_o \equiv ze\psi_o/kT$ is plotted as a function of the scaled

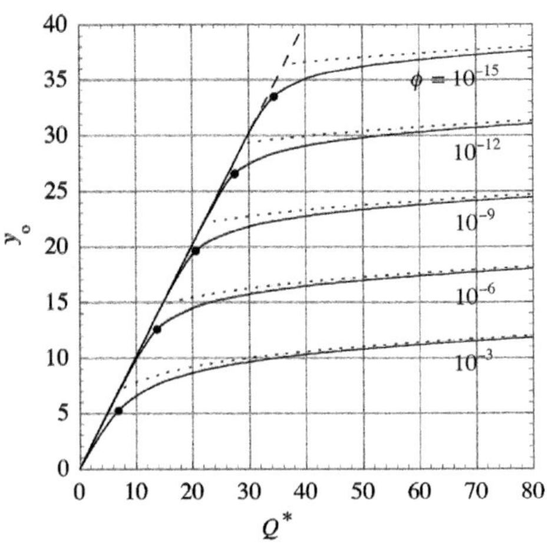

FIGURE 6.3 Scaled surface potential $y_o = ze\psi_o/kT$, defined by $y_o \equiv y(a) - y(R)$ of a sphere of radius a, as a function of the scaled total surface charge $Q^* = (ze/kT)Q/4\pi\varepsilon_r\varepsilon_o a$ for various values of the particle volume fraction ϕ. Solid lines represent exact numerical results. The dashed line, which passes through the origin ($y_o = 0$ and $Q^* = 0$), is approximate results calculated by an unscreened Coulomb potential, Eq. (6.37) (case 1: low surface charge densities) and dotted lines by Eq. (6.50) (case 2: high surface charge densities). Closed circles correspond to the approximate critical values of Q^* separating cases 1 and 2, given by Eq. (6.58).

total surface charge Q^* for various values of the particle volume fraction ϕ. Note that y_o is always positive. The agreement between the exact results and the approximate results is excellent with negligible errors except in the transition region between cases 1 and 2. It follows from the conditions for case 1 (Eq. (6.30)) and case 2 (Eq. (6.41)) that an approximate expression for the critical value of Q separating cases 1 and 2 for dilute suspensions is given by

$$Q^* = \ln(1/\phi) \tag{6.58}$$

6.3 CYLINDRICAL PARTICLE

Consider a dilute suspension of parallel cylindrical particles of radius a with a surface charge density σ or total surface charge $Q = 2\pi a\sigma$ per unit length in a salt-free medium containing only counterions. We assume that each cylinder is surrounded by a cylindrical free volume of radius R, within which counterions are distributed so that electroneutrality as a whole is satisfied. We define the particle volume fraction ϕ as

$$\phi = (a/R)^2 \tag{6.59}$$

The cylindrical particles are assumed to be parallel and equidistant. We treat the case of dilute suspensions, namely, $\phi \ll 1$ or $a/R \ll 1$. Let the average number density and the valence of counterions in the absence of the applied electric field be n and $-z$, respectively. Then we have

$$Q = 2\pi a\sigma = \pi(R^2 - a^2)zen \tag{6.60}$$

We set the equilibrium electric potential $\psi^{(0)}(r)$ (where r is the distance from the axis of one cylinder) to zero at points where the volume charge density resulting from counterions equals its average value ($-zen$). We assume that the distribution of counterions obeys a Boltzmann distribution (Eq. (6.4)). The cylindrical Poisson–Boltzmann equation thus becomes

$$\frac{d^2y}{dr^2} + \frac{1}{r}\frac{dy}{dr} = \kappa^2 e^y \tag{6.61}$$

with

$$\kappa = \left(\frac{z^2e^2n}{\varepsilon_r\varepsilon_o kT}\right)^{1/2} = \left[\frac{ze}{\varepsilon_r\varepsilon_o kT}\cdot\frac{2a\sigma}{(R^2 - a^2)}\right]^{1/2} = \left[\frac{zeQ}{\pi\varepsilon_r\varepsilon_o kT(R^2 - a^2)}\right]^{1/2} \tag{6.62}$$

where $y(r) = ze\psi(r)/kT$ is the scaled equilibrium potential and κ is the Debye–Hückel parameter in the present system. The boundary conditions are

$$\left.\frac{d\psi}{dr}\right|_{r=a} = -\frac{\sigma}{\varepsilon_r\varepsilon_o} = -\frac{Q}{2\pi\varepsilon_r\varepsilon_o a} \tag{6.63}$$

$$\left.\frac{d\psi}{dr}\right|_{r=R} = 0 \tag{6.64}$$

The Poisson–Boltzmann equation (6.61) subject to boundary conditions (6.63) and (6.64) has been solved independently by Fuoss et al. [1] and Afrey et al. [2]. The results are given below.

(a) *Case 1 (low surface charge case).* If the condition

$$0 < Q^* < \frac{\ln(a/R)}{\ln(a/R) - 1} \tag{6.65}$$

is satisfied, then we have

$$y(a) = -\ln\left[\frac{\kappa^2 a^2}{2}\cdot\frac{1}{(1-Q^*)^2 - \beta^2}\right] \tag{6.66}$$

$$y(R) = -\ln\left[\frac{\kappa^2 R^2}{2}\cdot\frac{1}{1-\beta^2}\right] \tag{6.67}$$

where Q^* is a scaled particle charge per unit length defined by

$$Q^* = \frac{Q}{4\pi\varepsilon_r\varepsilon_o}\cdot\frac{ze}{kT} \tag{6.68}$$

(Note that Q^* is always positive, since Q and z are always of the same sign) and β is given by the solution to the following transcendental equation:

$$Q^* = \frac{1-\beta^2}{1-\beta\coth\{\beta\ln(a/R)\}} \tag{6.69}$$

(b) *Case 2 (high surface charge case).* If the condition

$$Q^* \geq \frac{\ln(a/R)}{\ln(a/R) - 1} \tag{6.70}$$

is satisfied, then we have

$$y(a) = -\ln\left[\frac{\kappa^2 a^2}{2} \cdot \frac{1}{(1-Q^*)^2 + \beta^2}\right] \tag{6.71}$$

$$y(R) = -\ln\left[\frac{\kappa^2 R^2}{2} \cdot \frac{1}{1+\beta^2}\right] \tag{6.72}$$

where β is given by the solution to the following transcendental equation:

$$Q^* = \frac{1+\beta^2}{1-\beta\cot\{\beta\ln(a/R)\}} \tag{6.73}$$

Now we confine ourselves to the dilute case $a/R \ll 1$, in which case the ln $(a/R)/(\ln(a/R) - 1)$ on the right-hand side of Eqs. (6.65) and (6.70), which is the critical value separating cases 1 (low charge case) and 2 (high charge case), tends to unity. Also in this case we find that Eqs. (6.60) and (6.62) give

$$Q = \pi R^2 zen \tag{6.74}$$

$$Q^* = \frac{\kappa^2 R^2}{4} \tag{6.75}$$

and that the above results for cases 1 and 2 can be rewritten as follows [2].

(a) *Case 1 (low surface charge case).* If the condition

$$0 < Q^* < 1 \tag{6.76}$$

is satisfied, then we have that

$$y(a) = Q^* \ln(1/\phi) \tag{6.77}$$

$$y(R) = \ln\left(1 - \frac{Q^*}{2}\right) \tag{6.78}$$

(b) *Case 2 (high surface charge case).* If the condition

$$Q^* \geq 1 \tag{6.79}$$

is satisfied, then we have that

$$y(a) = \ln\left(\frac{Q^*}{2\phi}\right) \tag{6.80}$$

$$y(R) = -\ln(2Q^*) \tag{6.81}$$

Further, if we define the particle surface potential ψ_o as $\psi_o = \psi(a) - \psi(R)$ or the scaled surface potential $y_o \equiv (ze/kT)\psi_o$ as $y_o = y(a) - y(R)$, then we obtain for the dilute case

$$y_o = Q^* \ln(1/\phi), \quad \text{for } 0 < Q^* < 1 \text{ (case 1: low surface charge case)} \tag{6.82}$$

$$y_o = \ln\left(\frac{Q^{*2}}{\phi}\right), \quad \text{for } Q^* \geq 1 \text{ (case 2: high surface charge case)} \tag{6.83}$$

Equation (6.82) agrees with the unscreened Coulomb potential of a charged cylinder. That is, the surface potential in this case is the same as if the counterions were absent.

6.4 EFFECTS OF A SMALL AMOUNT OF ADDED SALTS

We consider a suspension of spherical colloidal particles of radius a with monovalent ionized groups on their surface in a medium which contains counterions produced by dissociation of the particle surface groups and a small amount of monovalent symmetrical electrolytes [8]. We may assume that the particles are positively charged without loss of generality. We treat the case in which N ionized groups of valence +1 are uniformly distributed on the surface of each particle and N counterions of valence -1 are produced by dissociation of the particle surface groups. Each particle thus carries a charge $Q = eN$, where e is the elementary electric charge. We also assume that each particle is surrounded by a spherical free volume of radius R (Fig. 6.1), within which electroneutrality as a whole is satisfied. The particle volume fraction ϕ is given by $\phi = (a/R)^3$. We treat the case of dilute (but not infinitely dilute) suspensions, namely, $\phi \ll 1$ or $a/R \ll 1$. We assume that the electrolyte is completely dissociated to give M cations of valence +1 and M anions of valence -1 in each free volume so that there are $N + M$ counterions and M coions in the free volume. Let the average concentration (number density) of total counterions be $n + m$ and that of anions be m, n and m being given by

$$n = \frac{N}{V}, \quad m = \frac{M}{V} \tag{6.84}$$

where

$$V = \frac{4}{3}\pi(R^3 - a^3) \tag{6.85}$$

is the volume available for counterions and coions within each free volume. From the condition of electroneutrality in each free volume, we have

$$Q = eN = Ven \tag{6.86}$$

We assume that the equilibrium distribution of ions obeys a Boltzmann distribution so that volume charge density $\rho(r)$ at position r, r being the distance from the particle center ($r \geq a$), is given by

$$\rho(r) = e\left[-(n+m)e^y \frac{V}{4\pi \int_a^R e^y r^2\,dr} + me^y \frac{V}{4\pi \int_a^R e^{-y} r^2\,dr}\right] \tag{6.87}$$

where $y(r) = e\psi(r)/kT$ is the scaled potential. Note that Eq. (6.87) satisfies the electroneutrality condition of a free volume, namely,

$$4\pi \int_a^R \rho(r) r^2 dr = e\{-(N+M) + M\} = -eN = -Q \tag{6.88}$$

We set the electric potential $\psi(r)$ to zero at points where the concentration of total counterions equals its average value $n + m$ so that

$$4\pi \int_a^R e^y r^2 dr = V = \frac{4\pi}{3}(R^3 - a^3) \tag{6.89}$$

and Eq. (6.87) becomes

$$\rho(r) = en\left[-(1+p)e^y + \frac{p}{W}e^{-y}\right] \tag{6.90}$$

with

$$W = \frac{4\pi \int_a^R e^{-y} r^2\,dr}{V} = \frac{\int_a^R e^{-y} r^2\,dr}{(R^3 - a^3)/3} \tag{6.91}$$

and

$$p = \frac{m}{n} \tag{6.92}$$

where p is the ratio of the concentration of counterions (or coions) resulting from the added electrolytes to that of counterions from the particle. The case of $p=0$ corresponds to the completely salt-free case.

By combining the Poisson equation (6.5) and Eq. (6.87), we obtain the following Poisson–Boltzmann equation for the equilibrium electric potential $\psi(r)$ at position r, r being the distance from the particle center ($r \geq a$):

$$\frac{d^2y}{dr^2} + \frac{2}{r}\frac{dy}{dr} = \frac{\kappa^2}{1+2p}\left[(1+p)e^y - \frac{p}{W}e^{-y}\right] \tag{6.93}$$

where κ is the Debye–Hückel parameter defined by

$$\kappa = \left[\frac{e^2(n+2m)}{\varepsilon_r\varepsilon_o kT}\right]^{1/2} = \left[\frac{3Q^*\phi(1+2p)}{a^2(1-\phi)}\right]^{1/2} \tag{6.94}$$

and

$$Q^* = \frac{Q}{4\pi\varepsilon_r\varepsilon_o a}\left(\frac{e}{kT}\right) = \frac{\kappa^2 a^2(1-\phi)}{3\phi(1+2p)} \tag{6.95}$$

is the scaled particle charge. Note that κ generally depends on Q^*, p, a, and ϕ and that when $n \ll m$, κ becomes the usual Debye–Hückel parameter of a 1–1 electrolyte of concentration m, given by

$$\kappa = \left(\frac{2e^2m}{\varepsilon_r\varepsilon_o kT}\right)^{1/2} \tag{6.96}$$

The boundary conditions are the same as Eqs. (6.11) and (6.12).

We derive some approximate solutions to Eq. (6.93). It can be shown that as in the case of completely salt-free case, there are three possible regions I, II, and III in the potential distribution around the particle surface for the dilute case. As will be seen later, region I (where $y(r) \approx y(R)$) is found to be much wider than regions II and III, and in region III the potential is very high so that there are essentially no coions. We may thus approximate Eq. (6.91) as

$$W \approx \frac{e^{-y(R)}\int_a^R r^2\,dr}{(R^3 - a^3)/3} = e^{-y(R)}, \quad \text{for } \phi \ll 1 \tag{6.97}$$

Thus Eq. (6.93) can be approximated by

$$\frac{d^2y}{dr^2} + \frac{2}{r}\frac{dy}{dr} = \frac{\kappa^2}{1+2p}\left[(1+p)e^{y(r)} - p\,e^{-\{y(r)-y(R)\}}\right] \tag{6.98}$$

or, equivalently

$$\frac{d^2y}{dr^2} + \frac{2}{r}\frac{dy}{dr} = \frac{3aQ^*}{R^3(1-\phi)}\left[(1+p)e^{y(r)} - p\,e^{-\{y(r)-y(R)\}}\right] \tag{6.99}$$

For $p = 0$, Eq. (6.98) tends to the Poisson–Boltzmann equation for the completely salt-free case (Eq. (6.61)). For large $p \gg 1$, $y(R)$ approaches 0 (as shown later by Eq. (6.123) and Eq. (6.99) tends to

$$\frac{d^2y}{dr^2} + \frac{2}{r}\frac{dy}{dr} = \kappa^2 \sinh y \tag{6.100}$$

with κ given by Eq. (6.96). Equation (6.100), which ignores counterions from the particle, is the familiar PB equation for media containing ample salts (Eq. (1.27)). In region I, $y(r) \approx y(R)$ so that Eq. (6.99) can further be approximated by, for the dilute case ($\phi \ll 1$),

$$\frac{d^2y}{dr^2} + \frac{2}{r}\frac{dy}{dr} = \frac{3aQ^*}{R^3}\left[(1+p)e^{y(R)} - p\right] \tag{6.101}$$

which, subject to Eq. (6.12), can be solved to give

$$y(r) = Q^*\left[(1+p)e^{y(R)} - p\right]\left(\frac{a}{r}\right)\left(1 + \frac{r^3}{2R^3} - \frac{3r}{2R}\right) + y(R), \quad \text{for } R^* \le r \le R \tag{6.102}$$

In the dilute case ($a/R \ll 1$), as r decreases to R^*, approaching region II, the term proportional to $1/r$ in Eq. (6.102) becomes dominant, that is, $y(r)$ becomes an unscreened Coulomb potential. Since $y(r)$ must be continuous at $r = R^*$, $y(r)$ in region II should be of the form

$$y(r) = Q^*\left[(1+p)e^{y(R)} - p\right]\left(\frac{a}{r}\right) + y(R), \quad \text{for } a^* \le r \le R^* \tag{6.103}$$

Note that Eq. (6.103) satisfies

$$\frac{d^2y}{dr^2} + \frac{2}{r}\frac{dy}{dr} = 0 \tag{6.104}$$

and that Eq. (6.98) becomes Eq. (6.104), if the following two conditions are both satisfied in addition to Eq. (6.2): (i) the terms on the right-hand side of Eq. (6.98) are very small as compared with a^2 and (ii) these terms are also much smaller than the second term of the left-hand side of Eq. (6.98). The first condition may be expressed as

$$\kappa a \ll 1 \tag{6.105}$$

or equivalently,

$$(1+2p)Q^*\phi \ll 1 \tag{6.106}$$

and the second condition as

$$\left|\frac{1}{r}\frac{dy}{dr}\right| \gg \frac{3aQ^*}{R^3}\left[(1+p)e^{y(r)} - p\,e^{-\{y(r)-y(R)\}}\right] \tag{6.107}$$

By evaluating Eq. (6.107) at $r = a^*$ using Eq. (6.103), we obtain

$$(1+p)e^{y(R)} - p \gg \frac{3a^{*3}}{R^3}\left[(1+p)e^{y(a^*)} - p\,e^{-\{y(a^*)-y(R)\}}\right] \tag{6.108}$$

which yields, by taking the logarithm of both sides,

$$y(a^*) - y(R) < \ln\left(1 - \frac{p}{1+p}e^{-y(R)}\right) + \ln\left(\frac{R^3}{a^{*3}}\right) \approx \ln(1/\phi) \tag{6.109}$$

Equation (6.109) implies that the value of $y(a^*) - y(R)$ cannot exceed $\ln(1/\phi)$, which is thus the possible maximum value of $y(a^*) - y(R)$. From Eq. (6.103), on the other hand, we obtain

$$y(a^*) - y(R) = -a^*\left|\frac{dy}{dr}\right|_{r=a^*} \tag{6.110}$$

By combining Eqs. (6.109) and (6.110), we obtain

$$-a^*\frac{dy}{dr}\bigg|_{r=a^*} < \ln(1/\phi), \quad \text{for } \phi \ll 1 \tag{6.111}$$

We must thus consider two cases 1 and 2 separately, depending on whether is larger or smaller than $\ln(1/\phi)$. This situation is the same as that for $p = 0$. The results are summarized below.

(a) *Case 1 (low surface charge case)*. Consider first case 1, in which the following condition holds:

$$Q^* \leq \ln(1/\phi) \tag{6.112}$$

In this case, region II can extend up to $r = a$, that is, a^* can reduce to a. Thus, the entire region of r can be covered only with regions I and II and one does not need region III. For the low-charge case, Eq. (6.101), which is now applicable essentially for the entire region of r, can further be simplified to give

$$y(r) = Q^* \frac{a}{r}\left(1 + \frac{r^3}{2R^3} - \frac{9r}{5R}\right), \quad a \leq r \leq R \tag{6.113}$$

which is independent of p. For the dilute case ($\phi \ll 1$), Eq. (6.113) gives

$$y(R) \approx 0 \tag{6.114}$$

and becomes essentially a Coulomb potential, namely,

$$y(r) = Q^* \frac{a}{r}, \quad a \leq r \leq R \tag{6.115}$$

The scaled particle surface potential $y_o = y(a) - y(R) = y(a)$ in case 1 is thus given by

$$y_o = Q^* \tag{6.116}$$

The above results do not depend on p and agree with those for the completely salt- free case.

(b) *Case 1 (high surface charge case)*. Consider next case 2, in which the condition

$$Q^* > \ln(1/\phi) \tag{6.117}$$

is satisfied so that

$$Q^* > -a^* \left.\frac{dy}{dr}\right|_{r=a*} \tag{6.118}$$

That is, region II cannot extend up to $r = a$. Therefore, the entire region of r cannot be covered only with regions I and II and thus one needs region III very near the particle surface between $r = a$ and $r = a^*$, where counterions are condensed (counter-ion condensation). In region III, $y(r)$ is very high so that Eq. (6.99) becomes

$$\frac{d^2y}{dr^2} = \frac{3\phi Q^*}{a^2}(1+p)e^{y(r)} \tag{6.119}$$

By integrating Eq. (6.119), we obtain for the dilute case

$$y(a) = \ln\left[\frac{Q^*}{6\phi(1+p)}\right] \tag{6.120}$$

The value of $y(R)$ can be obtained as follows. In case 2, $y(a^*) - y(R)$ has already reached its maximum $\ln(1/\phi)$ so we find that Eq. (6.102) becomes

$$y(r) = Q^*_{\text{eff}} \frac{a}{r} + y(R), \quad a^* \le r \le R^* \tag{6.121}$$

where

$$Q^*_{\text{eff}} = y(a^*) - y(R) = Q^*\left[(1+p)e^{y(R)} - p\right] = \ln(1/\phi) \tag{6.122}$$

is the scaled effective particle charge and the value of $y(R)$ is given by

$$y(R) = \ln\left[\frac{\ln(1/\phi) + pQ^*}{Q^*(1+\phi)}\right] \tag{6.123}$$

which is obtained from the last equation of Eq. (6.122). We see from Eq. (6.123) that $y(R)$ tends to zero as p increases. Equation (6.121) implies that in region II the potential distribution around the particle is a Coulomb potential produced by the scaled particle charge $Q^*_{\text{eff}} = \ln(1/\phi)$. In other words, the particle behaves as if the particle charge were Q^*_{eff} instead of Q^* for regions I and II, in which regions Eq. (6.102) may be expressed as

$$y(r) = Q^*_{\text{eff}} \frac{a}{r}\left(1 + \frac{r^3}{2R^3} - \frac{3r}{2R}\right) + y(R), \quad a^* \le r \le R \tag{6.124}$$

The particle surface potential $y_o = y(a) - y(R)$ in case 2 is found to be, from Eqs. (6.120) and (6.123),

$$y_o = \ln\left[\frac{(Q^*)^2}{6\phi\{\ln(1/\phi) + pQ^*\}}\right] \tag{6.125}$$

When $p = 0$, Eqs. (6.120), (6.123) and (6.125) all reduce back to results obtained for the completely salt-free case. Also note that the critical value separating cases 1 and 2 is $\ln(1/\phi)$, which is the same as that for the completely salt-free case.

6.5 SPHERICAL SOFT PARTICLE

Consider a dilute suspension of polyelectrolyte-coated spherical colloidal particles (soft particles) in a salt-free medium containing counterions only. We assume that the particle core of radius a (which is uncharged) is coated with an ion-penetrable layer of polyelectrolytes of thickness d. The polyelectrolyte-coated particle has thus an inner radius a and an outer radius $b \equiv a + d$ (Fig. 6.4). We also assume that ionized groups of valence Z are distributed at a uniform density N in the polyelectrolyte

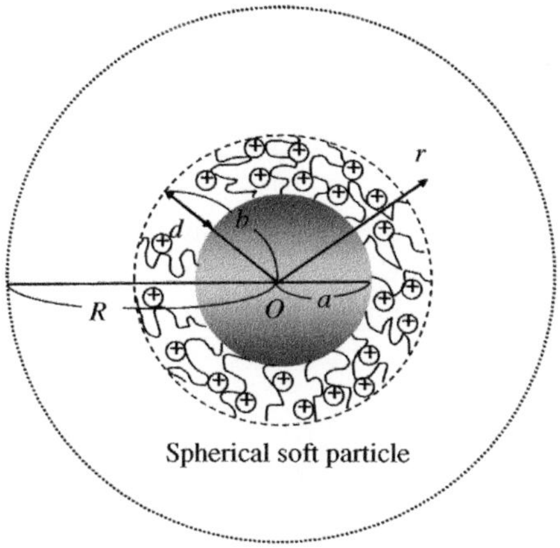

FIGURE 6.4 A polyelectrolyte-coated spherical particle (a spherical soft particle) in a free volume of radius R containing counterions only. a is the radius of the particle core. $b = a + d$. d is the thickness of the polyelectrolyte layer covering the particle core. $(b/R)^3$ equals the particle volume fraction ϕ. From Ref. [9].

layer so that the charge density in the polyelectrolyte layer is given by ZeN, where e is the elementary electric charge. The total charge amount Q of the particle is thus given by

$$Q = \frac{4}{3}\pi(b^3 - a^3)ZeN \tag{6.126}$$

We assume that each sphere is surrounded by a spherical free volume of radius R (Fig. 6.4), within which counterions are distributed so that electroneutrality as a whole is satisfied. The particle volume fraction ϕ is given by

$$\phi = (b/R)^3 \tag{6.127}$$

In the following we treat the case of dilute suspensions, namely,

$$\phi \ll 1 \quad \text{or} \quad b/R \ll 1 \tag{6.128}$$

We denote the electric potential at a distance r from the center O of one particle by $\psi(r)$. Let the average number density and the valence of counterions be n and $-z$, respectively. Then from the condition of electroneutrality in the free volume, we have

$$Q = \frac{4}{3}\pi(b^3 - a^3)ZeN = \frac{4}{3}\pi(R^3 - a^3)zen \tag{6.129}$$

which gives

$$\frac{ZN}{zn} = \frac{R^3 - a^3}{b^3 - a^3} \tag{6.130}$$

We set $\psi(r) = 0$ at points where the counter-ion concentration equals its average value n so that the volume charge density $\rho(r)$ resulting from counterions equals its average value $(-zen)$. We assume that the distribution of counterions obeys a Boltzmann distribution, namely,

$$\rho(r) = -zen \exp\left[-\frac{-ze\psi(r)}{kT}\right] = -zen \exp\left[\frac{ze\psi(r)}{kT}\right] \tag{6.131}$$

The Poisson equations for the electric potential $\psi(r)$ in the region inside and outside the polyelectrolyte layer are then given by

$$\frac{d^2\psi(r)}{dr^2} + \frac{2}{r}\frac{d\psi(r)}{dr} = -\frac{\rho(r)}{\varepsilon_r\varepsilon_o} - \frac{ZeN}{\varepsilon_r\varepsilon_o}, \quad a < r < b \tag{6.132}$$

$$\frac{d^2\psi(r)}{dr^2} + \frac{2}{r}\frac{d\psi(r)}{dr} = -\frac{\rho(r)}{\varepsilon_r\varepsilon_o}, \quad b < r < R \tag{6.133}$$

Here we have assumed that the relative permittivity ε_r takes the same value in the medium inside and outside the polyelectrolyte layer. By combining Eqs. (6.131)–(6.133), we obtain the following Poisson–Boltzmann equations:

$$\frac{d^2y}{dr^2} + \frac{2}{r}\frac{dy}{dr} = \kappa^2\left(e^y - \frac{ZN}{zn}\right), \quad a < r < b \tag{6.134}$$

$$\frac{d^2y}{dr^2} + \frac{2}{r}\frac{dy}{dr} = \kappa^2 e^y, \quad b < r < R \tag{6.135}$$

where $y(r) = ze\psi(r)/kT$ is the scaled potential and

$$\kappa = \left(\frac{z^2e^2n}{\varepsilon_r\varepsilon_o kT}\right)^{1/2} = \left[\frac{Zze^2N(b^3 - a^3)}{\varepsilon_r\varepsilon_o kT(R^3 - a^3)}\right]^{1/2} = \left[\frac{3zeQ}{4\pi\varepsilon_r\varepsilon_o kT(R^3 - a^3)}\right]^{1/2} = \left(\frac{3Q^*b}{R^3 - a^3}\right)^{1/2} \tag{6.136}$$

is the Debye–Hückel parameter in the present system. In Eq. (6.136), Q^* is the scaled particle charge defined by

$$Q^* = \frac{Q}{4\pi\varepsilon_r\varepsilon_o b}\left(\frac{ze}{kT}\right) \tag{6.137}$$

By using Eq. (6.130), we can rewrite Eq. (6.134) as

$$\frac{d^2y}{dr^2}+\frac{2}{r}\frac{dy}{dr}=\kappa^2\left[e^y-\left(\frac{R^3-a^3}{b^3-a^3}\right)\right]=\frac{3Q^*b}{R^3}\left[e^y-\left(\frac{R^3-a^3}{b^3-a^3}\right)\right],\quad a<r<b \tag{6.138}$$

which becomes for the dilute case

$$\frac{d^2y}{dr^2}+\frac{2}{r}\frac{dy}{dr}=\kappa^2\left(e^y-\frac{R^3}{b^3-a^3}\right)=\frac{3Q^*b}{R^3}\left(e^y-\frac{R^3}{b^3-a^3}\right),\quad a<r<b \tag{6.139}$$

The boundary conditions for y are expressed as

$$\left.\frac{dy}{dr}\right|_{r=a}=0 \tag{6.140}$$

$$\left.\frac{dy}{dr}\right|_{r=R}=0 \tag{6.141}$$

$$y(b^-)=y(b^+) \tag{6.142}$$

$$\left.\frac{dy}{dr}\right|_{r=b^-}=\left.\frac{dy}{dr}\right|_{r=b^+} \tag{6.143}$$

We derive approximate solutions to Eqs. (6.135) and (6.139) for the dilute case by using an approximation method described in Ref. [5], as follows.

(a) *Case 1 (low surface charge case)*. If the condition

$$Q^*\leq \ln(1/\phi) \tag{6.144}$$

is satisfied, Eq. (6.135) can be approximated by

$$\frac{d^2y}{dr^2}+\frac{2}{r}\frac{dy}{dr}=\kappa^2,\quad b<r<R \tag{6.145}$$

from which we have

$$y(r)=Q^*\frac{b}{r}\left(1+\frac{r^3}{2R^3}-\frac{9r}{5R}\right),\quad (b\leq r\leq R) \tag{6.146}$$

or

$$\psi(r) = \frac{Q}{4\pi\varepsilon_r\varepsilon_o r}\left(1 + \frac{r^3}{2R^3} - \frac{9r}{5R}\right), \quad (b \le r \le R) \tag{6.147}$$

For the dilute case ($\phi \ll 1$), Eq. (6.147) can be approximated well by an unscreened Coulomb potential of a sphere carrying charge Q, namely,

$$\psi(r) = \frac{Q}{4\pi\varepsilon_r\varepsilon_o r}, \quad (b \le r \le R) \tag{6.148}$$

That is, in the low charge case, the potential is essentially the same as if conterions were absent. This suggests that the potential inside the polyelectrolyte layer is also approximately the same as if conterions were absent. If counterions are absent, then Eq. (6.139) becomes

$$\frac{d^2y}{dr^2} + \frac{2}{r}\frac{dy}{dr} = -\frac{\kappa^2 R^3}{b^3 - a^3} = -\frac{3bQ^*}{b^3 - a^3}, \quad (a < r < b) \tag{6.149}$$

with the solution

$$y(r) = \frac{Q^* b^3}{(b^3 - a^3)}\left(\frac{3}{2} - \frac{r^2}{2b^2} - \frac{a^3}{b^2 r}\right), \quad (a \le r \le b) \tag{6.150}$$

or

$$\psi(r) = \frac{Qb^2}{4\pi\varepsilon_r\varepsilon_o(b^3 - a^3)}\left(\frac{3}{2} - \frac{r^2}{2b^2} - \frac{a^3}{b^2 r}\right), \quad (a \le r \le b) \tag{6.151}$$

If the surface potential ψ_s of the polyelectrolyte-coated particle is identified as $\psi_s = \psi(b) - \psi(R)$, then we find that for the dilute case

$$\psi(R) = 0 \tag{6.152}$$

$$\psi_o = \psi(b) = \frac{Q}{4\pi\varepsilon_r\varepsilon_o b} \tag{6.153}$$

or

$$y(R) = 0 \tag{6.154}$$

$$y_o = Q^* \tag{6.155}$$

where

$$y_o = \frac{ze\psi_o}{kT} \tag{6.156}$$

is the scaled surface potential ($y_o = y(b) - y(R)$).

(b) *Case 2 (high surface charge case).* If the condition

$$Q^* > \ln(1/\phi) \tag{6.157}$$

is satisfied, then as in the case of hard particles, we find that for the outer solution ($b \le r \le R$),

$$\frac{dy}{dr} = -\sqrt{2}\kappa e^{y/2} \quad (r \approx b) \tag{6.158}$$

$$y(r) = \ln\left(\frac{1}{\phi}\right)\left(\frac{b}{r}\right)\left(1 + \frac{r^3}{2R^3} - \frac{3r}{2R}\right) + y(R), \quad (b < r \le R) \tag{6.159}$$

$$y(R) = -\ln\left[\frac{Q^*}{\ln(1/\phi)}\right] \tag{6.160}$$

Note that Eqs. (6.158)–(6.160) are independent of a.

For the inner region ($a \le r \le b$), by noting that the potential variation is very small as compared with that in the outer region ($b \le r \le R$), we may replace the spherical Poisson–Boltzmann equation [6.139] with the following planar Poisson–Boltzmann equation:

$$\frac{d^2y}{dr^2} = \kappa^2\left[e^y - \left(\frac{R^3}{b^3 - a^3}\right)\right], \quad a < r < b \tag{6.161}$$

The solution to Eq. (6.161) is

$$\frac{dy}{dr} = -\sqrt{2}\kappa\sqrt{\exp(y) - \exp(y(a)) - \left(\frac{R^3}{b^3 - a^3}\right)\{y - y(a)\}} \tag{6.162}$$

By evaluating Eqs. (6.158) and (6.162) at $r = b$ and using Eq. (6.143), we have

$$y(b) = y(a) - \left(\frac{b^3 - a^3}{R^3}\right)\exp[y(a)] \tag{6.163}$$

Equation (6.162) can further be integrated to yield

$$-\sqrt{2}\kappa(r-a) = \int_{y(a)}^{y} \frac{dy}{\sqrt{\exp(y) - \exp(y(a)) - (R^3/(b^3-a^3))\{y - y(a)\}}} \approx \frac{-2\sqrt{y(a)-y}}{\sqrt{R^3/(b^3-a^3) - \exp[y(a)]}} \tag{6.164}$$

where we have expanded the integrand around $y = y(a)$ and carried out the integration, since the largest contribution to the integrand comes from the region near $y = y(a)$. Equation (6.164) yields

$$y(r) = y(a) + \left[e^{y(a)} - \frac{R^3}{(b^3-a^3)}\right]\kappa^2(r-a)^2 \tag{6.165}$$

which may also be obtained by directly expanding $y(r)$ around $r = a$ on the basis of Eq. (6.139). From Eqs. (6.163) and (6.165) (evaluated at $r = b$) we obtain

$$y(a) = \ln\left[\frac{3Q^* bR^3}{\{3Q^* b(b-a) + 2(b^2+ab+a^2)\}(b^2+ab+a^2)}\right] \tag{6.166}$$

and

$$y(b) = \ln\left[\frac{3Q^* bR^3}{\{3Q^* b(b-a) + 2(b^2+ab+a^2)\}(b^2+ab+a^2)}\right] - \frac{3Q^* b(b-a)}{\{3Q^* b(b-a) + 2(b^2+ab+a^2)\}} \tag{6.167}$$

Thus, we find that the scaled surface potential $y_o = y(b) - y(R)$ is given by

$$y_o = \ln\left[\frac{3Q^{*2} bR^3}{\ln(1/\phi)\{3Q^* b(b-a) + 2(b^2+ab+a^2)\}(b^2+ab+a^2)}\right] - \frac{3Q^* b(b-a)}{\{3Q^* b(b-a) + 2(b^2+ab+a^2)\}} \tag{6.168}$$

The Poisson–Boltzmann equations (6.135) and (6.139) subject to boundary conditions (6.140)–(6.143) can be solved numerically with Mathematica to obtain the potential distribution $y(x)$. The results of some calculations for the potential

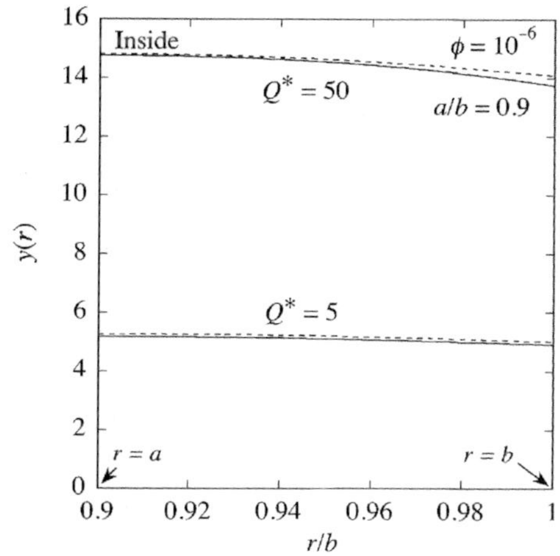

FIGURE 6.5 Distribution of the scaled electric potential $y(r)$ within the polyelectrolyte layer ($a \leq r \leq b$) at $\phi = 10^{-6}$ and $a/b = 0.9$ for $Q^* = 5$ (low charge case) and $Q^* = 50$ (high charge case). Solid curves, exact results; dotted curves, approximate results (Eq. (6.150) for $Q^* = 5$ and Eq. (6.165) 6. for $Q^* = 50$). From Ref. [9].

distribution are given in Figs 6.5 and 6.6, which demonstrate $y(r)$ for $Q^* = 5$ (low charge case) and 50 (high charge case) at $\phi = 10^{-6}$ and $a/b = 0.9$. Figure 6.5 shows the exact numerical results for the inner solution ($a \leq r \leq b$) in comparison with approximate results (Eq. (6.150) for $Q^* = 5$ and Eq. (6.165) for $Q^* = 50$), showing

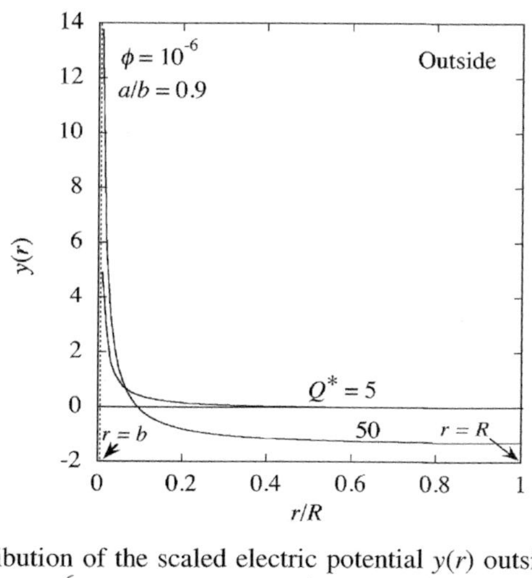

FIGURE 6.6 Distribution of the scaled electric potential $y(r)$ outside the polyelectrolyte layer ($b \leq r \leq R$) at $\phi = 10^{-6}$ and $a/b = 0.9$ for $Q^* = 5$ (low charge case) and $Q^* = 50$ (high charge case). Approximate results (Eq. (6.146) for $Q^* = 5$ and Eq. (6.159) for $Q^* = 50$) agree with exact results within the linewidth. From Ref. [9].

good agreement between numerical and approximate results. Figure 6.6 shows the exact numerical results for the outer region $b \leq r \leq R$, which agree with approximate results (Eq. (6.146) for $Q^* = 5$ and Eq. (6.159) for $Q^* = 50$) within the linewidth. It is seen that for the high charge case ($Q^* = 50$) the potential outside the polyelectrolyte layer decreases very sharply near the surface of the polyelectrolyte layer ($r = b$) and essentially constant ($y(r) \approx y(R)$) except in the region very near the polyelectrolyte layer.

In Fig. 6.7, the scaled surface potential $y_s = y(b) - y(R)$ is plotted as a function of Q^* at $a/b = 0.5$ for $\phi = 10^{-3}$, 10^{-6}, and 10^{-9}. We see that as in the case of a rigid particle, there is a certain critical value separating the low charge case and the high charge case. Approximate results (Eq. (6.155) for the low charge case and Eq. (6.168) for the high charge case) are also shown in Fig. 6.7. An approximate expression for the critical value Q^*_{cr} is given by (see Eqs. (6.144) and (6.157))

$$Q^*_{cr} = \ln(1/\phi) \tag{6.169}$$

When $Q^* \leq Q^*_{cr}$ (low charge case), the surface potential may be approximated by Eq. (6.155), which is the unscreened Coulomb potential for the case where counterions are absent. The surface potential y_s is proportional to Q^* in this case. When $Q^* > Q^*_{cr}$ (high charge case), the surface potential can be approximated by

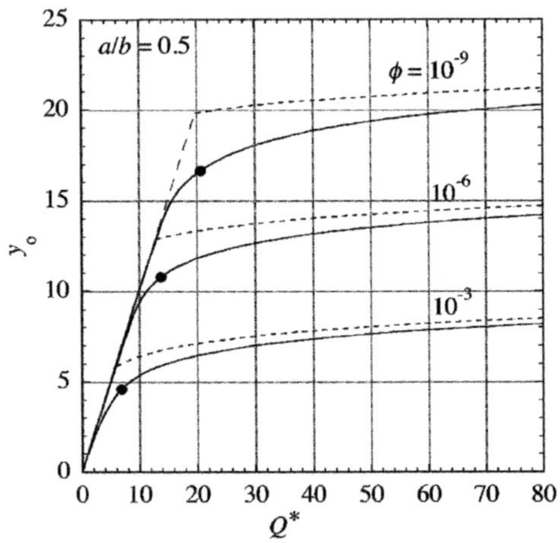

FIGURE 6.7 Scaled surface potential $y_o = ze\psi_o/kT$, defined by $y_o \equiv y(b) - y(R)$, as a function of the scaled charge $Q^* = (ze/kT)Q/4\pi\varepsilon_r\varepsilon_o b$ at $a/b = 0.5$ for various values of the particle volume fraction ϕ. Solid lines represent exact numerical results. The dashed line, which passes through the origin ($y_o = 0$ and $Q^* = 0$), is approximate results calculated by an unscreened Coulomb potential, Eq. (6.155) (low charge case) and dotted lines by Eq. (6.168) (high charge case). Closed circles correspond to the approximate critical values Q^*_{cr} of the scaled particle charge Q^* separating cases, the low and high charge cases, given by $Q^*_{cr} = \ln(1/\phi)$. From Ref. [9].

Eq. (6.168). The dependence of the surface potential y_s upon the particle charge Q^* is considerably suppressed since the counterions are accumulated within and near the polyelectrolyte layer (counter-ion condensation).

For the limiting case of $a = b$ (in which case the polyelectrolyte-coated particle becomes a rigid particle with no polyelectrolyte layer), Eq. (6.168) tends to

$$y_o = \ln\left[\frac{Q^{*2}}{6\phi \ln(1/\phi)}\right] \tag{6.170}$$

which agrees with Eq. (6.50) for the surface potential of a rigid particle in a salt-free medium. For the case where $a \cdot b$ and Q is high, Eq. (6.168) tends to

$$y_o = \ln\left[\frac{Q^*}{\exp(1)\phi \ln(1/\phi)(1 - a^3/b^3)}\right] \tag{6.171}$$

When $a = 0$, in particular, the polyelectrolyte-coated particle becomes a spherical polyelectrolyte with no particle core. In this case, Eq. (6.171) tends to

$$y_s = \ln\left[\frac{Q^*}{\exp(1)\phi \ln(1/\phi)}\right] \tag{6.172}$$

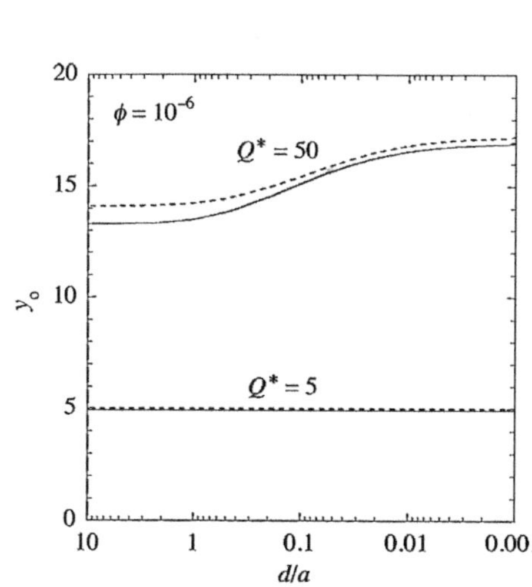

FIGURE 6.8 Scaled surface potential $y_o = ze\psi_o/kT$, defined by $y_o \equiv y(b) - y(R)$, as a function of the ratio d/a of the thickness $d = b - a$ of the polyelectrolyte layer to the radius a of the particle core at $\phi = 10^{-6}$ for $Q^* = 5$ (low charge case) and $Q^* = 50$ (high charge case). Solid lines represent exact numerical results. The dotted lines are approximate results calculated by an unscreened Coulomb potential (Eq. (6.155)) for the low charge case and by Eq. (6.168) for the high charge case. From Ref. [9]

Figure 6.8 shows y_o as a function of the ratio d/a of the polyelectrolyte layer thickness d to the core radius a for two values of Q^* (5 and 50) at $\phi = 10^{-6}$. Note that as d/a tends to zero, the polyelectrolyte-coated particle becomes a hard sphere with no polyelectrolyte layer, while as d/a tends to infinity, the particle becomes a spherical polyelectrolyte with no particle core. Approximate results calculated with Eq. (6.155) for $Q^* = 5$ (low charge case) and Eq. (6.168) for $Q^* = 50$ (high charge case) are also shown in Fig. 6.8. Agreement between exact and approximate results is good. For the low charge case, the surface potential is essentially independent of d and is determined only by the charge amount Q^*. In the example given in Fig. 6.8, for the high charge case, the particle behaves like a hard particle with no polyelectrolyte layer for $d/a \cdot 10^{-3}$ and the particle behaves like a spherical polyelectrolyte for $d/a \cdot 1$.

REFERENCES

1. R. M. Fuoss, A. Katchalsky, and S. Lifson, *Proc. Natl. Acad. Sci. USA* 37 (1951) 579.
2. T. Afrey, P. W. Berg, and H. Morawetz, *J. Polym. Sci.* 7 (1951) 543.
3. N. Imai and F. Oosawa, *Busseiron Kenkyu* 52 (1952) 42; 59 (1953) 99 (in Japanese).
4. F. Oosawa, *Polyelectrolytes*, Dekker, New York, 1971.
5. H. Ohshima, *J. Colloid Interface Sci.* 247 (2002) 18.
6. H. Ohshima, *J. Colloid Interface Sci.* 255 (2002) 202.
7. H. Ohshima, *Theory of Colloid and Interfacial Electric Phenomena*, Elsevier/Academic Press, Amsterdam, 2006.
8. H. Ohshima, *Colloid Polym. Sci.* 282 (2004) 1185.
9. H. Ohshima, *J. Colloid Interface Sci.* 268 (2003) 429.

PART II
Interaction Between Surfaces

7 Electrostatic Interaction of Point Charges in an Inhomogeneous Medium

7.1 INTRODUCTION

The electric properties of biological membranes and proteins depend on the potential distribution and the electrostatic interaction energy of their charges (see e.g., Refs [1–4]). The potential distribution $\psi(\boldsymbol{r})$ at position $\boldsymbol{r}$ around fixed point charges with a distribution $\rho(\boldsymbol{r})$ in a uniform medium of relative permittivity ε_r is described by the Poisson equation,

$$\Delta\psi(\boldsymbol{r}) = -\frac{\rho(\boldsymbol{r})}{\varepsilon_r \varepsilon_o} \tag{7.1}$$

Here we consider fixed charges so that we treat the Poisson equation (not the Poisson-Boltzmann equation). If a single point charge e_1 is located at position $\boldsymbol{r}_1$, then

$$\rho(\boldsymbol{r}) = e_1 \delta(\boldsymbol{r} - \boldsymbol{r}_1) \tag{7.2}$$

where $\delta(\boldsymbol{r})$ is Dirac's delta function, and Eq. (7.2) becomes

$$\Delta\psi(\boldsymbol{r}) = -\frac{e_1 \delta(\boldsymbol{r} - \boldsymbol{r}_1)}{\varepsilon_r \varepsilon_o} \tag{7.3}$$

with the solution

$$\psi(\boldsymbol{r}) = \frac{e_1}{4\pi\varepsilon_r \varepsilon_o |\boldsymbol{r} - \boldsymbol{r}_1|} \tag{7.4}$$

where we have used the following formula:

$$\Delta \frac{1}{|\boldsymbol{r} - \boldsymbol{r}_1|} = -4\pi\delta(\boldsymbol{r} - \boldsymbol{r}_1) \tag{7.5}$$

In this chapter, we generalize the above result to consider the potential created by point charges located in an inhomogeneous media and the interaction energy between these point charges.

7.2 PLANAR GEOMETRY

Consider three different isotropic regions 1–3 separated by planar boundaries at $z = 0$ and $z = d$, where the z-axis is taken to be perpendicular to the boundaries (Fig. 7.1). We suppose that M_i discrete point charges $e_k^{(i)}$ ($i = 1, 2, 3$; $k = 1, 2, \ldots, M_i$) are situated in region i of dielectric permittivity ε_i ($i = 1, 2, 3$). The position $\boldsymbol{r}_k^{(i)}$ of each charge $e_k^{(i)}$ is given by cylindrical coordinates $(\rho_k^{(i)}, \phi_k^{(i)}, \phi_k^{(i)})$. Let Ψ_i be the potential in region i ($i = 1$–3). The Poisson equations for regions 1–3 are given as follows.

For region 1

$$\Delta\Psi_1(\boldsymbol{r}) = -\frac{\rho_1(\boldsymbol{r})}{\varepsilon_r \varepsilon_o} \tag{7.6}$$

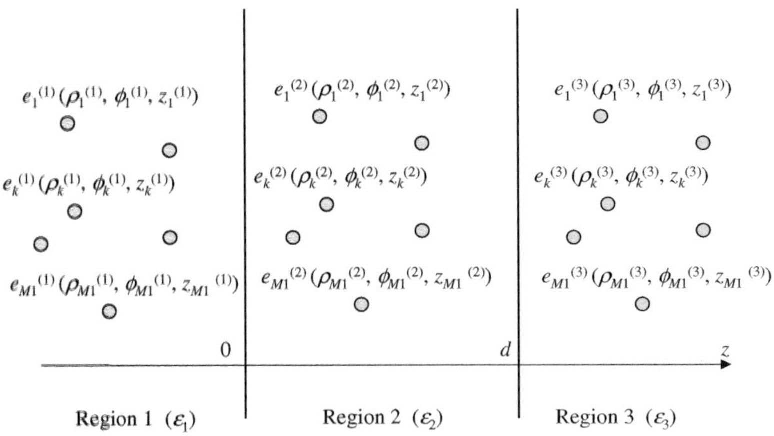

FIGURE 7.1 Three different isotropic regions 1–3 separated by planar boundaries at $z = 0$ and $z = d$, where the z-axis is taken to be perpendicular to the boundaries. There are M_i discrete point charges $e_k^{(i)}$ ($i = 1, 2, 3$; $k = 1, 2, \ldots, M_i$) in region i of dielectric permittivity ε_i ($i = 1, 2, 3$). The position $\boldsymbol{r}_k^{(i)}$ of each charge $e_k^{(i)}$ is given by cylindrical coordinates $(\rho_k^{(i)}, \phi_k^{(i)}, \phi_k^{(i)})$.

with

$$\rho_1(\boldsymbol{r}) = \sum_{k=1}^{M_1} e_k^{(1)} \delta(\boldsymbol{r} - \boldsymbol{r}_k^{(1)}) \tag{7.7}$$

For region 2

$$\Delta\Psi_2(\boldsymbol{r}) = -\frac{\rho_2(\boldsymbol{r})}{\varepsilon_r \varepsilon_o} \tag{7.8}$$

with

$$\rho_2(\boldsymbol{r}) = \sum_{k=1}^{M_2} e_k^{(2)} \delta(\boldsymbol{r} - \boldsymbol{r}_k^{(2)}) \tag{7.9}$$

For region 3

$$\Delta\Psi_3(\boldsymbol{r}) = -\frac{\rho_3(\boldsymbol{r})}{\varepsilon_r \varepsilon_o} \tag{7.10}$$

with

$$\rho_3(\boldsymbol{r}) = \sum_{k=1}^{M_3} e_k^{(3)} \delta(\boldsymbol{r} - \boldsymbol{r}_k^{(3)}) \tag{7.11}$$

The solution Ψ_i to each of the Poisson equations (7.6), (7.8), and (7.10) for region i ($i = 1$–3) can be expressed as the sum of Coulomb potentials created by all the charges $e_k^{(1)}$ ($i = 1$–3) in each region i and the solution ψ_i of the Laplace equations

$$\Delta\psi_i(\boldsymbol{r}) = 0, \quad i = 1, 2, 3 \tag{7.12}$$

The electric potential $\Psi_i(\rho, \phi, z)$ at position $\boldsymbol{r}(\rho, \phi, z)$ in region i ($i = 1$–3) can be expressed in cylindrical coordinates as

$$\Psi_1(\rho, \phi, z) = \sum_{k=1}^{M_1} \frac{e_k^{(1)}}{4\pi\varepsilon_1\varepsilon_o \left| \boldsymbol{r} - \boldsymbol{r}_k^{(1)} \right|} + \psi_1(\rho, \phi, z) \tag{7.13}$$

$$\Psi_2(\rho, \phi, z) = \sum_{k=1}^{M_2} \frac{e_k^{(2)}}{4\pi\varepsilon_2\varepsilon_o \left| \boldsymbol{r} - \boldsymbol{r}_k^{(2)} \right|} + \psi_2(\rho, \phi, z) \tag{7.14}$$

$$\Psi_3(\rho, \phi, z) = \sum_{k=1}^{M_3} \frac{e_k^{(3)}}{4\pi\varepsilon_r\varepsilon_o \left| \boldsymbol{r} - \boldsymbol{r}_k^{(3)} \right|} + \psi_3(\rho, \phi, z) \tag{7.15}$$

where the first and second terms on the right-hand side of each of Eqs. (7.13)–(7.15) are, respectively, the Coulomb potential created by the charge $e_k^{(1)}$ and the second term $\psi_i\,(\rho, \phi, z)$ is due to the electrostatic interaction with other charges. The potential $\psi_i(\rho, \phi, z)$ can be expressed in terms of cylindrical coordinates (ρ, ϕ, z) as

$$\psi_1(\rho, \phi, z) = \int_0^\infty dt \sum_{m=-\infty}^{+\infty} e^{zt} B_m(t) J_m(\rho t) e^{im\phi} \tag{7.16}$$

$$\psi_2(\rho, \phi, z) = \int_0^\infty dt \sum_{m=-\infty}^{+\infty} [e^{zt} B_m(t) + e^{-zt} D_m(t)] J_m(\rho t) e^{im\phi} \tag{7.17}$$

$$\psi_3(\rho, \phi, z) = \int_0^\infty dt \sum_{m=-\infty}^{+\infty} e^{-zt} E_m(t) J_m(\rho t) e^{im\phi} \tag{7.18}$$

where $\left| \boldsymbol{r} - \boldsymbol{r}_k^{(i)} \right|$ is the distance of charge $e_k^{(i)}$ from the point $\boldsymbol{r}$, ψ_i is the contribution to Ψ_i arising from regions other than region i, and $J_m(\rho t)$ is the Bessel function of order m. The functions $B_m(t)$, $C_m(t)$, $D_m(t)$, and $E_m(t)$ are to be determiners by the proper boundary conditions. The first term on the right-hand side of each of Eqs. (7.13)–(7.15) (the Coulomb potential) can further be expanded in Bessel functions as

$$\sum_{k=1}^{M_1} \frac{e_k^{(1)}}{4\pi\varepsilon_r\varepsilon_o \left| \boldsymbol{r} - \boldsymbol{r}_k^{(1)} \right|} = \int_0^\infty dt \sum_{m=-\infty}^{+\infty} e^{-zt} G_m(t) J_m(\rho t) e^{im\phi}, \quad z > z_1^{(1)}, z_2^{(1)}, \ldots, z_{M_1}^{(1)} \tag{7.19}$$

$$\sum_{k=1}^{M_2} \frac{e_k^{(2)}}{4\pi\varepsilon_r\varepsilon_o \left| \boldsymbol{r} - \boldsymbol{r}_k^{(2)} \right|} = \begin{cases} \displaystyle\int_0^\infty dt \sum_{m=-\infty}^{+\infty} e^{zt} H_m^{(1)}(t) J_m(\rho t) e^{im\phi}, & z < z_1^{(2)}, z_2^{(2)}, \ldots, z_{M_2}^{(2)} \\ \displaystyle\int_0^\infty dt \sum_{m=-\infty}^{+\infty} e^{-zt} H_m^{(2)}(t) J_m(\rho t) e^{im\phi}, & z > z_1^{(2)}, z_2^{(2)}, \ldots, z_{M_2}^{(2)} \end{cases} \tag{7.20}$$

$$\sum_{k=1}^{M_3} \frac{e_k^{(3)}}{4\pi\varepsilon_r\varepsilon_o \left| \boldsymbol{r} - \boldsymbol{r}_k^{(3)} \right|} = \int_0^\infty dt \sum_{m=-\infty}^{+\infty} e^{zt} L_m(t) J_m(\rho t) e^{im\phi}, \quad z > z_1^{(3)}, z_2^{(3)}, \ldots, z_{M_3}^{(3)} \tag{7.21}$$

where

$$G_m(t) = \sum_{k=1}^{M_1} \frac{e_k^{(1)}}{4\pi\varepsilon_1\varepsilon_o} e^{z_k^{(1)}t} J_m(\rho_k^{(1)}t) e^{-im\phi_k^{(1)}} \tag{7.22}$$

$$H_m^{(1)}(t) = \sum_{k=1}^{M_2} \frac{e_k^{(2)}}{4\pi\varepsilon_2\varepsilon_o} e^{-z_k^{(2)}t} J_m(\rho_k^{(2)}t) e^{-im\phi_k^{(2)}} \tag{7.23}$$

$$H_m^{(2)}(t) = \sum_{k=1}^{M_2} \frac{e_k^{(2)}}{4\pi\varepsilon_2\varepsilon_o} e^{z_k^{(2)}t} J_m(\rho_k^{(2)}t) e^{-im\phi_k^{(2)}} \tag{7.24}$$

$$L_m(t) = \sum_{k=1}^{M_3} \frac{e_k^{(3)}}{4\pi\varepsilon_1\varepsilon_o} e^{-z_k^{(3)}t} J_m(\rho_k^{(3)}t) e^{-im\phi_k^{(3)}} \tag{7.25}$$

The boundary conditions are

$$\Psi_1(\rho,\phi,0) = \Psi_2(\rho,\phi,0) \tag{7.26}$$

$$\Psi_2(\rho,\phi,d) = \Psi_3(\rho,\phi,d) \tag{7.27}$$

$$\varepsilon_1 \frac{\partial \Psi_1}{\partial z}\bigg|_{z=0} = \varepsilon_2 \frac{\partial \Psi_2}{\partial z}\bigg|_{z=0} \tag{7.28}$$

$$\varepsilon_2 \frac{\partial \Psi_2}{\partial z}\bigg|_{z=d} = \varepsilon_3 \frac{\partial \Psi_3}{\partial z}\bigg|_{z=d} \tag{7.29}$$

Substituting Eqs. (7.13)–(7.15) into Eqs. (7.26)–(7.29), we obtain

$$B_m(t) = \frac{(\Delta_{12} - \Delta_{32}e^{-2dt})G_m(t) + (1-\Delta_{12})H_m^{(1)}(t) - \Delta_{32}e^{-2dt}H_m^{(2)}(t) + (1+\Delta_{32})L_m(t)}{1 - \Delta_{12}\Delta_{32}e^{-2dt}}$$

$$= \left\{ \Delta_{12} + (\Delta_{12}^2 - 1)\Delta_{32} \sum_{n=1}^{\infty} (\Delta_{12}\Delta_{32})^{n-1} e^{-2ndt} \right\} G_m(t)$$

$$+(1-\Delta_{12})\left\{1+\sum_{n=1}^{\infty}(\Delta_{12}\Delta_{32})^n e^{-2ndt}\right\}H_m^{(1)}(t)$$

$$-(1-\Delta_{12})\Delta_{32}\sum_{n=1}^{\infty}(\Delta_{12}\Delta_{32})^{n-1}e^{-2ndt}H_m^{(2)}(t)$$

$$+(1-\Delta_{12})(1+\Delta_{32})\left\{1+\sum_{n=1}^{\infty}(\Delta_{12}\Delta_{32})^n e^{-2ndt}\right\}L_m(t) \quad (7.30)$$

$$C_m(t)=\frac{-(1+\Delta_{12})\Delta_{32}e^{-2dt}G_m(t)+\Delta_{12}\Delta_{32}e^{-2dt}H_m^{(1)}(t)-\Delta_{32}e^{-2dt}H_m^{(2)}(t)+(1+\Delta_{32})L_m(t)}{1-\Delta_{12}\Delta_{32}e^{-2dt}}$$

$$=-(1+\Delta_{12})\Delta_{32}\sum_{n=1}^{\infty}(\Delta_{12}\Delta_{32})^{n-1}e^{-2ndt}G_m(t)$$

$$+\sum_{n=1}^{\infty}(\Delta_{12}\Delta_{32})^n e^{-2ndt}H_m^{(1)}(t)$$

$$-\Delta_{32}\sum_{n=1}^{\infty}(\Delta_{12}\Delta_{32})^{n-1}e^{-2ndt}H_m^{(2)}(t)$$

$$+(1+\Delta_{32})\{1+\sum_{n=1}^{\infty}(\Delta_{12}\Delta_{32})^n e^{-2ndt}\}L_m(t) \quad (7.31)$$

$$D_m(t)=\frac{(1+\Delta_{12})G_m(t)-\Delta_{12}H_m^{(1)}(t)+\Delta_{12}\Delta_{32}e^{-2dt}H_m^{(2)}(t)-(1+\Delta_{32})\Delta_{12}L_m(t)}{1-\Delta_{12}\Delta_{32}e^{-2dt}}$$

$$=(1+\Delta_{12})\left\{1+\sum_{n=1}^{\infty}(\Delta_{12}\Delta_{32})^n e^{-2ndt}\right\}G_m(t)$$

$$-\Delta_{12}\left\{1+\sum_{n=1}^{\infty}(\Delta_{12}\Delta_{32})^n e^{-2ndt}\right\}H_m^{(1)}(t)$$

$$-\sum_{n=1}^{\infty}(\Delta_{12}\Delta_{32})^n e^{-2ndt}H_m^{(2)}(t)$$

$$-\Delta_{12}(1+\Delta_{32})\left\{1+\sum_{n=1}^{\infty}(\Delta_{12}\Delta_{32})^n e^{-2ndt}\right\}L_m(t) \quad (7.32)$$

$$E_m(t)=\frac{(1+\Delta_{12})(1+\Delta_{32})G_m(t)-\Delta_{12}(1-\Delta_{32})H_m^{(1)}(t)+(1-\Delta_{32})H_m^{(2)}(t)-(\Delta_{12}-\Delta_{32}e^{2dt})L_m(t)}{1-\Delta_{12}\Delta_{32}e^{-2dt}}$$

$$=(1+\Delta_{12})(1-\Delta_{32})\left\{1+\sum_{n=1}^{\infty}(\Delta_{12}\Delta_{32})^n e^{-2ndt}\right\}G_m(t)$$

$$-\Delta_{12}(1-\Delta_{32})\left\{1+\sum_{n=1}^{\infty}(\Delta_{12}\Delta_{32})^n e^{-2ndt}\right\}H_m^{(1)}(t)$$

$$+(1-\Delta_{32})\left\{1+\sum_{n=1}^{\infty}(\Delta_{12}\Delta_{32})^n e^{-2ndt}\right\}H_m^{(2)}(t)$$

$$-\left\{\Delta_{12}(1-\Delta_{32}^2)-\Delta_{32}e^{2dt}+\Delta_{12}(1-\Delta_{32}^2)\sum_{n=1}^{\infty}(\Delta_{12}\Delta_{32})^n e^{-2ndt}\right\}L_m(t) \tag{7.33}$$

where

$$\Delta_{ij}=\frac{\varepsilon_i-\varepsilon_j}{\varepsilon_i+\varepsilon_j}=-\Delta_{ji}\ \ (i,j=1,2,3) \tag{7.34}$$

The interaction energy V of charges $e_k^{(i)}$ ($i=1, 2, 3$; $k=1, 2, \ldots, M_i$), which is equal to the work of placing charges $e_k^{(i)}$, can be expressed as

$$V=\frac{1}{2}\sum_{k=1}^{M_1}e_k^{(1)}\psi_1(\rho_k^{(1)},\phi_k^{(1)},z_k^{(1)})+\frac{1}{2}\sum_{k=1}^{M_2}e_k^{(2)}\psi_2(\rho_k^{(2)},\phi_k^{(2)},z_k^{(2)})$$

$$+\frac{1}{2}\sum_{k=1}^{M_3}e_k^{(3)}\psi_3(\rho_k^{(3)},\phi_k^{(3)},z_k^{(3)})+V_{\text{self}} \tag{7.35}$$

where

$$V_{\text{self}}=\frac{1}{2}\sum_{k=1}^{M_1}\sum_{\substack{l=1\\l\neq k}}^{M_1}\frac{e_k^{(1)}e_l^{(1)}}{4\pi\varepsilon_1\varepsilon_o\left|\boldsymbol{r}_k^{(1)}-\boldsymbol{r}_l^{(1)}\right|}$$

$$+\frac{1}{2}\sum_{k=1}^{M_2}\sum_{\substack{l=1\\l\neq k}}^{M_2}\frac{e_k^{(2)}e_l^{(2)}}{4\pi\varepsilon_2\varepsilon_o\left|\boldsymbol{r}_k^{(2)}-\boldsymbol{r}_l^{(2)}\right|}+\frac{1}{2}\sum_{k=1}^{M_3}\sum_{\substack{l=1\\l\neq k}}^{M_3}\frac{e_k^{(3)}e_l^{(3)}}{4\pi\varepsilon_3\varepsilon_o\left|\boldsymbol{r}_k^{(3)}-\boldsymbol{r}_l^{(3)}\right|} \tag{7.36}$$

is the self energy.

We substitute Eqs. (7.30)–(7.33) into Eqs. (7.16)–(7.18) and calculate interaction energy V via Eq. (7.35) using Neuman's additional theorem

$$\sum_{m=-\infty}^{\infty} J_m(\rho_k^{(i)}t)J_m(\rho_l^{(j)}t)\exp(-im(\phi_k^{(i)}-\phi_l^{(j)}))$$
$$= J_0(t\sqrt{(\rho_k^{(i)})^2+(\rho_l^{(j)})^2-2\rho_k^{(i)}\rho_l^{(j)}\cos(\phi_k^{(i)}-\phi_l^{(j)})}) \tag{7.37}$$

and the integral of Lipschitz:

$$\int_0^{\infty} e^{-\alpha t}J_0(\beta t)dt = \frac{1}{\sqrt{\alpha^2+\beta^2}}, \qquad \alpha > 0 \tag{7.38}$$

The result is as follows.

$$V = \frac{1}{8\pi\varepsilon_1\varepsilon_o}\sum_{k=1}^{M_1}\sum_{l=1}^{M_1} e_k^{(1)}e_l^{(1)}\Big[\Delta_{12}f\left(\left|z_k^{(1)}\right|+\left|z_l^{(1)}\right|\right)$$
$$+\Delta_{32}(\Delta_{12}^2-1)\sum_{n=1}^{\infty}(\Delta_{12}\Delta_{32})^{n-1}f\left(\left|z_k^{(1)}\right|+\left|z_l^{(1)}\right|+2nd\right)\Big]$$
$$+\frac{1}{8\pi\varepsilon_3\varepsilon_o}\sum_{k=1}^{M_3}\sum_{l=1}^{M_3} e_k^{(3)}e_l^{(3)}\Big[\Delta_{32}f((z_k^{(3)}-d)+(z_l^{(3)}-d))$$
$$+\Delta_{12}(\Delta_{32}^2-1)\sum_{n=1}^{\infty}(\Delta_{12}\Delta_{32})^{n-1}f((z_k^{(3)}-d)+(z_l^{(3)}-d)+2nd)\Big]$$
$$+\frac{1}{8\pi\varepsilon_2\varepsilon_o}\sum_{k=1}^{M_2}\sum_{l=1}^{M_2} e_k^{(2)}e_l^{(2)}\Big[\sum_{n=0}^{\infty}(\Delta_{12}\Delta_{32})^{n}\Delta_{21}\{f(z_k^{(2)}+z_l^{(2)}+2nd)$$
$$+\Delta_{23}f((d-z_k^{(2)})+(d-z_l^{(2)})+2nd)\}+2\sum_{n=1}^{\infty}(\Delta_{12}\Delta_{23})^{n}f(z_k^{(2)}-z_l^{(2)}+2nd)\Big]$$
$$+\frac{1}{4\pi\varepsilon_2\varepsilon_o}\sum_{k=1}^{M_1}\sum_{l=1}^{M_3} e_k^{(1)}e_l^{(3)}(1-\Delta_{12})(1-\Delta_{32})\sum_{n=0}^{\infty}(\Delta_{12}\Delta_{32})^{n1}f\left(\left|z_k^{(0)}\right|+\left|z_l^{(3)}\right|+2nd\right)$$
$$+\frac{1}{4\pi\varepsilon_1\varepsilon_o}\sum_{k=1}^{M_2}\sum_{l=1}^{M_2} e_k^{(1)}e_l^{(2)}(1+\Delta_{12})\sum_{n=0}^{\infty}(\Delta_{12}\Delta_{32})^{n}\Big[f\left(\left|z_k^{(1)}\right|+\left|z_l^{(2)}\right|+2nd\right)$$
$$+\Delta_{23}f\left((d-\left|z_k^{(1)}\right|)+\left|z-z_l^{(2)}\right|+2nd\right)\Big]$$

$$+\frac{1}{4\pi\varepsilon_3\varepsilon_o}\sum_{k=1}^{M_3}\sum_{l=1}^{M_2}e_k^{(3)}e_l^{(2)}(1+\Delta_{13})\sum_{n=0}^{\infty}(\Delta_{12}\Delta_{32})^n\Big[f(z_k^{(3)}-z_l^{(2)}+2nd)$$

$$+\Delta_{21}f(z_k^{(3)}+z_l^{(2)}+2nd)\Big]+V_{\text{self}} \tag{7.39}$$

where $f(z_k^{(i)}+z_l^{(j)}+2nd)$ is defined by

$$f(z_k^{(i)}+z_l^{(j)}+2nd)=\frac{1}{\sqrt{(\rho_k^{(i)})^2+(\rho_l^{(j)})^2-2\rho_k^{(i)}\rho_l^{(j)}\cos(\phi_k^{(i)}-\phi_l^{(j)})+(z_k^{(i)}+z_l^{(j)}+2nd)^2}} \tag{7.40}$$

Equation (7.39) is the general result for the interaction energy V. We give below the explicit expressions for V for various cases.

(a) *Two regions of different permittivities*

(i) When a charge e_1 is located at position (ρ_1, ϕ_1, z_1) in region 1 of relative permittivity ε_1 (Fig. 7.2), Eq. (7.39) gives

$$V=\frac{\Delta_{12}e_1^2}{16\pi\varepsilon_1\varepsilon_o|z_1|} \tag{7.41}$$

with

$$\Delta_{12}=\frac{\varepsilon_1-\varepsilon_2}{\varepsilon_1+\varepsilon_2} \tag{7.42}$$

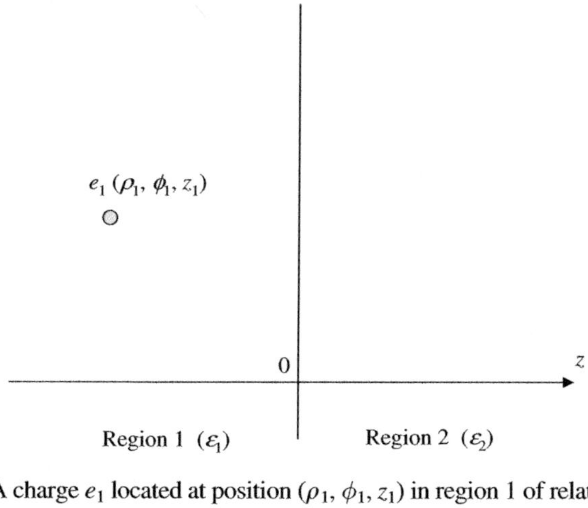

FIGURE 7.2 A charge e_1 located at position (ρ_1, ϕ_1, z_1) in region 1 of relative permittivity ε_1.

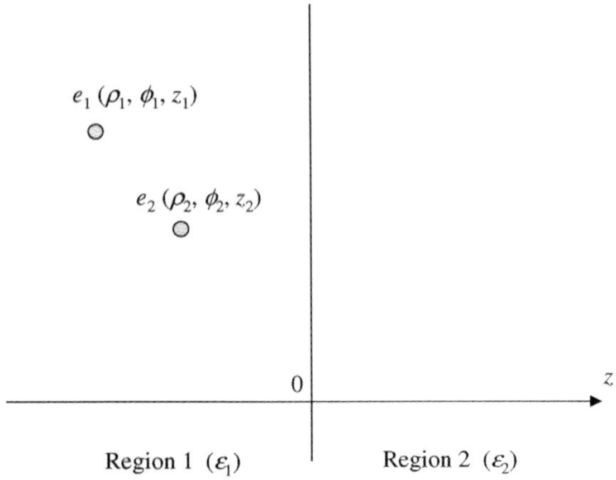

FIGURE 7.3 Two charges e_1 and e_2 located at positions (ρ_1, ϕ_1, z_1) and (ρ_2, ϕ_2, z_2), respectively, both being in region 1 of relative permittivity ε_1.

(ii) When two charges e_1 and e_2 are located at positions (ρ_1, ϕ_1, z_1) and (ρ_2, ϕ_2, z_2), respectively, both being in region 1 of relative permittivity ε_1 (Fig. 7.3), Eq. (7.39) gives

$$V = \frac{1}{16\pi\varepsilon_1\varepsilon_o}\left[\Delta_{12}\left(\frac{e_1^2}{|z_1|} + \frac{e_2^2}{|z_2|} + \frac{4e_1e_2}{R_{10}}\right) + \frac{4e_1e_2}{R_{00}}\right] \tag{7.43}$$

with

$$R_{00} = \sqrt{\rho_1^2 + \rho_2^2 - 2\rho_1\rho_2\cos(\phi_1 - \phi_2) + (z_1 - z_2)^2} \tag{7.44}$$

$$R_{10} = \sqrt{\rho_1^2 + \rho_2^2 - 2\rho_1\rho_2\cos(\phi_1 - \phi_2) + (z_1 + z_2)^2} \tag{7.45}$$

where R_{00} is the distance between two charges e_1 and e_2 and R_{10} is the distance between e_1 and the image of e_2 (= the distance between e_2 and the image of e_1).

(iii) When a charge e_1 is located at position (ρ_1, ϕ_1, z_1) in region 1 of relative permittivity ε_1 and a charge e_2 is located at position (ρ_2, ϕ_2, z_2) in region 2 of relative permittivity ε_2 (Fig. 7.4), Eq. (7.39) gives

$$V = \frac{1}{16\pi\varepsilon_o}\left[\frac{\Delta_{12}e_1^2}{\varepsilon_1|z_1|} + \frac{\Delta_{21}e_2^2}{\varepsilon_2|z_2|} + \left(\frac{8}{\varepsilon_1 + \varepsilon_2}\right)\frac{e_1e_2}{R_{00}}\right] \tag{7.46}$$

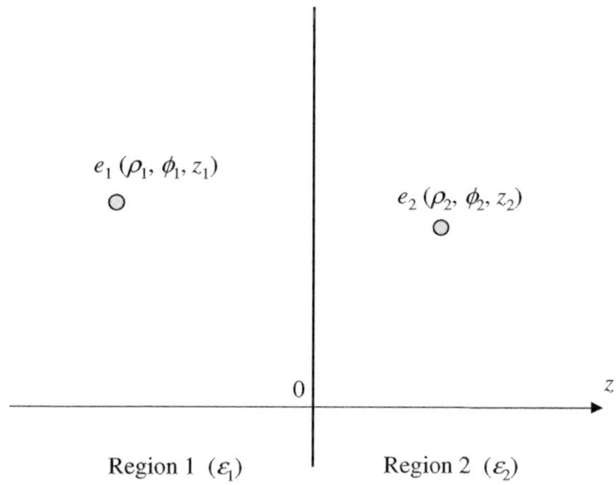

FIGURE 7.4 A charge e_1 at position (ρ_1, ϕ_1, z_1) in region 1 of relative permittivity ε_1 and a charge e_2 located at position (ρ_2, ϕ_2, z_2) in region 2 of relative permittivity ε_2.

with

$$\Delta_{21} = \frac{\varepsilon_2 - \varepsilon_1}{\varepsilon_2 + \varepsilon_1} = -\Delta_{12} \tag{7.47}$$

(b) *Three regions of different permittivities*

(i) When a charge e_1 is located at position (ρ_1, ϕ_1, z_1) in region 1 of relative permittivity ε_1 (Fig. 7.5), Eq. (7.39) gives

$$V = \frac{e_1^2}{16\pi\varepsilon_1\varepsilon_o}\left[\frac{\Delta_{12}}{|z_1|} + \Delta_{32}(\Delta_{12}^2 - 1)\sum_{n=1}^{\infty}\frac{(\Delta_{12}\Delta_{32})^{n-1}}{|z_1| + nd}\right] \tag{7.48}$$

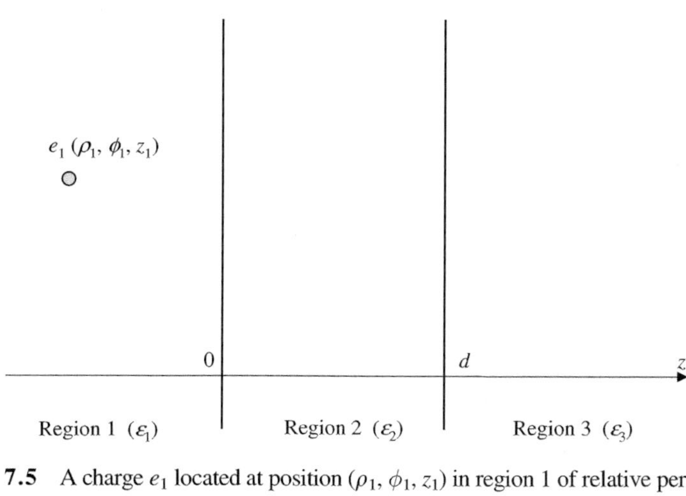

FIGURE 7.5 A charge e_1 located at position (ρ_1, ϕ_1, z_1) in region 1 of relative permittivity ε_1.

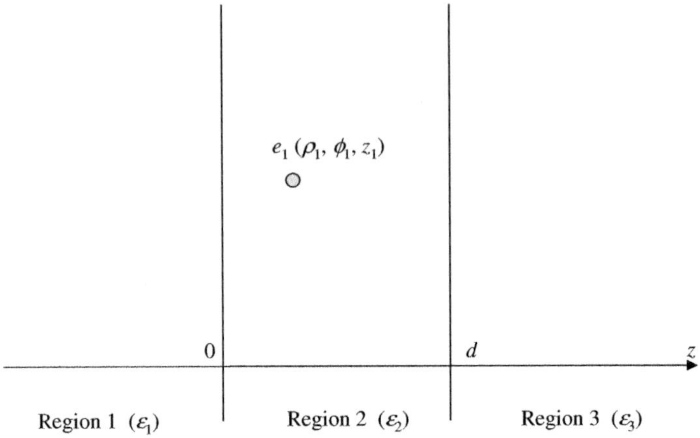

FIGURE 7.6 A charge e_1 located at position (ρ_1, ϕ_1, z_1) in region 2 of relative permittivity ε_2.

with

$$\Delta_{32} = \frac{\varepsilon_3 - \varepsilon_2}{\varepsilon_3 + \varepsilon_2} \tag{7.49}$$

(ii) When a charge e_1 is located at position (ρ_1, ϕ_1, z_1) in region 2 of relative permittivity ε_2 (Fig. 7.6), Eq. (7.39) gives

$$V = \frac{e_1^2}{16\pi\varepsilon_2\varepsilon_o}\left[\frac{\Delta_{21}}{z_1} + \frac{\Delta_{23}}{d - z_1} + \sum_{n=1}^{\infty}(\Delta_{12}\Delta_{32})^n\left\{\frac{\Delta_{21}}{z_1 + nd} + \frac{\Delta_{23}}{(d - z_1) + nd} + \frac{2}{nd}\right\}\right] \tag{7.50}$$

with

$$\Delta_{23} = \frac{\varepsilon_2 - \varepsilon_3}{\varepsilon_2 + \varepsilon_3} = -\Delta_{32} \tag{7.51}$$

(iii) When two charges e_1 and e_2 are located at positions (ρ_1, ϕ_1, z_1) and (ρ_2, ϕ_2, z_2), respectively, both being in region 1 of relative permittivity ε_1 (Fig. 7.7), Eq. (7.39) gives

$$V = \frac{1}{16\pi\varepsilon_1\varepsilon_o}\left[\Delta_{12}\left(\frac{e_1^2}{|z_1|} + \frac{e_2^2}{|z_2|} + \frac{4e_1e_2}{R_{10}}\right)\right.$$
$$\left. + \Delta_{32}(\Delta_{12}^2 - 1)\sum_{n=1}^{\infty}(\Delta_{12}\Delta_{32})^{n-1}\left(\frac{e_1^2}{|z_1| + nd} + \frac{e_2^2}{|z_2| + nd} + \frac{4e_1e_2}{R_{1n}}\right) + \frac{4e_1e_2}{R_{00}}\right] \tag{7.52}$$

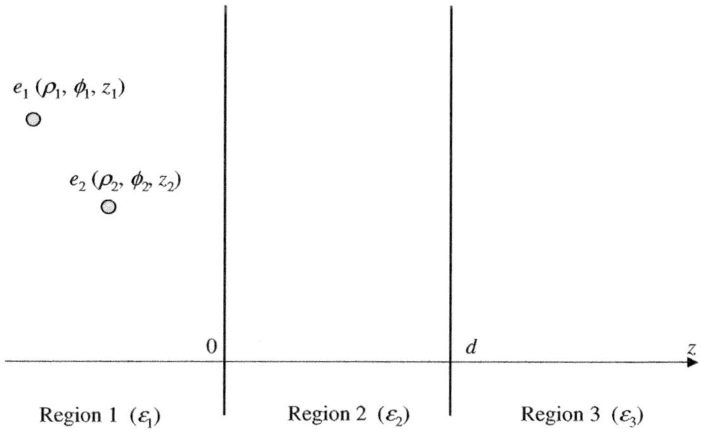

FIGURE 7.7 Two charges e_1 and e_2 located at positions (ρ_1, ϕ_1, z_1) and (ρ_2, ϕ_2, z_2), respectively, both being in region 1 of relative permittivity ε_1.

with

$$R_{1n} = \sqrt{\rho_1^2 + \rho_2^2 - 2\rho_1\rho_2\cos(\phi_1 - \phi_2) + (z_1 + z_2 - 2nd)^2} \qquad (7.53)$$

(iv) When two charges e_1 and e_2 are located at positions (ρ_1, ϕ_1, z_1) and (ρ_2, ϕ_2, z_2), respectively, both being in region 2 of relative permittivity ε_2 (Fig. 7.8), Eq. (7.39) gives

$$V = \frac{1}{16\pi\varepsilon_2\varepsilon_o}\left[e_1^2\left\{\frac{\Delta_{21}}{z_1} + \frac{\Delta_{23}}{d - z_1} + \sum_{n=1}^{\infty}(\Delta_{12}\Delta_{32})^n \times\left(\frac{\Delta_{21}}{z_1 + nd} + \frac{\Delta_{23}}{(d - z_1) + nd} + \frac{2}{nd}\right)\right\} + e_2^2\left\{\frac{\Delta_{21}}{z_2} + \frac{\Delta_{23}}{d - z_2} + \sum_{n=1}^{\infty}(\Delta_{12}\Delta_{32})^n\left(\frac{\Delta_{21}}{z_2 + nd} + \frac{\Delta_{23}}{(d - z_2) + nd} + \frac{2}{nd}\right)\right\} + 4e_1e_2\left\{\frac{1}{R_{00}} + \frac{\Delta_{21}}{R_{20}} + \frac{\Delta_{23}}{R_{30}} + \sum_{n=1}^{\infty}(\Delta_{12}\Delta_{32})^n\left(\frac{\Delta_{21}}{R_{2n}} + \frac{\Delta_{23}}{R_{3n}} + \frac{1}{R_{4n}} + \frac{1}{R_{5n}}\right)\right\}\right] \qquad (7.54)$$

with

$$R_{2n} = \sqrt{\rho_1^2 + \rho_2^2 - 2\rho_1\rho_2\cos(\phi_1 - \phi_2) + (z_1 + z_2 + 2nd)^2} \qquad (7.55)$$

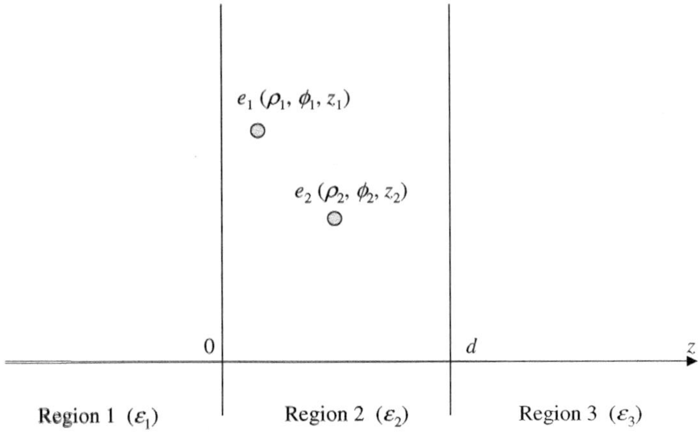

FIGURE 7.8 Two charges e_1 and e_2 located at positions (ρ_1, ϕ_1, z_1) and (ρ_2, ϕ_2, z_2), respectively, both being in region 2 of relative permittivity ε_2.

$$R_{3n} = \sqrt{\rho_1^2 + \rho_2^2 - 2\rho_1\rho_2\cos(\phi_1 - \phi_2) + \{(d_1 - z_1) + (d_1 - z_2) + 2nd\}^2} \tag{7.56}$$

$$R_{4n} = \sqrt{\rho_1^2 + \rho_2^2 - 2\rho_1\rho_2\cos(\phi_1 - \phi_2) + (z_1 - z_2 + 2nd)^2} \tag{7.57}$$

$$R_{5n} = \sqrt{\rho_1^2 + \rho_2^2 - 2\rho_1\rho_2\cos(\phi_1 - \phi_2) + (z_2 - z_1 + 2nd)^2} \tag{7.58}$$

(v) When a charge e_1 is located at position (ρ_1, ϕ_1, z_1) in region 1 of relative permittivity ε_1 and a charge e_2 is located at position (ρ_2, ϕ_2, z_2) in region 2 of relative permittivity ε_2 (Fig. 7.9), Eq. (7.39) gives

$$\begin{aligned} V = \frac{1}{16\pi\varepsilon_o}\Bigg[&\frac{e_1^2}{\varepsilon_1}\left\{\frac{\Delta_{12}}{|z_1|} + \Delta_{32}(\Delta_{12}^2 - 1)\sum_{n=1}^{\infty}\frac{(\Delta_{12}\Delta_{32})^{n-1}}{|z_1| + nd}\right\} \\ &+ \frac{e_2^2}{\varepsilon_2}\left\{\frac{\Delta_{21}}{z_2} + \frac{\Delta_{23}}{d - z_2} + \sum_{n=1}^{\infty}(\Delta_{12}\Delta_{32})^{n-1}\left(\frac{\Delta_{21}}{z_2 + nd} + \frac{\Delta_{23}}{(d - z_2) + nd} + \frac{2}{nd}\right)\right\} \\ &+ \frac{4e_1e_2}{\varepsilon_1}(1 + \Delta_{12})\sum_{n=1}^{\infty}(\Delta_{12}\Delta_{32})^n\left(\frac{\Delta_{23}}{R_{3n}} + \frac{1}{R_{5n}}\right)\Bigg] \end{aligned} \tag{7.59}$$

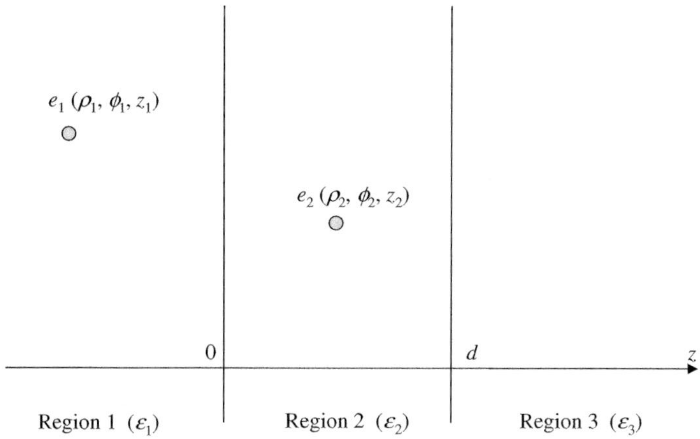

FIGURE 7.9 A charge e_1 located at position (ρ_1, ϕ_1, z_1) in region 1 of relative permittivity ε_1 and a charge e_2 located at position (ρ_2, ϕ_2, z_2) in region 2 of relative permittivity ε_2.

(vi) When a charge e_1 is located at position (ρ_1, ϕ_1, z_1) in region 1 of relative permittivity ε_1 and a charge e_2 is located at position (ρ_2, ϕ_2, z_2) in region 3 of relative permittivity ε_3 (Fig. 7.10), Eq. (7.39) gives

$$V = \frac{1}{16\pi\varepsilon_0}\left[\frac{e_1^2}{\varepsilon_1}\left\{\frac{\Delta_{12}}{|z_1|} + \Delta_{32}(\Delta_{12}^2 - 1)\sum_{n=1}^{\infty}\frac{(\Delta_{12}\Delta_{32})^{n-1}}{|z_1| + nd}\right\}\right.$$
$$+\frac{e_2^2}{\varepsilon_3}\left\{\frac{\Delta_{32}}{z_2 - d} + \Delta_{12}(\Delta_{32}^2 - 1)\sum_{n=1}^{\infty}\frac{(\Delta_{12}\Delta_{32})^{n-1}}{(z_2 - d) + nd}\right\}$$
$$\left.+\frac{e_1 e_2}{\varepsilon_2}(1 - \Delta_{12})(1 - \Delta_{32})\sum_{n=1}^{\infty}\frac{(\Delta_{12}\Delta_{32})^{n}}{R_{5n}}\right] \tag{7.60}$$

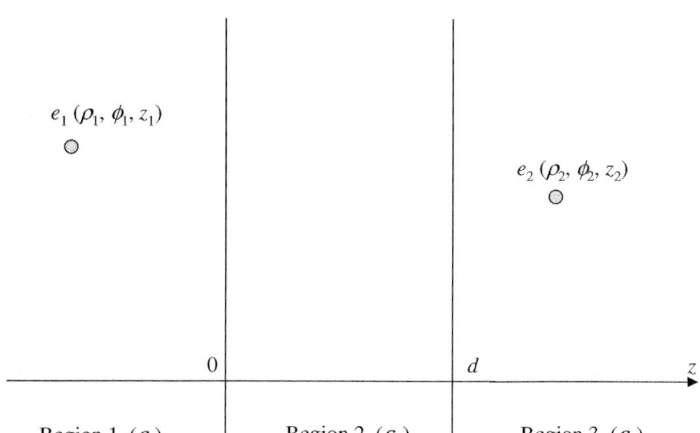

FIGURE 7.10 A charge e_1 located at position (ρ_1, ϕ_1, z_1) in region 1 of relative permittivity ε_1 and a charge e_2 located at position (ρ_2, ϕ_2, z_2) in region 3 of relative permittivity ε_3.

7.3 CYLINDRICAL GEOMETRY

Consider two regions 1 and 2 of different relative permittivites ε_1 and ε_2. Region 1 is the inside of a cylinder of radius a and region 2 is the outside of the cylinder. We take the z-axis to be the axis of the cylinder (Fig. 7.11). We suppose that there exist M_i point charges $e_k^{(i)}$ in region i ($i = 1, 2$; $k = 1, 2, \ldots, M_i$). The position $\boldsymbol{r}_k^{(i)}$ of each charge $e_k^{(i)}$ is given by cylindrical coordinates $(\rho_k^{(i)}, \phi_k^{(i)}, z_k^{(i)})$.

The electric potential $\Psi_i(\rho, \phi, z)$ at position $\boldsymbol{r}(\rho, \phi, z)$ in region i ($i = 1, 2$) can be expressed in cylindrical coordinates as

$$\Psi_1(\rho,\phi,z) = \sum_{k=1}^{M_1} \frac{e_k^{(1)}}{4\pi\varepsilon_1\varepsilon_o\left|\boldsymbol{r}-\boldsymbol{r}_k^{(1)}\right|} + \psi_1(\rho,\phi,z) \tag{7.61}$$

$$\Psi_2(\rho,\phi,z) = \sum_{k=1}^{M_2} \frac{e_k^{(2)}}{4\pi\varepsilon_2\varepsilon_o\left|\boldsymbol{r}-\boldsymbol{r}_k^{(2)}\right|} + \psi_2(\rho,\phi,z) \tag{7.62}$$

with

$$\psi_1(\rho,\phi,z) = \int_{-\infty}^{\infty} dt \sum_{m=-\infty}^{+\infty} e^{izt} B_m(t) J_m(\rho t) e^{im\phi} \tag{7.63}$$

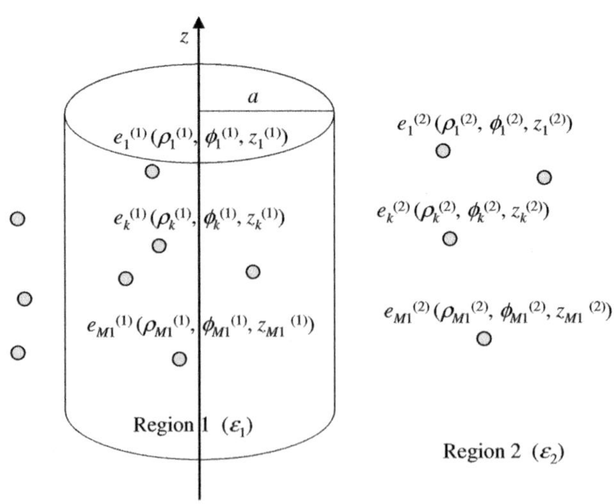

FIGURE 7.11 Two regions 1 and 2 of different relative permittivites ε_1 and ε_2. Region 1 is the inside of a cylinder of radius a and region 2 is the outside of the cylinder. The z-axis is taken to be the axis of the cylinder. There exist M_i point charges $e_k^{(i)}$ in region i ($i = 1, 2$; $k = 1, 2, \ldots, M_i$). The position $\boldsymbol{r}_k^{(i)}$ of each charge $e_k^{(i)}$ is given by cylindrical coordinates $(\rho_k^{(i)}, \phi_k^{(i)}, \phi_k^{(i)})$.

$$\psi_2(\rho, \phi, z) = \int_{-\infty}^{\infty} dt \sum_{m=-\infty}^{+\infty} e^{zt} B_m(t) K_m(\rho t) e^{im\phi} \tag{7.64}$$

where $I_m(z)$ and $K_m(z)$ are the modified Bessel functions of order m, ψ_i is the contribution from region j ($j \neq i$). The functions $B_m(t)$ and $C_m(t)$ are to be determined by the proper boundary conditions. The first term on the right-hand side of each of Eqs. (7.61) and (7.62) (the Coulomb potential) can further be expanded in Bessel functions as

$$\sum_{k=1}^{M_1} \frac{e_k^{(1)}}{4\pi\varepsilon_r\varepsilon_o \left| \boldsymbol{r} - \boldsymbol{r}_k^{(1)} \right|} = \int_{-\infty}^{\infty} dt \sum_{m=-\infty}^{+\infty} e^{izt} G_m(t) K_m(\rho t) e^{im\phi} \tag{7.65}$$

$$\sum_{k=1}^{M_2} \frac{e_k^{(2)}}{4\pi\varepsilon_r\varepsilon_o \left| \boldsymbol{r} - \boldsymbol{r}_k^{(2)} \right|} = \int_{-\infty}^{\infty} dt \sum_{m=-\infty}^{+\infty} e^{izt} H_m(t) I_m(\rho t) e^{im\phi} \tag{7.66}$$

where

$$G_m(t) = \sum_{k=1}^{M_1} \frac{e_k^{(1)}}{4\pi^2 \varepsilon_1 \varepsilon_o} e^{-iz_k^{(1)} t} I_m(\rho_k^{(1)} t) e^{-im\phi_k^{(1)}} \tag{7.67}$$

$$H_m(t) = \sum_{k=1}^{M_2} \frac{e_k^{(1)}}{4\pi^2 \varepsilon_2 \varepsilon_o} e^{-iz_k^{(2)} t} K_m(\rho_k^{(2)} t) e^{-im\phi_k^{(2)}} \tag{7.68}$$

Thus, Eqs. (7.61) and (7.62) become

$$\Psi_2(\rho, \phi, z) = \int_{-\infty}^{\infty} dt \sum_{m=-\infty}^{+\infty} e^{izt} e^{im\phi} [B_m(t) I_m(\rho t) + G_m(t) K_m(\rho t)] \tag{7.69}$$

$$\Psi_2(\rho, \phi, z) = \int_{-\infty}^{\infty} dt \sum_{m=-\infty}^{+\infty} e^{izt} e^{im\phi} [C_m(t) K_m(\rho t) + H_m(t) I_m(\rho t)] \tag{7.70}$$

which are subject to the following boundary conditions at $\rho = a$:

$$\Phi_1(a, \phi, z) = \Phi_2(a, \phi, z) \tag{7.71}$$

$$\varepsilon_1 \frac{\partial \Phi_1}{\partial \rho}\bigg|_{\rho=a} = \varepsilon_2 \frac{\partial \Phi_2}{\partial \rho}\bigg|_{\rho=a} \tag{7.72}$$

Substituting Eqs. (7.69) and (7.70) into Eqs. (7.71) and (7.72), we obtain

$$B_m(t) = -\frac{\Delta_{12} at K_m(at)\{K_{m-1}(at) + K_{m+1}(at)\}G_m(t) + (1-\Delta_{12})H_m(t)}{\Delta_{12} at I_m(at)\{K_{m-1}(at) + K_{m+1}(at)\} - (1+\Delta_{12})} \tag{7.73}$$

$$C_m(t) = \frac{\Delta_{12} at I_m(at)\{I_{m-1}(at) + I_{m+1}(at)\}H_m(t) - (1+\Delta_{12})G_m(t)}{\Delta_{12} at I_m(at)\{K_{m-1}(at) + K_{m+1}(at)\} - (1+\Delta_{12})} \tag{7.74}$$

The interaction energy V of charges $e_k^{(i)}$ ($i = 1, 2$; $k = 1, 2, \ldots, M_i$), which is equal to the work of placing charges $e_k^{(i)}$, can be expressed as

$$V = \frac{1}{2}\sum_{k=1}^{M_1} e_k^{(1)}\psi_1(\rho_k^{(1)}, \phi_k^{(1)}, z_k^{(1)}) + \frac{1}{2}\sum_{k=1}^{M_2} e_k^{(2)}\psi_2(\rho_k^{(2)}, \phi_k^{(2)}, z_k^{(2)}) + V_{\text{self}} \tag{7.75}$$

where

$$V_{\text{self}} = \frac{1}{2}\sum_{k=1}^{M_1}\sum_{\substack{l=1 \\ l\neq k}}^{M_1} \frac{e_k^{(1)} e_l^{(1)}}{4\pi\varepsilon_1\varepsilon_o\left|\boldsymbol{r}_k^{(1)} - \boldsymbol{r}_l^{(1)}\right|} + \frac{1}{2}\sum_{k=1}^{M_2}\sum_{\substack{l=1 \\ l\neq k}}^{M_2} \frac{e_k^{(2)} e_l^{(2)}}{4\pi\varepsilon_2\varepsilon_o\left|\boldsymbol{r}_k^{(2)} - \boldsymbol{r}_l^{(2)}\right|} \tag{7.76}$$

is the self energy.

By substituting Eqs. (7.73) and (7.74) into Eqs. (7.69) and (7.70) and calculating interaction energy V via Eq. (7.75), we obtain

$$\begin{aligned}
V &= \frac{1}{4\pi^2\varepsilon_1\varepsilon_o}\sum_{k=1}^{M_1}\sum_{l=1}^{M_1} e_k^{(1)} e_l^{(1)} \int_0^\infty dt \sum_{m=0}^{\infty} \varepsilon_m \cos(m(\phi_k^{(1)} - \phi_l^{(1)}))\cos(t(z_k^{(1)} - z_l^{(1)})) \\
&\times \frac{\Delta_{12} at K_m(at)\{K_{m-1}(at) + K_{m+1}(at)\} I_m(\rho_k^{(1)} t) I_m(\rho_l^{(1)} t)}{1 + \Delta_{12} - \Delta_{12} at I_m(at)\{K_{m-1}(at) + K_{m+1}(at)\}} \\
&+ \frac{1}{4\pi^2\varepsilon_2\varepsilon_o}\sum_{k=1}^{M_2}\sum_{l=1}^{M_2} e_k^{(2)} e_l^{(2)} \int_0^\infty dt \sum_{m=0}^{\infty} \varepsilon_m \cos(m(\phi_k^{(2)} - \phi_l^{(2)}))\cos(t(z_k^{(2)} - z_l^{(2)})) \\
&\times \frac{\Delta_{21} at I_m(at)\{I_{m-1}(at) + I_{m+1}(at)\} K_m(\rho_k^{(2)} t) K_m(\rho_l^{(2)} t)}{1 + \Delta_{12} - \Delta_{12} at I_m(at)\{K_{m-1}(at) + K_{m+1}(at)\}} \\
&+ \frac{1}{\pi^2(\varepsilon_1 + \varepsilon_2)\varepsilon_o}\sum_{k=1}^{M_2}\sum_{l=1}^{M_2} e_k^{(1)} e_l^{(2)} \int_0^\infty dt \sum_{m=0}^{\infty} \varepsilon_m \cos(m(\phi_k^{(1)} - \phi_l^{(2)}))\cos(t(z_k^{(1)} - z_l^{(2)})) \\
&\times \frac{I_m(\rho_k^{(1)} t) I_m(\rho_l^{(2)} t)}{1 + \Delta_{12} - \Delta_{12} at I_m(at)\{K_{m-1}(at) + K_{m+1}(at)\}} + V_{\text{self}}
\end{aligned} \tag{7.77}$$

where

$$\varepsilon_m = \begin{cases} 2, & m \neq 0 \\ 1, & m = 0 \end{cases} \tag{7.78}$$

Equation (7.77) is the general result for the interaction energy V. We give below the explicit expressions for V for various cases.

(i) When two charges e_1 and e_2 are located at positions (ρ_1, ϕ_1, z_1) and (ρ_2, ϕ_2, z_2), respectively, both being in region 1 (inside the cylinder) of relative permittivity ε_1 (Fig. 7.12), Eq. (7.77) gives

$$V = \frac{1}{4\pi^2\varepsilon_1\varepsilon_o}\int_0^\infty dt \sum_{m=0}^{\infty} \varepsilon_m \frac{F_{2m}(at)}{F_{1m}(at)} \left[e_1^2 I_m^2(\rho_1 t) + e_2^2 I_m^2(\rho_2 t)\right.$$
$$\left. +2e_1e_2\cos(m(\phi_1 - \phi_2))\cos(t(z_1 - z_2))I_m(\rho_1 t)I_m(\rho_2 t)\right] + \frac{e_1 e_2}{4\pi\varepsilon_1\varepsilon_o R_{00}} \tag{7.79}$$

with

$$F_{1m}(at) = 1 + \Delta_{12} - \Delta_{12} at I_m(at)[K_{m-1}(at) + K_{m+1}(at)] \tag{7.80}$$

$$F_{2m}(at) = \Delta_{12} at K_m(at)[K_{m-1}(at) + K_{m+1}(at)] \tag{7.81}$$

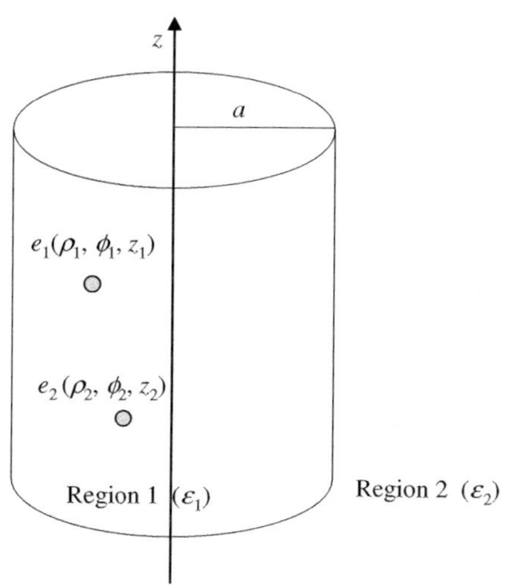

FIGURE 7.12 Two charges e_1 and e_2 located at positions (ρ_1, ϕ_1, z_1) and (ρ_2, ϕ_2, z_2), respectively, both being in region 1 (inside the cylinder) of relative permittivity ε_1.

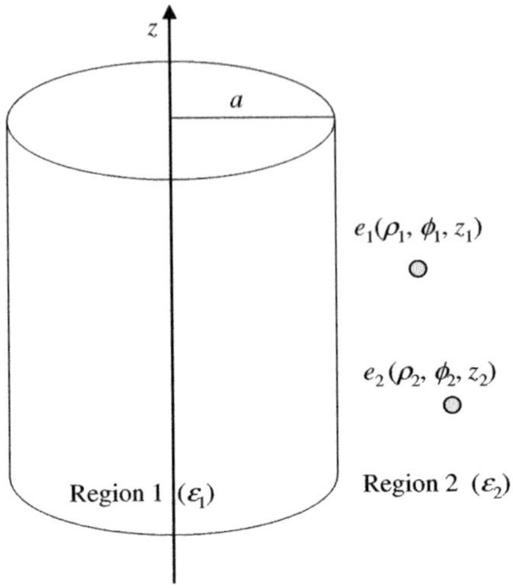

FIGURE 7.13 Two charges e_1 and e_2 located at positions (ρ_1, ϕ_1, z_1) and (ρ_2, ϕ_2, z_2), respectively, both being in region 2 (outside the cylinder) of relative permittivity ε_2.

(ii) When two charges e_1 and e_2 are located at positions (ρ_1, ϕ_1, z_1) and (ρ_2, ϕ_2, z_2), respectively, both being in region 2 (outside the cylinder) of relative permittivity ε_2 (Fig. 7.13), Eq. (7.77) gives

$$V = \frac{1}{4\pi^2\varepsilon_2\varepsilon_o}\int_0^\infty dt \sum_{m=0}^\infty \varepsilon_m \frac{F_{3m}(at)}{F_{1m}(at)}\left[e_1^2K_m^2(\rho_1 t) + e_2^2K_m^2(\rho_2 t)\right. \tag{7.82}$$
$$\left. + 2e_1e_2\cos(m(\phi_1 - \phi_2))\cos(t(z_1 - z_2))K_m(\rho_1 t)K_m(\rho_2 t)\right] + \frac{e_1e_2}{4\pi\varepsilon_2\varepsilon_o R_{00}}$$

with

$$F_{3m}(at) = \Delta_{21} at I_m(at)[I_{m-1}(at) + I_{m+1}(at)] \tag{7.83}$$

(iii) When a charge e_1 is located at position (ρ_1, ϕ_1, z_1) in region 1 (inside the cylinder) of relative permittivity ε_1 and a charge e_2 is located at position (ρ_2, ϕ_2, z_2) in region 2 (outside the cylinder) of relative permittivity ε_2 (Fig. 7.14), Eq. (7.39) gives

$$V = \frac{1}{4\pi^2\varepsilon_o}\int_0^\infty dt \sum_{m=0}^\infty \varepsilon_m \frac{1}{F_{1m}(at)}\left[\frac{e_1^2}{\varepsilon_1}F_{1m}(at)I_m^2(\rho_1 t) + \frac{e_2^2}{\varepsilon_2}F_{3m}(at)K_m^2(\rho_2 t)\right.$$
$$\left. + \frac{4e_1e_2}{\varepsilon_1 + \varepsilon_2}\cos(m(\phi_1 - \phi_2))\cos(t(z_1 - z_2))I_m(\rho_1 t)K_m(\rho_2 t)\right] \tag{7.84}$$

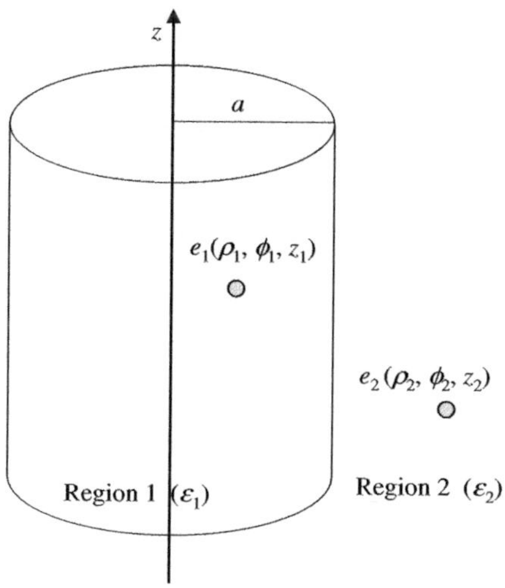

FIGURE 7.14 A charge e_1 located at position (ρ_1, ϕ_1, z_1) in region 1 (inside the cylinder) of relative permittivity ε_1 and a charge e_2 located at position (ρ_2, ϕ_2, z_2) in region 2 (outside the cylinder) of relative permittivity ε_2.

REFERENCES

1. J. G. Kirkwood, *J. Chem. Phys.* 2 (1934) 351.
2. C. Tanford and J. G. Kirkwood, *J. Am. Chem. Soc.* 79 (1957) 5333.
3. J. B. Matthew, *Ann. Rev. Biophys. Biophys. Chem.* 14 (1985) 387.
4. B. H. Honig, W. L. Hubbell, and R. F. Flewelling, *Ann. Rev. Biophys. Biophys. Chem.* 15 (1986) 163.

8 Force and Potential Energy of the Double-Layer Interaction Between Two Charged Colloidal Particles

8.1 INTRODUCTION

When two charged particles are approaching each other in an electrolyte solution, the electrical diffuse double layers around the particles overlap, resulting in electrostatic force between the approaching particles [1–8]. The theory of the double-layer interaction between two charged particles has been developed by Derjaguin and Landau [1] and Verwey and Overbeek [2] and their theory is called the DLVO theory. In this theory, the balance between the electrostatic force and the van der Waals force (see Chapter 19) acting between two particles determines the behavior of a suspension of charged colloidal particles, as will be discussed in Chapter 20. That is, the DLVO theory assumes that the behavior of a colloidal suspension is determined by the interaction potential between two particles, that is, the pair potential. In the DLVO theory it is also essential to assume that the sizes of colloidal particles are very large compared with those of electrolyte ions. Because of this size difference, the rates of motion of the particles are negligibly small compared with those of the electrolyte ions. This makes it possible to regard the motion of the electrolyte ions as being about fixed particles placed at given distances from one another. As a result, the interaction force and energy between the particles are expressed as a function of the interparticle distance. This assumption corresponds to the adiabatic approximation employed in quantum mechanics to describe diatomic molecules. The pair potential is thus the adiabatic pair potential. In this chapter, we give general expressions for the force and potential energy of the electrical double-layer interaction and analytic approximations for the interaction between two parallel plates.

8.2 OSMOTIC PRESSURE AND MAXWELL STRESS

Hydrostatic pressure p_o is acting on a single uncharged particle in a liquid in the absence of electrolyte ions. When the particle is immersed in a liquid containing

Biophysical Chemistry of Biointerfaces By Hiroyuki Ohshima

N ionic species with valence z_i and bulk concentration (number density) n_i^∞ ($i = 1, 2, \ldots, N$), in addition to the hydrostatic pressure p_o, the osmotic pressure is acting on the particle, which is obtained by integrating the osmotic pressure tensor over an arbitrary surface surrounding the particle. The osmotic pressure is given by the bulk osmotic pressure tensor Π_o,

$$\Pi_o = kT \sum_{i=1}^{N} n_i^\infty \tag{8.1}$$

If, further, the particle is charged, the particle charge and electrolyte ions (mainly counterions) form an electrical double layer around the particle, as shown in Chapter 1. The osmotic pressure becomes

$$\begin{aligned} \Pi(\boldsymbol{r}) &= kT \sum_{i=1}^{N} n_i(\boldsymbol{r}) \\ &= kT \sum_{i=1}^{N} n_i^\infty \exp\left(-\frac{z_i e \psi(\boldsymbol{r})}{kT}\right) \end{aligned} \tag{8.2}$$

The electrolyte ions in the electrical double layer thus exert an excess osmotic pressure $\Delta\Pi(\boldsymbol{r})$ on the particle, which is given by

$$\begin{aligned} \Delta\Pi(\boldsymbol{r}) &= \Pi(\boldsymbol{r}) - \Pi_o \\ &= kT \sum_{i=1}^{N} n_i^\infty \left[\exp\left(-\frac{z_i e \psi(\boldsymbol{r})}{kT}\right) - 1\right] \end{aligned} \tag{8.3}$$

At the same time, the coulomb attraction acts between the charges on the particle surface and the counterions within the electrical double layer, which is obtained by integrating the Maxell stress tensor over an arbitrary surface surrounding the particle. The Maxwell stress tensor is given by

$$\mathbf{T}(\boldsymbol{r}) = \varepsilon_r \varepsilon_o \left(\boldsymbol{E}\boldsymbol{E} - \frac{1}{2} E^2 \mathbf{I}\right) \tag{8.4}$$

where $\mathbf{I}$ is the unit tensor and $\boldsymbol{E} = -\,\mathrm{grad}\psi$ is the electrostatic field vector with magnitude E. Thus, the electrical double layer exerts two additional forces on the charged particle: the excess osmotic pressure and the Maxwell stress (Fig. 8.1).

When two charged colloidal particles approach each other, their electrical double layers overlap so that the concentration of counterions in the region between the particles increases, resulting in electrostatic forces between them (Fig. 8.2). There are two methods for calculating the potential energy of the double-layer interaction between two charged colloidal particles [1,2]: In the first method, one directly calculates the interaction force $\boldsymbol{P}$ from the excess osmotic pressure tensor $\Delta\boldsymbol{\Pi}$ and

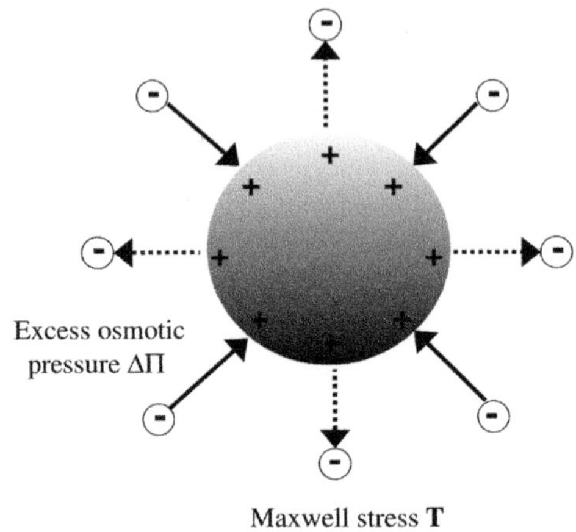

FIGURE 8.1 Electrical double layer around a charged particle exert the excess osmotic pressure $\Delta\Pi$ and the Maxwell stress **T** on the particle.

the Maxwell stress tensor **T** (Section 8.3). The potential energy V of the double--layer interaction is then obtained by integrating the force $\boldsymbol{P}$ with respect to the particle separation. In the second method, the interaction energy V is obtained from the difference between the Helmholtz free energy F of the system of two interacting particles at a given separation and those at infinite separation (Section 8.4).

8.3 DIRECT CALCULATION OF INTERACTION FORCE

The interaction force $\boldsymbol{P}$ can be calculated by integrating the excess osmotic pressure $\Delta\Pi$ and the Maxwell stress tensor **T** over an arbitrary closed surface Σ enclosing either one of the two interacting particles (Fig. 8.3), which is written as [8]

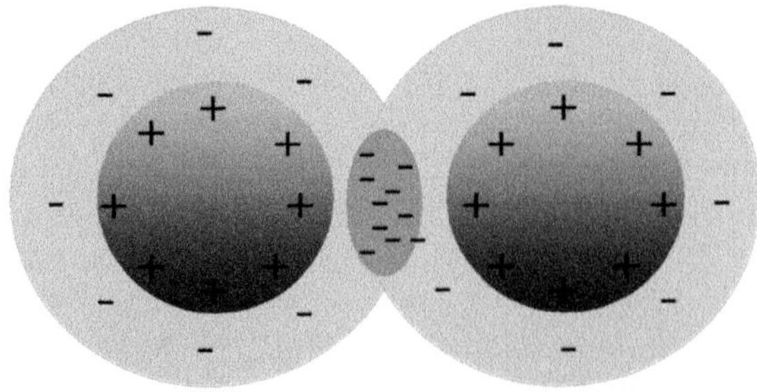

FIGURE 8.2 Overlapping of the electrical double layers around two interacting particles.

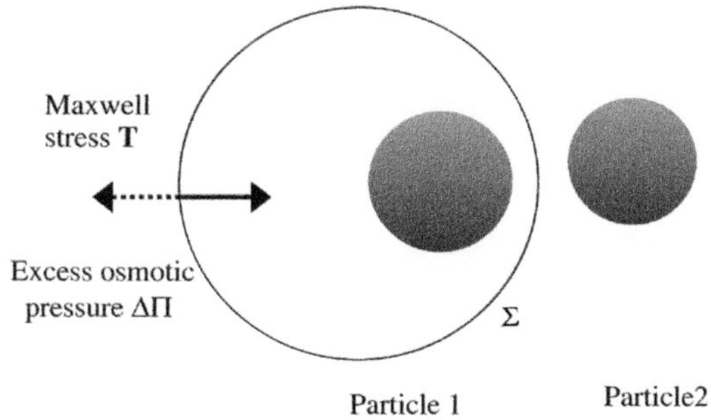

FIGURE 8.3 Calculation of the interaction force between two particles by integrating the excess osmotic pressure $\Delta\Pi$ and the Maxwell stress $\mathbf{T}$ over an arbitrary surface Σ enclosing one of the particles.

$$\boldsymbol{P} = \int_{\Sigma} \left[\left(\Delta\Pi + \frac{1}{2}\varepsilon_r\varepsilon_o E^2 \mathbf{I} \right) \cdot \boldsymbol{n} - \varepsilon_r\varepsilon_o(\boldsymbol{E}\cdot\boldsymbol{n})E \right] dS \tag{8.5}$$

The potential energy V of the double-layer interaction is then obtained by integrating the force $\boldsymbol{P}$ with respect to the particle separation.

Consider several cases.

(i) *Two parallel plates.* Consider two parallel dissimilar plates 1 and 2 of thicknesses d_1 and d_2, respectively, separated by a distance h between their surfaces (Fig. 8.4). We take an x-axis perpendicular to the plates with its origin

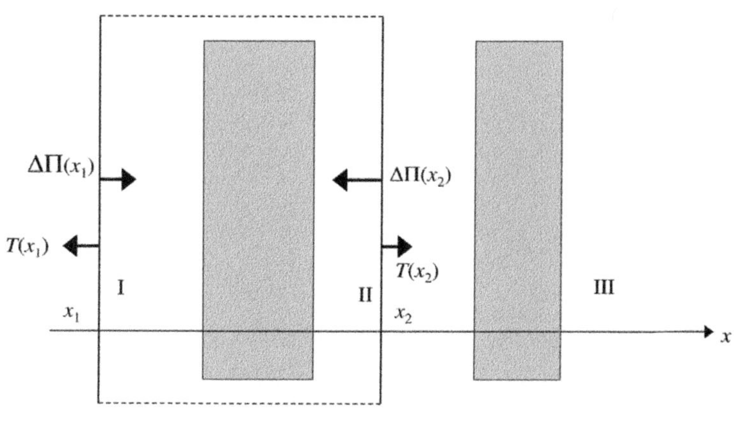

FIGURE 8.4 Two parallel plates 1 and 2 at separation h. The interaction force P between the plates can be calculated from $\{\Delta\Pi(x_2) - T(x_2)\} - \{\Delta\Pi(x_1) - \mathrm{T}(x_1)\}$, where x_1 is an arbitrary point in region I ($-\infty < x \leq 0$) and x_2 is an arbitrary point in region II ($0 \leq x \leq h$).

0 at the surface of plate 1. We denote the regions ($-\infty < x \leq -d_1$), ($0 \leq x \leq h$), and ($h + d_2 \leq x < \infty$) by regions I, II, and III, respectively. As an arbitrary surface Σ enclosing plate 1, we choose a plane $x = x_1$ in region I ($-\infty < x \leq -d_1$) and a plane $x = x_2$ in region III. Here we have assumed that the plates are infinitely large so that end effects may be neglected. The interaction force $P(h)$ per unit area between plates 1 and 2 per unit area can be expressed as

$$\begin{aligned} P(h) &= [T(x_2) + \Delta\Pi(x_2)] - [T(x_1) + \Delta\Pi(x_1)] \\ &= [T(x_2) + \Pi(x_2)] - [T(x_1) + \Pi(x_1)] \end{aligned} \tag{8.6}$$

with

$$T(x) = -\frac{1}{2}\varepsilon_r\varepsilon_o\left(\frac{d\psi}{dx}\right)^2 \tag{8.7}$$

$$\Pi(x) = kT\sum_{i=1}^{N} n_i^{\infty}\exp\left(-\frac{z_i e\psi(x)}{kT}\right) \tag{8.8}$$

$$\Delta\Pi(x) = kT\sum_{i=1}^{N} n_i^{\infty}\left[\exp\left(-\frac{z_i e\psi(x)}{kT}\right) - 1\right] \tag{8.9}$$

where $T(x)$ is the Maxwell stress and $\Delta\Pi(x)$ is the excess osmotic pressure at position x. By substituting Eqs. (8.7) and (8.8) into Eq. (8.6), we obtain

$$\begin{aligned} P(h) = &\left[kT\sum_{i=1}^{N} n_i^{\infty}\exp\left(-\frac{z_i e\psi(x_2)}{kT}\right) - \frac{1}{2}\varepsilon_r\varepsilon_o\left(\left.\frac{d\psi}{dx}\right|_{x=x_2}\right)^2\right] \\ &- \left[kT\sum_{i=1}^{N} n_i^{\infty}\left\{\exp\left(-\frac{z_i e\psi(x_1)}{kT}\right) - 1\right\} - \frac{1}{2}\varepsilon_r\varepsilon_o\left(\left.\frac{d\psi}{dx}\right|_{x=x_1}\right)^2\right] \end{aligned} \tag{8.10}$$

We assume that the potential $\psi(x)$ in the region outside the plates obeys the one-dimensional planar Poisson–Boltzmann equation:

$$\frac{d^2\psi(x)}{dx^2} = -\frac{1}{\varepsilon_r\varepsilon_o}\sum_{i=1}^{N} z_i e n_i^{\infty}\exp\left(-\frac{z_i e\psi(x)}{kT}\right) \tag{8.11}$$

On multiplying $d\psi/dx$ on both sides of Eq. (8.11), we have

$$\frac{d\psi}{dx}\cdot\frac{d^2\psi}{dx^2} = -\frac{1}{\varepsilon_r\varepsilon_o}\sum_{i=1}^{N} z_i e n_i^{\infty} \exp\left(-\frac{z_i e\psi(x)}{kT}\right)\frac{d\psi}{dx} \tag{8.12}$$

which is rewritten as

$$\frac{d}{dx}\left[\frac{1}{2}\varepsilon_r\varepsilon_o\left(\frac{d\psi}{dx}\right)^2\right] = \frac{d}{dx}\left[kT\sum_{i=1}^{N} n_i^{\infty} \exp\left(-\frac{z_i e\psi(x)}{kT}\right)\right] \tag{8.13}$$

By noting Eqs. (8.7) and (8.8), Eq.(8.13) can be rewritten as

$$\frac{d}{dx}[\Pi(x) + T(x)] = 0 \tag{8.14}$$

or

$$\frac{d}{dx}[\Delta\Pi(x) + T(x)] = 0 \tag{8.15}$$

That is,

$$\Delta\Pi(x) + T(x) = C \tag{8.16}$$

or

$$kT\sum_{i=1}^{N} n_i^{\infty}\left[\exp\left(-\frac{z_i e\psi(x)}{kT}\right) - 1\right] - \frac{1}{2}\varepsilon_r\varepsilon_o\left(\frac{d\psi}{dx}\right)^2 = C \tag{8.17}$$

where C is an integration constant independent of x but the value of C in the region between the plates (region II) is different from that in regions I and III. We denote the values of C in regions I, II, and III by C_{I}, C_{II}, and C_{III} $(= C_{\mathrm{I}})$ respectively. We thus see that $P(h)$ is given by

$$P(h) = C_{\mathrm{II}} - C_{\mathrm{I}} \tag{8.18}$$

It is convenient to set $x_1 = -\infty$, since $\psi(x) = d\psi(x)/dx = 0$ at $x = -\infty$ so that $\Delta\Pi(-\infty) = 0$ and $T(-\infty) = 0$ (Fig. 8.5). The left-hand side of Eq. (8.16) (or Eq. (8.17)) becomes zero for region I, namely,

$$C_{\mathrm{I}} = 0 \tag{8.19}$$

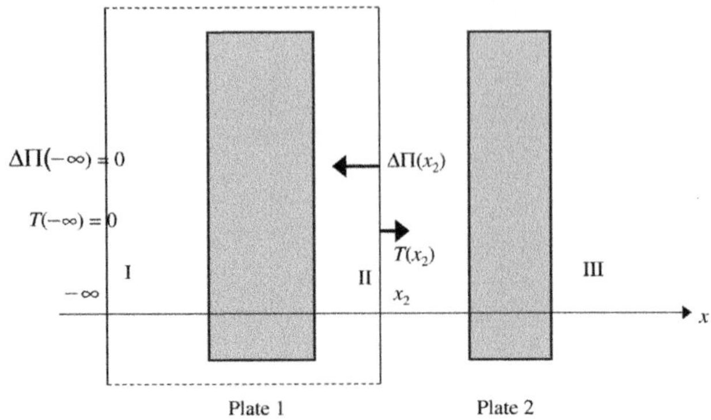

FIGURE 8.5 Two parallel plates 1 and 2 at separation h. If $x_1 = -\infty$, then $\Delta\Pi(x_1) = 0$ and $T(x_1) = 0$.

Consequently, Eq. (8.18) reduces to

$$\begin{aligned} P(h) &= C_{\mathrm{II}} \\ &= \Delta\Pi(x_2) + T(x_2) \\ &= kT\sum_{i=1}^{N} n_i^{\infty}\left[\exp\left(-\frac{z_i e\psi(x_2)}{kT}\right) - 1\right] - \frac{1}{2}\varepsilon_r\varepsilon_o\left(\left.\frac{d\psi}{dx}\right|_{x=x_2}\right)^2 \end{aligned} \tag{8.20}$$

where x_2 is an arbitrary point between plates 1 and 2 (region II).

For low potentials, Eq. (8.9) can be linearized to give

$$\begin{aligned} \Delta\Pi(x) &= kT\sum_{i=1}^{N} n_i^{\infty}\left[\exp\left(-\frac{z_i e\psi(x)}{kT}\right) - 1\right] \\ &= kT\sum_{i=1}^{N} n_i^{\infty}\left[\left\{1 - \frac{z_i e\psi(x)}{kT} + \frac{1}{2}\left(\frac{z_i e\psi(x)}{kT}\right)^2 + \cdots\right\} - 1\right] \\ &= -\sum_{i=1}^{N} n_i^{\infty} z_i e\psi(x) + kT\sum_{i=1}^{N} n_i^{\infty}\frac{1}{2}\left(\frac{z_i e\psi(x)}{kT}\right)^2 + \cdots \end{aligned} \tag{8.21}$$

Here we use the condition of electroneutrality

$$\sum_{i=1}^{N} z_i n_i^{\infty} = 0 \tag{8.22}$$

and introduce the Debye–Hückel parameter

$$\kappa = \left(\frac{1}{\varepsilon_r \varepsilon_o kT} \sum_{i=1}^{N} z_i^2 e^2 n_i^\infty \right)^{1/2} \tag{8.23}$$

Then, Eq. (8.21) reduces to

$$\Delta\Pi(x) = \frac{1}{2} \varepsilon_r \varepsilon_o \kappa^2 \psi^2(x) \tag{8.24}$$

where higher-order terms of $\psi(x)$ has been neglected. Thus, Eq. (8.20) becomes

$$P(h) = \frac{1}{2} \varepsilon_r \varepsilon_o \left[\kappa^2 \psi^2(x_2) - \left(\left. \frac{d\psi}{dx} \right|_{x=x_2} \right)^2 \right] \tag{8.25}$$

For further calculation of $P(h)$, one needs to solve the Poisson–Boltzmann equation (8.11) under appropriate boundary conditions on the plate surfaces to obtain the value of C_{II} or $\psi(x_2)$ at an arbitrary point $x = x_2$ in region II between the plates, as shown in the next chapters.

If we approximate $\psi(x)$ in region II ($0 \leq x \leq h$) as the simple sum of the unperturbed potentials of the two plates (see Eq. (1.25)), namely,

$$\psi(r, \theta) = \psi_{o1} e^{-\kappa x} + \psi_{o2} e^{-\kappa(h - x_1)} \tag{8.26}$$

This is only correct in the limit of large particle separations, since Eq. (8.26) does not always satisfy the boundary conditions on the particles surface. The interaction force $P(h)$ per unit area between two parallel plates at separation h is thus given by

$$P(h) = 2\varepsilon_r \varepsilon_o \kappa^2 \psi_{o1} \psi_{o2} e^{-\kappa h} \tag{8.27}$$

The potential energy $V(h)$ of the double-layer interaction per unit area between two parallel plates at separation h is given by

$$\begin{aligned} V(h) &= -\int_\infty^h P(h) dh \\ &= 2\varepsilon_r \varepsilon_o \kappa \psi_{o1} \psi_{o2} e^{-\kappa h} \end{aligned} \tag{8.28}$$

Equations (8.27) and (8.28) are correct in the limit of large separations.

(ii) *Two spheres.* Consider two interacting dissimilar spheres of radii a_1 and a_2 at separation R between their centers (Fig. 8.6). As an arbitrary surface Σ enclosing sphere 1, we choose the surface of sphere 1. We consider the

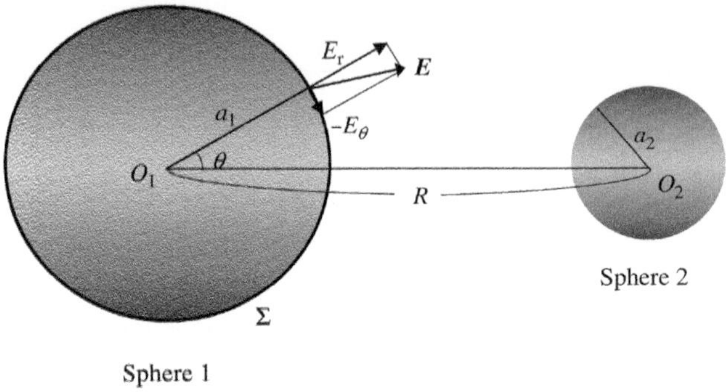

FIGURE 8.6 Two spheres 1 and 2 at separation R.

low potential case and use the low potential form of Eq. (8.3)

$$\begin{aligned}\Delta\Pi(\boldsymbol{r}) &= \Pi(\boldsymbol{r}) - \Pi_o \\ &= \frac{1}{2}\varepsilon_r\varepsilon_o\kappa^2\psi^2\end{aligned} \tag{8.29}$$

It follows from Eq.(8.5) that the interaction force f is given by

$$P = \int_0^\pi \left[\left(\Delta\Pi + \frac{1}{2}\varepsilon_r\varepsilon_o(E_r^2 + E_\theta^2)\right)\cos\theta - \varepsilon_r\varepsilon_o E_r(E_r\cos\theta - E_\theta\sin\theta)\right]_{r=a_1} \times 2\pi a_1^2 \sin\theta\, d\theta \tag{8.30}$$

with

$$\Delta\Pi = \frac{1}{2}\varepsilon_r\varepsilon_o\kappa^2\psi^2 \tag{8.31}$$

$$E_r(r,\theta) = -\frac{\partial\psi(r,\theta)}{\partial r} \tag{8.32}$$

$$E_\theta(r,\theta) = -\frac{1}{r}\frac{\partial\psi(r,\theta)}{\partial\theta} \tag{8.33}$$

where $P > 0$ corresponds to repulsion and $P < 0$ to attraction. The potential distribution is derived by solving the spherical Poisson–Boltzmann equation. Here we consider the simple case where the potential is assumed to be simply given by the sum of the unperturbed potentials of

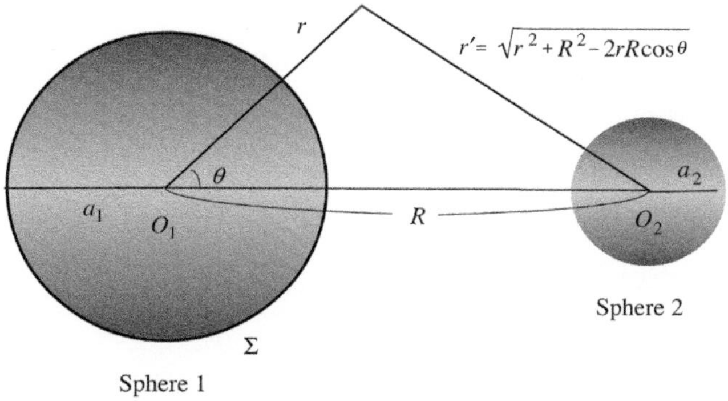

FIGURE 8.7 Spheres 1 and 2. r is the distance from the center O_1 of sphere 1 and $r' = \sqrt{r^2 + R^2 - 2rR\cos\theta}$ is the distance from the center O_2 of sphere 2.

the two spheres (Eq. (1.72)) (see Fig. 8.7).

$$\psi(r,\theta) = \psi_{o1}\frac{a_1}{r}e^{-\kappa(r-a_1)} + \psi_{o1}\frac{a_1}{r'}e^{-\kappa(r'-a_1)} \tag{8.34}$$

with

$$r' = \sqrt{r^2 + R^2 - 2rR\cos\theta} \tag{8.35}$$

where ψ_{o1} and ψ_{o2} are, respectively, the unperturbed surface potentials of spheres 1 and 2, respectively. Equation (8.34) is only correct in the limit of large particle separations, since Eq. (8.34) does not always satisfy the boundary conditions on the particles surface. The interaction force $P(R)$ between two spheres 1 and 2 at separation R is thus given by

$$P(R) = \frac{4\pi\varepsilon_r\varepsilon_o\psi_{o1}\psi_{o2}a_1a_2}{R^2}(1+\kappa R)e^{-\kappa(R-a_1-a_2)} \tag{8.36}$$

The potential energy $V(R)$ of the double-layer interaction between two spheres at separation R is given by

$$\begin{aligned} V(R) &= -\int_\infty^R P(R)dR \\ &= \frac{4\pi\varepsilon_r\varepsilon_o\psi_{o1}\psi_{o2}a_1a_2}{R}e^{-\kappa(R-a_1-a_2)} \end{aligned} \tag{8.37}$$

Equations (8.36) and (8.37) are correct in the limit of large separations.

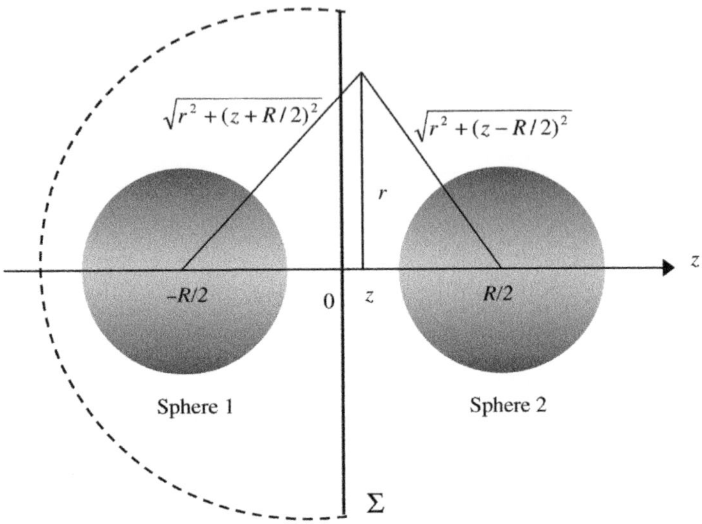

FIGURE 8.8 Two identical spheres 1 and 2 with the intermediate plane Σ.The integral taken on the surface represented by a dashed line (far from the sphere 1) may be neglected.

For two identical spheres 1 and 2 of radius a carrying unperturbed surface potential ψ_o separated by R, one may choose the intermediate plane at $z = 0$ between the spheres as an arbitrary plane enclosing sphere 1 (Fig. 8.8). Here we use the cylindrical coordinate system (r, z) and take the z-axis to be the axis connecting the centers of the spheres and r to be the radial distance from the z-axis. Equation (8.34) can be rewritten by using the cylindrical coordinate (r, z) as

$$\psi(r,z) = \psi_o \frac{a}{\sqrt{r^2+(z+R/2)^2}} \exp\left(-\kappa\left[\sqrt{r^2+(z+R/2)^2}-a\right]\right) + \psi_o \frac{a}{\sqrt{r^2+(z-R/2)^2}} \exp\left(-\kappa\left[\sqrt{r^2+(z-R/2)^2}-a\right]\right) \tag{8.38}$$

and Eq. (8.30) as

$$P = \int_0^\infty \left[\Delta\Pi + \frac{1}{2}\varepsilon_r\varepsilon_o\left\{\left(\frac{\partial\psi}{\partial r}\right)^2 - \left(\frac{\partial\psi}{\partial z}\right)^2\right\}\right]_{z=0} 2\pi r dr$$
$$= \frac{1}{2}\varepsilon_r\varepsilon_o \int_0^\infty \left[\kappa^2\psi^2(r,0) + \left\{\left(\frac{\partial\psi}{\partial r}\right)^2 - \left(\frac{\partial\psi}{\partial z}\right)^2\right\}\right]_{z=0} 2\pi r dr \tag{8.39}$$

for the low potential case. By substituting Eq. (8.38) into Eq. (8.39), we obtain

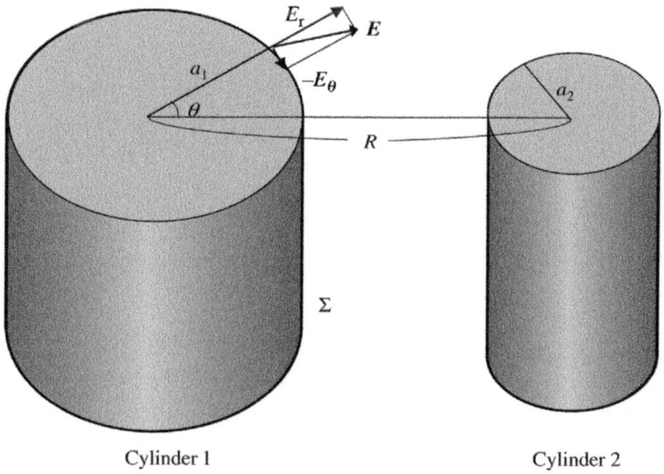

FIGURE 8.9 Two cylinders at separation R.

$$P(R) = \frac{4\pi\varepsilon_r\varepsilon_o\psi_o^2 a^2}{R^2}(1+\kappa R)e^{-\kappa(R-2a)} \tag{8.40}$$

(iii) *Two cylinders.* Consider two interacting infinitely long dissimilar cylinders of radii a_1 and a_2 at separation R between their axes. As an arbitrary surface Σ enclosing cylinder 1, we choose the surface of cylinder 1 (Fig. 8.9). We consider the low potential case and use the cylindrical coordinate system (r, z), where the z-axis is taken to be parallel to the cylinder axes and r is the distance measured from the cylinder axis. In this case, the interaction force $P(R)$ per unit length becomes

$$P = \int_0^{2\pi} \left[\left(\Delta\Pi + \frac{1}{2}\varepsilon_r\varepsilon_o(E_r^2 + E_\theta^2)\right)\cos\theta - \varepsilon_r\varepsilon_o E_r(E_r\cos\theta - E_\theta\sin\theta)\right]_{r=a_1} \cdot a_1 d\theta \tag{8.41}$$

with

$$\Delta\Pi = \frac{1}{2}\varepsilon_r\varepsilon_o\kappa^2\psi^2 \tag{8.42}$$

$$E_r(r,\theta) = -\frac{\partial\psi(r,\theta)}{\partial r} \tag{8.43}$$

$$E_\theta(r,\theta) = -\frac{1}{r}\frac{\partial\psi(r,\theta)}{\partial\theta} \tag{8.44}$$

where $P > 0$ corresponds to repulsion and $P < 0$ to attraction. We consider the simple case where the potential is assumed to be simply given by the sum of the unperturbed potentials of the two cylinders (Eq. (1.143))

$$\psi(r, \theta) = \psi_{o1} \frac{K_0(\kappa r)}{K_0(\kappa a_1)} + \psi_{o2} \frac{K_0(\kappa r')}{K_0(\kappa a_2)} \tag{8.45}$$

$$r' = \sqrt{r^2 + R^2 - 2rR\cos\theta} \tag{8.46}$$

where ψ_{o1} and ψ_{o2} are, respectively, the unperturbed surface potentials of cylinders 1 and 2, respectively. $K_n(z)$ is the modified Bessel function of the second kind of order n. Equation (8.45) is only correct in the limit of large particle separations, since Eq. (8.45) does not always satisfy the boundary conditions on the particles surface. The interaction force $P(R)$ per unit length between two spheres 1 and 2 at separation R is thus given by

$$P(R) = 2\pi\varepsilon_r\varepsilon_o\kappa\psi_{o1}\psi_{o2} \frac{K_1(\kappa R)}{K_0(\kappa a_1)K_0(\kappa a_2)} \tag{8.47}$$

The potential energy $V(R)$ of the double-layer interaction per unit length between two cylinders at separation R is given by

$$\begin{aligned} V(R) &= -\int_\infty^R P(R)dR \\ &= 2\pi\varepsilon_r\varepsilon_o\psi_{o1}\psi_{o2} \frac{K_0(\kappa R)}{K_0(\kappa a_1)K_0(\kappa a_2)} \end{aligned} \tag{8.48}$$

Equations (8.47) and (8.48) are correct in the limit of large separations.

8.4 FREE ENERGY OF DOUBLE-LAYER INTERACTION

In this section, we calculate the interaction energy V from the difference between the Helmholtz free energy F of the system of two interacting particles at a given separation and those at infinite separation (Fig. 8.10), namely,

$$V = F - F(\infty) \tag{8.49}$$

The form of the Helmholtz free energy F depends on the type of the origin of surface charges on the interacting particles. The following two types of interaction, that is (i) interaction at constant surface charge density and (ii) interaction at constants surface potential are most frequently considered. We denote the free energy F for the constant surface potential case by F^{ψ} and that for the constant surface

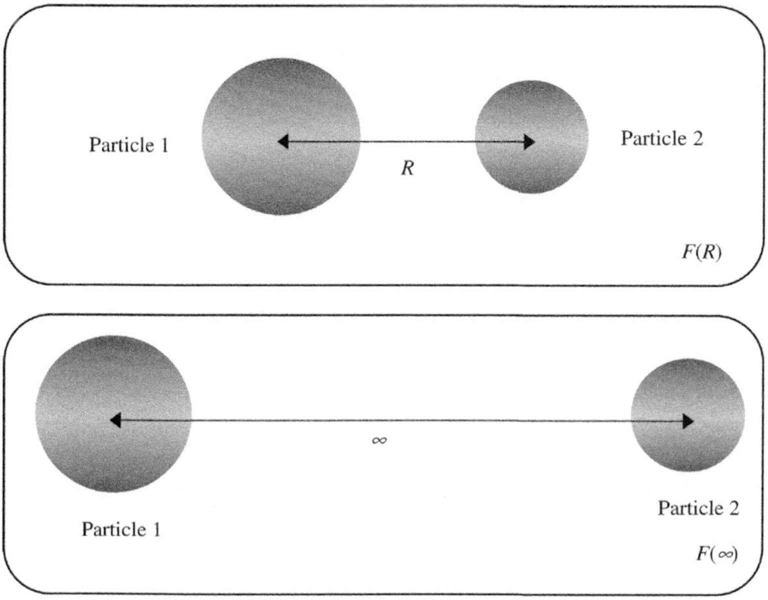

FIGURE 8.10 Difference between the free energy $F(R)$ of two interacting particles at separation R and that for infinite separation gives the potential energy $V(R)$ of the double--layer interaction between the particles.

charge density case by F^{σ}. The expression for the interaction force (Eq. (8.5)), on the other hand, does not depend on the type of the double-layer interaction.

8.4.1 Interaction at Constant Surface Charge Density

The first case corresponds to the situation in which the surface charge densities of the interacting particles 1 and 2 remain constant during interaction. If the particle surface charge is caused by dissociation of ionizable groups, and further, if the dissociation is complete, then the free energy F^{σ} for the interaction between two particles 1 and 2 with constant surface charge densities σ_1 and σ_2, respectively, can be given by the electric part F_{el} of the double-layer free energy (Eq. (5.48)), namely,

$$F^{\sigma} = F_{\mathrm{el}} = \int_{S_1}\int_0^{\sigma_1} \psi_1(\sigma')d\sigma' dS_1 + \int_{S_2}\int_0^{\sigma_2} \psi_2(\sigma')d\sigma' dS_2 \tag{8.50}$$

where surface integration is carried out over the surface S_i of particle i ($i = 1, 2$) and integration with respect to σ' is the electric work of charging the surface of each particle and $\psi_i(\sigma')$ is the surface potential of particle i at a stage at which the surface charge density is σ' during the charging process. For the low potential case where $\psi(\sigma')$ is proportional to σ',

$$\int_0^{\sigma} \psi_i(\sigma')d\sigma' = \frac{1}{2}\sigma_i\psi_{oi} \quad (i = 1, 2) \tag{8.51}$$

and thus Eq. (8.50) reduces to

$$F^{\sigma} = \frac{1}{2}\sigma_1 \int_{S_1} \psi_{o1} dS_1 + \frac{1}{2}\sigma_2 \int_{S_2} \psi_{o2} dS_2 \tag{8.52}$$

Here ψ_{oi} is the actual surface potential of particle i (i.e., the surface potential at the final stage) and may differ at different points on the particle surfaces.

8.4.2 Interaction at Constants Surface Potential

In the second case, the surface potentials of the interacting particles remain constant during interaction. Consider two interacting particles 1 and 2 whose surface charges are due to adsorption of N_i ions (potential-determining ions) of valence Z adsorb onto the surface of particle i ($i = 1$, 2). If the configurational entropy S_c of the adsorbed ions does not depend on N_i, then the surface potential ψ_{oi} of particle i is given by Eq. (5.10), namely,

$$\psi_{oi} = \frac{kT}{Ze} \ln\left(\frac{n}{n_{oi}}\right) \tag{8.53}$$

where n_{oi} is the value of the concentration n of potential-determining ions at which ψ_{oi} becomes zero, and the double-layer free energy is given by Eq. (5.13).

The free energy F^{ψ} of two interacting particles carrying constant surface potentials ψ_{o1} and ψ_{o2} can thus be expressed by

$$\begin{aligned} F^{\psi} &= F_{\text{el}} - \int_{S_1} \sigma_1\psi_{o1} dS_1 - \int_{S_2} \sigma_2\psi_{o2} dS \\ &= \int_{S_1} \left\{ \int_0^{\sigma_1} \psi_1(\sigma'_1)d\sigma'_1 - \sigma_1\psi_{o1} \right\} dS_1 + \int_{S_2} \left\{ \int_0^{\sigma_2} \psi_2(\sigma'_2)d\sigma'_2 - \sigma_2\psi_{o2} \right\} dS_2 \\ &= -\int_{S_1}\int_0^{\psi_{o1}} \sigma_1(\psi')d\psi' dS_1 - \int_{S_2}\int_0^{\psi_{o2}} \sigma_2(\psi')d\psi' dS_2 \end{aligned} \tag{8.54}$$

where surface integration is carried out over the surface S_i of particle i ($i = 1$, 2) and $\sigma_i(\psi')$ is the surface charge density of particle i at a stage at which the surface potential is ψ' during the charging process.

For the low potential case, Eq. (8.54) reduces to

$$F^{\psi} = -\frac{1}{2}\psi_{1o} \int_{S_1} \sigma_1 dS_1 - \frac{1}{2}\psi_{2o} \int_{S_2} \sigma_2 dS_2 \tag{8.55}$$

where σ_{oi} is the actual surface charge density of particle i (i.e., the surface charge density at the final stage) and may differ at different points on the particle surfaces. If the thermodynamic equilibrium with respect to adsorption of potential-determining ions does not attain, the constant surface charge model (instead of the constant surface potential model) should be used [9, 10].

8.5 ALTERNATIVE EXPRESSION FOR THE ELECTRIC PART OF THE FREE ENERGY OF DOUBLE-LAYER INTERACTION

The electric part F_{el} of the free energy of double-layer interaction (Eq. (5.4)) can be expressed in a different way on the basis of the Debye charging process in which all the ions and the particles are charged simultaneously from zero to their full charge. Let λ be a parameter that expresses a stage of the charging process and varies from 0 to 1. Then F_{el} can be expressed by

$$F_{el} = 2\int_0^{\lambda} \frac{1}{\lambda} E(\lambda) d\lambda \tag{8.56}$$

with

$$E(\lambda) = \frac{1}{2}\int_V \rho(\lambda)\psi(\lambda)dV + \frac{1}{2}\int_{S_1} \lambda\sigma_1\psi(\lambda)dS_1 + \frac{1}{2}\psi_{2o}\int_{S_2} \lambda\sigma_2\psi(\lambda)dS_2 \tag{8.57}$$

where $E(\lambda)$, $\rho(\lambda)$, and $\psi(\lambda)$ are, respectively, the internal energy, the volume charge density and the electric potential at stage λ in the charging process, and volume integration is carried out over both regions outside and inside the particles.

8.6 CHARGE REGULATION MODEL

If the dissociation of the ionizable groups on the particle surface is not complete, or the configurational entropy S_c of adsorbed potential-determining ions depends on N, then neither of ψ_o nor of σ remain constant during interaction. This type of double--layer interaction is called charge regulation model. In this model, we should use Eqs. (8.35) and (5.44) for the double-layer free energy [11–13].

Namely, if there are N_{maxi} binding sites for ions of valence Z on the surface of particle i ($i = 1, 2$) and we assume 1:1 binding of the Langmuir type, then the double-layer free energy is given by

$$\begin{aligned} F = &-\int_{S_1}\int_0^{\psi_{o1}} \sigma(\psi'_{o1})d\psi'_{o1}dS_1 - \int_{S_1} N_{max1}kT\ln(1 + K_{a1}ne^{-y_{o1}})dS_1 \\ &-\int_{S_2}\int_0^{\psi_{o2}} \sigma(\psi'_{o2})d\psi'_{o2}dS_2 - \int_{S_2} N_{max2}kT\ln(1 + K_{a2}ne^{-y_{o2}})dS_2 \end{aligned} \tag{8.58}$$

with

$$y_{oi} = \frac{Ze\psi_{oi}}{kT} \quad (i = 1, 2) \tag{8.59}$$

where ψ_{oi} and K_{ai} are, respectively, the surface potential and the adsorption constant of ions onto the surface of particle i ($i = 1, 2$).

If, on the other hand, there are $N_{\max i}$ dissociable sites on the surface of particle i, then the double-layer free energy is given by

$$\begin{aligned} F = &-\int_{S_1}\int_0^{\psi_{o1}} \sigma(\psi'_{o1})d\psi'_{o1}dS_1 - \int_{S_1} N_{\max 1}kT \ln\left(1 + \frac{n_{\mathrm{H}}e^{-y_{o1}}}{K_{d1}}\right)dS_1 \\ &-\int_{S_2}\int_0^{\psi_{o2}} \sigma(\psi'_{o2})d\psi'_{o2}dS_2 - \int_{S_2} N_{\max 2}kT \ln\left(1 + \frac{n_{\mathrm{H}}e^{-y_{o1}}}{K_{d1}}\right)dS_2 \end{aligned} \tag{8.60}$$

$$y_{oi} = \frac{e\psi_{oi}}{kT} \quad (i = 1, 2) \tag{8.61}$$

where ψ_{oi} and K_{di} are, respectively, the surface potential and the dissociation constant of dissociable groups on the surface of particle i ($i = 1, 2$).

REFERENCES

1. B. V. Derjaguin and L. D. Landau, *Acta Physicochim.* 14 (1941) 633.
2. E. J. W. Verwey and J. Th. G. Overbeek, *Theory of the Stability of Lyophobic Colloids*, Elsevier/Academic Press, Amsterdam, 1948.
3. B. V. Derjaguin, *Theory of Stability of Colloids and Thin Films*, Consultants Bureau, New York, London, 1989.
4. J. N. Israelachvili, *Intermolecular and Surface Forces, 2nd edition*, Academic Press, New York, 1992.
5. J. Lyklema, *Fundamentals of Interface and Colloid Science, Solid–Liquid Interfaces*, Vol. 2, Academic Press, New York, 1995.
6. H. Ohshima,in: H. Ohshima and K. Furusawa(Eds.), *Electrical Phenomena at Interfaces, Fundamentals, Measurements, and Applications, 2nd edition*, revised and expanded, Dekker, New York, 1998, Chapter 3.
7. H. Ohshima, *Theory of Colloid and Interfacial Electric Phenomena*, Elsevier/Academic Press, Amsterdam, 1968.
8. N. E. Hoskin and S. Levine, *Phil. Trans. Roy. Soc. London A* 248 (1956) 499.
9. G. Frens,Thesis, The reversibility of irreversible colloids, Utrecht, 1968.
10. G. Frens and J. Th. G. Overbeek, *J. Colloid Interface Sci.* 38 (1972) 376.
11. B. W. Ninham and V. A. Parsegian, *J. Theor. Biol.* 31 (1971) 405.
12. D. Y. C. Chan, T. W. Healy, and L. R. White, *J. Chem. Soc., Faraday Trans. 1*, 72 (1976) 2845.
13. H. Ohshima, Y. Inoko, and T. Mitsui, *J. Colloid Interface Sci.* 86 (1982) 57.

9 Double-Layer Interaction Between Two Parallel Similar Plates

9.1 INTRODUCTION

In this chapter, we give exact expressions and various approximate expressions for the force and potential energy of the electrical double-layer interaction between two parallel similar plates. Expressions for the double-layer interaction between two parallel plates are important not only for the interaction between plate-like particles but also for the interaction between two spheres or two cylinders, because the double-interaction between two spheres or two cylinders can be approximately calculated from the corresponding interaction between two parallel plates via Derjaguin's approximation, as shown in Chapter 12. We will discuss the case of two parallel dissimilar plates in Chapter 10.

9.2 INTERACTION BETWEEN TWO PARALLEL SIMILAR PLATES

Consider two parallel similar plates 1 and 2 of thickness d separated by a distance h immersed in a liquid containing N ionic species with valence z_i and bulk concentration (number density) n_i^∞ ($i = 1, 2, \ldots, N$). Without loss of generality, we may assume that plates 1 and 2 are positively charged. We take an x-axis perpendicular to the plates with its origin at the right surface of plate 1, as in Fig. 9.1. From the symmetry of the system we need consider only the region $-\infty < x \le h/2$. We assume that the electric potential $\psi(x)$ outside the plate ($-\infty < x < -d$ and $0 < x \le h/2$) obeys the following one-dimensional planar Poisson–Boltzmann equation:

$$\frac{d^2\psi(x)}{dx^2} = -\frac{1}{\varepsilon_r\varepsilon_o}\sum_{i=1}^{N} z_i e n_i^\infty \exp\left(-\frac{z_i e\psi(x)}{kT}\right), \quad 0 < x \le h/2 \qquad (9.1)$$

$$\frac{d^2\psi(x)}{dx^2} = -\frac{1}{\varepsilon_r\varepsilon_o}\sum_{i=1}^{N} z_i e n_i^\infty \exp\left(-\frac{z_i e\psi(x)}{kT}\right), \quad x < -d \qquad (9.2)$$

Biophysical Chemistry of Biointerfaces By Hiroyuki Ohshima

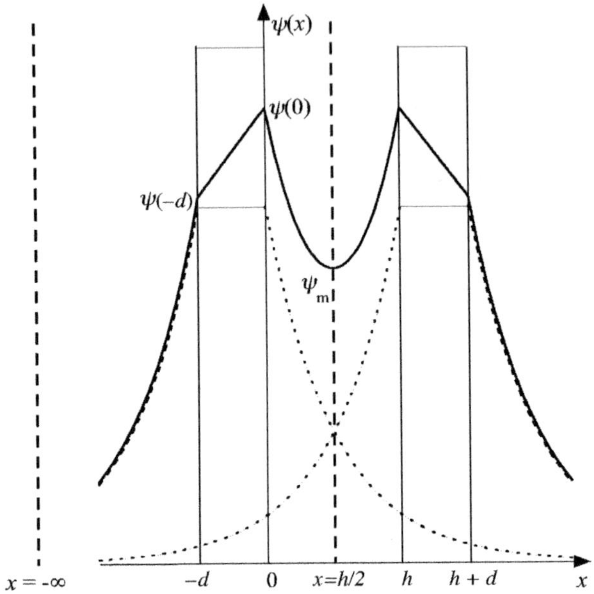

FIGURE 9.1 Schematic representation of the double-layer interaction between two parallel plates 1 and 2 separated by h. ψ_m is the potential at the midpoint between the plates.

For the region inside the plate, the potential satisfies the Laplace equation,

$$\frac{d^2\psi(x)}{dx^2} = 0, \qquad -d < x < 0 \tag{9.3}$$

By symmetry of the system, we have

$$\left.\frac{d\psi}{dx}\right|_{x=h/2} = 0 \tag{9.4}$$

and $\psi(x)$ and $d\psi/dx$ must vanish as $x \rightarrow -\infty$, namely,

$$\psi(x) \rightarrow 0 \quad \text{as } x \rightarrow -\infty \tag{9.5}$$

$$\frac{d\psi}{dx} \rightarrow 0 \quad \text{as } x \rightarrow -\infty \tag{9.6}$$

The boundary conditions at the plate surface depends on the type of the double--layer interaction between plates 1 and 2. If the surface potential of the plates remains constant at ψ_o, then

$$\psi(0) = \psi(-d) = \psi_o \tag{9.7}$$

On the other hand, if the surface charge density of the plates remains constant at σ, then

$$\psi(-d^-) = \psi(-d^+) \tag{9.8}$$

$$\psi(0^-) = \psi(0^+) \tag{9.9}$$

$$\varepsilon_r \left.\frac{d\psi}{dx}\right|_{x=-d^-} - \varepsilon_p \left.\frac{d\psi}{dx}\right|_{x=-d^+} = \frac{\sigma}{\varepsilon_o} \tag{9.10}$$

$$\varepsilon_p \left.\frac{d\psi}{dx}\right|_{x=0^-} - \varepsilon_r \left.\frac{d\psi}{dx}\right|_{x=0^+} = \frac{\sigma}{\varepsilon_o} \tag{9.11}$$

where ε_p is the relative permittivity of the plate. For the constant surface charge density case, one must solve not only the Poisson–Boltzmann equations (9.1) and (9.2) for the region outside the plates but also that for the region inside the plates (Eq. (9.3)) [1–5], the latter of which is integrated to give

$$\frac{d\psi}{dx} = \frac{\psi(0) - \psi(-d)}{d} \tag{9.12}$$

Further integration of Eq. (9.12) yields

$$\psi(x) = \frac{\psi(0) - \psi(-d)}{d} x + \psi(0) \tag{9.13}$$

With the help of Eq. (9.12), we have

$$\left.\frac{d\psi}{dx}\right|_{x=-d^+} = \left.\frac{d\psi}{dx}\right|_{x=0^-} = \frac{\psi(0) - \psi(-d)}{d} \tag{9.14}$$

By using Eq. (9.14), we can rewrite Eqs. (9.10) and (9.11) as

$$\varepsilon_r \left.\frac{d\psi}{dx}\right|_{x=-d^-} - \varepsilon_p \frac{\psi(0) - \psi(-d)}{d} = \frac{\sigma}{\varepsilon_o} \tag{9.15}$$

$$\varepsilon_p \frac{\psi(0) - \psi(-d)}{d} - \varepsilon_r \left.\frac{d\psi}{dx}\right|_{x=0^+} = \frac{\sigma}{\varepsilon_o} \tag{9.16}$$

Note that if the electric fields inside the plates can be neglected, then Eqs. (9.10) and (9.11) (or Eqs. (9.15) and (9.16)) may be replaced by

$$\left.\frac{d\psi}{dx}\right|_{x=-d^-} = +\frac{\sigma}{\varepsilon_r \varepsilon_o} \tag{9.17}$$

$$\left.\frac{d\psi}{dx}\right|_{x=0^+} = -\frac{\sigma}{\varepsilon_r \varepsilon_o} \tag{9.18}$$

which are often employed as an approximate boundary conditions at the plate surface for the constant surface charge density case.

As shown in Chapter 8, the interaction force P between plates 1 and 2 can be calculated by integrating the excess osmotic pressure $\Delta\Pi$ and the Maxwell stress T over an arbitrary closed surface Σ enclosing either one of the two interacting plates (Eq. (8.6)). As an arbitrary surface Σ enclosing plate 1, it is convenient to choose the plane $x = -\infty$ and the midplane at $x = h/2$, since $\psi(x)$ and $d\psi/dx = 0$ at $x = -\infty$ (Eqs. (9.5) and (9.6)) and $d\psi/dx = 0$ at $x = h/2$ (Eq. (9.4)) so that the excess osmotic pressure $\Delta\Pi$ and the Maxwell stress T are both zero at $x = -\infty$ and the Maxwell stress T is zero at $x = h/2$. Thus, the force of the double-layer interaction per unit area between plates 1 and 2 can be expressed as

$$\begin{aligned} P(h) &= [T(h/2) + \Delta\Pi(h/2)] - [T(-\infty) + \Delta\Pi(-\infty)] \\ &= [T(h/2) + \Pi(h/2)] - [T(-\infty) + \Pi(-\infty)] \end{aligned} \tag{9.19}$$

with

$$T(x) = -\frac{1}{2}\varepsilon_r \varepsilon_o \left(\frac{d\psi}{dx}\right)^2 \tag{9.20}$$

$$\Pi(x) = kT \sum_{i=1}^{N} n_i^\infty \exp\left(-\frac{z_i e \psi(x)}{kT}\right) \tag{9.21}$$

$$\Delta\Pi(x) = kT \sum_{i=1}^{N} n_i^\infty \left[\exp\left(-\frac{z_i e \psi(x)}{kT}\right) - 1\right] \tag{9.22}$$

where $T(x)$ is the Maxwell stress and $\Delta\Pi(x)$ is the excess osmotic pressure at position x. By substituting Eqs. (9.21) and (9.22) into Eq. (9.19) and noting that

$$T(-\infty) = 0 \tag{9.23}$$

$$\Delta\Pi(-\infty) = 0 \tag{9.24}$$

$$T(h/2) = 0 \tag{9.25}$$

we obtain

$$P(h) = kT\sum_{i=1}^{N} n_i^{\infty}\left[\exp\left(-\frac{z_i e\psi_{\mathrm{m}}}{kT}\right) - 1\right] \tag{9.26}$$

where $\psi_{\mathrm{m}} = \psi(h/2)$ is the potential at the midpoint between plates 1 and 2. Thus, one needs to find only the value of $\psi_{\mathrm{m}} = \psi(h/2)$ that is obtained by solving the Poisson–Boltzmann equation (9.1) for the electric potential $\psi(x)$.

9.3 LOW POTENTIAL CASE

For the low potential case, simple analytic expressions for the force and potential energy of the double-layer interaction between two plates can be derived. In this case Eq. (9.26) for the interaction force $P(h)$ per unit area between the plates at separation h reduces to

$$P(h) = \frac{1}{2}\varepsilon_{\mathrm{r}}\varepsilon_{\mathrm{o}}\kappa^2\psi_{\mathrm{m}}^2 \tag{9.27}$$

with

$$\kappa = \left(\frac{1}{\varepsilon_{\mathrm{r}}\varepsilon_{\mathrm{o}}kT}\sum_{i=1}^{N} z_i^2 e^2 n_i^{\infty}\right)^{1/2} \tag{9.28}$$

being the Debye–Hückel parameter.

For the low potential case, Eqs. (9.1) and (9.2) can be linearized to give

$$\frac{d^2\psi}{dx^2} = \kappa^2\psi, \quad 0 < x \leq h/2 \tag{9.29}$$

$$\frac{d^2\psi}{dx^2} = \kappa^2\psi, \quad -\infty < x < -d \tag{9.30}$$

9.3.1 Interaction at Constant Surface Charge Density

For the constant surface charge density case, one must consider the internal field within the plates, since the plate surface becomes no longer equipotential as the plates approach each other. The solutions to Eqs. (9.29), (9.30) and (9.3) subject to Eqs. (9.4), (9.8), (9.9), (9.15) and (9.16) are [1]

$$\psi(x) = \psi_o \frac{(1+\alpha)\cosh[\kappa(h/2 - x)]}{\{\alpha + \tanh(\kappa h/2)\}\cosh(\kappa h/2)}, \quad 0 \leq x \leq h/2 \tag{9.31}$$

$$\psi(x) = \psi_o \left\{ \frac{2\alpha + (1-\alpha)\tanh(\kappa h/2)}{\alpha + \tanh(\kappa h/2)} \right\} e^{\kappa(x+d)}, \quad -\infty < x \leq -d \tag{9.32}$$

$$\psi(x) = \psi(0) + \left\{ \frac{\psi(0) - \psi(-d)}{d} \right\} x, \quad -d \leq x \leq 0 \tag{9.33}$$

with

$$\psi_o = \frac{\sigma}{\varepsilon_r \varepsilon_o \kappa} \tag{9.34}$$

$$\alpha = \frac{1}{1 + (\varepsilon_r/\varepsilon_p)\kappa d} \tag{9.35}$$

where ψ_o is the unperturbed surface potential of the plates at $\kappa h = \infty$, and α characterizes the influence of the internal electric field within the plates. In the limit of $\varepsilon_p \to 0$ and/or $d \to \infty$, α tends to 0. This situation, where the influence of the internal electric fields within the plates may be ignored, corresponds to $\alpha = 0$.

The surface potentials $\psi(0)$ and $\psi(-d)$ are functions of plate separation h. We denote $\psi(0)$ and $\psi(-d)$ at plate separation h by $\psi_0(h)$ and $\psi_{-d}(h)$, respectively. These surface potentials can be calculated from Eqs. (9.31) and (9.32)

$$\psi_0(h) = \psi_o \left\{ \frac{1+\alpha}{\alpha + \tanh(\kappa h/2)} \right\} \tag{9.36}$$

$$\psi_{-d}(h) = \psi_o \left\{ \frac{2\alpha + (1-\alpha)\tanh(\kappa h/2)}{\alpha + \tanh(\kappa h/2)} \right\} \tag{9.37}$$

Equations (9.36) and (9.37) show that as the two plates approach each other, $\psi_0(h)$ and $\psi_{-d}(h)$ increase from their unperturbed values for a single plate, that is,

$$\psi_0(\infty) = \psi_{-d}(\infty) = \psi_o \tag{9.38}$$

down to

$$\psi_0(0) = \psi_o\left(\frac{1+\alpha}{\alpha}\right) \tag{9.39}$$

$$\psi_{-d}(0) = 2\psi_o \tag{9.40}$$

at $h = 0$.

Figure 9.2 gives the potential distribution $\psi(x)$ across two interacting plates 1 and 2 at constant surface charge density σ calculated with Eqs. (9.31)–(9.33) for $\kappa h = 1$, 2, and ∞ at $\kappa d = 1$ and $\alpha = 0.1$, showing how $\psi(x)$ changes as the plates approach each other. We see that as plates 1 and 2 approach each other, the surface potentials $\psi_0(h)$ and $\psi_{-d}(h)$ changes, inducing the internal electric fields within the plates.

For the special case where $\varepsilon_p = 0$ and/or $d = \infty$, that is, $\alpha = 0$ so that the influence of the internal fields may be neglected. In this case, Eqs. (9.31)–(9.33) become

$$\psi(x) = \psi_o \frac{\cosh[\kappa(h/2 - x)]}{\sinh(\kappa h/2)}, \quad 0 \le x \le h/2 \tag{9.41}$$

$$\psi(x) = \psi_o\, e^{\kappa(x+d)}, \quad -\infty < x \le 0 \tag{9.42}$$

$$\psi(x) = \psi(0) + \left\{\frac{\psi(0) - \psi(-d)}{d}\right\}x, \quad -d \le x \le 0 \tag{9.43}$$

and the surface potentials (Eqs. (9.36) and (9.37)) become

$$\psi_0(h) = \psi_o \frac{1}{\tanh(\kappa h/2)} \tag{9.44}$$

$$\psi_{-d}(h) = \psi_o \tag{9.45}$$

By evaluating $\psi_m = \psi(h/2)$ from Eq. (9.31), namely,

$$\psi_m = \psi(h/2) = \psi_o \frac{1+\alpha}{\{\alpha + \tanh(\kappa h/2)\}\cosh(\kappa h/2)} \tag{9.46}$$

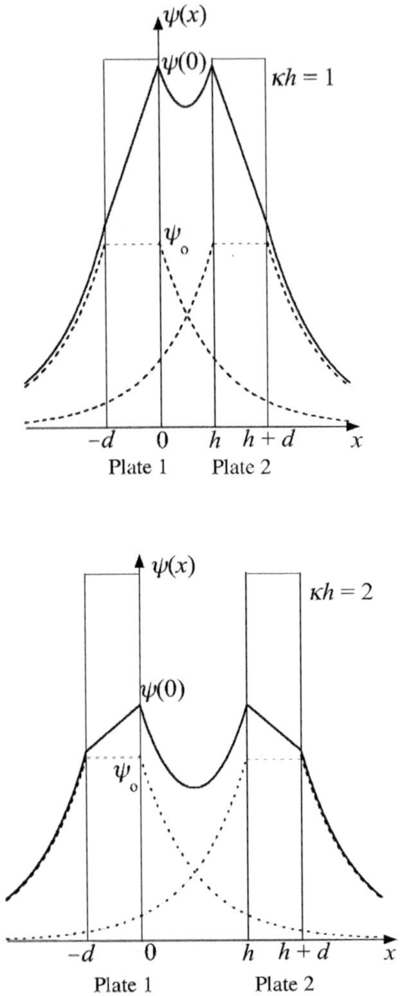

FIGURE 9.2 Potential distribution $\psi(x)$ across two interacting plates 1 and 2 at constant surface charge density σ calculated with Eqs. (9.31)–(9.33) for $\kappa h = 1$ (upper) and 2 (lower). The dotted lines, which stand for unperturbed potentials, correspond to the case of infinite separation ($\kappa h = \infty$).

and substituting this result into Eq. (9.27), we obtain the interaction force $P^{\sigma}(h)$ per unit area between plates 1 and 2 carrying constant surface charge density σ, namely,

$$P^{\sigma}(h) = \frac{1}{2}\varepsilon_r\varepsilon_o\kappa^2\psi_o^2\left\{\frac{(1+\alpha)\mathrm{sech}(\kappa h/2)}{\alpha + \tanh(\kappa h/2)}\right\}^2 \tag{9.47}$$

The potential energy per unit area of the double-layer interaction between plates 1 and 2 is obtained by integrating Eq. (9.47) with respect to h with the result that

$$
\begin{aligned}
V^{\sigma}(h) &= \int_h^{\infty} P(h)dh \\
&= \varepsilon_r\varepsilon_o\kappa^2\psi_o^2\left[\frac{(1+\alpha)\{1-\tanh(\kappa h/2)\}}{\alpha+\tanh(\kappa h/2)}\right]
\end{aligned}
\tag{9.48}
$$

Equation (9.48) can also be derived from the low potential approximate from for the double-layer free energy between two parallel plates at constant surface charge density σ (Eq. (8.52))

$$
V^{\sigma}(h) = F^{\sigma}(h) - F^{\sigma}(\infty) \tag{9.49}
$$

where

$$
\begin{aligned}
F^{\sigma}(h) &= \sigma\psi_0(h) + \sigma\psi_{-d}(h) \\
&= \sigma\psi_o\left\{\frac{1+\alpha}{\alpha+\tanh(\kappa h/2)}\right\} + \sigma\psi_o\left\{\frac{2\alpha+(1-\alpha)\tanh(\kappa h/2)}{\alpha+\tanh(\kappa h/2)}\right\} \\
&= \varepsilon_r\varepsilon_o\kappa\psi_o^2\left[\left\{\frac{1+\alpha}{\alpha+\tanh(\kappa h/2)}\right\} + \left\{\frac{2\alpha+(1-\alpha)\tanh(\kappa h/2)}{\alpha+\tanh(\kappa h/2)}\right\}\right] \\
&= \varepsilon_r\varepsilon_o\kappa\psi_o^2\left\{\frac{1+3\alpha+(1-\alpha)\tanh(\kappa h/2)}{\alpha+\tanh(\kappa h/2)}\right\}
\end{aligned}
\tag{9.50}
$$

$$
\begin{aligned}
F^{\sigma}(\infty) &= \sigma\psi_0(\infty) + \sigma\psi_{-d}(\infty) \\
&= 2\varepsilon_r\varepsilon_o\kappa\psi_o^2
\end{aligned}
\tag{9.51}
$$

Substituting Eqs. (9.50) and (9.51) into Eq. (9.49) gives Eq. (9.48). For the special case where $\varepsilon_p = 0$ and/or $d = \infty$, where $\alpha = 0$ so that the influence of the internal fields may be neglected, Eqs. (9.47) and (9.48) become

$$
P^{\sigma}(h) = \frac{1}{2}\varepsilon_r\varepsilon_o\kappa^2\psi_o^2\,\mathrm{cosech}^2(\kappa h/2) \tag{9.52}
$$

$$
V^{\sigma}(h) = \varepsilon_r\varepsilon_o\kappa\psi_o^2\{\coth(\kappa h/2) - 1\} \tag{9.53}
$$

9.3.2 Interaction at Constant Surface Potential

Consider next the case where the surface potentials of plates 1 and 2 both remain constant at ψ_o during interaction. The solution to Eqs. (9.29), (9.30) and (9.3) subject to Eqs. (9.4)–(9.7) is

$$\psi(x) = \psi_o \frac{\cosh[\kappa(h/2 - x)]}{\cosh(\kappa h/2)}, \quad 0 \leq x \leq h/2 \tag{9.54}$$

$$\psi(x) = \psi_o, \quad -d \leq x \leq 0 \tag{9.55}$$

$$\psi(x) = \psi_o\, e^{\kappa(x+d)}, \quad -\infty < x \leq -d \tag{9.56}$$

Figure 9.3 gives the potential distribution $\psi(x)$ across two interacting plates 1 and 2 at constant surface potential ψ_o calculated with Eqs. (9.54)–(9.56) for $\kappa h = 1$, 2, and ∞, showing how $\psi(x)$ depends on the plate separation h. In this case, $\psi(x)$ does not depend on the plate thickness d, since there is no electric field inside the plates and $\psi(x)$ in the region $x < 0$ remain unchanged during interaction.

The surface charge density $\sigma_0(h)$ at $x = 0$ of plate 1, which, in this case, is a function of plate separation h, is given by the boundary condition at the plate surface $x = 0$, namely,

$$\sigma_0(h) = -\varepsilon_r \varepsilon_o \left.\frac{d\psi}{dx}\right|_{x=0^+} \tag{9.57}$$

where we have used the fact that there is no electric field within the plate and thus $d\psi/dx$ at $x = 0^-$ is zero. By substituting Eq. (9.54) into Eq. (9.57), we obtain

$$\sigma_0(h) = \varepsilon_r \varepsilon_o \kappa \psi_o \tanh(\kappa h/2) \tag{9.58}$$

Equation (9.58) shows that as the two plates approach each other, $\sigma_0(h)$ decreases from its value for a single plate, that is,

$$\sigma_0(\infty) = \varepsilon_r \varepsilon_o \kappa \psi_o \tag{9.59}$$

down to zero at $h = 0$, at which the plate charge is completely discharged,

$$\sigma_0(0) = 0 \tag{9.60}$$

The charge density $\sigma_{-d}(h)$ at $x = -d$, on the other hand, is given by

$$\sigma_{-d}(h) = -\varepsilon_r \varepsilon_o \left.\frac{d\psi}{dx}\right|_{x=-d^-} \tag{9.61}$$

By substituting Eq. (9.56) into Eq. (9.61), we obtain

$$\sigma_{-d}(h) = \sigma_{-d}(\infty) = \varepsilon_r \varepsilon_o \kappa \psi_o \tag{9.62}$$

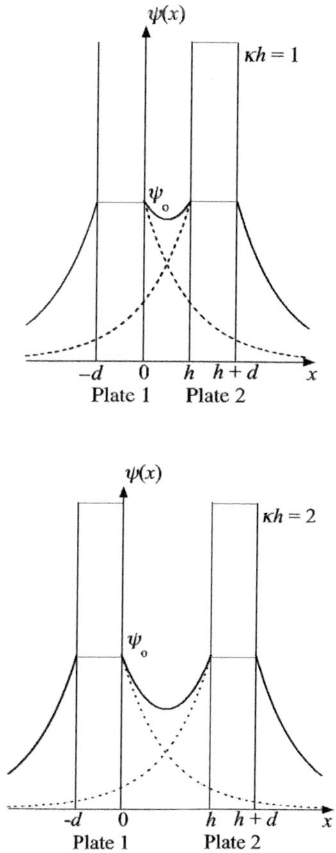

FIGURE 9.3 Potential distribution $\psi(x)$ across two interacting plates 1 and 2 at constant surface potential ψ_o calculated with Eqs. (9.54) and (9.56) for $\kappa h = 1$ (upper) and 2 (lower). The dotted lines, which stand for unperturbed potentials, correspond to the case of infinite separation ($\kappa h = \infty$).

That is, $\sigma_{-d}(h)$ remains unchanged at its unperturbed value $\sigma_{-d}(\infty) = \varepsilon_r \varepsilon_o \kappa \psi_o$ (Eq. (9.34)).

From Eq. (9.54) we can obtain simple analytic approximate expressions for the force and potential energy of the electrostatic interaction between two parallel plates 1 and 2. It follows from Eq. (9.54) that the potential $\psi_m = \psi(h/2)$ at the midpoint $x = h/2$ is given by

$$\psi_m = \psi(h/2) = \psi_o \frac{1}{\cosh(\kappa h/2)} \tag{9.63}$$

By substituting Eq. (9.63) into Eq. (9.27), we obtain the following expression $P^{\psi}(h)$ for the interaction force at constant potential per unit area between two parallel similar plates 1 and 2 at separation h:

$$P^{\psi}(h) = \frac{1}{2}\varepsilon_r\varepsilon_o\kappa^2\psi_o^2\frac{1}{\cosh^2(\kappa h/2)} \tag{9.64}$$

The corresponding potential energy $V^{\psi}(h)$ per unit area is obtained by integrating Eq. (9.64) with respect to h

$$V^{\psi}(h) = \int_h^{\infty} P(h)dh = \varepsilon_r\varepsilon_o\kappa\psi_o^2\{1 - \tanh(\kappa h/2)\} \tag{9.65}$$

Equation (9.65) can also be derived from the low-potential approximate expression (Eq. (8.55)) for the double-layer free energy $F^{\psi}(h)$ of two parallel plates 1 and 2 at constant surface potential ψ_o, namely,

$$V^{\psi}(h) = F^{\psi}(h) - F^{\psi}(\infty) \tag{9.66}$$

with

$$\begin{aligned} F^{\psi}(h) &= -\sigma_0(h)\psi_o - \sigma_{-d}(h)\psi_o \\ &= -\varepsilon_r\varepsilon_o\kappa\psi_o^2\tanh(\kappa h/2) - \varepsilon_r\varepsilon_o\kappa\psi_o^2 \end{aligned} \tag{9.67}$$

$$\begin{aligned} F^{\psi}(\infty) &= -\sigma_0(\infty)\psi_o - \sigma_{-d}(\infty)\psi_o \\ &= -2\varepsilon_r\varepsilon_o\kappa\psi_o^2 \end{aligned} \tag{9.68}$$

where $\sigma_0(h)$ and $\sigma_{-d}(h)$ are, respectively, given by Eqs. (9.58) and (9.62). Substituting Eqs. (9.67) and (9.68) into Eq. (9.66) gives Eq. (9.65).

9.4 ARBITRARY POTENTIAL CASE

For simplicity, we consider the case where two parallel plates 1 and 2 at separation h are immersed in a symmetrical electrolyte solution of valence z and bulk concentration n.

9.4.1 Interaction at Constant Surface Charge Density

First we consider the double-layer interaction between two parallel plates with constant surface charge density σ [3].

The Poisson–Boltzmann equation (9.1) for the electric potential $\psi(x)$ in the region outside the plates becomes

$$\frac{d^2\psi}{dx^2} = -\frac{zen}{\varepsilon_r\varepsilon_o}\left[\exp\left(-\frac{ze\psi}{kT}\right) - \exp\left(\frac{ze\psi}{kT}\right)\right], \quad -\infty < x < -d \text{ and } 0 < x < -h/2 \tag{9.69}$$

or

$$\frac{d^2\psi}{dx^2} = \frac{2zen}{\varepsilon_r\varepsilon_o}\sinh\left(\frac{ze\psi}{kT}\right), \quad -\infty < x < -d \text{ and } 0 < x < -h/2 \tag{9.70}$$

If the scaled potential

$$y(x) = \frac{ze\psi(x)}{kT} \tag{9.71}$$

is introduced, then Eq. (9.70) is further transformed into

$$\frac{d^2y}{dx^2} = \kappa^2 \sinh y \tag{9.72}$$

with

$$\kappa = \sqrt{\frac{2nz^2e^2n}{\varepsilon_r\varepsilon_o kT}} \tag{9.73}$$

being the Debye–Hückel parameter. We rewrite Eq. (9.72) as

$$\frac{1}{\kappa^2}\frac{d^2y}{dx^2} = \frac{1}{\kappa}\frac{dw}{dx} = \sinh y, \quad 0 < x \leq h/2 \text{ and } -\infty < x < -d \tag{9.74}$$

with

$$w = \frac{1}{\kappa}\frac{dy}{dx} \tag{9.75}$$

For the internal region of plate 1 ($-d < x < 0$), the potential satisfies the Laplace equation

$$\frac{d^2y}{dx^2} = \frac{dw}{dx} = 0 \tag{9.76}$$

which is integrated to give

$$\frac{dy}{dx} = \frac{y(0) - y(-d)}{d} \tag{9.77}$$

The boundary conditions given by Eqs. (9.4), (9.5), (9.6), (9.15) and (9.16) become

$$w(h/2) = 0 \tag{9.78}$$

$$y(-\infty) = 0 \tag{9.79}$$

$$w(-\infty) = 0 \tag{9.80}$$

$$|w(-d)| - a\left\{|y(0)| - y|(-d)|\right\} = |\sigma'| \tag{9.81}$$

$$|w(0)| + a\left\{|y(0)| - y|(-d)|\right\} = |\sigma'| \tag{9.82}$$

with

$$\sigma' = \frac{ze}{kT} \cdot \frac{\sigma}{\varepsilon_r \varepsilon_o \kappa} \tag{9.83}$$

$$a = \frac{\varepsilon_p}{\varepsilon_r \kappa d} \tag{9.84}$$

where σ' is the scaled surface charge density and a is related to α (Eq. (9.35)) by

$$\alpha = \frac{a}{1 + a} \tag{9.85}$$

The interaction force between the plates per unit area is given by (9.26), which becomes in this case

$$P(h) = 2nkT(\cosh y_m - 1) = 4nkT \sinh^2(y_m/2) \tag{9.86}$$

with

$$y_m = \frac{ze\psi_m}{kT} \tag{9.87}$$

where $\psi_m = \psi(h/2)$ is the scaled potential at the midpoint between plates 1 and 2, and $y_m = y(h/2)$ is its scaled quantity. The value of y_m can be obtained as follows. By integrating Eq. (9.74) under Eqs. (9.78)–(9.80), we obtain

$$\begin{aligned}\left(\frac{1}{\kappa}\frac{dy}{dx}\right)^2 &= w^2 = 2\cosh y - 1 \\ &= 4\sinh^2\left(\frac{y}{2}\right) \quad \text{for } x \leq -d\end{aligned} \tag{9.88}$$

$$\begin{aligned}\left(\frac{1}{\kappa}\frac{dy}{dx}\right)^2 &= w^2 = 2(\cosh y - \cosh y_m) \\ &= 4\left\{\cosh^2\left(\frac{y}{2}\right) - \cosh^2\left(\frac{y_m}{2}\right)\right\} \quad \text{for } 0 \leq x \leq h/2\end{aligned} \tag{9.89}$$

From Eq. (9.88), we have

$$|w(-d)| = 2\sinh\left(\frac{y(-d)}{2}\right) \tag{9.90}$$

By using Eqs. (9.81), (9.82) and (9.90), we obtain

$$|y(-d)| = 2\,\text{arcsinh}\left[2\sinh\left(\frac{|y_o|}{2}\right) - \frac{|z(-d)|}{2}\right] \tag{9.91}$$

where we have introduced the unperturbed surface potential ψ_o, that is, the potential of the plates at infinite separation.

$$\psi_o = 2\left(\frac{kT}{ze}\right)\text{arcsinh}\left(\frac{\sigma'}{2}\right) \tag{9.92}$$

and y_o is the scaled unperturbed surface potential, namely,

$$y_o = 2\,\text{arcsinh}\left(\frac{\sigma'}{2}\right) \tag{9.93}$$

which is obtained by integrating Eq. (9.74) under the boundary conditions for an isolated plate (see Eq. (1.42)). From Eq. (9.89), we obtain

$$\kappa\,dx = -\frac{dy}{2\sqrt{\{\cosh^2(y/2) - \cosh^2(y_m/2)\}}} \tag{9.94}$$

Further integration of Eq. (9.94) gives

$$\kappa\left(\frac{h}{2}-x\right) = \int_{y_{\rm m}}^{y(x)} \frac{dy}{2\sqrt{\{\cosh^2(y/2)-\cosh^2(y_{\rm m}/2)\}}}$$
$$= k\left\{K(k) - F\left[\arcsin\left(\frac{1}{k\cosh(y(x)/2)}\right), k\right]\right\} \tag{9.95}$$

where $F(\phi, k)$ is an elliptic integral of the first kind with a modulus k given by

$$k = \frac{1}{\cosh(y_{\rm m}/2)} \tag{9.96}$$

and $K(k) = F(\pi/2, k)$ is the complete elliptic integral of the first kind with a modulus k. By using a Jacobian elliptic function $dc\ (u, k)\ (= dn(u, k)/cn(u, k))$ with a modulus k, Eq. (9.95) can be rewritten as

$$y(x) = 2\ \mathrm{arccosh}\left[dc\left(\frac{\kappa(h/2-x)}{k}, k\right)\right] \tag{9.97}$$

Evaluating Eqs. (9.95) and (9.97) at $x = 0$, we obtain

$$\kappa h = 2k\left\{K(k) - F\left[\arcsin\left(\frac{1}{k\cosh(y(0)/2)}\right), k\right]\right\} \tag{9.98}$$

$$y(0) = 2\ \mathrm{arccosh}\left(\frac{dc(u,k)}{k}\right) \tag{9.99}$$

where

$$u = \frac{\kappa h}{2k} \tag{9.100}$$

and the following inequality must be satisfied:

$$0 \le u \le K(k) \tag{9.101}$$

From Eq. (9.89) we also have

$$|w(0)| = 2\sqrt{\cosh^2(y(0)/2) - \cosh^2(y_{\rm m}/2)} \tag{9.102}$$

Substituting Eq. (9.99), we have

$$|w(0)| = \frac{2k'}{k} tn(u, k) \tag{9.103}$$

where k' is the complementary modulus given by

$$k' = \sqrt{1 - k^2} = \tanh\left(\frac{|y_m|}{2}\right) \tag{9.104}$$

and $tn(u, k)$ (= $sn(u, k)/cn(u, k)$) is also a Jacobian elliptic function with a modulus k. Substituting Eqs. (9.90), (9.91), (9.99) and (9.103) into Eqs. (9.81) and (9.82), we obtain the following transcendental equation for y_m:

$$\begin{aligned}\sinh\left(\frac{|y_o|}{2}\right) &- \frac{k'}{k} tn(u, k) \\ &+ a\left[\operatorname{arcsinh}\left(2\sinh\left(\frac{|y_o|}{2}\right) - \frac{k'}{k} tn(u, k)\right) - \operatorname{arccosh}\left(\frac{1}{k} dc(u, k)\right)\right] = 0\end{aligned} \tag{9.105}$$

Thus, if, for given values of y_o (or σ'), a, and κh, the solution of Eq. (9.105) satisfying Eq. (9.101) is found, the interaction force P can be calculated as a function of y_o (or σ'), a, and κh by using Eq. (9.86).

The potential energy $V^\sigma(h)$ of the double-layer interaction per unit area between two parallel plates 1 and 2 with constant surface charge density σ at separation h is given by

$$V^\sigma(h) = F^\sigma(h) - F^\sigma(\infty) \tag{9.106}$$

where $F^\sigma(h)$ is the free energy of the double layer. In the present case, it is more convenient to use Eq. (8.56) for F^σ than Eq. (8.50)

$$F^\sigma = F_{el} = \int_0^1 d\lambda \frac{2}{\lambda} E(\lambda) \tag{9.107}$$

with

$$\begin{aligned}E(\lambda) = &\int_{-\infty}^{-d} \psi(x, \lambda)\rho(x, \lambda)dx + \int_0^{h/2} \psi(x, \lambda)\rho(x, \lambda)dx \\ &+ \lambda\sigma\psi(0, \lambda) + \lambda\sigma\psi(-d, \lambda)\end{aligned} \tag{9.108}$$

where $\rho(x, \lambda)$ is the volume charge density at stage λ. Then

$$F^\sigma = 2I(-\infty, -d) + 2I(-d, 0) + 2\sigma\int_0^1 \psi(-d, \lambda)d\lambda + 2\sigma\int_0^1 \psi(0, \lambda)d\lambda \tag{9.109}$$

where

$$I(a,b) = \int_a^b dx \int_0^1 \frac{d\lambda}{\lambda} \psi(x,\lambda)\rho(x,\lambda) \tag{9.110}$$

By using the Poisson–Boltzmann equation at stage λ,

$$\begin{aligned} \frac{\partial^2 \psi(x,\lambda)}{\partial x^2} &= -\frac{\rho(x,\lambda)}{\varepsilon_r \varepsilon_o} \\ &= \frac{2z(\lambda e)n}{\varepsilon_r \varepsilon_o} \sinh\left(\frac{z(\lambda e)\psi}{kT}\right) \end{aligned} \tag{9.111}$$

For the region outside plate 1, that is, $-\infty < x \leq -d$ and $0 \leq x \leq -d$, Eq. (9.110) can be transformed into

$$\begin{aligned} I(a,b) = & -\int_a^b 2nkT\left\{\cosh\left(\frac{ze\psi(x,1)}{kT}\right) - 1\right\}dx - \frac{1}{2}\varepsilon_r\varepsilon_o \int_a^b \left(\frac{\partial\psi(x,1)}{\partial x}\right)^2 dx \\ & + \varepsilon_r\varepsilon_o \int_0^1 \left[\left(\frac{\partial\psi(x,\lambda)}{\partial x}\right)_{x=b} \cdot \frac{\partial\psi(b,\lambda)}{\partial\lambda} - \left(\frac{\partial\psi(x,\lambda)}{\partial x}\right)_{x=a} \cdot \frac{\partial\psi(a,\lambda)}{\partial\lambda}\right] d\lambda \\ & \text{for } -\infty < x \leq -d \text{ and } 0 \leq x \leq -d \end{aligned} \tag{9.112}$$

On the other hand, for the internal region $-\infty < x \leq -d$, there are no charges, that is,

$$\frac{\partial^2 \psi(x,\lambda)}{\partial x^2} = -\frac{\rho(x,\lambda)}{\varepsilon_r \varepsilon_o} = 0, \quad \text{for } -\infty < x \leq -d \tag{9.113}$$

Thus

$$I(-d,0) = \int_{-d}^0 dx \int_0^1 \frac{d\lambda}{\lambda} \psi(x,\lambda)\rho(x,\lambda) = 0 \tag{9.114}$$

That is,

$$\begin{aligned} 0 = & -\frac{1}{2}\varepsilon_r\varepsilon_o \int_{-d}^0 \left(\frac{\partial\psi(x,1)}{\partial x}\right)^2 dx \\ & + \varepsilon_r\varepsilon_o \int_0^1 \left[\left(\frac{\partial\psi(x,\lambda)}{\partial x}\right)_{x=0} \cdot \frac{\partial\psi(b,\lambda)}{\partial\lambda} - \left(\frac{\partial\psi(x,\lambda)}{\partial x}\right)_{x=-d} \cdot \frac{\partial\psi(a,\lambda)}{\partial\lambda}\right] d\lambda \end{aligned} \tag{9.115}$$

By using Eqs. (9.88), (9.99), (9.112), and (9.115) and the boundary conditions (9.8)–(9.11) at stage λ, we have the following expression for the double-layer free energy per unit area of plates 1 and 2.

$$F^{\sigma}(h) = -P(h)h - 2\varepsilon_r\varepsilon_o\left[\int_{-\infty}^{-d}\left(\frac{d\psi}{dx}\right)^2 dx + \int_0^{h/2}\left(\frac{d\psi}{dx}\right)^2 dx\right] - \varepsilon_p\varepsilon_o\int_{-d}^{0}\left(\frac{d\psi}{dx}\right)^2 dx + 2\sigma\{\psi(-d,h)+\psi(0,h)\} \tag{9.116}$$

Here with the help of Eq. (9.12), we have

$$\int_{-d}^{0}\left(\frac{d\psi}{dx}\right)^2 dx = \frac{1}{d}\{\psi(0)-\psi(-d)\}^2 \tag{9.117}$$

By using Eqs. (9.86) and (9.117), Eq. (9.116) becomes

$$F^{\sigma}(h) = -4nkT\sinh^2\left(\frac{ze\psi_m}{2kT}\right)\cdot h - 2\varepsilon_r\varepsilon_o\left[\int_{-\infty}^{-d}\left(\frac{d\psi}{dx}\right)^2 dx + \int_0^{h/2}\left(\frac{d\psi}{dx}\right)^2 dx\right] - \frac{\varepsilon_p\varepsilon_o}{d}\{\psi(0)-\psi(-d)\}^2 + 2\sigma\{\psi(-d,h)+\psi(0,h)\} \tag{9.118}$$

which can be rewritten in terms of $y(x)$ and $w(x)$ as

$$F^{\sigma}(h) = -\frac{4nkT}{\kappa}\left[\sinh^2\left(\frac{|y_m|}{2}\right)\cdot\kappa h + \kappa\int_{-\infty}^{-d} w^2\,dx + \kappa\int_0^{h/2} w^2\,dx + \frac{a}{2}\{|y(0)|-|y(-d)|\}^2 - 2\sinh\left(\frac{|y_o|}{2}\right)\{|y(0)|+|y(-d)|\}\right] \tag{9.119}$$

The second term in the brackets is integrated by using Eq. (9.88) to give

$$\int_{-\infty}^{-d} w^2\,dx = \int_{-\infty}^{-d}\left(\frac{1}{\kappa}\frac{dy}{dx}\right)^2 dx = \frac{1}{\kappa^2}\int_0^{y(-d)}\left(\frac{dy}{dx}\right)dy = \frac{1}{\kappa}\left\{\cosh\left(\frac{|y(-d)|}{2}\right)-1\right\} \tag{9.120}$$

and the third term is transformed into, by using Eqs. (9.89), (9.99) and (9.102),

$$\begin{aligned}\int_0^{h/2} w^2\,dx &= \int_0^{h/2}\left(\frac{1}{\kappa}\frac{dy}{dx}\right)^2 dx = \frac{1}{\kappa^2}\int_{|y_m|}^{|y(0)|}\left(\frac{dy}{dx}\right)dy\\ &= \frac{2}{\kappa}\int_{|y_m|}^{|y(0)|}\sqrt{\cosh^2(y/2)-\cosh^2(y_m/2)}dy \\ &= \frac{1}{\kappa}\left\{-\frac{4}{k}\varepsilon(u,k)+\frac{2|w(0)|}{\tanh(|y(0)|/2)}\right\}\end{aligned} \tag{9.121}$$

with

$$\varepsilon(u,k) = E(\arcsin(sn(u),k) = \int_0^u dn^2(u,k)du \tag{9.122}$$

where $E(\phi, k)$ is an elliptic integral of the second kind with a modulus k. Thus, Eq. (9.118) can be rewritten as

$$\begin{aligned}F^{\sigma}(h) = -\frac{4nkT}{\kappa}\Bigg[&\frac{k'^2}{k^2}\cdot\kappa h - \frac{4}{k}E(u,k) + \frac{2|w(0)|}{\tanh(|y(0)|/2)} + 4\left\{\cosh\left(\frac{|y(-d)|}{2}\right)-1\right\}\\ &+\frac{a}{2}\{|y(0)|-|y(-d)|\}^2 - 2\sinh\left(\frac{|y_o|}{2}\right)\{|y(0)|+|y(-d)|\}\Bigg]\end{aligned} \tag{9.123}$$

For $h \rightarrow \infty$, Eq. (9.123) becomes

$$F^{\sigma}(\infty) = \frac{16nkT}{\kappa}\left\{|y_o|\sinh\left(\frac{|y_o|}{2}\right) - 2\cosh\left(\frac{|y_o|}{2}\right)+2\right\} \tag{9.124}$$

The potential energy $V(h)$ of the double-layer interaction per unit area between plates 1 and 2 is thus given by Eq. (9.118), namely,

$$\begin{aligned}V^{\sigma}(h) &= F^{\sigma}(h) - F^{\sigma}(\infty)\\ &= -\frac{4nkT}{\kappa}\Bigg[\frac{k'^2}{k^2}\cdot\kappa h - \frac{4}{k}\varepsilon(u,k) + \frac{2|w(0)|}{\tanh(|y(0)|/2)} + 4\left\{\cosh\left(\frac{|y(-d)|}{2}\right)-1\right\}\\ &\quad+\frac{a}{2}\{|y(0)|-|y(-d)|\}^2 - 2\sinh\left(\frac{|y_o|}{2}\right)\{|y(0)|+|y(-d)|-2|y_o|\}\\ &\quad-8\left\{\cosh\left(\frac{|y_o|}{2}\right)-1\right\}\Bigg]\end{aligned} \tag{9.125}$$

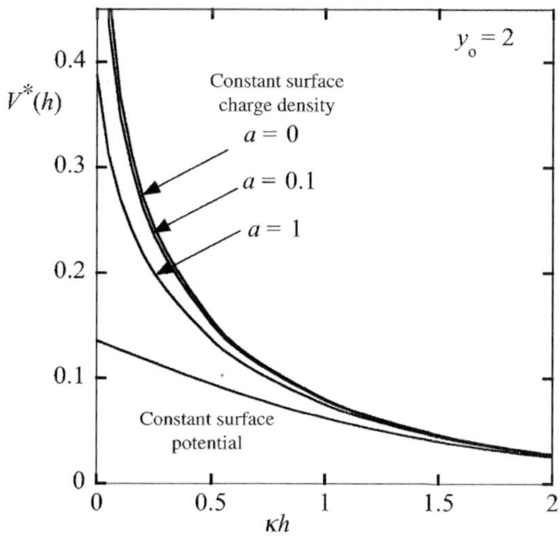

FIGURE 9.4 Reduced potential energy $V^*(h) = (\kappa/64nkT)V^\sigma(h)$ as a function of scaled plate separation κh for the constant surface charge density model calculated with Eq. (9.125) for $a = 0$, 0.1, and 1 in comparison with $V^*(h) = (\kappa/64nkT)V^\psi(h)$ for the constant surface potential model calculated with Eq. (9.141).

This is the require expression for $V^\sigma(h)$. Figure 9.4 shows V^σ with $a = 0$, 0.1, and 1 calculated at $y_o = 2$, showing the strong dependence of V^σ upon a.

If the influence of the internal electric fields within the plates may be neglected, then $a = 0$ so that Eqs. (9.105) and (9.125) reduce to

$$\sinh\left(\frac{|y_o|}{2}\right) - \frac{k'}{k} tn(u, k) = 0 \tag{9.126}$$

and

$$\begin{aligned} V^\sigma(h) &= F^\sigma(h) - F^\sigma(\infty) \\ &= -\frac{4nkT}{\kappa}\left[\frac{k'^2}{k^2}\cdot \kappa h - \frac{4}{k}\varepsilon(u,k) + \frac{2|w(0)|}{\tanh(|y(0)|/2)} + 4\left\{\cosh\left(\frac{|y(-d)|}{2}\right) - 1\right\}\right] \end{aligned} \tag{9.127}$$

Finally, we give alternative expressions for $y(x)$ and $V^\sigma(h)$ by changing the definition of k for the special case where the influence of the external fields may be neglected ($a = 0$) [6,7]. If we define

$$k = \exp(-y_m) \tag{9.128}$$

then Eq. (9.95) is changed into

$$\kappa\left(\frac{h}{2}-x\right)=2\sqrt{k}\left\{K(k)-F\left[\arcsin\left(\frac{1}{\sqrt{k}}\exp\left(\frac{y(x)}{2}\right)\right),k\right]\right\} \tag{9.129}$$

which can be rewritten as

$$y(x)=y_{\mathrm{m}}+2\ln\left[dc\left(\frac{\kappa(h/2-x)}{2\sqrt{k}},k\right)\right] \tag{9.130}$$

Evaluating Eq. (9.130) at $x=0$, we obtain

$$y(0)=y_{\mathrm{m}}+2\ln[dc(u,k)] \tag{9.131}$$

where

$$u=\frac{\kappa h}{4\sqrt{k}} \tag{9.132}$$

and the following inequality must be satisfied:

$$0\leq u\leq K(k) \tag{9.133}$$

If the internal fields within the plates may be neglected, then the transcendental equation for y_{m} (Eq. (9.126)) is changed into

$$2\sinh\left(\frac{|y_{\mathrm{o}}|}{2}\right)=\sqrt{\frac{\{dc^2(u,k)-1\}\{dc^2(u,k)-k^2\}}{k\,dc^2(u,k)}} \tag{9.134}$$

With this change of k, Eq. (9.127) becomes

$$\begin{aligned}V^{\sigma}(h) = \frac{64nkT}{\kappa}\Bigg[&\frac{1}{4}(e^{-y_{\mathrm{o}}/2}-1)-\frac{kh}{64}(3\,e^{y_{\mathrm{m}}}-2-e^{-y_{\mathrm{m}}})\\ &+\frac{1}{8}\{y(0)-y_{\mathrm{o}}\}\sinh\left(\frac{y_{\mathrm{o}}}{2}\right)\\ &+\frac{1}{4}e^{y_{\mathrm{m}}/2}\{E(k)-E(\arcsin\left[e^{y_{\mathrm{m}}-y(0)}\right],k)\}\Bigg]\end{aligned} \tag{9.135}$$

where $E(\pi/2, k)=E(k)$ is the complete elliptic integral of the second kind.

9.4.2 Interaction at Constant Surface Potential

Here we use Eq. (9.128) (not Eq. (9.96)) for k and Eq. (9.132) (not Eq. (9.100)) for u. For the interaction between two parallel plates with constant surface potential ψ_{o},

$\psi(0)$ is always equal to ψ_o so that Eq. (9.131), which holds also for the constant surface potential case, becomes

$$\psi_o = \psi_m + \frac{2kT}{ze}\ln[dc(u,k)] \tag{9.136}$$

or

$$y_o = y_m + 2\ln[dc(u,k)] \tag{9.137}$$

with

$$k = \exp(-y_m) \tag{9.138}$$

$$u = \frac{\kappa h}{4\sqrt{k}} \tag{9.139}$$

Equation (9.136) (or Eq. (9.137)) is a transcendental equation for ψ_m, which can be solved numerically. By substituting the obtained value ψ_m into Eq. (9.86), which holds irrespective of the type of double-layer interaction (constant surface potential or constant surface charge density models), we can calculate the interaction force $P(h)$.

The potential energy $V^{\psi}(h)$ of the double-layer interaction per unit area between two parallel plates 1 and 2 at separation h is given by Eq. (9.66). The double-layer free energy $F^{\psi}(h)$ of two parallel plates 1 and 2 at constant surface potential ψ_o is given by Eq. (8.54), namely,

$$F^{\psi}(h) = -P(h)h - 2\varepsilon_r\varepsilon_o\int_0^{h/2}\left(\frac{d\psi}{dx}\right)^2 dx \tag{9.140}$$

Here the terms relating to the region $-\infty < x \leq 0$ have been dropped, since only the potential distribution $\psi(x)$ in the region $0 \leq x \leq h$ between plates 1 and 2 changes depending on plate separation h, while the potential distribution in the region $-\infty < x \leq 0$ remains unchanged independent of h. It can be shown that Eq. (9.140) gives [6,7]

$$\begin{aligned} V^{\psi}(h) = \frac{64nkT}{\kappa}\Bigg[&\frac{1}{4}\left\{\cosh\left(\frac{y_o}{2}\right) - 1\right\} - \frac{\kappa h}{64}(3e^{y_m} - 2 - e^{-y_m}) \\ &-\frac{1}{8}\sqrt{2(\cosh y_o - \cosh y_m)} \\ &+\frac{1}{4}e^{y_m/2}\{E(k) - E(\arcsin[e^{y_m - y_o}], k)\}\Bigg] \end{aligned} \tag{9.141}$$

Frens [6–8] found that the following simple relation holds between V^{ψ} and V^{σ}:

$$V^{\sigma}(h) = V^{\psi(0)}(h) + \frac{8nkT}{\kappa}\left[\{y(0) - y_{o}\}\sinh\left(\frac{y_{o}}{2}\right) -2\left\{\cosh\left(\frac{y(0)}{2}\right) - \cosh\left(\frac{y_{o}}{2}\right)\right\}\right] \tag{9.142}$$

In Fig. 9.4, V^{ψ} is plotted as a function of κh in comparison with V^{σ}, showing that V^{ψ} is much lower than V^{σ}.

9.5 COMPARISON BETWEEN THE THEORY OF DERJAGUIN AND LANDAU AND THE THEORY OF VERWEY AND OVERBEEK

The force and potential energy of the double-layer interaction between two parallel plates were calculated by the theory of Derjaguin and Landau [9] and that of Verwey and Overbeek [10], independently. As mentioned earlier, the force of the double-layer interaction between two parallel plates 1 and 2 can be obtained by integrating the osmotic pressure and the Maxwell stress over an arbitrary surface Σ enclosing either one of the two plates. These two theories both treat the double-layer interaction between two similar plates with constant surface potential ψ_{o} at separation h. It is interesting to be noted that these two theories chose different surfaces Σ (Fig. 9.5). In the case of two parallel plates as S, one can choose two arbitrary planes $x = x_1$ $(-\infty < x_1 \leq -d)$ and $x = x_2$ $(0 \leq x_1 \leq h/2)$ enclosing the left plate (plate 1). In the DLVO theory, the interaction between two similar plates at constant surface potential are considered [9,10]. Derjaguin and Landau [9] chose $x_1 = 0$ and $x_2 = -d$, that is, the two surfaces of plate 1 itself. Since $\psi(-d) = \psi\ (0) = \psi_{o}$ (=constant) at the constant surface potential condition, $\Delta\Pi(0) = \Delta\Pi(-d)$ so that $P(h)$ is given by

$$\begin{aligned} P(h) &= [T(x_2) + \Delta\Pi(x_2)] - [T(x_1) + \Delta\Pi(x_1)] \\ &= T(-d) - T(0) \end{aligned} \tag{9.143}$$

Verwey and Overbeek [10], on the other hand, chose $x_1 = -\infty$ and $x_2 = h/2$. Since $\psi = d\psi/dx = 0$ at $x_1 = -\infty$ so that $T(-\infty) = 0$ and $d\psi/dx = 0$ at $x_2 = h/2$ so that $T(h/2) = 0$, $P(h)$ is given by

$$\begin{aligned} P(h) &= [T(x_2) + \Delta\Pi(x_2)] - [T(x_1) + \Delta\Pi(x_1)] \\ &= \Delta\Pi(h/2) - \Delta\Pi(-\infty) \end{aligned} \tag{9.144}$$

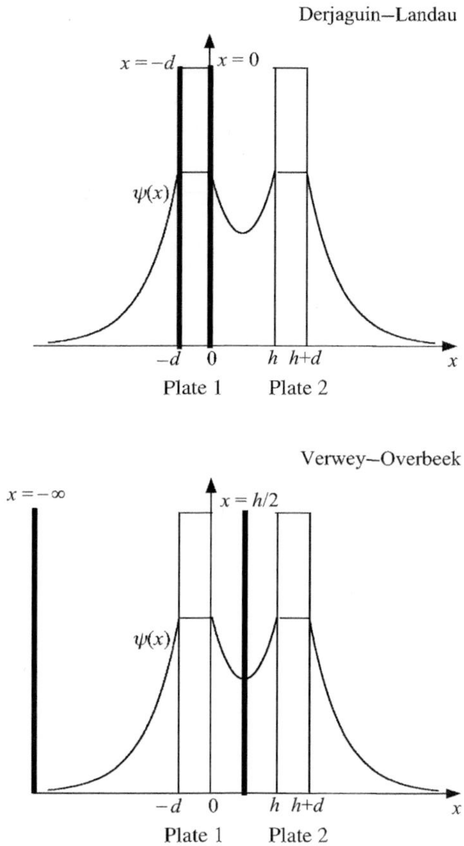

FIGURE 9.5 Comparison between the Derjaguin–Landau theory [9] and the Verwey–Overbeek theory [10]. These two theories give the same result.

Although the choice of Σ is different for the Derjaguin–Landau theory and the Verwey and Overbeek theories, Eqs. (9.143) and (9.144) both give the identical result (Eq. (9.86)).

9.6 APPROXIMATE ANALYTIC EXPRESSIONS FOR MODERATE POTENTIALS

The calculation of the potential energies $V^{\sigma}(h)$ and $V^{\psi}(h)$ of the double-layer interaction between two parallel plates requires numerical solutions to transcendental equations (9.105) and (9.137), respectively. In the following we give approximate analytic expressions for $V^{\psi}(h)$, which does not require numerical calculation. The obtained results, which are correct to the order of the sixth power of the unperturbed surface potential ψ_o, are applicable for low and moderate potentials.

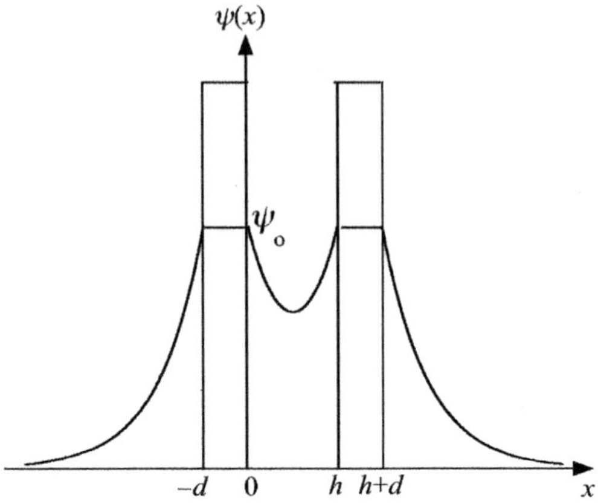

FIGURE 9.6 Schematic representation of the double-layer interaction at constant surface potential between two parallel plates 1 and 2 separated by h. ψ_o is the potential at the mid-point between the plates.

Consider two parallel plates 1 and 2 at separation h in a symmetrical electrolyte solution of valence z and bulk concentration n. We take an x-axis perpendicular to the plate with its origin at the surface of plate 1 (Fig. 9.6). We expand the right-hand side of Eq. (9.72) in a power series of the scaled potential $y = ze\psi/kT$ up to y^5,

$$\begin{aligned}\frac{d^2y}{dx^2} &= \kappa^2 \sinh y \\ &= \kappa^2\left(y + \frac{y^3}{6} + \frac{y^5}{120} + \cdots\right)\end{aligned} \tag{9.145}$$

We express y as a power series of the scaled surface potential $y_o = ze\psi_o/kT$,

$$y(x) = y_o A_1(x) + y_o^3 A_2(x) + y_o^5 A_3(x) + \cdots \tag{9.146}$$

where the functions $A_i(x)$ remain to be determined.

The double-layer free energy can be expressed as (see Eq. (8.54))

$$F^{\psi} = -2\int_0^{\psi_o} \sigma(\psi')d\psi' = -2\psi_o \int_0^1 \sigma(\lambda)d\lambda \tag{9.147}$$

where ψ' $(= \lambda\psi_o)$ and $\sigma(\lambda)$ are, respectively, the surface potential and the surface charge density of plate 1 at stage λ in the charging process (λ varying from 0 to 1). The factor 2 in Eq. (9.147) corresponds to two plate surfaces at $x = 0$ and $x = h$.

The surface charge density σ of plate 1 is related to the potential by

$$\sigma = -\varepsilon_r \varepsilon_o \left. \frac{d\psi}{dx} \right|_{x=0^+} \tag{9.148}$$

According to Eq. (9.146), we expand σ as

$$\sigma = \sigma^{(1)} + \sigma^{(2)} + \sigma^{(3)} + \cdots \tag{9.149}$$

Since $\sigma^{(1)}$, $\sigma^{(2)}$, and $\sigma^{(3)}$ are, respectively, first-, third-, and fifth-degree functions of y_o, their values at stage λ are proportional to λ^1, λ^3, and λ^5, respectively. Thus, we can write the double-layer free energy F^{ψ} as

$$F^{\psi} = -\psi_o \sigma^{(1)} - \frac{1}{2}\psi_o \sigma^{(2)} - \frac{1}{3}\psi_o \sigma^{(3)} + \cdots \tag{9.150}$$

Determining $A_i(x)$ so as to satisfy the boundary conditions on the plate surface, we finally obtain [11]

$$\begin{aligned}
V^{\psi}(h) &= \frac{2nkT}{\kappa} y_o^2 \{1 - \tanh(\kappa h/2)\} \\
&+ \frac{2nkT}{\kappa} y_o^4 \left[\frac{1}{48}\{1 - \tanh(\kappa h/2)\} - \frac{1}{32}\frac{\tanh(\kappa h/2)}{\cosh^2(\kappa h/2)} - \frac{(\kappa h/2)}{32}\frac{1}{\cosh 4(\kappa h/2)} \right] \\
&+ \frac{2nkT}{\kappa} y_o^6 \left[\frac{1}{5760}\{1 - \tanh(\kappa h/2)\} + \frac{1}{1536}\frac{\tanh(\kappa h/2)}{\cosh^2(\kappa h/2)} \right. \\
&- \frac{1}{1024}\frac{\tanh(\kappa h/2)}{\cosh^4(\kappa h/2)} - \frac{7(\kappa h/2)}{3072}\frac{1}{\cosh^6(\kappa h/2)} \\
&\left. + \frac{(\kappa h/2)}{384}\frac{\tanh^2(\kappa h/2)}{\cosh^4(\kappa h/2)} + \frac{(\kappa h/2)^2}{256}\frac{\tanh(\kappa h/2)}{\cosh^6(\kappa h/2)} \right]
\end{aligned} \tag{9.151}$$

where the first term on the right-hand side of Eq. (9.151) agrees with Eq. (9.65). A more detailed mathematical derivation of Eq. (9.151) will be described in the next chapter for the interaction between two parallel dissimilar plates.

Honig and Mul [7] suggested that better approximations can be obtained if the interaction energy is expressed as a series of $\tanh(ze\psi_o/kT)$ instead of ψ_o and derived the interaction energy correct to $\tanh^4(ze\psi_o/kT)$. The interaction energy $V^{\psi}(h)$ per unit area correct to $\tanh^6(ze\psi_o/kT)$ for the interaction between two parallel similar plates at constant surface potential ψ_o separated by a distance h in a symmetrical electrolyte solution of valence z and bulk concentration n can be derived as follows [12]. That is, by introducing a new variable

$$Y(x) = \tanh\left(\frac{y(x)}{4}\right) \tag{9.152}$$

we rewrite the Poisson–Boltzmann equation (9.72) as

$$\begin{aligned} \frac{d^2Y}{dx^2} - \kappa^2 Y &= \frac{2Y}{1-Y^2}\left[\kappa^2 Y^2 - \left(\frac{dY}{dx}\right)^2\right] \\ &= 2Y(1+Y^2+Y^4+\cdots)\left[\kappa^2 Y^2 - \left(\frac{dY}{dx}\right)^2\right] \end{aligned} \tag{9.153}$$

We treat the interaction between two parallel similar plates placed at $x=0$ and h (Fig. 9.6) (i.e., the plates are at separation h) for the case where ψ_o remains constant during interaction. The boundary conditions are

$$Y(0) = Y(h) = \gamma \tag{9.154}$$

where

$$\gamma = \tanh(y_o/4) \tag{9.155}$$

and $y_o = ze\psi_o/kT$ is the scaled surface potential. We can expand $Y(h)$ as

$$Y(h) = \gamma A_1(h) + \gamma^3 A_2(h) + \gamma^5 A_3(h) + \cdots \tag{9.156}$$

From Eq. (9.154) it follows that $A_n(h)$ ($n = 1, 2, \ldots$) must satisfy

$$A_1(0) = A_1(h) = 1 \tag{9.157}$$

$$A_n(0) = A_n(h) = 0 \quad (n = 1, 2, \ldots) \tag{9.158}$$

Substituting Eq. (9.156) into Eq. (9.153) and equating terms of like powers of γ on both sides of Eq. (9.153), we have successive equations for A_n ($n = 1, 2, \ldots$).

Equation (9.86) for the interaction force $P(h)$ per unit area of plates 1 and 2 can be rewritten as

$$\begin{aligned} P(h) &= 4nkT\sinh^2(y_m/2) \\ &= 16nkT\frac{Y_m^2}{(1-Y_m^2)^2} \\ &= 16nkTY_m^2(1+2Y_m^2+3Y_m^4+\cdots) \end{aligned} \tag{9.159}$$

with

$$Y_m = \tanh(y_m/4) = \tanh(ze\psi_m/4kT)$$

where ψ_m is the potential at the midpoint between plates 1 and 2 and y_m is its scaled quantity. Integration of Eq. (9.159) gives the interaction energy per unit area $V^{\psi}(h)$, with the result that

$$\begin{aligned} V^{\psi}(h) = \frac{32nkT}{\kappa} \Bigg[& \gamma^2(1+\gamma^2+\gamma^4)\{1-\tanh(\kappa h/2)\} \\ & - \gamma^4 \left\{ (1+\gamma^2)\frac{\sinh(\kappa h/2)}{2\cosh^3(\kappa h/2)} + \frac{(\kappa h/2)}{2\cosh^4(\kappa h/2)} \right\} \\ & - \gamma^6 \left\{ \frac{\sinh(\kappa h/2)}{4\cosh^5(\kappa h/2)} + \frac{5(\kappa h/2)}{4\cosh^6(\kappa h/2)} - \frac{(\kappa h/2)^2 \sinh(\kappa h/2)}{\cosh^7(\kappa h/2)} \right\} \Bigg] \end{aligned} \tag{9.160}$$

Agreement of Eq. (9.160) with the exact numerical results is excellent. For $y_o \leq 2$, the relative error ε is less than 1% at all κh. For $y_o = 3$, $\varepsilon < 1\%$ at $\kappa h \geq 0.6$ and increases as $\kappa h \to 0$, tending to its maximum 6.6% at $\kappa h = 0$. Even for $y_o = \varepsilon$ $\leq 1\%$ at $\kappa h \geq 0.9$. Note that Eq. (9.160) gives the correct limiting form at large κh, that is, $V^{\psi}(h) \to (64nkT\gamma^2/\kappa)e^{-\kappa h}$ (see Chapter 11).

9.7 ALTERNATIVE METHOD OF LINEARIZATION OF THE POISSON–BOLTZMANN EQUATION

In this section, we present a novel linearization method for simplifying the nonlinear Poisson–Boltzmann equation to derive an accurate analytic expression for the interaction energy between two parallel similar plates in a symmetrical electrolyte solution [13, 14]. This method is different from the usual linearization method (i.e., the Debye–Hückel linearization approximation) in that the Poisson–Boltzmann equation in this method is linearized with respect to the deviation of the electric potential from the surface potential so that this approximation is good for small particle separations, while in the usual method, linearization is made with respect to the potential itself so that this approximation is good for low potentials.

9.7.1 Interaction at Constant Surface Potential

We first consider two interacting parallel similar plates 1 and 2 with constant surface potential ψ_o at separation h in a symmetrical electrolyte solution of valence z and bulk concentration n[13]. We assume that the surface potential ψ_o remains unchanged independent of h during interaction. We take an x-axis perpendicular to the plates with its origin 0 at the surface of plate 1 (Fig. 9.6). The Poisson–Boltzmann equation for the electric potential $\psi(x)$ (which is a function of x only) relative to the bulk solution phase, where $\psi(x)$ is zero, is given by

$$\frac{d^2\psi}{dx^2} = -\frac{zen}{\varepsilon_r\varepsilon_o}\left[\exp\left(-\frac{ze\psi}{kT}\right) - \exp\left(\frac{ze\psi}{kT}\right)\right] \tag{9.161}$$

or in terms of the scaled potential $y = ze\psi/kT$

$$\frac{d^2y}{dx^2} = \kappa^2 \sinh y \tag{9.162}$$

where the Debye–Hückel parameter κ being given by Eq. (9.73). The boundary condition at the plate surface $x = 0$ is

$$y(0) = y_o \tag{9.163}$$

where $y_o = ze\psi_o/kT$ is the scaled surface potential and as a result of the symmetry of the system

$$\left.\frac{dy}{dx}\right|_{x=h/2} = 0 \tag{9.164}$$

In the usual Debye–Hückel linearization approximation, Eq. (9.162) is linearized with respect to y itself, namely,

$$\frac{d^2y}{dx^2} = \kappa^2 y \tag{9.165}$$

which is not a good approximation for large y. Instead, let us now linearize Eq. (9.162) with respect to the deviation Δy of the potential y from the surface potential y_o

$$\Delta y = y - y_o \tag{9.166}$$

By substituting Eq. (9.166) into Eq. (9.162) and linearizing with respect to Δy, we obtain

$$\begin{aligned}\frac{d^2\,\Delta y}{dx^2} &= \kappa^2 \sinh(y_o + \Delta y)\\ &= \kappa^2(\sinh y_o + \cosh_{yo} \cdot \Delta y)\\ &= \kappa^2 \cosh y_o \cdot (\tanh y_o + \Delta y)\end{aligned} \tag{9.167}$$

which has a solution in the form

$$\Delta y = -\tanh y_o + A\exp(-\sqrt{\cosh y_o} \cdot \kappa x) + B\exp(\sqrt{\cosh y_o} \cdot \kappa x) \tag{9.168}$$

where A and B are integration constants. After determining A and B so as to satisfy the boundary conditions (9.163) and (9.164), we obtain for the potential distribution $y(x)$

$$y(x) = y_o - \tanh y_o \cdot \left\{ 1 - \frac{\cosh[\sqrt{\cosh y_o} \cdot \kappa(h/2 - x)]}{\cosh[\sqrt{\cosh y_o} \cdot \kappa h/2]} \right\}, \quad 0 \le x \le \frac{h}{2} \tag{9.169}$$

Note that Eq. (9.169) is only accurate near the plate surface, since Eq. (9.169) is obtained by linearizing the Poisson–Boltzmann equation with respect to $\Delta y = y - y_o$.

Now we calculate the double-layer free energy $F^{\psi}(h)$ per unit area of two interacting parallel similar plates at separation h by Eq. (8.54),

$$F^{\psi}(h) = 2\int_0^{\sigma} \psi_o(\sigma)d\sigma - 2\sigma\psi_o = -2\int_0^{\psi_o} \sigma(\psi_o)d\psi_o \tag{9.170}$$

where $\psi_o(\sigma)$ is the surface potential when the surface charge density is σ during the charging process. The relation between the surface potential ψ_o (or y_o) and the surface charge density σ is obtained from

$$\sigma = -\varepsilon_r\varepsilon_o \left.\frac{d\psi}{dx}\right|_{x=0^+} \tag{9.171}$$

Substituting Eq. (9.169) into Eq. (9.171) yields

$$\sigma = \frac{\varepsilon_r\varepsilon_o\kappa kT}{ze} \tanh y_o \cdot \sqrt{\cosh y_o} \cdot \tanh\left[\sqrt{\cosh y_o} \cdot \kappa h/2\right] \tag{9.172}$$

Then substituting Eq. (9.172) into Eq. (9.170) and integrating, we obtain

$$F^{\psi}(h) = \frac{16nkT}{\kappa^2 h} \ln\left[\frac{1 + \exp(-\kappa h)}{1 + \exp(-\sqrt{\cosh y_o} \cdot \kappa h)}\right] - \frac{8nkT}{\kappa}(\sqrt{\cosh y_o} - 1) \tag{9.173}$$

Note that Eq. (9.173) has the correct form at separation $h = 0$, namely, $F^{\psi}(0) = 0$, but in the limit $\kappa h \to \infty$, Eq. (9.173) tends to

$$F^{\psi}(\infty) = -\frac{8nkT}{\kappa}(\sqrt{\cosh y_o} - 1) \tag{9.174}$$

which disagrees with the correct limiting expression (see Eq. (5.53))

$$F^{\psi}_{\text{correct}}(\infty) = -\frac{16nkT}{\kappa}\{\cosh(y_o/2) - 1\} \tag{9.175}$$

The reason for this discrepancy between Eqs. (9.174) and (9.175) (which agree with each other only for small y_o) is that Eq. (9.169) is only accurate for small h but becomes less accurate as h increases. Thus, in order to calculate the double-layer interaction energy one must use the correct expression $F^{\psi}_{\mathrm{correct}}(\infty)$ given by Eq. (9.175) instead of $F^{\psi}(\infty)$ given by Eq. (9.174).

The potential energy $V^{\psi}(h)$ of the double-layer interaction per unit area of the plates at separation h is thus given by

$$V^{\psi}(h) = F^{\psi}(h) - F^{\psi}_{\mathrm{correct}}(\infty) \tag{9.176}$$

Substituting Eqs. (9.173) and (9.175) into Eq. (9.176), we obtain

$$V^{\psi}(h) = \frac{8nkT}{\kappa}\left\{\frac{2}{\kappa h}\ln\left[\frac{1+\exp(-\kappa h)}{1+\exp(-\sqrt{\cosh y_o}\cdot\kappa h)}\right] + 2\cosh(y_o/2) - \sqrt{\cosh y_o} - 1\right\} \tag{9.177}$$

This is the required expression for the interaction energy per unit area between two parallel similar plates with constant surface potential. From the nature of this linearization, the obtained potential distribution (9.177) is only accurate near the plate surface and thus the interaction energy expression (9.177) is also only accurate for small plate separations. Indeed, at $h=0$, Eq. (9.177) gives the following correct expression, regardless of the value of y_o,

$$V^{\psi}(0) = \frac{16nkT}{\kappa}\{\cosh(y_o/2) - 1\} \tag{9.178}$$

As h increases, on the other hand, Eq. (9.177) becomes less accurate and in the limit $h \to \infty$, Eq. (9.177) does not become zero.

Note, however, that in the limit of low surface potential, Eq. (9.177) reduces to the correct limiting form, regardless of the value of κh,

$$V^{\psi}(h) = \frac{2nkT}{\kappa} y_o^2\{1 - \tanh(\kappa h/2)\} \tag{9.179}$$

which agrees with Eq. (9.65).

In Fig. 9.7, we compare exact numerical values and approximate results obtained from an approximate expression (9.177). It is seen that the present linearization approximation works quite well for small separations for all values of the reduced surface potential y_o but becomes less accurate at larger separations.

9.7.2 Interaction at Constant Surface Charge Density

Consider next two interacting parallel similar plates 1 and 2 with surface charge density σ at separation h in a symmetrical electrolyte solution of valence z and bulk concentration n [14]. We assume that the surface charge density σ remains unchanged independent of h during interaction. We take an x-axis perpendicular to

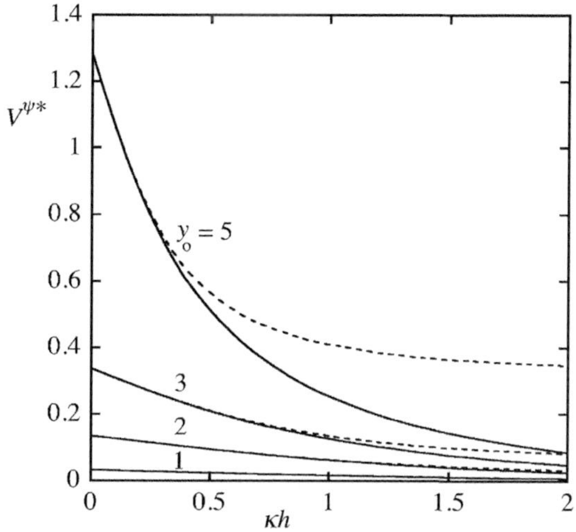

FIGURE 9.7 Reduced potential energy $V^{\psi *} \equiv (\kappa/64nkT)V^{\psi}$ of the double-layer interaction per unit area between two parallel similar plates with constant surface potential ψ_o as a function of the reduced distance κh between the plates for several values of the scaled unperturbed surface potential $y_o \equiv ze\psi_o/kT$. Solid lines are exact values and dotted lines represent approximate results calculated by Eq. (9.177). The exact and approximate results for $y_o = 1$ agree with each other within the linewidth. (From Ref. 13.)

the plates with its origin 0 at the surface of plate 1 (Fig. 9.1). The Poisson–Boltzmann equation is given by Eq. (9.162). As a result of the symmetry of the system, we need consider only the region $0 \leq x \leq h/2$. We treat the case in which the influence of the internal electric fields can be neglected. The boundary condition at the plate surface $x = 0$ is therefore

$$\left.\frac{d\psi}{dx}\right|_{x=0^+} = -\frac{\sigma}{\varepsilon_r \varepsilon_o} \tag{9.180}$$

or

$$\left.\frac{dy}{dx}\right|_{x=0^+} = -\frac{ze\sigma}{\varepsilon_r \varepsilon_o kT} \tag{9.181}$$

and by symmetry

$$\left.\frac{dy}{dx}\right|_{x=h/2} = 0 \tag{9.182}$$

We introduce the unperturbed surface potential ψ_o, that is, the surface potential $\psi(0)$ in the absence of interaction at $h \to \infty$. The surface charge density σ is related to ψ_o by Eq. (1.41), namely,

$$\sigma = \frac{2\varepsilon_r\varepsilon_o\kappa kT}{ze}\sinh\left(\frac{y_o}{2}\right) = \frac{4nze}{\kappa}\sinh\left(\frac{y_o}{2}\right) \tag{9.183}$$

We linearize Eq. (9.162) with respect to the deviation Δy of the potential y from the surface potential $y(0)$

$$\Delta y = y(x) - y(0). \tag{9.184}$$

By substituting Eq. (9.184) into Eq. (9.162) and linearizing with respect to Δy, we obtain

$$\begin{aligned} \frac{d^2\,\Delta y}{dx^2} &= \kappa^2 \sinh[y(0) + \Delta y] \\ &= \kappa^2[\sinh y(0)\cdot\cosh\Delta y + \cosh y(0)\cdot\sinh\Delta y] \\ &\approx \kappa^2[\sinh y(0) + \cosh y(0)\cdot\Delta y] \\ &= \kappa^2\cosh y(0)\cdot[\tanh y(0) + \Delta y] \end{aligned} \tag{9.185}$$

where we have considered the deviation Δy to be small so that $\cosh\Delta y \approx 1$ and $\sinh \Delta y \approx \Delta y$. Equation (9.185) has a solution in the form

$$\Delta y = -\tanh y(0) + A\exp(-\sqrt{\cosh y(0)}\cdot\kappa x) + B\exp(\sqrt{\cosh y(0)}\cdot\kappa x) \tag{9.186}$$

where A and B are integration constants. After determining A and B so as to satisfy the boundary conditions (9.181) and (9.182), we obtain for the potential distribution $y(x)$

$$\begin{aligned} y(x) &= y(0) + \Delta y \\ &= y(0) - \tanh y(0) + \frac{\cosh[\sqrt{\cosh y(0)}\cdot\kappa(h/2 - x)]}{\sinh[\sqrt{\cosh y(0)}\cdot\kappa h/2]}\frac{ze\sigma}{\varepsilon_r\varepsilon_o\kappa kT\sqrt{\cosh y(0)}}, \\ &0 \le x \le \frac{h}{2} \end{aligned} \tag{9.187}$$

On setting $x = 0$ in Eq. (9.187), we obtain the following relation between $y(0)$ and σ

$$\sigma = \frac{\varepsilon_r\varepsilon_o\kappa kT}{ze}\tanh y(0)\cdot\sqrt{\cosh y(0)}\cdot\tanh\left[\sqrt{\cosh y(0)}\cdot\kappa h/2\right] \tag{9.188}$$

which gives $\sigma(\psi(0))$ as a function of $\psi(0)$. Note that Eq. (9.188) has the correct limiting form as $h \to 0$, namely,

$$\sinh y(0) \to \frac{\sigma}{nzeh} \quad \text{or} \quad y(0) \to 2\ln\left(\frac{1}{h}\right) \quad \text{as } h \to 0 \tag{9.189}$$

but Eq. (9.188) does not give the correct relation (9.183) in the limit $\kappa h \to \infty$.

The criterion for the present approximation to be valid is $|\Delta y| \ll 1$. Since

$$|\Delta y| = |y(x) - y(0)| < \left| \left.\frac{dy}{dx}\right|_{x=0^+} \right| \cdot \frac{h}{2} = 2\kappa \sinh\left(\frac{|y_o|}{2}\right) \cdot \frac{h}{2} \tag{9.190}$$

the criterion may be expressed as

$$\sinh\left(\frac{|y_o|}{2}\right) \cdot \frac{\kappa h}{2} \ll 1 \tag{9.191}$$

The double-layer free energy $F(h)$ per unit area of two interacting parallel similar plates with constant surface charge density σ at separation h is given by Eq. (8.50), namely,

$$F^\sigma(h) = 2\int_0^\sigma \psi(0) d\sigma = -2\int_0^{\psi(0)} \sigma(\psi(0)) d\psi(0) + 2\sigma\psi(0) \tag{9.192}$$

Then substituting Eq. (9.188) for $\sigma(\psi(0))$ and Eq. (9.183) for σ into Eq. (9.192) and integrating, we obtain

$$F^\sigma(h) = \frac{8nkT}{\kappa}\left\{\frac{2}{\kappa h}\ln\left[\frac{1+\exp(-\kappa h)}{1+\exp(-\sqrt{\cosh y(0)}\cdot\kappa h)}\right] - \sqrt{\cosh y(0)} + 1 + y(0)\sinh\left(\frac{y_o}{2}\right)\right\} \tag{9.193}$$

Note that Eq. (9.193) has the correct limiting form at separation $h \to 0$ [15, 16], namely,

$$F^\sigma(h) \to \frac{2|\sigma|kT}{ze}\ln\left(\frac{1}{h}\right) \quad \text{as } h \to 0 \tag{9.194}$$

(which follows from Eqs. (9.183) and (9.189)), but in the limit $\kappa h \to \infty$, Eq. (9.193), which is correct only for small κh, does not tend to the following correct limiting expression:

$$F^\sigma_{\text{correct}}(\infty) = \frac{8nkT}{\kappa}\left\{y_o \sinh\left(\frac{y_o}{2}\right) - 2\left(\cosh\left(\frac{y_o}{2}\right) - 1\right)\right\} \tag{9.195}$$

which is obtained from Eqs. (9.183) and (9.192).

The potential energy $V^\sigma(h)$ of the double-layer interaction per unit area between plates 1 and 2 at separation h is thus given by

$$V^\sigma(h) = F^\sigma(h) - F^\sigma_{\text{correct}}(\infty) \tag{9.196}$$

Substituting Eqs. (9.193) and (9.195) into Eq. (9.196), we obtain

$$V^{\sigma}(h) = \frac{8nkT}{\kappa}\left\{\frac{2}{\kappa h}\ln\left[\frac{1+\exp(-\kappa h)}{1+\exp(-\sqrt{\cosh y(0)}\cdot\kappa h)}\right] - \sqrt{\cosh y(0)} - 1 + 2\cosh\left(\frac{y_o}{2}\right) + [y(0) - y_o]\sinh\left(\frac{y_o}{2}\right)\right\} \tag{9.197}$$

Here $y(0)$ is related to y_o by

$$\tanh y(0) = \frac{2\sinh(y_o/2)}{\tanh[\sqrt{\cosh y(0)}\cdot\kappa h/2]\cdot\sqrt{\cosh y(0)}} \tag{9.198}$$

which is obtained from Eqs. (9.183) and (9.188). Note that for small κh (i.e., under the condition given by Eq. (9.191)), Eq. (9.198) reduces to

$$\sinh y(0) = \frac{2\sinh(y_o/2)}{\kappa h/2} \tag{9.199}$$

while for small $y(0)$, $\sqrt{\cosh y(0)}$ in Eq. (9.198) may be replaced by 1 and $\tanh y(0) \approx \sinh y(0)$ so that Eq. (9.198) is approximated by

$$\sinh y(0) = \frac{2\sinh(y_o/2)}{\tanh(\kappa h/2)} \tag{9.200}$$

Since for small κh, Eq. (9.199) tends to Eq. (9.200), Eq. (9.188) is a good approximation for Eq. (9.198) even for large $y(0)$. It is thus found that Eq. (9.198) can be approximated fairly well by

$$y(0) = \text{arcsinh}\left[\frac{2\sinh(y_o/2)}{\tanh(\kappa h/2)}\right] = \ln\left[\frac{2\sinh(y_o/2)}{\tanh(\kappa h/2)} + \sqrt{\left\{\frac{2\sinh(y_o/2)}{\tanh(\kappa h/2)}\right\}^2 + 1}\right] \tag{9.201}$$

Equation (9.197) as combined with Eq. (9.201) is the required expression for the potential energy of the double-layer interaction per unit area between two parallel similar plates with constant surface charge density σ. From the nature of this linearization, the obtained potential distribution (Eq. (9.187)) is only accurate near the plate surface and thus the interaction energy expression (9.197) is also only accurate for small plate separations h. Indeed, as $h \to 0$, Eq. (9.197) gives the following correct limiting expression [15, 16]:

$$V^{\sigma}(h) \to \frac{2|\sigma|kT}{ze}\ln\left(\frac{1}{h}\right) \quad \text{as } h \to 0 \tag{9.202}$$

which in turn gives a correct limiting form of the interaction force $P(h) = -dV(h)/dh$ per unit area between the two plates as $h \to 0$

$$P^{\sigma}(h) \to \frac{2|\sigma|kT}{zeh} \quad \text{as } h \to 0 \tag{9.203}$$

As h increases, on the other hand, Eq. (9.197) becomes less accurate and thus in the limit $h \to \infty$, Eq. (9.197) does not become zero.

Note, however, that in the limit of low surface potential y_o, Eq. (9.197) reduces to the correct limiting form,

$$V^{\sigma}(h) = \frac{2nkT}{\kappa} y_o^2 \{\coth(\kappa h/2) - 1\} \tag{9.204}$$

In Fig. 9.8, we compare exact values and approximate results obtained from Eq. (9.197) as combined with Eq. (9.201). It is seen that the present linearization approximation works quite well for small separations for all values of the surface charge density σ or the reduced surface potential y_o but becomes less accurate at larger separations. For $y_o = 1$, the relative error ε is 0.02% at $\kappa h = 1.2$ and 2.7% at $\kappa h = 2$. For $y_o = 2$, $\varepsilon = -0.9\%$ at $\kappa h = 0.6$ and -3.6% at $\kappa h = 1.2$. For $y_o = 3$, $\varepsilon = -1.3\%$ at $\kappa h = 0.4$ and -7.8% at $\kappa h = 0.9$. For $y_o = 5$, $\varepsilon = -2.3\%$ at $\kappa h = 0.2$.

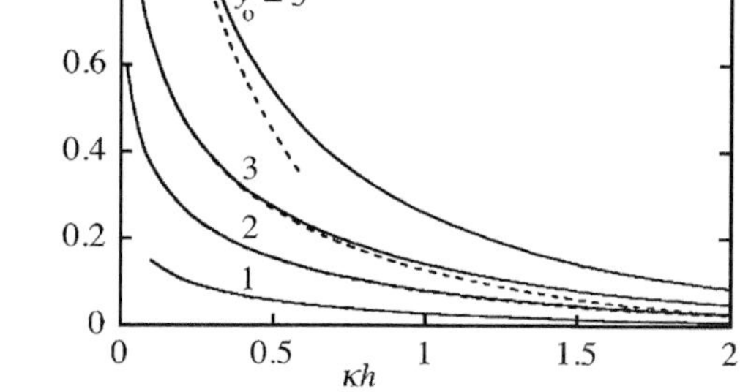

FIGURE 9.8 Reduced potential energy $V^{\sigma *} \equiv (\kappa/64nkT)V^{\sigma}$ of the double-layer interaction per unit area between two parallel similar plates with constant surface charge density σ as a function of the reduced distance κh between the plates for several values of the scaled unperturbed surface potential $y_o \equiv ze\psi_o/kT$. Solid lines are exact values and dotted lines represent approximate results calculated by Eq. (9.197) as combined with Eq. (9.201). The exact and approximate results for $y_o = 1$ agree with each other within the linewidth. (From Ref. 14.)

Here $\varepsilon > 0$ corresponds to overestimation and $\varepsilon < 0$ to underestimation. We see that for larger y_o, the approximation is good only for smaller separations. This can be explained by noting that the criterion for this approximation to be valid is given by Eq. (9.191). That is, as y_o increases, the range of κh in which the approximation is good becomes smaller.

REFERENCES

1. H. Ohshima, *Colloid Polym. Sci.* 252 (1974) 158.
2. H. Ohshima, *Colloid Polym. Sci.* 253 (1975) 150.
3. H. Ohshima, *Colloid Polym. Sci.* 254 (1976) 484.
4. H. Ohshima,in: H. Ohshima and K. Furusawa (Eds.), *Electrical Phenomena at Interfaces, Fundamentals, Measurements, and Applications, 2nd edition*, revised and expanded, Dekker, New York, 1998, Chapter 3.
5. H. Ohshima, *Theory of Colloidal and Interfacial Electrical Phenomena*, Elsevier/ Academic Press, Amsterdam, 2006.
6. G. Frens,Thesis, The reversibility of irreversible colloids, Utrecht, 1968.
7. E. P. Honig and P. M. Mul, *J. Colloid Interface Sci.* 36 (1971) 258.
8. G. Frens and J. Th. G. Overbeek, *J. Colloid Interface Sci.* 38 (1972) 376.
9. B. V. Derjaguin and L. Landau, *Acta Physicochim.* 14 (1941) 633.
10. E. J. W. Verwey and J. Th. G. Overbeek, *Theory of the Stability of Lyophobic Colloids*, Elsevier, Amsterdam, 1948.
11. H. Ohshima, T. W. Healy, and L. R. White, *J. Colloid Interface Sci.* 89 (1982) 484.
12. H. Ohshima and T. Kondo, *J. Colloid Interface Sci.* 122 (1988) 591.
13. H. Ohshima, *Colloids Surf. A: Physicochem. Eng. Sci.* 146 (1999) 213.
14. H. Ohshima, *J. Colloid Interface Sci.* 212 (1999) 130.
15. J. N. Israelachvili, *Intermolecular and Surface Forces, 2nd edition*, Academic Press, New York, 1992.
16. D. McCormack, S. L. Carnie, and D. Y. C. Chan, *J. Colloid Interface Sci.* 169 (1995) 177.

10 Electrostatic Interaction Between Two Parallel Dissimilar Plates

10.1 INTRODUCTION

In this chapter, we discuss two models for the electrostatic interaction between two parallel dissimilar hard plates, that is, the constant surface charge density model and the surface potential model. We start with the low potential case and then we treat with the case of arbitrary potential.

10.2 INTERACTION BETWEEN TWO PARALLEL DISSIMILAR PLATES

Consider two parallel dissimilar plates 1 and 2 having thicknesses d_1 and d_2, respectively, separated by a distance h immersed in a liquid containing N ionic species with valence z_i and bulk concentration (number density) n_i^∞ ($i = 1, 2, \ldots, N$). We take an x-axis perpendicular to the plates with its origin at the right surface of plate 1, as in Fig. 10.1. We assume that the electric potential $\psi(x)$ outside the plates ($-\infty < x < -d$, $0 < x \leq h$, and $h + d_2 < x < \infty$) obeys the following one-dimensional planar Poisson–Boltzmann equation:

$$\frac{d^2\psi(x)}{dx^2} = -\frac{1}{\varepsilon_r\varepsilon_o}\sum_{i=1}^{N} z_i e n_i^\infty \exp\left(-\frac{z_i e\psi(x)}{kT}\right), \quad \text{for } -\infty < x < -d_1,\ 0 < x < h,$$
$$\text{and } h + d_2 < x < \infty \tag{10.1}$$

For the region inside the plates ($-d_1 < x < 0$ and $h < x < h + d_2$), the potential satisfies the Laplace equation

$$\frac{d^2\psi(x)}{dx^2} = 0, \quad \text{for } -d_1 < x < 0 \quad \text{and} \quad h < x < h + d_2 \tag{10.2}$$

Biophysical Chemistry of Biointerfaces By Hiroyuki Ohshima

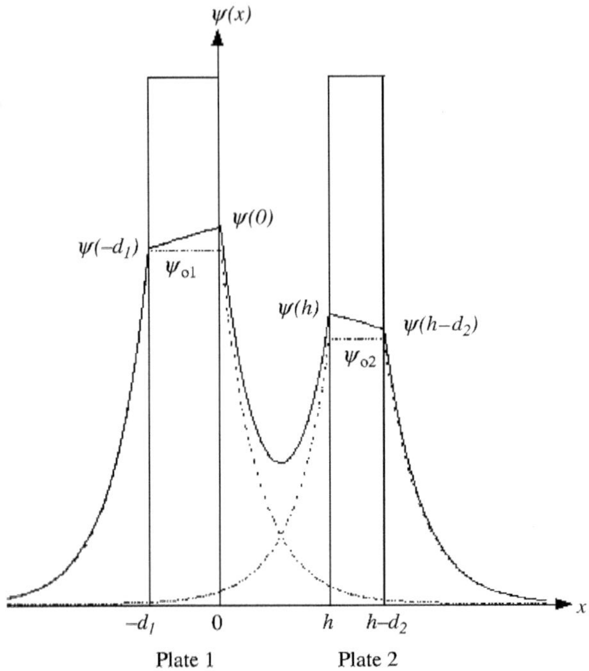

FIGURE 10.1 Schematic representation of the potential distribution $\psi(x)$ (solid line) across two interacting parallel dissimilar plates 1 and 2 at separation h. Dotted line is the unperturbed potential distribution at $h=\infty$. ψ_{o1} and ψ_{o2} are the unperturbed surface potentials of plate 1 and 2, respectively.

The boundary conditions at $x \to \pm\infty$ are

$$\psi(x) \to 0 \quad \text{as } x \to \pm\infty \tag{10.3}$$

$$\frac{d\psi}{dx} \to 0 \quad \text{as } x \to \pm\infty \tag{10.4}$$

The boundary conditions at the plate surface depends on the type of the double--layer interaction between plates 1 and 2. If the surface potentials of the plates 1 and 2 remain constant at ψ_{o1} and ψ_{o2} during interaction respectively, then

$$\psi(-d_1) = \psi(0) = \psi_{o1} \tag{10.5}$$

$$\psi(h) = \psi(h + d_2) = \psi_{o2} \tag{10.6}$$

On the other hand, if the surface charge densities of plates 1 and 2 remain constant at σ_1 and σ_2, then

$$\psi(-d_1^-) = \psi(-d_1^+) \tag{10.7}$$

$$\psi(0^-) = \psi(0^+) \tag{10.8}$$

$$\psi(h^-) = \psi(h^+) \tag{10.9}$$

$$\psi(h + d_2{}^-) = \psi(h + d_2{}^+) \tag{10.10}$$

$$\varepsilon_r \left.\frac{d\psi}{dx}\right|_{x=-d_1{}^-} - \varepsilon_{p1} \left.\frac{d\psi}{dx}\right|_{x=-d_1{}^+} = \frac{\sigma_1}{\varepsilon_o} \tag{10.11}$$

$$\varepsilon_r \left.\frac{d\psi}{dx}\right|_{x=0^-} - \varepsilon_{p1} \left.\frac{d\psi}{dx}\right|_{x=0^+} = \frac{\sigma_1}{\varepsilon_o} \tag{10.12}$$

$$\varepsilon_r \left.\frac{d\psi}{dx}\right|_{x=h^-} - \varepsilon_{p2} \left.\frac{d\psi}{dx}\right|_{x=h^+} = \frac{\sigma_2}{\varepsilon_o} \tag{10.13}$$

$$\varepsilon_r \left.\frac{d\psi}{dx}\right|_{x=h+d_2{}^-} - \varepsilon_{p2} \left.\frac{d\psi}{dx}\right|_{x=h+d_2{}^+} = \frac{\sigma_2}{\varepsilon_o} \tag{10.14}$$

where ε_{pi} is the relative permittivity of plate i ($i = 1$ and 2). For the mixed case where plate 1 is at constant potential y_{o1} and plate 2 is at constant surface charge density σ_2 ($= \varepsilon_r\varepsilon_o\kappa\psi_{o2}$), Eqs. (10.5) for plate 1, and Eqs. (10.9), (10.13), and (10.14) must be employed.

As shown in Chapter 8, the interaction force P can be calculated by integrating the excess osmotic pressure $\Delta\Pi$ and the Maxwell stress T over an arbitrary closed surface Σ enclosing either one of the two interacting plates (Eq. (8.6)). As an arbitrary surface Σ enclosing plate 1, we choose two planes $x = -\infty$ and $x = 0$, since $\psi(x) = d\psi/dx = 0$ at $x = -\infty$ (Eqs. (10.3) and (10.4)) so that the excess osmotic pressure $\Delta\Pi$ and the Maxwell stress T are both zero at $x = -\infty$. Thus, the force $P(h)$ of the double-layer interaction per unit area between plates 1 and 2 can be expressed as

$$\begin{aligned} P(h) &= [T(0) + \Delta\Pi(0)] - [T(-\infty) + \Delta\Pi(-\infty)] \\ &= T(0) + \Delta\Pi(0) \end{aligned} \tag{10.15}$$

with

$$T(0) = -\frac{1}{2}\varepsilon_r\varepsilon_o \left(\left.\frac{d\psi}{dx}\right|_{x=0^+} \right)^2 \tag{10.16}$$

$$\Delta\Pi(0) = kT\sum_{i=1}^{N} n_i^{\infty}\left[\exp\left(-\frac{z_i e\psi(0)}{kT}\right) - 1\right] \tag{10.17}$$

That is,

$$P(h) = kT\sum_{i=1}^{N} n_i^{\infty}\left[\exp\left(-\frac{z_i e\psi(0)}{kT}\right) - 1\right] - \frac{1}{2}\varepsilon_r\varepsilon_o\left(\left.\frac{d\psi}{dx}\right|_{x=0^+}\right)^2 \tag{10.18}$$

Here $P(h) > 0$ corresponds to repulsion and $P(h) < 0$ to attraction.

10.3 LOW POTENTIAL CASE

For the low potential case, Eq. (10.18) for the interaction force $P(h)$ per unit area between plates 1 and 2 at separation h reduces to

$$P(h) = \frac{1}{2}\varepsilon_r\varepsilon_o\left[\kappa^2\psi^2(0) - \left(\left.\frac{d\psi}{dx}\right|_{x=0^+}\right)^2\right] \tag{10.19}$$

with

$$\kappa = \left(\frac{1}{\varepsilon_r\varepsilon_o kT}\sum_{i=1}^{N} z_i^2 e^2 n_i^{\infty}\right)^{1/2} \tag{10.20}$$

being the Debye–Hückel parameter. For the low potential case, Eq. (10.1) can be linearized to give

$$\frac{d^2\psi}{dx^2} = \kappa^2\psi, \quad \text{for } -\infty < x < -d_1,\ 0 < x < h, \quad \text{and} \quad h + d_2 < x < \infty \tag{10.21}$$

10.3.1 Interaction at Constant Surface Charge Density

For the constant surface charge density case, the solutions to Eqs. (10.2) and (10.21) subject to Eqs. (10.3), (10.4), and and (10.7)–(10.14) are [1]

$$\psi(x) = \psi(-d_1)e^{\kappa(x+d_1)}, \quad -\infty < x \le -d_1 \tag{10.22}$$

$$\psi(x) = \frac{\psi(0) - \psi(-d_1)}{d_1}x + \psi(0), \quad \text{for } -d_1 \le x \le 0 \tag{10.23}$$

$$\psi(x) = \frac{\psi(0)\sinh(\kappa(h-x)) + \psi(h)\sinh(\kappa x)}{\sinh(\kappa h)}, \qquad -0_1 \leq x \leq \mathrm{h} \tag{10.24}$$

$$\psi(x) = \frac{\psi(h+d_2) - \psi(h)}{d_2}(x-h) + \psi(h), \quad \text{for } h \leq x \leq h + d_1 \tag{10.25}$$

$$\psi(x) = \psi(h+d_2)e^{-\kappa(x-h-d_2)}, \quad h + d_2 \leq x < \infty \tag{10.26}$$

with

$$\psi(-d_1) = \frac{\psi_{o1}\{2\alpha_1 + \alpha_2 - \alpha_1\alpha_2 + (1 - \alpha_1 + 2\alpha_1\alpha_2)\tanh(\kappa h)\} + \psi_{o2}\alpha_1(1+\alpha_2)\mathrm{sech}(\kappa h)}{\alpha_1 + \alpha_2 + (1+\alpha_1\alpha_2)\tanh(\kappa h)} \tag{10.27}$$

$$\psi(0) = \frac{\psi_{o1}(1+\alpha_1)\{1 + \alpha_2\tanh(\kappa h)\} + \psi_{o2}(1+\alpha_2)\mathrm{sech}(\kappa h)}{\alpha_1 + \alpha_2 + (1+\alpha_1\alpha_2)\tanh(\kappa h)} \tag{10.28}$$

$$\psi(h) = \frac{\psi_{o2}(1+\alpha_2)\{1 + \alpha_1\tanh(\kappa h)\} + \psi_{o1}(1+\alpha_1)\mathrm{sech}(\kappa h)}{\alpha_1 + \alpha_2 + (1+\alpha_1\alpha_2)\tanh(\kappa h)} \tag{10.29}$$

$$\psi(h+d_2) = \frac{\psi_{o2}\{2\alpha_2 + \alpha_1 - \alpha_1\alpha_2 + (1 - \alpha_2 + 2\alpha_1\alpha_2)\tanh(\kappa h)\} + \psi_{o1}\alpha_2(1+\alpha_1)\mathrm{sech}(\kappa h)}{\alpha_1 + \alpha_2 + (1+\alpha_1\alpha_2)\tanh(\kappa h)} \tag{10.30}$$

$$\psi_{o1} = \frac{\sigma_1}{\varepsilon_r\varepsilon_o\kappa} \tag{10.31}$$

$$\psi_{o2} = \frac{\sigma_2}{\varepsilon_r\varepsilon_o\kappa} \tag{10.32}$$

$$\alpha_1 = \frac{1}{1 + (\varepsilon_r/\varepsilon_{p1})\kappa d} \tag{10.33}$$

$$\alpha_2 = \frac{1}{1 + (\varepsilon_r/\varepsilon_{p2})\kappa d} \tag{10.34}$$

where ψ_{oi} is the unperturbed surface potential of plate i at $\kappa h = \infty$, and α_i characterizes the influence of the internal electric field within plate i. In the limit of $\varepsilon_{pi} \to 0$ and/or $d_i \to \infty$, α_i tends to 0. This situation, where the influence of the internal electric fields within plate i may be ignored, corresponds to $\alpha_i = 0$.

By using Eqs. (10.19) and (10.24), we obtain the following expression for the interaction force $P^{\sigma}(h)$ per unit area between plates 1 and 2 [1]

$$P^{\sigma}(h) = \frac{1}{2}\varepsilon_r\varepsilon_o\kappa^2(1+\alpha_1)(1+\alpha_2)$$
$$\times \frac{\psi_{o1}^2(1+\alpha_1)(1-\alpha_2) + \psi_{o2}^2(1-\alpha_1)(1+\alpha_2) + 2\psi_{o1}\psi_{o2}\{(1+\alpha_1\alpha_2)\cosh(\kappa h) + (\alpha_1+\alpha_2)\sinh(\kappa h)\}}{\{(1+\alpha_1\alpha_2)\sinh(\kappa h) + (\alpha_1+\alpha_2)\cosh(\kappa h)\}^2} \tag{10.35}$$

We see that if σ_1 and σ_2 are of like sign (i.e., ψ_{o1} and ψ_{o2} are of like sign), or if either of σ_1 or σ_2 is zero, then P is always positive (repulsive). If σ_1 and σ_2 are of opposite sign (i.e., ψ_{o1} and ψ_{o2} are of opposite sign), P is negative (attractive) at $\kappa h \to \infty$, but in some cases there exists a minimum of P and P may change from negative (attractive) to positive (repulsive), as summarized below.

(i) If the condition

$$\left|\frac{\sigma_1}{\sigma_2}\right| \geq \frac{1+\alpha_2}{1-\alpha_2} \tag{10.36}$$

is satisfied, then P reaches a minimum P_m

$$P_m = -\frac{1}{2}\varepsilon_r\varepsilon_o\kappa^2\left(\frac{1+\alpha_2}{1-\alpha_2}\right)\psi_{o2}^2 \tag{10.37}$$

at $h = h_m$, given by

$$\kappa h_m = \ln\left[\left|\frac{\sigma_1}{\sigma_2}\right|\left(\frac{1-\alpha_2}{1+\alpha_2}\right) + \sqrt{\left(\frac{\sigma_1}{\sigma_2}\cdot\frac{1-\alpha_2}{1+\alpha_2}\right)^2 - \frac{(1-\alpha_1)(1-\alpha_2)}{(1+\alpha_1)(1+\alpha_2)}}\right] \tag{10.38}$$

and becomes zero at $h = h_0$, given by

$$\kappa h_0 = \ln\left[\left|\frac{\sigma_1}{\sigma_2}\right|\left(\frac{1-\alpha_2}{1+\alpha_2}\right)\right] \tag{10.39}$$

For $h < h_0$, P is repulsive (positive). At $h = h_o$, at which the interaction from P becomes zero, the potential distribution near the higher charged plate (plate 1) equals the unperturbed potential distribution in the absence of plate 2 (see Fig. 10.2) so that the net force acting plate 1 becomes zero.

(ii) If

$$\frac{1}{2}\left(\frac{1-\alpha_1}{1+\alpha_1}+\frac{1+\alpha_2}{1-\alpha_2}\right) \leq \left|\frac{\sigma_1}{\sigma_2}\right| < \frac{1+\alpha_2}{1-\alpha_2} \tag{10.40}$$

then P is attractive for all κh and also has a minimum P_m given by Eq. (10.37) at $h = h_m$.

(iii) If

$$2\left(\frac{1+\alpha_1}{1-\alpha_1}+\frac{1-\alpha_2}{1+\alpha_2}\right)^{-1} < \left|\frac{\sigma_1}{\sigma_2}\right| < \frac{1}{2}\left(\frac{1-\alpha_1}{1+\alpha_1}+\frac{1+\alpha_2}{1-\alpha_2}\right) \tag{10.41}$$

then P is attractive for all κh and has its minimum at $h = 0$.

(iv) If

$$\left|\frac{\sigma_1}{\sigma_2}\right| \leq 2\left(\frac{1+\alpha_1}{1-\alpha_1}+\frac{1-\alpha_2}{1+\alpha_2}\right)^{-1} \tag{10.42}$$

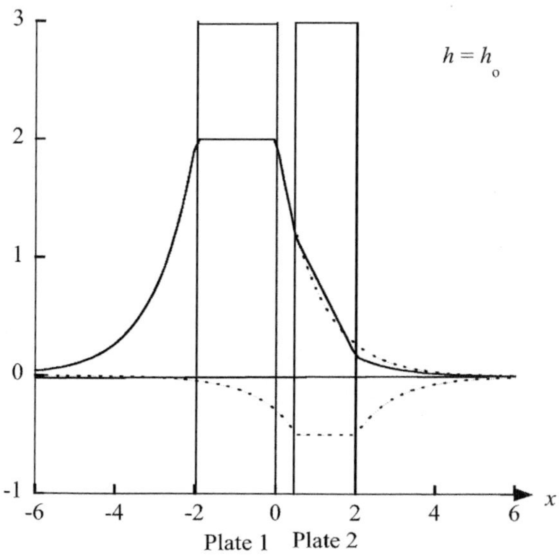

FIGURE 10.2 Potential distribution $\psi(x)$ when at $h = h_o = 0.539$ (Eq. (10.38), at which $P(h) = 0$. Calculated with $a_1 = 0.1$ and $a_2 = 0.4$ and $\sigma_1/\sigma_2 = -8$ (Case (i)).

then P is attractive for all κh and has a minimum P'_{m},

$$P'_{\mathrm{m}} = -\frac{1}{2}\varepsilon_r\varepsilon_o\kappa^2\left(\frac{1+\alpha_1}{1-\alpha_1}\right)\psi_{o1}^2 \tag{10.43}$$

at $h = h'_{\mathrm{m}}$, given by

$$\kappa h'_{\mathrm{m}} = \ln\left[\left|\frac{\psi_{o2}}{\psi_{o1}}\right|\left(\frac{1-\alpha_1}{1+\alpha_1}\right) + \sqrt{\left(\frac{\psi_{o2}}{\psi_{o1}}\cdot\frac{1-\alpha_1}{1+\alpha_1}\right)^2 - \frac{(1-\alpha_1)(1-\alpha_2)}{(1+\alpha_1)(1+\alpha_2)}}\right] \tag{10.44}$$

Figure 10.3 shows $P(h)$ plotted as a function of κh for various values of the ratio of $\sigma_2/\sigma_1 = \psi_{o2}/\psi_{o1}$ at $\alpha_1 = 0.1$ and $\alpha_2 = 0.4$. As shown in Fig. 10.3, for the interaction between likely charged plates, $P(h)$ is always positive (repulsive) (curve with $\sigma_2/\sigma_1 = 10$) but for oppositely charged plates, curve with $\sigma_2/\sigma_1 = -4$, which corresponds to case (i), shows that $P(h)$ is negative (attractive) at large separations, exhibits a minimum, and tends to zero at some point, becoming positive (repulsive) beyond this point. Curve with $\sigma_2/\sigma_1 = -2$, which corresponds to case (ii), shows that $P(h)$ exhibits a minimum but always negative (attractive). Curve with $\sigma_2/\sigma_1 = -1.3$, which corresponds to case (iii), shows that $P(h)$ has no minimum.

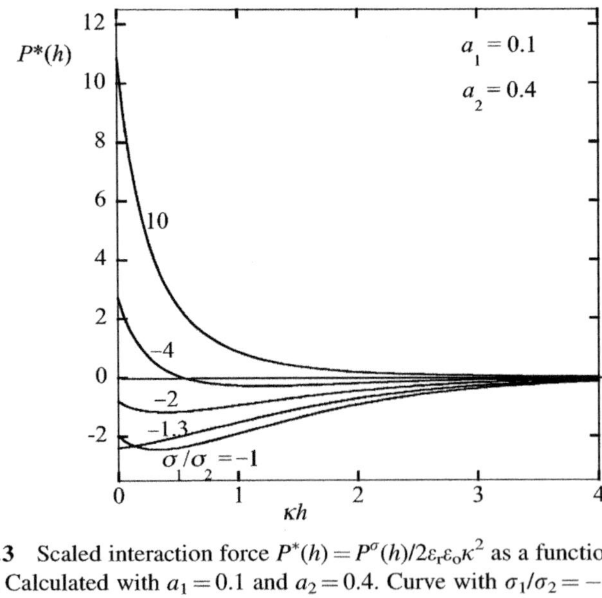

FIGURE 10.3 Scaled interaction force $P^*(h) = P^\sigma(h)/2\varepsilon_r\varepsilon_o\kappa^2$ as a function of scaled plate separation κh. Calculated with $a_1 = 0.1$ and $a_2 = 0.4$. Curve with $\sigma_1/\sigma_2 = -4$ corresponds to case (i), curve with $\sigma_1/\sigma_2 = -2$ to case (ii), curve with $\sigma_1/\sigma_2 = -1.3$ to case (iii), curve with $\sigma_1/\sigma_2 = -1$ to case (iv).

Curve with $\sigma_2/\sigma_1 = -1$, which corresponds to case (iv), shows that $P(h)$ exhibits a minimum but always negative (attractive).

The potential energy $V^{\sigma}(h)$ of the double-layer interaction per unit area is obtained by integrating $P^{\sigma}(h)$ as [1]

$$V^{\sigma}(h) = \frac{1}{2}\varepsilon_r\varepsilon_o\kappa \times \frac{\{\psi_{o1}^2(1+\alpha_1)(1-\alpha_2)+\psi_{o2}^2(1-\alpha_1)(1+\alpha_2)\}\{1-\tanh(\kappa h)\}+2\psi_{o1}\psi_{o2}(1+\alpha_1)(1+\alpha_2)\mathrm{sech}(\kappa h)}{\alpha_1+\alpha_2+(1+\alpha_1\alpha_2)\tanh(\kappa h)} \tag{10.45}$$

We see that if σ_1 and σ_2 are of like sign (i.e., ψ_{o1} and ψ_{o2} are of like sign), or if either of σ_1 or σ_2 is zero, then V^{σ} is always positive. If σ_1 and σ_2 are of opposite sign (i.e., ψ_{o1} and ψ_{o2} are of opposite sign), V^{σ} is negative at large κh, but may become either positive or negative as decreasing κh, as summarized below.

(i) If the condition

$$\left|\frac{\psi_{o1}}{\psi_{o2}}\right| \geq \frac{1+\alpha_2}{1-\alpha_2}\left\{1+\sqrt{1-\frac{(1-\alpha_1)(1-\alpha_2)}{(1+\alpha_1)(1+\alpha_2)}}\right\} \tag{10.46}$$

is satisfied, then V^{σ} has a minimum V^{σ}_{m} given by

$$V^{\sigma}_{\mathrm{m}} = -\varepsilon_r\varepsilon_o\kappa\frac{1+\alpha_2}{1-\alpha_2}\psi_{o2}^2 \tag{10.47}$$

at $h = h_o$ (Eq. (10.39)) and becomes zero at $h = h'$, given by

$$\kappa h' = \ln\left[\frac{1}{2}\left(\left|\frac{\psi_{o2}}{\psi_{o1}}\right|\frac{1-\alpha_1}{1+\alpha_1}+\left|\frac{\psi_{o1}}{\psi_{o2}}\right|\frac{1+\alpha_2}{1-\alpha_2}\right)\right] \tag{10.48}$$

For $h < h'$, V becomes positive.

(ii) If

$$\frac{1+\alpha_2}{1-\alpha_2} \leq \left|\frac{\psi_{o1}}{\psi_{o2}}\right| < \frac{1+\alpha_2}{1-\alpha_2}\left\{1+\sqrt{1-\frac{(1-\alpha_1)(1-\alpha_2)}{(1+\alpha_1)(1+\alpha_2)}}\right\} \tag{10.49}$$

then V^{σ} is negative for all κh and has a minimum V^{σ}_{m} at $h = h_o$.

(iii) If

$$\left|\frac{\psi_{o1}}{\psi_{o2}}\right| \leq \frac{1+\alpha_2}{1-\alpha_2} \tag{10.50}$$

then V^{σ} is negative for all κh and has no minimum.

Equation (10.45) for $V^{\sigma}(h)$ can also be derived from the double-layer free energy $F^{\sigma}(h)$, namely, $V^{\sigma}(h) = F^{\sigma}(h) - F^{\sigma}(\infty)$, where

$$F^{\sigma}(h) = \frac{1}{2}\sigma_1\{\psi(-d_1) + \psi(0)\} + \frac{1}{2}\sigma_2\{\psi(h) + \psi(h + d_2)\} \tag{10.51}$$

$$F^{\sigma}(\infty) = \sigma_1\psi_{o1} + \sigma_2\psi_{o2} \tag{10.52}$$

Figure 10.4 shows $V^{\sigma}(h)$ plotted as a function of κh for various values of the ratio of $\sigma_2/\sigma_1 = \psi_{o2}/\psi_{o1}$ at $\alpha_1 = 0.1$ and $\alpha_2 = 0.4$. As shown in Fig. 10.4, for the interaction between likely charged plates, $V^{\sigma}(h)$ is always positive (curve with $\sigma_2/\sigma_1 = 10$) but for oppositely charged plates, curve with $\sigma_2/\sigma_1 = -8$, which corresponds to case (i), shows that $V^{\sigma}(h)$ is negative at large separations, exhibits a minimum, and tends to zero at some point, becoming positive beyond this point. Curve with $\sigma_2/\sigma_1 = -4$, which corresponds to case (ii), shows that $V^{\sigma}(h)$ exhibits a minimum but always negative. Curve with $\sigma_2/\sigma_1 = -2.5$, which corresponds to case (iii), shows that $V^{\sigma}(h)$ has no minimum.

If the influence of the internal fields within the interacting plates can be neglected ($a_1 = a_2 = 0$), then Eqs. (10.24) reduce to

$$\psi(x) = \frac{\psi_{o1}\cosh[\kappa(h - x)] + \psi_{o2}\cosh(\kappa x)}{\sinh(\kappa h)}, \quad 0 \leq x \leq h \tag{10.53}$$

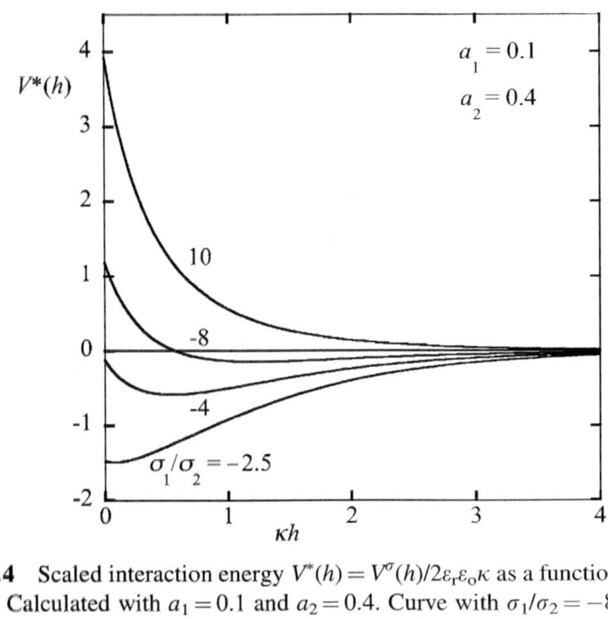

FIGURE 10.4 Scaled interaction energy $V^{*}(h) = V^{\sigma}(h)/2\varepsilon_r\varepsilon_o\kappa$ as a function of scaled plate separation κh. Calculated with $a_1 = 0.1$ and $a_2 = 0.4$. Curve with $\sigma_1/\sigma_2 = -8$ corresponds to case (i), curve with $\sigma_1/\sigma_2 = -4$ to case (ii), and curve with $\sigma_1/\sigma_2 = -2.5$ to case (iii).

and Eqs. (10.35) and (10.45) to

$$\begin{aligned} P^{\sigma}(h) &= \frac{1}{2}\varepsilon_r\varepsilon_o\kappa^2\left[\left(\frac{\psi_{o1}+\psi_{o2}}{2}\right)^2\frac{1}{\sinh^2(\kappa h/2)} - \left(\frac{\psi_{o1}-\psi_{o2}}{2}\right)^2\frac{1}{\cosh^2(\kappa h/2)}\right] \\ &= \frac{1}{2}\varepsilon_r\varepsilon_o\kappa^2\left[\frac{2\psi_{o1}\psi_{o2}\cosh(\kappa h)+(\psi_{o1}^2+\psi_{o2}^2)}{\sinh^2(\kappa h)}\right] \end{aligned} \tag{10.54}$$

$$V^{\sigma}(h) = \varepsilon_r\varepsilon_o\kappa\left[\left(\frac{\psi_{o1}+\psi_{o2}}{2}\right)^2\left\{\coth\left(\frac{\kappa h}{2}\right)-1\right\} - \left(\frac{\psi_{o1}-\psi_{o2}}{2}\right)^2\left\{1-\tanh\left(\frac{\kappa h}{2}\right)\right\}\right] \tag{10.55}$$

which was derived by Wiese and Healy.

10.3.2 Interaction at Constant Surface Potential

If the interacting plates 1 and 2 are at constant surface potentials ψ_{o1} and ψ_{o2}, respectively, then the solution $\psi(x)$ to (10.21) subject to Eqs. (10.5) and (10.6) is

$$\psi(x) = \frac{\psi_{o1}\sinh[\kappa(h-x)] + \psi_{o2}\sinh(\kappa x)}{\sinh(\kappa h)}, \quad 0 \le x \le h \tag{10.56}$$

The interaction force $P^{\psi}(h)$ per unit area between the two plates unit area is calculated by substituting Eq. (10.56) into Eq. (10.19) with the result that

$$\begin{aligned} P^{\psi}(h) &= \frac{1}{2}\varepsilon_r\varepsilon_o\kappa^2\left[\left(\frac{\psi_{o1}+\psi_{o2}}{2}\right)^2\frac{1}{\cosh^2(\kappa h/2)} - \left(\frac{\psi_{o1}-\psi_{o2}}{2}\right)^2\frac{1}{\sinh^2(\kappa h/2)}\right] \\ &= \frac{1}{2}\varepsilon_r\varepsilon_o\kappa^2\left[\frac{2\psi_{o1}\psi_{o2}\cosh(\kappa h)-(\psi_{o1}^2+\psi_{o2}^2)}{\sinh^2(\kappa h)}\right] \end{aligned} \tag{10.57}$$

By integrating Eq. (10.57) with respect to h, we obtain the following expression for the interaction energy between the two plates per unit area:

$$V^{\psi}(h) = \varepsilon_r\varepsilon_o\kappa\left[\left(\frac{\psi_{o1}+\psi_{o2}}{2}\right)^2\left\{1-\tanh\left(\frac{\kappa h}{2}\right)\right\} - \left(\frac{\psi_{o1}-\psi_{o2}}{2}\right)^2\left\{\coth\left(\frac{\kappa h}{2}\right)-1\right\}\right] \tag{10.58}$$

Equation (10.58) can also be derived from the double-layer free energy $F^{\psi}(h)$ of the system of two interacting plates 1 and 2, namely,

$$V^{\psi} = F^{\psi}(h) - F^{\psi}(\infty) \tag{10.59}$$

with

$$F^{\psi} = -\frac{1}{2}\sigma_1(h)\psi_{o1} - \frac{1}{2}\sigma_2(h)\psi_{o2} \tag{10.60}$$

$$\sigma_1(h) = -\varepsilon_r\varepsilon_o \left.\frac{d\psi}{dx}\right|_{x=0^+} = \varepsilon_r\varepsilon_o\kappa \frac{\psi_{o1}\cosh(kh) - \psi_{o1}}{\sinh(\kappa h)} \tag{10.61}$$

$$\sigma_2(h) = +\varepsilon_r\varepsilon_o \left.\frac{d\psi}{dx}\right|_{x=h^-} = -\varepsilon_r\varepsilon_o\kappa \frac{\psi_{o1} - \psi_{o2}\cosh(kh)}{\sinh(\kappa h)} \tag{10.62}$$

Here $\sigma_1(h)$ and $\sigma_2(h)$ are, respectively, the surface charge densities of plates 1 and 2, which are not constant but depend on the value of plate separation h. Equation (10.58), which was derived by Hogg et al. , is called the Hogg–Healy–Fuerstenau (HHF) formula.

10.3.3 Mixed Case

The double-layer interaction energy $V^{\psi-\sigma}(h)$ for the case where plate 1 is at constant surface potential ψ_{o1} and plate 2 is at constant surface charge density $\sigma_2 = \varepsilon_r\varepsilon_o\kappa\psi_{o2}$ is given by [4]

$$V^{\psi-\sigma} = \varepsilon_r\varepsilon_o\kappa\left[\frac{\psi_{o1}\psi_{o2}}{\cosh(\kappa h)} + \frac{1}{2}(\psi_{o1}^2 - \psi_{o2}^2)\{1 - \tanh(\kappa h)\}\right] \tag{10.63}$$

10.4 ARBITRARY POTENTIAL: INTERACTION AT CONSTANT SURFACE CHARGE DENSITY

10.4.1 Isodynamic Curves

Consider the electrostatic interaction between two parallel dissimilar plates 1 and 2 separated by a distance h in a symmetrical electrolyte solution of valence z and bulk concentration n for the case where the potential magnitude is arbitrary. For the interaction at constant surface potential, detailed description is given in Refs [5–7]. We consider the constant surface charge density case in detail. We treat the case where the influence of the internal fields within the interacting plates is negligible. In such a case, we need consider only the region $0 < x < h$, since the potential distribution in the region $x < 0$ and $x > h$ remain unchanged. We consider two parallel plates 1 and 2 carrying surface charge densities σ_1 and σ_2, respectively [1].

We start with the nonlinear planar Poisson–Boltzmann equation (see Eq. (9.74)),

$$\frac{d^2y}{dx^2} = \kappa^2 \sinh y, \quad \text{for } 0 < x < h \tag{10.64}$$

which can be rewritten as

$$\frac{dw}{dx} = \kappa \sinh y, \quad \text{for} \quad 0 < x < h \tag{10.65}$$

with

$$w = \frac{dy}{dx} \tag{10.66}$$

The boundary conditions are

$$w(0^+) = \sigma_1' \tag{10.67}$$

$$w(h^-) = \sigma_2' \tag{10.68}$$

with

$$\sigma_1' = \frac{ze}{kT} \cdot \frac{\sigma_1}{\varepsilon_r \varepsilon_o \kappa} \tag{10.69}$$

$$\sigma_2' = \frac{ze}{kT} \cdot \frac{\sigma_2}{\varepsilon_r \varepsilon_o \kappa} \tag{10.70}$$

where σ_1' and σ_2' are the scaled surface charge densities of plates 1 and 2, respectively. Integration of Eq. (10.65) gives

$$w^2 = 2 \cosh y + C \tag{10.71}$$

where C is an integration constant.

The force of the electrostatic interaction per unit area between plates 1 and 2, which is calculated from Eq. (10.18), becomes for the case of symmetrical electrolyte solution

$$P(h) = nkT\{2(\cosh y - 1) - w^2\} \tag{10.72}$$

which can be rewritten with the help of Eq. (10.71) as

$$P(h) = -nkT(C + 2) \tag{10.73}$$

From Eqs. (10.65) and (10.71), we have

$$\left(\frac{w^2 - C}{2}\right)^2 - \left(\frac{1}{\kappa}\frac{dw}{dx}\right)^2 = 1 \tag{10.74}$$

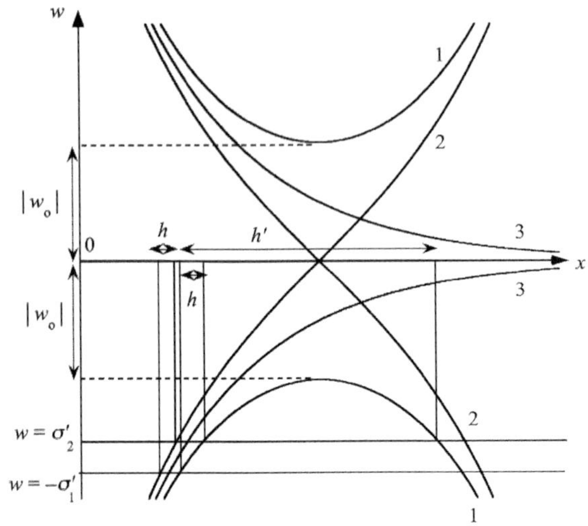

FIGURE 10.5 Schematic representation of a set of integral (isodynamic) curves for the dissimilar double-layer interaction. (1) $C > -2$, (2) $C < -2$, (3) $C = 2$. From Ref. [10].

Considering x as a function of C (for given values of σ'_1 and σ'_2), we plot schematically in Fig. 10.5 a set of integral curves (isodynamic curves) $w = w(\kappa x, C)$ according to Eq. (10.74) for various values of C, which have the w-axis as an asymptote. The intersections of one of these integral curves with two straight lines $w = -\sigma'_1$ and $w = \sigma'_2$ give the possible values of h. The idea of isodynamic curves has been introduced by Derjaguin [5].

When σ' and σ'_2 are of like sign, or when either of them is zero, only the case $C < -2$ is possible. Hence the force is always repulsive and the integration of Eq. (10.74) is given by

$$\kappa h = 2 \int_{-|\sigma'_2|}^{|\sigma'_1|} \frac{dw}{\sqrt{(w^2 - C + 2)(w^2 - C - 2)}} \tag{10.75}$$

This can be expressed in terms of elliptic integrals as

$$\kappa h = \frac{2}{\sqrt{-C+2}} \left\{ F\left[\arctan \frac{\sigma'_1}{\sqrt{-C-2}}, \frac{2}{\sqrt{-C+2}}\right] + F\left[\arctan \frac{\sigma'_2}{\sqrt{-C-2}}, \frac{2}{\sqrt{-C+2}}\right] \right\} \tag{10.76}$$

where $F(\phi, k)$ is an elliptic integral of the first kind of modulus k. The magnitude of y has a minimum $|y_0|$, given by

$$\cosh y_0 = -C/2 \tag{10.77}$$

at a certain distance $h = h_0$. In terms of y_0 and h_0, Eq. (10.75) can be rewritten as

$$\kappa h = 2\int_0^{|\sigma'_1|} \frac{dw}{\sqrt{\left(w^2 + 4\cosh^2(y_0/2)\right)\left(w^2 + 4\sinh^2(y_0/2)\right)}} \tag{10.78}$$

$$\kappa(h - h_0) = 2\int_0^{|\sigma'|} \frac{dw}{\sqrt{\left(w^2 + 4\cosh^2(y_0/2)\right)\left(w^2 + 4\sinh^2(y_0/2)\right)}} \tag{10.79}$$

and Eq. (10.73) for the force $P(h)$ can be written as

$$P = 4nkT\sinh^2(y_0/2) \tag{10.80}$$

When σ'_1 and σ'_2 are of opposite sign and $\sigma'_1 \neq \sigma'_2$, the interaction force may be either attractive ($C > -2$) or repulsive ($C < -2$). For $C = -2$, where the integral curve has the x-axis as an asymptote, the force P is zero and $h = h_0$, given by

$$\begin{aligned} \kappa h_0 &= 2\int_{|\sigma'_2|}^{|\sigma'_1|} \frac{dw}{w\sqrt{w^2 + 4}} \\ &= \ln\left[\frac{|\sigma'_1|\left(2 + \sqrt{\sigma'^2_2 + 4}\right)}{|\sigma'_2|\left(2 + \sqrt{\sigma'^2_1 + 4}\right)}\right] \end{aligned} \tag{10.81}$$

which corresponds to Eq. (10.39) for the low potential case. In the region $h < h_0$, which corresponds to $C < -2$, integration of Eq. (10.74) is

$$\kappa h = 2\int_{|\sigma'_2|}^{|\sigma'_1|} \frac{dw}{\sqrt{(w^2 - C + 2)(w^2 - C - 2)}} \tag{10.82}$$

or

$$\begin{aligned} \kappa h = \frac{2}{\sqrt{-C + 2}}\Bigg\{&F\left[\arctan\frac{\sigma'_1}{\sqrt{-C - 2}}, \frac{2}{\sqrt{-C + 2}}\right] \\ &- F\left[\arctan\frac{\sigma'_2}{\sqrt{-C - 2}}, \frac{2}{\sqrt{-C + 2}}\right]\Bigg\} \end{aligned} \tag{10.83}$$

In the region $h > h_0$, which corresponds to $C > -2$, the integral curve has a minimum or a maximum w_0, given by

$$w_0 = \pm\sqrt{C + 2} \tag{10.84}$$

Each line $w = \text{constant}$ ($|w| > |w_o|$) intersects the integral curve twice and then there exist two different values of h. When the line $w = \sigma'_2$ is tangent to the integral curve, that is,

$$C = C_m = \sigma'^2_2 - 2 \tag{10.85}$$

or

$$P = P_m = -nkT\sigma'^2_2 = -\frac{\sigma_2^2}{2\varepsilon_r\varepsilon_o} \tag{10.86}$$

the two values of h have the same value $h = h_m$, which is

$$\kappa h_m = 2\int_{|\sigma'_2|}^{|\sigma'_1|} \frac{dw}{\sqrt{\left(w^2 - \sigma'^2_2\right)\left(w^2 - \sigma'^2_2 + 4\right)}} \tag{10.87}$$

or

$$\kappa h_m = F\left(\arccos\left|\frac{\sigma'_2}{\sigma'_1}\right|, \sqrt{1 - \left(\frac{\sigma'_2}{2}\right)^2}\right), \quad \text{for } |\sigma'_2| \le 2 \tag{10.88}$$

$$\kappa h_m = \frac{2}{|\sigma'_2|}\left\{K\left(\sqrt{1 - \left(\frac{2}{\sigma'_2}\right)^2}\right) - F\left(\arcsin\left|\frac{\sigma'_2}{\sigma'_1}\right|, \sqrt{1 - \left(\frac{2}{\sigma'_2}\right)^2}\right)\right\}, \quad \text{for } |\sigma'_2| \ge 2 \tag{10.89}$$

where $K(k)$ is a complete elliptic integral of the first kind defined by $K(k) = F(\pi/2, k)$. The smaller value of h corresponds to $h_0 < h < h_m$ and is given by

$$\kappa h = 2\int_{|\sigma'_2|}^{|\sigma'_1|} \frac{dw}{\sqrt{\left(w^2 - C + 2\right)\left(w^2 - C - 2\right)}} \tag{10.90}$$

or

$$\begin{aligned}\kappa h = &\left\{F\left[\arccos\left(\frac{\sqrt{C+2}}{|\sigma'_1|}\right), \frac{\sqrt{-C+2}}{2}\right]\right.\\ &\left. - F\left[\arcsin\left(\frac{\sqrt{C+2}}{|\sigma'_2|}\right), \frac{\sqrt{-C+2}}{2}\right]\right\}, \quad \text{for } -2 < C \le 2\end{aligned} \tag{10.91}$$

$$\kappa h = \frac{2}{\sqrt{C+2}}\left\{F\left[\arcsin\left(\frac{\sqrt{C+2}}{|\sigma'_2|}\right), \sqrt{\frac{C-2}{C+2}}\right]\right.$$
$$\left. - F\left[\arcsin\left(\frac{\sqrt{C+2}}{|\sigma'_1|}\right), \sqrt{\frac{C-2}{C+2}}\right]\right\}, \quad \text{for } C \geq 2 \qquad (10.92)$$

The larger value of h corresponds to $h > h_\mathrm{m}$ and is given by

$$\kappa h = 2\int_{\sqrt{C+2}}^{|\sigma'_1|} \frac{dw}{\sqrt{(w^2 - C + 2)(w^2 - C - 2)}} + 2\int_{\sqrt{C+2}}^{|\sigma'_2|} \frac{dw}{\sqrt{(w^2 - C + 2)(w^2 - C - 2)}} \qquad (10.93)$$

or

$$\kappa h = \left\{F\left[\arccos\left(\frac{\sqrt{C+2}}{|\sigma'_1|}\right), \frac{\sqrt{-C+2}}{2}\right]\right.$$
$$\left. + F\left[\arcsin\left(\frac{\sqrt{C+2}}{|\sigma'_2|}\right), \frac{\sqrt{-C+2}}{2}\right]\right\}, \quad \text{for } -2 < C \leq 2 \qquad (10.94)$$

$$\kappa h = \frac{2}{\sqrt{C+2}}\left\{2K\left(\sqrt{\frac{C-2}{C+2}}\right) - F\left[\arcsin\left(\frac{\sqrt{C+2}}{|\sigma'_2|}\right), \sqrt{\frac{C-2}{C+2}}\right]\right.$$
$$\left. - F\left[\arcsin\left(\frac{\sqrt{C+2}}{|\sigma'_1|}\right), \sqrt{\frac{C-2}{C+2}}\right]\right\}, \quad \text{for } C \geq 2 \qquad (10.95)$$

When $C > C_\mathrm{m}$, the line $w = \sigma'_2$ cannot intersect the integral curve, that is, such values of C can occur at no values of h. Hence we see that P_m is the largest value of the attractive force. In the region $h \geq h_\mathrm{m}$, y has an inflection point at a certain distance $h = \bar{h}$, where $dw/dx = 0$ and $y = 0$ and $w = w_0$. In terms of w_0 and x, Eq. (10.93) can be written as

$$\kappa \bar{h} = 2\int_{|w_0|}^{|\sigma'_1|} \frac{dw}{\sqrt{(w^2 - w_0^2)(w^2 - w_0^2 + 4)}} \qquad (10.96)$$

$$\kappa(h - \bar{h}) = 2\int_{|w_0|}^{|\sigma'_2|} \frac{dw}{\sqrt{(w^2 - w_0^2)(w^2 - w_0^2 + 4)}} \qquad (10.97)$$

and Eq. (10.72) for the interaction force P can be written as

$$P(h) = -nkTw_0^2 \tag{10.98}$$

At $h = h_m$, y has an inflection point on the surface of the more weakly (in magnitude) charged plate. For $h < h_m$, y has no inflection point.

When $\sigma_1' = -\sigma_2'$, the lines $z = \sigma_1'$ and $z = \sigma_2'$ coincide each other so that only the case $C > -2$ is possible and thus the interaction force $P(h)$ is always attractive. Integration of Eq. (10.74) is also given by Eq. (10.93) and the magnitude of P has its largest value P_m given by Eq. (10.86) at $h = h_m$.

10.4.2 Interaction Energy

We calculate the potential energy of the double-layer interaction per unit area between two parallel dissimilar plates 1 and 2 at separation h carrying constant surface charge densities σ_1 and σ_2, as shown in Fig. 10.1[8]. Equation (9.116) for the interaction energy $F^\sigma(h)$ is generalized to cover the interaction between two parallel dissimilar plates as

$$\begin{aligned} F^\sigma(h) &= nkT(C+2)h - \varepsilon_r\varepsilon_o\left[\int_{-\infty}^{-d_1}\left(\frac{d\psi}{dx}\right)^2 dx + \int_0^h \left(\frac{d\psi}{dx}\right)^2 dx + \int_{h+d_2}^{\infty}\left(\frac{d\psi}{dx}\right)^2 dx\right] \\ &\quad - \frac{1}{2}\varepsilon_{p1}\varepsilon_o\int_{-d_1}^{0}\left(\frac{d\psi}{dx}\right)^2 dx - \frac{1}{2}\varepsilon_{p2}\varepsilon_o\int_{h}^{h+d_2}\left(\frac{d\psi}{dx}\right)^2 dx \\ &\quad + \sigma_1\{\psi(-d_1,h) + \psi(0,h)\} + \sigma_2\{\psi(h,h) + \psi(h+d_2,h)\} \end{aligned} \tag{10.99}$$

where $\psi(-d_1, h)$, $\psi(0, h)$, $\psi(h, h)$, and $\psi(h + d_2, h)$ are, respectively, the potentials at the surfaces at $x = -d_1$ and $x = 0$ of plate 1 and those at the surfaces at $x = h$ and $x = h + d_2$ of plate 2. These potentials are not constant but depend on h. The interaction energy $V^\sigma(h)$ can be calculated as follows.

$$\begin{aligned} V^\sigma(h) &= F^\sigma(h) - F^\sigma(\infty) \\ &= nkT(C+2)h + 2\left(\sqrt{\sigma_1'^2 + 4} + \sqrt{\sigma_2'^2 + 4} - 4\right) - J_1 + J_2 \end{aligned} \tag{10.100}$$

with

$$J_1 = \int_0^{\kappa h} w^2 d(\kappa x) = \int_{-\sigma_1'}^{\sigma_2'} w^2 \frac{d(\kappa x)}{dw} dw \tag{10.101}$$

and

$$J_2 = \sigma'_1\{y(0) - 2\,\text{arcsinh}(\sigma_1'/2)\} + \sigma_2'\{y(h) - 2\,\text{arcsinh}(\sigma_2'/2)\} \tag{10.102}$$

Now we derive expressions for J_1 and J_2.

(i) When σ'_1 and σ'_2 are of like sign, or when either of them is zero, only the case $C < -2$ is possible (the interaction force P is always repulsive). Then J_1 is given by

$$J_1 = 2\int_{-|\sigma'_2|}^{|\sigma'_1|} \frac{w^2 dw}{\sqrt{(w^2 - C + 2)(w^2 - C - 2)}} \tag{10.103}$$

which can be expressed in terms of elliptic integrals as

$$\begin{aligned} J_1 = {} & -2\sqrt{-C+2}\left\{E\left[\arctan\left(\frac{\sigma'_1}{\sqrt{-C-2}}\right), \frac{2}{\sqrt{-C+2}}\right]\right. \\ & \left. + E\left[\arctan\left(\frac{\sigma'_2}{\sqrt{-C-2}}\right), \frac{2}{\sqrt{-C+2}}\right]\right\} \\ & +2\left(|\sigma'_1|\sqrt{\frac{\sigma'^2_1 - C + 2}{\sigma'^2_1 - C - 2}} - |\sigma'_2|\sqrt{\frac{\sigma'^2_2 - C + 2}{\sigma'^2_2 - C - 2}}\right) \end{aligned} \tag{10.104}$$

where $E(\phi, k)$ is an elliptic integral of the second kind. It can be shown that J_2 can be written as

$$\begin{aligned} J_2 = {} & 2|\sigma'_1|\left\{\operatorname{arcsinh}\left(\frac{\sqrt{\sigma'^2_1 - C - 2}}{2}\right) - \operatorname{arcsinh}\left(\frac{|\sigma'_1|}{2}\right)\right\} \\ & +2|\sigma'_2|\left\{\operatorname{arcsinh}\left(\frac{\sqrt{\sigma'^2_2 - C - 2}}{2}\right) - \operatorname{arcsinh}\left(\frac{|\sigma'_2|}{2}\right)\right\} \end{aligned} \tag{10.105}$$

(ii) When σ'_1 and σ'_2 are of opposite sign and $|\sigma'_1| \neq |\sigma'_2|$, $C < -2$ for $h < h_0$ and $C > -2$ for $h > h_0$, h_0 being given by Eq. (10.81)

$$\kappa h_0 = \ln\left[\frac{|\sigma'_1|(2 + \sqrt{\sigma'^2_2 + 4})}{|\sigma'_2|(2 + \sqrt{\sigma'^2_1 + 4})}\right] \tag{10.106}$$

For $h = h_0$, $C = -2$ (the interaction force is zero) and thus the potential energy V has a minimum V_m. We have for J_1

$$J_1 = 2\int_{|\sigma'_2|}^{|\sigma'_1|} \frac{w\,dw}{\sqrt{w^2 + 4}} = 2\left(\sqrt{\sigma'^2_1 + 4} - \sqrt{\sigma'^2_2 + 4}\right) \tag{10.107}$$

and for J_2

$$J_2 = -4|\sigma'_2| \operatorname{arcsinh}\left(\frac{|\sigma'_2|}{2}\right) \tag{10.108}$$

Then, V_m is given by

$$V_m = \frac{8nkT}{\kappa}\left\{\sqrt{\sigma'^2_2 + 4} - 2|\sigma'_2| \operatorname{arcsinh}\left(\frac{|\sigma'_2|}{2}\right)\right\} \tag{10.109}$$

For $h < h_0$, where the interaction force is repulsive, J_1 is given by

$$J_1 = 2\int_{|\sigma'_2|}^{|\sigma'_1|} \frac{w^2 dw}{\sqrt{(w^2 - C + 2)(w^2 - C - 2)}} \tag{10.110}$$

or

$$\begin{aligned} J_1 = & -2\sqrt{-C+2}\left\{E\left[\arctan\left(\frac{\sigma'_1}{\sqrt{-C-2}}\right), \frac{2}{\sqrt{-C+2}}\right]\right. \\ & \left. - E\left[\arctan\left(\frac{\sigma'_2}{\sqrt{-C-2}}\right), \frac{2}{\sqrt{-C+2}}\right]\right\} \\ & + 2\left(|\sigma'_1|\sqrt{\frac{\sigma'^2_1 - C + 2}{\sigma'^2_1 - C - 2}} - |\sigma'_2|\sqrt{\frac{\sigma'^2_2 - C + 2}{\sigma'^2_2 - C - 2}}\right) \end{aligned} \tag{10.111}$$

and J_2 is given by

$$\begin{aligned} J_2 = & \ 2|\sigma'_1|\left\{\operatorname{arcsinh}\left(\frac{\sqrt{\sigma'^2_1 - C - 2}}{2}\right) - \operatorname{arcsinh}\left(\frac{|\sigma'_1|}{2}\right)\right\} \\ & - 2|\sigma'_2|\left\{\operatorname{arcsinh}\left(\frac{\sqrt{\sigma'^2_2 - C - 2}}{2}\right) - \operatorname{arcsinh}\left(\frac{|\sigma'_2|}{2}\right)\right\} \end{aligned} \tag{10.112}$$

For $h_0 < h \leq h_m$, where the interaction force is attractive, J_1 is given by

$$J_1 = 2\int_{|\sigma'_2|}^{|\sigma'_1|} \frac{w^2 dw}{\sqrt{(w^2 - C + 2)(w^2 - C - 2)}} \tag{10.113}$$

which can be expressed in terms of elliptic integrals as

$$J_1 = (C+2)\kappa h - 4\left\{E\left[\arctan\left(\frac{\sqrt{C+2}}{\sigma'_1}\right), \frac{\sqrt{-C+2}}{2}\right]\right.$$
$$- E\left[\arctan\left(\frac{\sqrt{C+2}}{\sigma'_2}\right), \frac{\sqrt{-C+2}}{2}\right]\bigg\}$$
$$+ 2\left\{\frac{1}{|\sigma'_1|}\sqrt{(\sigma'^2_1 - C + 2)(\sigma'^2_1 - C - 2)}\right.$$
$$\left. - \frac{1}{|\sigma'_2|}\sqrt{(\sigma'^2_2 - C + 2)(\sigma'^2_2 - C - 2)}\right\},$$
$$\text{for } -2 < C \leq 2 \tag{10.114}$$

$$J_1 = (C+2)\kappa h + 2\sqrt{C+2}\left\{E\left[\arcsin\left(\frac{\sqrt{C+2}}{|\sigma'_1|}\right), \sqrt{\frac{C-2}{C+2}}\right]\right.$$
$$- E\left[\arcsin\left(\frac{\sqrt{C+2}}{|\sigma'_2|}\right), \sqrt{\frac{C-2}{C+2}}\right]\bigg\}$$
$$+ 2\left\{\frac{1}{|\sigma'_1|}\sqrt{(\sigma'^2_1 - C + 2)(\sigma'^2_1 - C - 2)}\right.$$
$$\left. - \frac{1}{|\sigma'_2|}\sqrt{(\sigma'^2_2 - C + 2)(\sigma'^2_2 - C - 2)}\right\}, \quad \text{for } C \geq 2 \tag{10.115}$$

Here h_m is the separation at which the attractive force has its largest value, and is given by Eqs. (10.87). Expression for J_2 is also given by Eq. (10.112)

For $h \geq h_m$, where the interaction is also attractive, J_1 given by

$$J_1 = 2\int_{\sqrt{C+2}}^{|\sigma'_1|} \frac{w^2\,dw}{\sqrt{(w^2 - C + 2)(w^2 - C - 2)}}$$
$$+2\int_{\sqrt{C+2}}^{|\sigma'_2|} \frac{w^2\,dw}{\sqrt{(w^2 - C + 2)(w^2 - C - 2)}} \tag{10.116}$$

which can be rewritten in terms of elliptic integrals as

$$J_1 = (C+2)\kappa h - 4\left\{E\left[\arccos\left(\frac{\sqrt{C+2}}{|\sigma'_1|}\right), \frac{\sqrt{-C+2}}{2}\right]\right.$$

$$+E\left[\arcsin\left(\frac{\sqrt{C+2}}{|\sigma'_2|}\right),\frac{\sqrt{-C+2}}{2}\right]\Bigg\}$$

$$+2\left\{\frac{1}{|\sigma'_1|}\sqrt{(\sigma'^2_1-C+2)(\sigma'^2_1-C-2)}\right.$$

$$\left.+\frac{1}{|\sigma'_2|}\sqrt{(\sigma'^2_2-C+2)(\sigma'^2_2-C-2)}\right\},\quad \text{for } -2<C\leq 2 \qquad (10.117)$$

and

$$J_1=(C+2)\kappa h-2\sqrt{C+2}\left\{2E\left(\sqrt{\frac{C-2}{C+2}}\right)\right.$$

$$\left.-E\left[\arcsin\left(\frac{\sqrt{C+2}}{|\sigma'_1|}\right),\sqrt{\frac{C-2}{C+2}}\right]-E\left[\arcsin\left(\frac{\sqrt{C+2}}{|\sigma'_2|}\right),\sqrt{\frac{C-2}{C+2}}\right]\right\}$$

$$+2\left\{\frac{1}{|\sigma'_1|}\sqrt{(\sigma'^2_1-C+2)(\sigma'^2_1-C-2)}\right.$$

$$\left.+\frac{1}{|\sigma'_2|}\sqrt{(\sigma'^2_2-C+2)(\sigma'^2_2-C-2)}\right\},\quad \text{for } C\geq 2 \qquad (10.118)$$

(iii) When $\sigma'_1=-\sigma'_2$, only the case $C>-2$ is possible and $h_m=0$. Thus, J_1 is given by Eqs. (10.116) and J_2 is by Eq. (10.105).

10.5 APPROXIMATE ANALYTIC EXPRESSIONS FOR MODERATE POTENTIALS

The method for obtaining an analytic expression for the interaction energy at constant surface potential given in Chapter 9 (see Section 9.6) can be applied to the interaction between two parallel dissimilar plates. The results are given below [9].

$$V^{\psi}(h)=\frac{2nkT}{\kappa}\left[Y_+^2\left\{1-\tanh(\kappa h/2)\right\}-Y_-^2\left\{\coth(\kappa h/2)-1\right\}\right]$$

$$+\frac{2nkT}{\kappa}\left[\frac{1}{48}(Y_+^4+3Y_+^2Y_-^2)\left\{1-\tanh(\kappa h/2)\right\}\right.$$

$$-\frac{1}{48}(Y_-^4+3Y_+^2Y_-^2)\left\{\coth(\kappa h/2)-1\right\}$$

$$-\frac{Y_+^4}{32}\frac{\tanh(\kappa h/2)}{\cosh^2(\kappa h/2)}+\frac{Y_-^4}{32}\frac{\coth(\kappa h/2)}{\sinh^2(\kappa h/2)}$$

$$
\begin{aligned}
&- \frac{(\kappa h/2)}{32}\left\{\frac{Y_+^2}{\cosh^2(\kappa h/2)} - \frac{Y_-^2}{\sinh^2(\kappa h/2)}\right\}^2\Bigg] \\
&+ \frac{2nkT}{\kappa}\left[\frac{Y_+^2}{5760}\left\{Y_+^4 + \frac{15}{8}Y_-^2(7Y_+^2 + Y_-^2)\right\}\left\{1 - \tanh(\kappa h/2)\right\}\right. \\
&- \frac{Y_-^2}{5760}\left\{Y_-^4 + \frac{15}{8}Y_+^2(7Y_-^2 + Y_+^2)\right\}\left\{\coth(\kappa h/2) - 1\right\} \\
&+ \frac{Y_+^4}{1536}\left(Y_+^2 + \frac{3}{2}Y_-^2\right)\frac{\tanh(\kappa h/2)}{\cosh^2(\kappa h/2)} - \frac{Y_-^4}{1536}\left(Y_-^2 + \frac{3}{2}Y_+^2\right)\frac{\coth(\kappa h/2)}{\sinh^2(\kappa h/2)} \\
&- \frac{Y_+^6}{1024}\frac{\tanh(\kappa h/2)}{\cosh^4(\kappa h/2)} - \frac{Y_-^6}{1024}\frac{\coth(\kappa h/2)}{\sinh^4(\kappa h/2)} \\
&- \frac{7(\kappa h/2)}{3072}\left\{\frac{Y_+^2}{\cosh^2(\kappa h/2)} - \frac{Y_-^2}{\sinh^2(\kappa h/2)}\right\}^3 \\
&+ \frac{(\kappa h/2)}{384}\left\{\frac{Y_+^4}{\cosh^4(\kappa h/2)} - \frac{Y_-^4}{\sinh^4(\kappa h/2)}\right\} \\
&\times [Y_+^2\tanh^2(\kappa h/2) - Y_-^2\coth^2(\kappa h/2)] \\
&+ \frac{(\kappa h/2)^2}{256}\left\{\frac{Y_+^2}{\cosh^2(\kappa h/2)} - \frac{Y_-^2}{\sinh^2(\kappa h/2)}\right\}^2 \\
&\left.\times \left\{Y_+^2\frac{\tanh(\kappa h/2)}{\cosh^2(\kappa h/2)} - Y_-^2\frac{\coth(\kappa h/2)}{\sinh^2(\kappa h/2)}\right\}\right] \qquad (10.119)
\end{aligned}
$$

with

$$
Y_+ = \frac{y_{o1} + y_{o2}}{2}, \quad Y_- = \frac{y_{o1} - y_{o2}}{2} \qquad (10.120)
$$

where the first term on the right-hand side of Eq. (10.119) agrees with the HHF formula (10.58).

REFERENCES

1. H. Ohshima, *Colloid Polym. Sci.* 253 (1975) 150.
2. G. R. Wiese and D. W. Healy, *Trans. Faraday Soc.* 66 (1970) 490.
3. R. Hogg, T. W. Healy, and D. W. Fuerstenau, *Trans. Faraday Soc.* 62 (1966) 1638.
4. G. Kar, S. Chander, and T. S. Mika, *J. Colloid Interface Sci.* 44 (1973) 347.
5. B. V. Derjaguin, *Disc. Faraday Soc.* 18 (1954) 85.

6. O. F. Devereux and P. L. de Bruyn, *Interaction of Plane-Parallel Double Layers*, The MIT Press, Cambridge, 1963.
7. D. McCormack, S. L. Carnie, and D. Y. C. Chan, *J. Colloid Interface Sci.* 169 (1995) 177.
8. H. Ohshima, *Colloid Polym. Sci.* 253 (1975) 150.
9. H. Ohshima, T. W. Healy, and L. R. White, *J. Colloid Interface Sci.* 89 (1982) 484.

11 Linear Superposition Approximation for the Double-Layer Interaction of Particles at Large Separations

11.1 INTRODUCTION

A method is given for obtaining exact limiting expressions for the force and potential energy of the double-layer interaction between two charged particles at large separations as compared with the Debye length $1/\kappa$. This method is based on the idea that the potential distribution in the region between the two interacting particles at large separations can be approximated by the sum of the unperturbed potential distributions around the respective particles in the absence of interaction (or, when the particles are at infinite separation). This method, which gives the correct limiting expressions for the interaction force and energy, is called the linear superposition approximation (LSA) [1–5]. The expressions for the interaction force and energy involves only the product of the unperturbed surface potentials of the interacting particles and thus can be applied irrespective of the interaction models (that is, the constant surface potential model, constant surface charge density model, or the Donnan potential regulation model) and irrespective of whether the particles are soft or hard.

11.2 TWO PARALLEL PLATES

11.2.1 Similar Plates

Consider first the case of two similar plates 1 and 2 immersed in a liquid containing N ionic species with valence z_i and bulk concentration (number density) n_i^∞ ($i = 1$, 2, . . . , N). We take an x-axis perpendicular to the plates with its origin at the surface of plate 1 (Fig. 11.1). At large separations h as compared with the Debye length, the potential $\psi(h/2)$ at the midpoint between the plates can be approximated

Biophysical Chemistry of Biointerfaces By Hiroyuki Ohshima

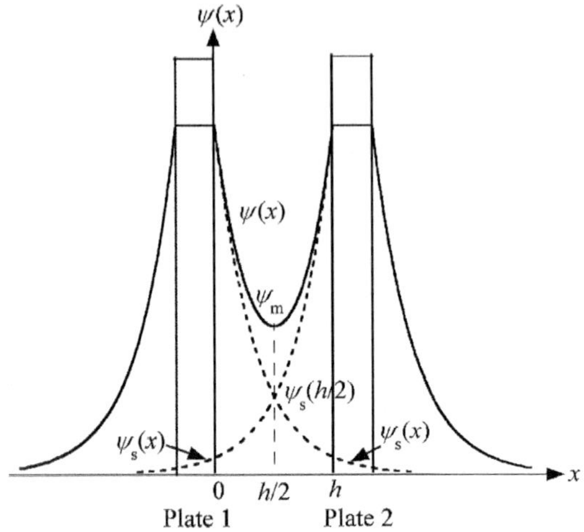

FIGURE 11.1 Schematic representation of the potential distribution $\psi(x)$ across two parallel similar plates 1 and 2.

by the sum of the asymptotic values of the two unperturbed potentials that is produced by the respective plates in the absence of interaction. For two similar plates, the interaction force $P(h)$ per unit area between plates 1 and 2 is given by Eq. (9.26), namely,

$$P(h) = kT\sum_{i=1}^{N} n_i^{\infty}\left[\exp\left(-\frac{z_i e\psi_{\rm m}}{kT}\right) - 1\right] \tag{11.1}$$

where $\psi_{\rm m} = \psi(h/2)$ is the potential at the midpoint between plates 1 and 2. Since at large κh the potential at the midpoint $x = h/2$ between the two plates is small so that Eq. (11.1) can be linearized with respect to $\psi_{\rm m}$ to give

$$P(h) = \frac{1}{2}\varepsilon_{\rm r}\varepsilon_{\rm o}\kappa^2\psi_{\rm m}^2 \tag{11.2}$$

with

$$\kappa = \left(\frac{1}{\varepsilon_{\rm r}\varepsilon_{\rm o}kT}\sum_{i=1}^{N} z_i^2 e^2 n_i^{\infty}\right)^{1/2} \tag{11.3}$$

being the Debye–Hückel parameter. In the linear superposition approximation, the potential $\psi_{\rm m} = \psi(h/2)$ is approximated by the sum of two unperturbed potential $\psi_{\rm s}(x)$ around a single plate [1, 2], namely,

$$\psi_{\rm m} = \psi(h/2) \approx 2\psi_{\rm s}(h/2) \tag{11.4}$$

This approximation holds good for large κh.

For a plate in a symmetrical electrolyte solution of valence z and bulk concentration n, $\psi_s(x)$ is given by Eq. (1.38), namely,

$$\psi_s(x) = \frac{2kT}{ze} \ln\left(\frac{1+\gamma\, e^{-\kappa x}}{1-\gamma\, e^{-\kappa x}}\right) \tag{11.5}$$

with

$$\gamma = \tanh\left(\frac{ze\psi_o}{4kT}\right) = \frac{\exp(ze\psi_o/2kT)-1}{\exp(ze\psi_o/2kT)+1} = \frac{\exp(y_o/2)-1}{\exp(y_o/2)+1} \tag{11.6}$$

where ψ_o is the unperturbed surface potential of the plate and $y_o = ze\psi_o/kT$ is the scaled surface potential. Far from the plate, Eq. (11.5) asymptotes to

$$\psi_s(x) = \psi_{\text{eff}}\, e^{-\kappa x} \tag{11.7}$$

with

$$\psi_{\text{eff}} = \frac{kT}{ze} Y = \frac{4kT}{ze}\gamma \tag{11.8}$$

where ψ_{eff} and $Y = 4\gamma$ are, respectively, the effective surface potential and scaled effective potential. In the low potential limit, ψ_{eff} tends to ψ_o. Equation (11.7) gives

$$\psi_s(h/2) = \psi_{\text{eff}}\, e^{-\kappa h/2} = \frac{kT}{ze} Y\, e^{-\kappa h/2} = \frac{4kT}{ze}\gamma\, e^{-\kappa h/2} \tag{11.9}$$

At large separations h, we have

$$\begin{aligned}\psi_m &= \psi(h/2) \approx 2\psi_s(h/2)\\ &\approx 2\psi_{\text{eff}}\, e^{-\kappa h/2} = \frac{8kT}{ze}\gamma\, e^{-\kappa h/2}\end{aligned} \tag{11.10}$$

By substituting Eq. (11.10) into Eq. (11.2). we obtain

$$P(h) = 32\varepsilon_r\varepsilon_o\kappa^2\left(\frac{kT}{ze}\right)^2 \gamma^2\, e^{-\kappa h} = 2\varepsilon_r\varepsilon_o\kappa^2\psi_{\text{eff}}^2\, e^{-\kappa h} = 2\varepsilon_r\varepsilon_o\kappa^2\left(\frac{kT}{ze}\right)^2 Y^2\, e^{-\kappa h} \tag{11.11}$$

or

$$P(h) = 64nkT\gamma^2\, e^{-\kappa h} \tag{11.12}$$

The corresponding interaction energy per unit area is obtained by integrating $P(h)$,

$$V(h) = 32\varepsilon_r\varepsilon_o\kappa\left(\frac{kT}{ze}\right)^2\gamma^2 e^{-\kappa h} = 2\varepsilon_r\varepsilon_o\kappa\psi_{\text{eff}}^2 e^{-\kappa h} = 2\varepsilon_r\varepsilon_o\kappa\left(\frac{kT}{ze}\right)^2 Y^2 e^{-\kappa h} \quad (11.13)$$

or

$$V(h) = \frac{64\gamma^2 nkT}{\kappa} e^{-\kappa h} \quad (11.14)$$

When a charged plate is immersed in a 2-1 electrolyte of bulk concentration n, $\psi_s(x)$ is given by Eq. (1.44), namely,

$$\psi_s(x) = \frac{kT}{e}\ln\left[\frac{3}{2}\left(\frac{1+\frac{2}{3}\gamma' e^{-\kappa x}}{1-\frac{2}{3}\gamma' e^{-\kappa x}}\right)^2 - \frac{1}{2}\right] \quad (11.15)$$

with

$$\gamma' = \frac{3}{2}\left\{\frac{\left(\frac{2}{3}e^{y_o} + \frac{1}{3}\right)^{1/2} - 1}{\left(\frac{2}{3}e^{y_o} + \frac{1}{3}\right)^{1/2} + 1}\right\} \quad (11.16)$$

where $y_o = e\psi_o/kT$ is the scaled surface potential and κ is given by Eq. (1.13). Far from the plate, Eq. (11.15) asymptotes to

$$\psi_s(x) = \psi_{\text{eff}}\, e^{-\kappa x} \quad (11.17)$$

where

$$\psi_{\text{eff}} = \frac{kT}{e}Y = \frac{4kT}{e}\gamma' \quad (11.18)$$

Eq. (11.15) gives at large κh,

$$\psi_m = \psi(h/2) \approx 2\psi_s(h/2) = 2\psi_{\text{eff}}\, e^{-\kappa h/2} = \frac{2kT}{e}Y e^{-\kappa h/2} = \frac{8kT}{e}\gamma' e^{-\kappa h/2} \quad (11.19)$$

By substituting Eq. (11.19) into Eq. (11.2), we obtain

$$P(h) = 32\varepsilon_r\varepsilon_o\kappa^2\left(\frac{kT}{ze}\right)^2\gamma'^2 e^{-\kappa h} = 2\varepsilon_r\varepsilon_o\kappa^2\psi_{\text{eff}}^2 e^{-\kappa h} = 2\varepsilon_r\varepsilon_o\kappa^2\left(\frac{kT}{ze}\right)^2 Y^2 e^{-\kappa h}$$

$$(11.20)$$

or

$$P(h) = 192nkT\gamma'^2 e^{-\kappa h} \tag{11.21}$$

The corresponding interaction energy per unit area is obtained by integrating $P(h)$,

$$V(h) = 32\varepsilon_r\varepsilon_o\kappa\left(\frac{kT}{ze}\right)^2 \gamma'^2 e^{-\kappa h} = 2\varepsilon_r\varepsilon_o\kappa\psi_{\mathrm{eff}}^2 e^{-\kappa h} = 2\varepsilon_r\varepsilon_o\kappa\left(\frac{kT}{ze}\right)^2 Y^2 e^{-\kappa h} \tag{11.22}$$

or

$$V(h) = \frac{192\gamma'^2 nkT}{\kappa} e^{-\kappa h} \tag{11.23}$$

For a plate in a mixed solution of 1-1 electrolyte of bulk concentration n_1 and 2-1 electrolyte of bulk concentration n_2. $\psi_s(x)$ is given by Eq. (1.49), namely,

$$\psi(x) = \frac{kT}{e}\ln\left[\left(\frac{1}{1-\eta/3}\right)\left\{\frac{1+(1-\eta/3)\gamma'' e^{-\kappa x}}{1-(1-\eta/3)\gamma'' e^{-\kappa x}}\right\}^2 - \left(\frac{\eta/3}{1-\eta/3}\right)\right] \tag{11.24}$$

with

$$\eta = \frac{3n_2}{n_1 + 3n_2} \tag{11.25}$$

$$\gamma'' = \left(\frac{1}{1-\eta/3}\right)\left[\frac{\{(1-\eta/3)e^{y_o} + \eta/3\}^{1/2} - 1}{\{(1-\eta/3)e^{y_o} + \eta/3\}^{1/2} + 1}\right] \tag{11.26}$$

where $y_o = e\psi_o/kT$ is the scaled surface potential and κ is given by Eq. (1.14). Eq. (11.24) gives at large κh,

$$\psi_m = \psi(h/2) \approx 2\psi_s(h/2) = 2\psi_{\mathrm{eff}}\, e^{-\kappa h/2} = \frac{2kT}{e} Y e^{-\kappa h/2} = \frac{8kT}{e}\gamma'' e^{-\kappa h/2} \tag{11.27}$$

By substituting Eq. (11.27) into Eq. (11.2), we obtain

$$P(h) = 32\varepsilon_r\varepsilon_o\kappa^2\left(\frac{kT}{ze}\right)^2 \gamma'^2 e^{-\kappa h} = 2\varepsilon_r\varepsilon_o\kappa^2\psi_{\mathrm{eff}}^2 e^{-\kappa h} = 2\varepsilon_r\varepsilon_o\kappa^2\left(\frac{kT}{ze}\right)^2 Y^2 e^{-\kappa h} \tag{11.28}$$

or

$$P(h) = 64(n_1 + 3n_2)kT\gamma''^2\, e^{-\kappa h} \tag{11.29}$$

The corresponding interaction energy per unit area is obtained by integrating $P(h)$,

$$V(h) = 32\varepsilon_r\varepsilon_o\kappa\left(\frac{kT}{ze}\right)^2\gamma'^2\, e^{-\kappa h} = 2\varepsilon_r\varepsilon_o\kappa\psi_{\text{eff}}^2\, e^{-\kappa h} = 2\varepsilon_r\varepsilon_o\kappa\left(\frac{kT}{ze}\right)^2 Y^2\, e^{-\kappa h} \tag{11.30}$$

or

$$V(h) = \frac{64\gamma''^2(n_1 + 3n_2)kT}{\kappa}\, e^{-\kappa h} \tag{11.31}$$

11.2.2 Dissimilar Plates

Consider two parallel plates 1 and 2 at separation h immersed in a liquid containing N ionic species with valence z_i and bulk concentration (number density) n_i^∞ ($i = 1, 2, \ldots, N$). We take an x-axis perpendicular to the plates with its origin at the surface of plate 1 (Fig. 11.2 for likely charged plates and Fig. 11.3 for oppositely charged plates). The interaction force P can be obtained by integrating the excess osmotic pressure and the Maxwell stress over an arbitrary surface Σ enclosing one of the plates Chapter 8). As Σ, we choose two planes located at $x = -\infty$ and $x = x\prime$ ($0 < x\prime < h$) enclosing plate 1. Here $x\prime$ is an arbitrary point near the midpoint in the region $0 < x < h$ between plates 1 and 2. The force P acting between two plates 1 and 2 per unit area is given by Eq. (8.20), namely,

$$P(h) = \sum_{i=1}^{N} n_i^\infty kT\left[\exp\left(-\frac{z_i e\psi(x')}{kT}\right) - 1\right] - \frac{1}{2}\varepsilon_r\varepsilon_o\left(\left.\frac{d\psi}{dx}\right|_{x=x'}\right)^2 \tag{11.32}$$

Since the potential around the midpoint between the two plates is small so that Eq. (11.28) can be linearized with respect to $\psi(x')$ to give

$$P(h) = \frac{1}{2}\varepsilon_r\varepsilon_o\left\{\kappa^2\psi^2(x') - \left(\left.\frac{d\psi}{dx}\right|_{x=x'}\right)^2\right\} \tag{11.33}$$

Here it must be stressed that the above linearization is not for the surface potential $\psi(0)$ and $\psi(h)$ but only for $\psi(x')$. That is, this linearization approximation holds good even when the surface potentials are arbitrary, provided that the particle separation h is large compared with the Debye length $1/\kappa$.

Consider first the case where plates 1 and 2 are immersed in a symmetrical electrolyte solution of valence z and bulk concentration n. The asymptotic forms of the

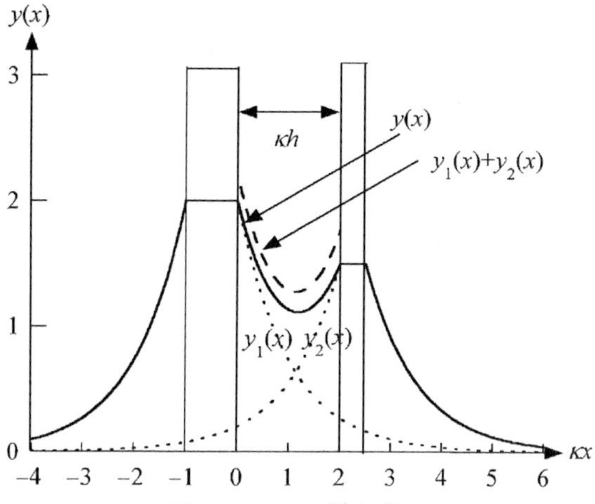

FIGURE 11.2 Scaled potential distribution $y(x)$ across two likely charged plates 1 and 2 at scaled separation $\kappa h = 2$ with unperturbed surface potentials $y_{o1} = 2$ and $y_{o2} = 1.5$. $y_1(x)$ and $y_2(x)$ are, respectively, scaled unperturbed potential distributions around plates 1 and 2 in the absence of interaction at $\kappa h = \infty$.

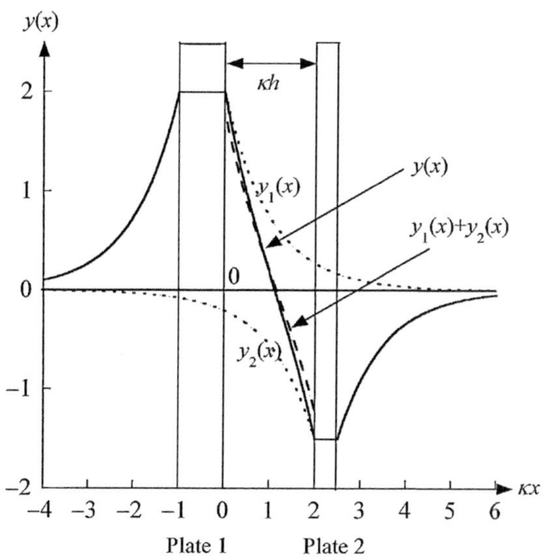

FIGURE 11.3 Scaled potential distribution $y(x)$ across two oppositely charged plates 1 and 2 at scaled separation $\kappa h = 2$ with unperturbed surface potentials $y_{o1} = 2$ and $y_{o2} = -1.5$. $y_1(x)$ and $y_2(x)$ are, respectively, scaled unperturbed potential distributions around plates 1 and 2 in the absence of interaction at $\kappa h = \infty$.

unperturbed potentials $\psi_1(x)$ and $\psi_2(x)$ produced at large distances by plates 1 and 2, respectively, in the absence of interaction are given by Eq. (1.177), namely,

$$\psi_1(x) = \psi_{\mathrm{eff1}}\, e^{-\kappa x} = \frac{kT}{ze} Y_1\, e^{-\kappa x} \tag{11.34}$$

$$\psi_2(x) = \psi_{\mathrm{eff2}}\, e^{-\kappa(h-x)} = \left(\frac{kT}{ze}\right) Y_2\, e^{-\kappa(h-x)} \tag{11.35}$$

where ψ_{eff1} and ψ_{eff2} are, respectively, the effective surface potentials of plates 1 and 2 and Y_1 and Y_2 are the corresponding scaled effective surface potentials. These potentials are given by

$$\psi_{\mathrm{eff1}} = \frac{kT}{ze} Y_1 = \frac{4kT}{ze}\gamma_1 \tag{11.36}$$

$$\psi_{\mathrm{eff2}} = \frac{kT}{ze} Y_2 = \frac{4kT}{ze}\gamma_2 \tag{11.37}$$

with

$$\gamma_1 = \tanh\left(\frac{ze\psi_{\mathrm{o1}}}{4kT}\right) \tag{11.38}$$

$$\gamma_2 = \tanh\left(\frac{ze\psi_{\mathrm{o2}}}{4kT}\right) \tag{11.39}$$

where ψ_{o1} and ψ_{o2} are, respectively, the unperturbed surface potentials of plates 1 and 2. The potential distribution $\psi(x)$ around the midpoint between plates 1 and 2 can thus be approximated by

$$\begin{aligned}\psi(x) &= \psi_1(x) + \psi_2(x) \\ &= \psi_{\mathrm{eff1}}\, e^{-\kappa x} + \psi_{\mathrm{eff2}}\, e^{-\kappa(h-x)} \\ &= \frac{4kT}{ze}\gamma_1\, e^{-\kappa x} + \frac{4kT}{ze}\gamma_2\, e^{-\kappa(h-x)}\end{aligned} \tag{11.40}$$

from which

$$\frac{d\psi}{dx} = -\kappa\frac{4kT}{ze}\left\{\gamma_1\, e^{-\kappa x} - \gamma_2\, e^{-\kappa(h-x)}\right\} \tag{11.41}$$

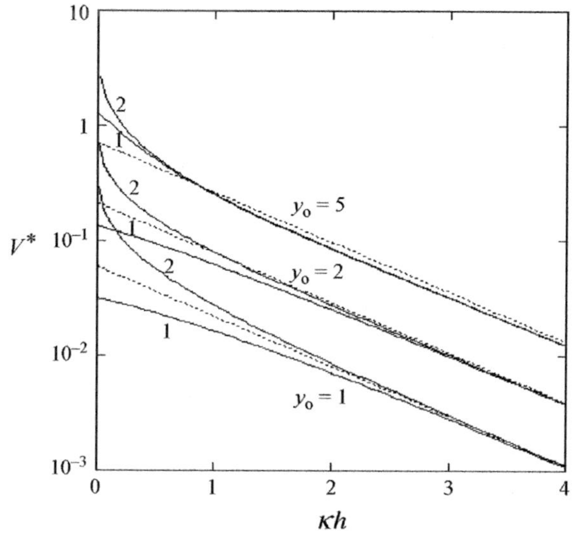

FIGURE 11.4 Scaled double-layer interaction energy $V^*(h) = (\kappa/64nkT)V(h)$ per unit area between two parallel similar plates as a function of scaled separation κh at the scaled unperturbed surface potential $y_o = 1$, 2, and 5 calculated with Eq. (11.14) (dotted lines) in comparison with the exact results under constant surface potential (curves 1) and constant surface charge density (curves 2). From Ref. [5].

By substituting Eqs. (11.40) and (11.41) into Eq. (11.33), we obtain

$$\begin{aligned} P(h) &= \frac{1}{2}\varepsilon_r\varepsilon_o\kappa^2\left(\frac{4kT}{ze}\right)^2\left\{\left(\gamma_1 e^{-\kappa x} + \gamma_2 e^{-\kappa(h-x)}\right)^2 - \left(\gamma_1 e^{-\kappa x} - \gamma_2 e^{-\kappa(h-x)}\right)^2\right\} \\ &= 2\varepsilon_r\varepsilon_o\kappa^2\left(\frac{4kT}{ze}\right)^2\gamma_1\gamma_2 e^{-\kappa h} \\ &= 2\varepsilon_r\varepsilon_o\kappa^2\psi_{\text{eff}1}\psi_{\text{eff}2}e^{-\kappa h} = 2\varepsilon_r\varepsilon_o\kappa^2\left(\frac{kT}{ze}\right)^2 Y_1 Y_2 e^{-\kappa h} \end{aligned} \tag{11.42}$$

or

$$P(h) = 64\pi nkT\gamma_1\gamma_2\, e^{-\kappa h} \tag{11.43}$$

Note that $P(h)$ is independent of the value of $x\prime$. The interaction energy $V(h)$ per unit area of plates 1 and 2 is thus

$$V(h) = \int_h^\infty P(h)dh$$
$$= 2\varepsilon_r\varepsilon_o\kappa\left(\frac{4kT}{ze}\right)^2 \gamma_1\gamma_2\, e^{-\kappa h} \tag{11.44}$$
$$= 2\varepsilon_r\varepsilon_o\kappa\psi_{\text{eff1}}\psi_{\text{eff2}}\, e^{-\kappa h} = 2\varepsilon_r\varepsilon_o\kappa\left(\frac{kT}{ze}\right)^2 Y_1Y_2\, e^{-\kappa h}$$

or

$$V(h) = \frac{64\pi nkT\gamma_1\gamma_2}{\kappa} e^{-\kappa h} \tag{11.45}$$

If plates 1 and 2 are immersed in a 2-1 electrolyte solution of bulk concentration n, then the effective surface potential $\psi_{\text{eff}i}$ and the scaled effective surface potential Y_i of plate i ($i = 1$ and 2) are given by Eqs. (1.182) and (1.183),

$$\psi_{\text{eff}i} = \frac{kT}{e}Y_i = \frac{4kT}{e}\gamma_i' = \frac{kT}{e}\cdot 6\left\{\frac{\left(\frac{2}{3}e^{y_{oi}}+\frac{1}{3}\right)^{1/2}-1}{\left(\frac{2}{3}e^{y_{oi}}+\frac{1}{3}\right)^{1/2}+1}\right\}, \quad (i = 1, 2) \tag{11.46}$$

where $y_{oi} = e\psi_{oi}/kT$ is the scaled unperturbed surface potential of plate i. The interaction force $P(h)$ per unit area between plates 1 and 2 are thus given by

$$P(h) = 32\varepsilon_r\varepsilon_o\kappa^2\left(\frac{kT}{e}\right)^2 \gamma_1'\gamma_2'\, e^{-\kappa h}$$
$$= 2\varepsilon_r\varepsilon_o\kappa^2\psi_{\text{eff1}}\psi_{\text{eff2}}\, e^{-\kappa h} = 2\varepsilon_r\varepsilon_o\kappa^2\left(\frac{kT}{e}\right)^2 Y_1Y_2\, e^{-\kappa h} \tag{11.47}$$

or

$$P(h) = 192nkT\gamma_1'\gamma_2'\, e^{-\kappa h} \tag{11.48}$$

The corresponding interaction energy $V(h)$ per unit area is

$$V(h) = 32\varepsilon_r\varepsilon_o\kappa\left(\frac{kT}{e}\right)^2 \gamma_1'\gamma_2'\, e^{-\kappa h}$$
$$= 2\varepsilon_r\varepsilon_o\kappa\psi_{\text{eff1}}\psi_{\text{eff2}}\, e^{-\kappa h} = 2\varepsilon_r\varepsilon_o\kappa\left(\frac{kT}{e}\right)^2 Y_1Y_2\, e^{-\kappa h} \tag{11.49}$$

or

$$V(h) = \frac{192nkT\gamma_1'\gamma_2'}{\kappa} e^{-\kappa h} \tag{11.50}$$

If plates 1 and 2 are immersed in a mixed solution of 1-1 electrolyte of bulk concentration n_1 and 2-1 electrolyte of bulk concentration n_2, then the effective surface potential ψ_{oi} and the scaled effective surface potential Y_i of plate i ($i = 1$ and 2) are given by Eqs. (1.185) and (1.186),

$$\psi_{\mathrm{eff}i} = \frac{kT}{e} Y_i = \frac{4kT}{e}\gamma_i''$$
$$= \frac{4kT}{e} \cdot \left(\frac{1}{1-\eta/3}\right)\left[\frac{\{(1-\eta/3)e^{y_{oi}}+\eta/3\}^{1/2}-1}{\{(1-\eta/3)e^{y_{oi}}+\eta/3\}^{1/2}+1}\right], \quad (i = 1, 2) \tag{11.51}$$

with

$$\eta = \frac{3n_2}{n_1 + 3n_2} \tag{11.52}$$

The interaction force $P(h)$ per unit area between plates 1 and 2 at separation h are thus given by

$$P(h) = 32\varepsilon_r\varepsilon_o\kappa^2\left(\frac{kT}{e}\right)^2\gamma_1''\gamma_2''\, e^{-\kappa h}$$
$$= 2\varepsilon_r\varepsilon_o\kappa^2\psi_{\mathrm{eff}1}\psi_{\mathrm{eff}2}\, e^{-\kappa h} = 2\varepsilon_r\varepsilon_o\kappa^2\left(\frac{kT}{e}\right)^2 Y_1 Y_2\, e^{-\kappa h} \tag{11.53}$$

or

$$P(h) = 64(n + 3n_2)kT\gamma_1'\gamma_2'\, e^{-\kappa h} \tag{11.54}$$

The corresponding interaction energy $V(h)$ per unit area is

$$V(h) = 32\varepsilon_r\varepsilon_o\kappa\left(\frac{kT}{e}\right)^2\gamma_1''\gamma_2''\, e^{-\kappa h}$$
$$= 2\varepsilon_r\varepsilon_o\kappa\psi_{\mathrm{eff}1}\psi_{\mathrm{eff}2}\, e^{-\kappa h} = 2\varepsilon_r\varepsilon_o\kappa\left(\frac{kT}{e}\right)^2 Y_1 Y_2\, e^{-\kappa h} \tag{11.55}$$

or

$$V(h) = \frac{64(n_1 + 3n_2)kT\gamma_1'\gamma_2'}{\kappa} e^{-\kappa h} \tag{11.56}$$

If plates 1 and 2 are in a general electrolyte solution containing N ionic species with valence z_i and bulk concentration (number density) n_i^∞ ($i = 1, 2, \ldots, N$), the effective surface potential $\psi_{\text{eff}i}$ and scaled effective surface potential Y_i of plate i ($i = 1, 2$) are given by Eq. (1.193),

$$Y_i = y_{oi}\exp\left[\int_0^{y_{oi}} \left\{\frac{1}{f(y)} - \frac{1}{y}\right\}dy\right] \tag{11.57}$$

with

$$\begin{aligned} f(y) &= sgn(y_{oi})\left[\frac{2\sum_{i=1}^{N} n_i(e^{-z_i y} - 1)}{\sum_{i=1}^{N} z_i^2 n_i}\right]^{1/2} \\ &= (1 - e^{-y})\left[\frac{2\sum_{i=1}^{N} n_i(e^{-z_i y} - 1)}{(1 - e^{-y})^2 \sum_{i=1}^{N} z_i^2 n_i}\right]^{1/2} \end{aligned} \tag{11.58}$$

The interaction force $P(h)$ per unit area between plates 1 and 2 are thus given by

$$P(h) = 2\varepsilon_r\varepsilon_o\kappa\psi_{\text{eff}1}\psi_{\text{eff}2}\, e^{-\kappa h} = 2\varepsilon_r\varepsilon_o\kappa\left(\frac{kT}{e}\right)^2 Y_1 Y_2\, e^{-\kappa h} \tag{11.59}$$

The corresponding interaction energy $V(h)$ per unit area is

$$V(h) = 2\varepsilon_r\varepsilon_o\kappa\psi_{\text{eff}1}\psi_{\text{eff}2}\, e^{-\kappa h} = 2\varepsilon_r\varepsilon_o\kappa\left(\frac{kT}{e}\right)^2 Y_1 Y_2\, e^{-\kappa h} \tag{11.60}$$

11.2.3 Hypothetical Charge

An alternative method for calculating the asymptotic expressions for the double--layer interaction between two parallel plates at large separations is given below. This method, which was introduced by Brenner and Parsegian [6] to obtain an asymptotic expression for the interaction energy between two parallel cylinders, is based on the concept of hypothetical charge, as shown below.

Consider two parallel plates at separation h in a general electrolyte solution. According to the method of Brenner and Parsegian [6], the asymptotic expression for the interaction energy $V(h)$ per unit area is given by

$$V(h) = \sigma_{\text{eff}1}\psi_2(h) = \sigma_{\text{eff}2}\psi_1(h) \tag{11.61}$$

Here $\sigma_{\text{eff}i}$ ($i = 1, 2$) is the hypothetical charge per unit area of an infinitely thin plate that would produce the same asymptotic potential as produced by plate i and $\psi_i(h)$

is the asymptotic form of the unperturbed potential at a large distance h from the surface of plate i, when each plate exists separately.

Consider the electric field produced by the above hypothetical infinitely thin plate having a surface charge density $\sigma_{\mathrm{eff}i}$. We take an x-axis perpendicular to the plate, which is placed at the origin $x=0$. The electric field produced by this hypothetical plate at a distance h from its surface is thus given by

$$E_i(h) = \frac{\sigma_{\mathrm{eff}i}}{2\varepsilon_r\varepsilon_o} e^{-\kappa h} \tag{11.62}$$

The corresponding electric potential $\psi_i(h)$ is given by

$$\psi_i(h) = \frac{\sigma_{\mathrm{eff}i}}{2\varepsilon_r\varepsilon_o\kappa} e^{-\kappa h} \tag{11.63}$$

The asymptotic expression for the unperturbed potential of plate i ($i=1, 2$) at a large distance h from the surface of plate i is given by

$$\psi_i(h) = \left(\frac{kT}{e}\right) Y_i\, e^{-\kappa h} \tag{11.64}$$

where Y_i is the effective surface potential or the asymptotic constant. If we introduce the scaled potential $y(x)=e\psi(x)/kT$ of plate i, then Eq. (11.64) reduces to

$$y_i(h) = Y_i e^{-\kappa h} \tag{11.65}$$

Comparison of Eqs. (11.63) and (11.64) leads to an expression for the hypothetical charge $\sigma_{\mathrm{eff}i}$ for plate i, namely,

$$\sigma_{\mathrm{eff}i} = 2\varepsilon_r\varepsilon_o\kappa\left(\frac{kT}{e}\right) Y_i \tag{11.66}$$

By combining Eqs. (11.61), (11.64), and (11.66), we obtain

$$V(h) = 2\varepsilon_r\varepsilon_o\kappa\left(\frac{kT}{e}\right)^2 Y_1 Y_2\, e^{-\kappa h} \tag{11.67}$$

which agrees with Eq. (11.60). This method can be applied to the sphere–sphere or cylinder–cylinder interaction, as shown below.

11.3 TWO SPHERES

Consider next two interacting spherical colloidal particles 1 and 2 of radii a_1 and a_2 having surface potentials ψ_{o1} and ψ_{o2}, respectively, at separation $R = H + a_1 + a_2$ between their centers in a general electrolyte solution. According to the method of Brenner and Parsegian [6], the asymptotic expression for the interaction energy $V(R)$ is given by

$$V(R) = Q_{\text{eff1}}\psi_2(R) = Q_{\text{eff2}}\psi_1(R) \tag{11.68}$$

Here $Q_{\text{eff}i}$ ($i = 1, 2$) is the hypothetical point charge that would produce the same asymptotic potential as produced by sphere i and $\psi_i(h)$ is the asymptotic form of the unperturbed potential at a large distance R from the center of sphere i in the absence of interaction. The hypothetical point charge $Q_{\text{eff}i}$ gives rise to the potential

$$\psi_i(R) = \frac{Q_{\text{eff}i}}{4\pi\varepsilon_r\varepsilon_o R}e^{-\kappa R} \tag{11.69}$$

The asymptotic expression for the unperturbed potential of sphere i ($i = 1, 2$) at a large distance R from the center of sphere i may be expressed as

$$\psi_i(R) = \left(\frac{kT}{e}\right)Y_i\frac{a_i}{R}e^{-\kappa(R-a_i)} \tag{11.70}$$

where Y_i is the scaled effective surface potential, which reduces to the scaled surface potential $y_{oi} = e\psi_{oi}/kT$ in the low potential limit. Comparison of Eqs. (11.69) and (11.70) leads to

$$Q_{\text{eff}i} = 4\pi\varepsilon_r\varepsilon_o a_i a_2\, e^{\kappa a_i}\left(\frac{kT}{e}\right)Y_i \tag{11.71}$$

By combining Eqs. (11.68),(11.70) and (1.71), we obtain [8, 9]

$$V(R) = 4\pi\varepsilon_r\varepsilon_o a_1 a_2\left(\frac{kT}{e}\right)^2 Y_1 Y_2\frac{e^{-\kappa(R-a_1-a_2)}}{R} \tag{11.72}$$

which is the required result for the asymptotic expression for the interaction energy of two spheres 1 and 2 at large separations. For low potentials Eq. (11.72) reduces to

$$V(R) = 4\pi\varepsilon_r\varepsilon_o a_1 a_2\psi_{o1}\psi_{o2}\frac{e^{-\kappa(R-a_1-a_2)}}{R} \tag{11.73}$$

Note that Bell et al. [7] derived Eq. (11.73) by the usual method, that is, by integrating the osmotic pressure and the Maxwell tensor over an arbitrary closed surface enclosing either sphere. Equation (11.73) agrees with the leading term of the exact expression for the double-layer interaction energy between two spheres in the low potential limit (see Chapter 14).

For a sphere of radius a having a surface potential ψ_o in a symmetrical electrolyte solution of valence z and bulk concentration n, we find that Y is approximately given by Eq. (1.197) [10], namely,

$$Y = \frac{8\tanh(y_o/4)}{1 + [1 - (2\kappa a + 1/(\kappa a + 1)^2)\tanh^2(y_o/4)]^{1/2}} \tag{11.74}$$

where $y_o = ze\psi_o/kT$ is the scaled surface potential of the sphere and κ is given by Eq. (1.20). By substituting Eq. (11.74) into Eq. (11.72), we obtain the following expression for the interaction energy $V(R)$ between two similar spheres of radius a at separation R between their centers carrying surface potential ψ_o:

$$V(R) = \frac{256\pi\varepsilon_r\varepsilon_o\gamma^2(kT/ze)^2}{[1 + \{1 - (2\kappa a + 1/(\kappa a + 1)^2)\gamma^2\}^{1/2}]^2}\frac{a^2\, e^{-\kappa(R-2a)}}{R} \tag{11.75}$$

or equivalently

$$V(R) = \frac{512\pi\gamma^2 nkT}{\kappa^2[1 + \{1 - (2\kappa a + 1/(\kappa a + 1)^2)\gamma^2\}^{1/2}]^2}\frac{a^2\, e^{-\kappa(R-2a)}}{R} \tag{11.75a}$$

where $H = R - 2a$ is the distance between the surfaces of the two spheres and $\gamma = \tanh(y_o/4)$. When large $\kappa a \gg 1$ and $H \ll a$, Eq. (11.75) reduces to Derjaguin's result, Eq. (12.17).

Equation (11.75) can be generalized to the case of two dissimilar spheres 1 and 2 of radii a_1 and a_2 carrying surface potentials ψ_{o1} and ψ_{o2}, respectively,

$$V(R) = \frac{256\pi\varepsilon_r\varepsilon_o\gamma_1\gamma_2(kT/ze)^2}{\left[1 + \{1 - (2\kappa a_1 + 1/(\kappa a_1 + 1)^2)\gamma_1^2\}^{1/2}\right]\left[1 + \{1 - (2\kappa a_2 + 1/(\kappa a_2 + 1)^2)\gamma_2^2\}^{1/2}\right]}$$
$$\times \frac{a_1 a_2\, e^{-\kappa(R-a_1-a_2)}}{R} \tag{11.76}$$

where $\gamma_i = \tanh(y_{oi}/4)$ $(i = 1, 2)$.

11.4 TWO CYLINDERS

Consider two parallel interacting cylindrical colloidal particles 1 and 2 of radii a_1 and a_2 having surface potentials ψ_{o1} and ψ_{o2}, respectively, at separation $R = H +$

a_1a_2 between their axes in a general electrolyte solution. According to the method of Brenner and Parsegian [6], the asymptotic expression for the interaction energy $V(R)$ per unit length is given by

$$V(R) = \sigma_{\text{eff}1}\psi_2(R) = \sigma_{\text{eff}2}\psi_1(R) \tag{11.77}$$

where $\sigma_{\text{eff}i}$ ($i = 1, 2$) is the charge density per unit length of the hypothetical cylinder of infinitesimal thickness that would produce the same asymptotic potential as produced by cylinder i and $\psi_i(R)$ ($i = 1, 2$) is the asymptotic form of the unperturbed potential at a large distance R from the axis of cylinder j, when each cylinder exists separately. The hypothetical line charge of density $\sigma_{\text{eff}i}$ gives rise to a potential

$$\psi_i(R) = \frac{Q_{\text{eff}i}}{2\pi\varepsilon_r\varepsilon_o} K_0(\kappa R) \tag{11.78}$$

where $K_0(\kappa R)$ is the zeroth order Bessel function of the second kind. The asymptotic expression for the unperturbed potential of cylinder i ($i = 1, 2$) at a large distance R from the axis of cylinder i may be expressed by Eq. (1.202), namely,

$$\psi_i(R) = \left(\frac{kT}{e}\right) Y_j \frac{K_0(\kappa R)}{K_0(\kappa a)} \tag{11.79}$$

where Y_i is the scaled effective surface potential [10]. Comparison of Eq. (11.78) with (11.79) leads to

$$\sigma_{\text{eff}i} = 2\pi\varepsilon_r\varepsilon_o \left(\frac{kT}{e}\right) Y_i \frac{1}{K_0(\kappa R)} \tag{11.80}$$

By combining Eqs. (11.77)–(11.79), we find that the double-layer interaction energy per unit length between two parallel cylinders at large separations is given by [11]

$$V(R) = 2\pi\varepsilon_r\varepsilon_o \left(\frac{kT}{e}\right)^2 Y_1 Y_2 \frac{K_0(\kappa R)}{K_0(\kappa a_1)K_0(\kappa a_2)} \tag{11.81}$$

Equation (11.81) becomes in the low ψ_{oi} limit

$$V(R) = 2\pi\varepsilon_r\varepsilon_o \psi_{o1}\psi_{o2} \frac{K_0(\kappa R)}{K_0(\kappa a_1)K_0(\kappa a_2)} \tag{11.82}$$

which agrees with Brenner and Parsegian's result [6] and also with the leading term of the exact expression for the double-layer interaction energy between two parallel cylinders per unit length in the low potential limit (see Chapter 14).

An approximate expression for the scaled effective surface potential Y for a cylinder of radius a having a surface potential ψ_o immersed in a symmetrical electrolyte solution of valence z and bulk concentration n is given by Eq. (1.203), namely,

$$Y = \frac{8\gamma}{1 + \{1 - (1 - \beta^2)\gamma^2\}^{1/2}} \tag{11.83}$$

where $\beta = K_0(\kappa a)/K_1(\kappa a)$, $\gamma = \tanh(y_o/4)$, and $y_o = ze\psi_o/kT$ is the scaled surface potential of the cylinder.

By substituting Eq. (11.83) into Eq. (11.81), we obtain the following expression for the interaction energy between two similar cylinders of radius a carrying surface potential ψ_o[1]:

$$V(R) = 128\pi\varepsilon_r\varepsilon_o \left(\frac{kT}{ze}\right)^2 \frac{\gamma^2}{[1 + \{1 - (1 - \beta^2)\gamma^2\}^{1/2}]^2} \frac{K_0(\kappa R)}{K_0^2(\kappa a)} \tag{11.84}$$

Equation (11.84) can be generalized to the case of two dissimilar cylinders 1 and 2 of radii a_1 and a_2 carrying surface potentials ψ_{o1} and ψ_{o2}, respectively,

$$V(R) = \frac{128\pi\varepsilon_r\varepsilon_o\gamma_1\gamma_2(kT/ze)^2}{[1 + \{1 - (1 - \beta_1^2)\gamma_1^2\}^{1/2}][1 + \{1 - (1 - \beta_2^2)\gamma_2^2\}^{1/2}]} \frac{K_0(\kappa R)}{K_0(\kappa a_1)K_0(\kappa a_2)} \tag{11.85}$$

where $\beta_i = K_0(\kappa a_i)/K_1(\kappa a_i)$ $(i = 1, 2)$, $\gamma_i = \tanh(y_{oi}/4)$, and $y_{oi} = ze\psi_{oi}/kT$ $(i = 1, 2)$.

REFERENCES

1. B. V. Derjaguin and L. Landau, *Acta Physicochim.* 14 (1941) 633.
2. E. J. W. Verwey and J. Th. G. Overbeek, *Theory of the Stability of Lyophobic Colloids*, Elsevier, Amsterdam, 1948.
3. B. V. Derjaguin, *Theory of Stability of Colloids and Thin Films*, Consultants Bureau, New York, London, 1989.
4. H. Ohshima and K. Furusawa (Eds.), *Electrical Phenomena at Interfaces, Fundamentals, Measurements, and Applications, 2nd edition*, revised and expanded, Dekker, New York, 1998.
5. H. Ohshima, *Theory of Colloid and Interfacial Electric Phenomena*, Elsevier/Academic Press, Amsterdam, 2006.
6. S. L. Brenner and V. A. Parsegian, *Biophys. J.* 14 (1974) 327.

7. G. M. Bell, S. Levine, and L. N. McCartney, *J. Colloid Interface Sci.* 33 (1970) 335.
8. H. Ohshima, T. W. Healy, and L. R. White, *J. Colloid Interface Sci.* 90 (1982) 17.
9. H. Ohshima, *J. Colloid Interface Sci.* 174 (1995) 45.
10. H. Ohshima, *J. Colloid Interface Sci.* 200 (1998) 291.
11. H. Ohshima, *Colloid Polym. Sci.* 274 (1996) 1176.

12 Derjaguin's Approximation at Small Separations

12.1 INTRODUCTION

With the help of Derjaguin's approximation, one can calculate the interaction energy between two spheres or two cylinders by integrating the interaction energy between the corresponding two parallel plates [1–9]. This approximation holds good for large particles with thin double layers at small separations as compared with the particle size. It should also be mentioned that Derjaguin's approximation can be applied not only to the electrostatic interaction between colloidal particles but also to the van der Waals interaction between particles at small particle separations, as will be seen in Chapter 19.

12.2 TWO SPHERES

Consider the interaction energy $V_{sp}(H)$ between two spheres 1 and 2 of radii a_1 and a_2 separated by a distance H between their surfaces (Fig. 12.1). The spherical Poisson–Boltzmann equation for the two interacting spheres has been not solved. If, however, the following conditions are satisfied,

$$\kappa a_1 \gg 1,\ \kappa a_2 \gg 1,\ H \ll a_1, \quad \text{and} \quad H \ll a_2 \tag{12.1}$$

then with the help of Derjaguin's approximation [1–4], one can calculate $V_{sp}(H)$ via the corresponding interaction energy $V_{pl}(h)$ between two parallel dissimilar plates, namely,

$$V_{sp}(H) = \frac{2\pi a_1 a_2}{a_1 + a_2} \int_H^\infty V_{pl}(h)dh \tag{12.2}$$

For the special case of two identical spheres of radius $a_1 = a_2 = a$ at separation H, Eq. (12.2) becomes

Biophysical Chemistry of Biointerfaces By Hiroyuki Ohshima

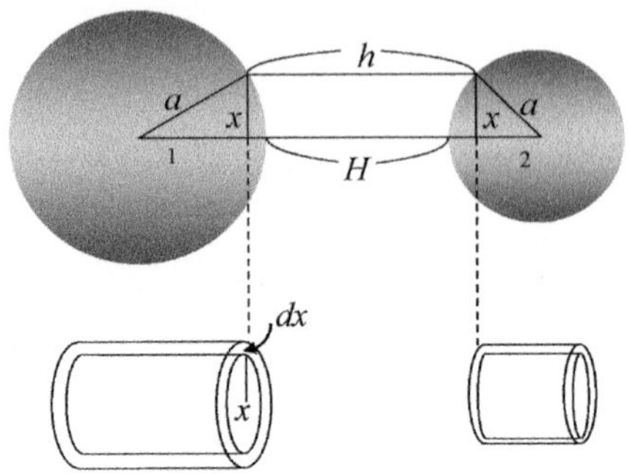

FIGURE 12.1 Derjaguin's approximation for the two interacting spheres 1 and 2 at separation H, having radii a_1 and a_2, respectively.

$$V_{\rm sp}(H) = \pi a \int_H^\infty V_{\rm pl}(h)dh \tag{12.3}$$

Equation (12.2) can be derived as follows (Fig. 12.1). The interaction energy $V_{\rm sp}(H)$ is considered to be formed by the contributions of parallel hollow cylinders of thickness dx, each pair of cylinders (having a planar intersection area of $2\pi x\,dx$) contributing to the interaction energy an amount equal to $2\pi x V_{\rm pl}(h)\,dx$, where $V_{\rm pl}(h)$ is the interaction energy per unit area between two parallel plates at a distance of h. The interaction energy $V_{\rm sp}(H)$ between two spheres 1 and 2, having radii a_1 and a_2, at separation H can be approximated by

$$V_{\rm sp}(H) = \int_{x=0}^{x=\infty} 2\pi x V_{\rm pl}(h)dx \tag{12.4}$$

where

$$h = H + \left(a_1 - \sqrt{a_1^2 - x^2}\right) + \left(a_2 - \sqrt{a_2^2 - x^2}\right) \tag{12.5}$$

from which

$$x\,dx = \frac{a_1 a_2}{a_1 + a_2} dh \tag{12.6}$$

Substituting Eq. (12.6) into Eq. (12.4) gives Eq. (12.2).

It follows from Eq. (12.2) that the interaction force $P_{\rm sp}(H)$ between two spheres at separation H is directly proportional to the interaction energy $V_{\rm pl}(H)$ per unit area between two parallel plates at separation H, namely,

$$P_{\rm sp}(H) = -\frac{\partial V_{\rm sp}(H)}{\partial H} = \frac{2\pi a_1 a_2}{a_1 + a_2} V_{\rm pl}(H) \tag{12.7}$$

12.2.1 Low Potentials

We apply the Derjaguin's approximation (Eq. (12.3)) to the low-potential approximate expression for the plate–plate interaction energy, that is, Eqs. (9.53) and (9.65), obtaining the following two formulas for the interaction between two similar spheres 1 and 2 of radius a carrying unperturbed surface potential ψ_o at separation H at constant surface potential, $V^{\psi}(H)$, and that for the constants surface charged density case, $V^{\sigma}(H)$:

$$V^{\psi}(H) = 2\pi\varepsilon_r\varepsilon_o a\psi_o^2 \ln(1 + e^{-\kappa H}) \tag{12.8}$$

and

$$V^{\sigma}(H) = 2\pi\varepsilon_r\varepsilon_o a\psi_o^2 \ln\left(\frac{1}{1 - e^{-\kappa H}}\right) \tag{12.9}$$

In Eq. (12.9) the unperturbed surface potential ψ_o is related to the surface charge density σ by $\psi_o = \sigma/\varepsilon_r\varepsilon_o\kappa$. Note that Eq. (12.9) ignores the influence of the internal electric fields induced within the interacting particles.

For the interaction between two dissimilar spheres 1 and 2 of radii a_1 and a_2 at separation H carrying constant surface potentials ψ_{o1} and ψ_{o2}, respectively, Hogg et al. [10] obtained the following expressions for $V^{\psi}(H)$ from Eq. (10.58):

$$V^{\psi}(H) = \pi\varepsilon_r\varepsilon_o \frac{a_1 a_2}{a_1 + a_2}\left\{(\psi_{o1} + \psi_{o2})^2 \ln(1 + e^{-\kappa H}) + (\psi_{o1} - \psi_{o2})^2 \ln(1 - e^{-\kappa H})\right\} \tag{12.10}$$

which is called the HHF formula for the sphere–sphere interaction. For the case of the interaction energy between two dissimilar spheres 1 and 2 of radii a_1 and a_2 carrying unperturbed surface potentials ψ_{o1} and ψ_{o2} at separation H at constant surface charge density, Wiese and Healy [11] derived the following expressions for $V^{\psi}(H)$ from Eq. (10.55):

$$V^{\sigma}(H) = \pi\varepsilon_r\varepsilon_o \frac{a_1 a_2}{a_1 + a_2}\left\{-(\psi_{o1} + \psi_{o2})^2 \ln(1 - e^{-\kappa H}) - (\psi_{o1} - \psi_{o2})^2 \ln(1 + e^{-\kappa H})\right\} \tag{12.11}$$

which again ignores the influence of the internal electric fields within the interacting particles.

For the mixed case where sphere 1 carries a constant surface potential ψ_{o1} and sphere 2 carries a constant surface charge density σ (or the corresponding unperturbed surface potential ψ_{o2}), Kar et al. [12] derived the following expression for the interaction energy $V^{\psi-\sigma}(H)$ between spheres 1 and 2 from Eq. (10.63):

$$V^{\psi-\sigma}(H) = 4\pi\varepsilon_r\varepsilon_o \frac{a_1 a_2}{a_1 + a_2}\left\{\psi_{o1}\psi_{o2}\arctan\left(e^{-\kappa H}\right) + \tfrac{1}{4}\left(\psi_{o1}^2 - \psi_{o2}^2\right)\ln\left(1 + e^{-2\kappa H}\right)\right\} \tag{12.12}$$

12.2.2 Moderate Potentials

Consider the double layer between two similar spheres of radius a at separation H between their centers carrying constant surface potential ψ_o in a symmetrical electrolyte solution of valence z and bulk concentration n between The expression for the interaction energy $V^{\psi}(H)$ correct to the sixth power in the surface potential ψ_o can be obtained by applying Derjaguin's approximation (Eq. (12.3)) to Eq. (9.151) with the result that [13]

$$\begin{aligned} V^{\psi}(H) &= \frac{4\pi ankT}{\kappa^2} y_o^2 \ln\left(1 + e^{-\kappa H}\right) + \frac{4\pi ankT}{\kappa^2}\left[-\frac{1}{48} y_o^4 \left(\frac{\kappa H}{2}\right)\left\{1 - \tanh\left(\frac{\kappa H}{2}\right)\right\}\right. \\ &\quad - \frac{y_o^4}{96}\frac{1 - (\kappa H/2)\tanh\left(\kappa H/2\right)}{\cosh^2\left(\kappa H/2\right)} - \frac{4\pi ankT}{\kappa^2}\frac{y_o^6}{5760}\left(\frac{\kappa H}{2}\right)\left\{1 - \tanh\left(\frac{\kappa H}{2}\right)\right\} \\ &\quad + y_o^6 \frac{17 + 4(\kappa H/2)\tanh(\kappa H/2)}{46080\cosh^2(\kappa H/2)} - y_o^6 \frac{1 - 11(\kappa H/2)\tanh(\kappa H/2)}{15360\cosh^4(\kappa H/2)} \\ &\quad \left. + \frac{(\kappa H/2)^2}{1536}\frac{y_o^6}{\cosh^6(\kappa H/2)}\right] \end{aligned} \tag{12.13}$$

where $y_o = ze\psi_{o1}/kT$ is the scaled surface potential and κ is the Debye–Hückel parameter given by Eq. (1.11) The first term on the right-hand side of Eq. (12.13) agrees with Eq. (12.8).

For the interaction between two parallel dissimilar spheres of radii a_1 and a_2 at separation H between their centers carrying constant surface potentials ψ_{o1} and ψ_{o2}, respectively, we obtain by applying Eq. (12.2) to Eq. (10.119) [13]

$$\begin{aligned} V^{\psi}(H) &= \frac{8\pi a_1 a_2 nkT}{\kappa^2(a_1 + a_2)}\left[Y_+^2 \ln\left(1 + e^{-\kappa H}\right) + Y_-^2 \ln\left(1 - e^{-\kappa H}\right)\right] \\ &\quad + \frac{8\pi a_1 a_2 nkT}{\kappa^2(a_1 + a_2)}\left[-\frac{1}{48}\left(Y_+^4 + 3Y_+^2 Y_-^2\right)\left(\frac{\kappa H}{2}\right)\left\{1 - \tanh\left(\frac{\kappa H}{2}\right)\right\}\right. \end{aligned}$$

$$+\frac{1}{48}\left(Y_-^4+3Y_+^2Y_-^2\right)\left(\frac{\kappa H}{2}\right)\left\{\coth\left(\frac{\kappa H}{2}\right)-1\right\}$$

$$\left.-\frac{Y_+^4}{96}\frac{1-(\kappa H/2)\tanh(\kappa H/2)}{\cosh^2(\kappa H/2)}-\frac{Y_-^4}{96}\frac{(\kappa H/2)\coth(\kappa H/2)-1}{\sinh^2(\kappa H/2)}\right]$$

$$+\frac{8\pi a_1 a_2 nkT}{\kappa^2(a_1+a_2)}\left[-\frac{Y_+^2}{5760}\left\{Y_+^4+\frac{15}{8}Y_-^2\left(7Y_+^2+Y_-^2\right)\right\}\left(\frac{\kappa H}{2}\right)\left\{1-\tanh\left(\frac{\kappa H}{2}\right)\right\}\right.$$

$$+\frac{Y_-^2}{5760}\left\{Y_-^4+\frac{15}{8}Y_+^2\left(7Y_-^2+Y_+^2\right)\right\}\left(\frac{\kappa H}{2}\right)\left\{\coth\left(\frac{\kappa H}{2}\right)-1\right\}$$

$$+Y_+^6\frac{17+4(\kappa H/2)\tanh(\kappa H/2)}{46080\cosh^2(\kappa H/2)}-Y_-^6\frac{4(\kappa H/2)\coth(\kappa H/2)+17}{46080\sinh^2(\kappa H/2)}$$

$$+Y_+^4Y_-^2\frac{1+(\kappa H/2)\tanh(\kappa H/2)}{1024\cosh^2(\kappa H/2)}-Y_+^2Y_-^4\frac{(\kappa H/2)\coth(\kappa H/2)+1}{1024\sinh^2(\kappa H/2)}$$

$$-Y_+^6\frac{1-11(\kappa H/2)\tanh(\kappa H/2)}{15360\cosh^4(\kappa H/2)}+Y_-^6\frac{11(\kappa H/2)\coth(\kappa H/2)-1}{15360\sinh^4(\kappa H/2)}$$

$$\left.+\frac{(\kappa H/2)^2}{1536}\left\{\frac{Y_+^2}{\cosh^2(\kappa H/2)}-\frac{Y_-^2}{\sinh^2(\kappa H/2)}\right\}^3\right] \tag{12.14}$$

with

$$Y_+=\frac{y_{o1}+y_{o2}}{2},\quad Y_-=\frac{y_{o1}-y_{o2}}{2} \tag{12.15}$$

where y_{o1} $(=ze\psi_{o1}/kT)$ and y_{o2} $(=ze\psi_{o2}/kT)$ are, respectively, the scaled surface potentials of spheres 1 and 2.The first term on the right-hand side of Eq. (12.14) agrees with the HHF formula (15.10).

Better approximations than Eq. (12.14) can be obtained if the interaction energy is expressed as a series of $\gamma=\tanh(ze\psi_o/kT)$ instead of ψ_o, as suggested by Honig and Mul [14]. For the case of two similar spheres of radius a carrying constant scaled surface potential $y_o=ze\psi_o/kT$ at separation H, by applying Derjaguin's approximation to Eq. (9.160) we obtain [15]

$$V^{\psi}(H)=\frac{64\pi ankT}{\kappa^2}\left[\gamma^2\left(1+\frac{2}{3}\gamma^2+\frac{23}{45}\gamma^4\right)\ln\left(1+e^{-\kappa H}\right)\right.$$

$$-\frac{1}{3}\gamma^4\left(1+\frac{22}{15}\gamma^2\right)\left(\frac{\kappa H}{2}\right)\left\{1-\tanh\left(\frac{\kappa H}{2}\right)-\frac{\tanh(\kappa H/2)}{2\cosh^2(\kappa H/2)}\right\}$$

$$-\frac{1}{6}\gamma^4\left(1+\frac{23}{30}\gamma^2\right)\frac{1}{\cosh^2(\kappa H/2)}$$

$$-\frac{\gamma^6}{60}\frac{1-11(\kappa H/2)\tanh(\kappa H/2)}{\cosh^4(\kappa H/2)}+\frac{\gamma^6}{6}\frac{(\kappa H/2)^2}{\cosh^6(\kappa H/2)}\Bigg] \tag{12.16}$$

12.2.3 Arbitrary Potentials: Derjaguin's Approximation Combined with the Linear Superposition Approximation

In Chapter 11, we derived the double-layer interaction energy between two parallel plates with arbitrary surface potentials at large separations compared with the Debye length $1/\kappa$ with the help of the linear superposition approximation. These results, which do not depend on the type of the double-layer interaction, can be applied both to the constant surface potential and to the constant surface charge density cases as well as their mixed case. In addition, the results obtained on the basis of the linear superposition approximation can be applied not only to hard particles but also to soft particles. We now apply Derjaguin's approximation to these results to obtain the sphere–sphere interaction energy, as shown below.

For the case where two similar spheres carrying unperturbed surface potential ψ_o at separation H are immersed in a symmetrical electrolyte of valence z and bulk concentration n, we obtain from Eq. (11.14)

$$V(H)=\frac{64\pi a\gamma^2 nkT}{\kappa^2}e^{-\kappa H} \tag{12.17}$$

or, equivalently

$$V(H)=32\pi\varepsilon_r\varepsilon_o a\gamma^2\left(\frac{kT}{ze}\right)^2 e^{-\kappa H} \tag{12.17a}$$

where $\gamma=\tanh(ze\psi_o/4kT)$. For two dissimilar spheres 1 and 2 of radii a_1 and a_2 carrying unperturbed surface potentials ψ_{o1} and ψ_{o2}, respectively, we obtain

$$V(H)=\frac{64\pi\gamma_1\gamma_2 nkT}{\kappa^2}\left(\frac{2a_1a_2}{a_1+a_2}\right)e^{-\kappa H} \tag{12.18}$$

or

$$V(H)=64\pi\varepsilon_r\varepsilon_o\gamma_1\gamma_2\left(\frac{a_1a_2}{a_1+a_2}\right)\left(\frac{kT}{ze}\right)^2 e^{-\kappa H} \tag{12.18a}$$

where $\gamma_1=\tanh(ze\psi_{o1}/4kT)$ and $\gamma_2=\tanh(ze\psi_{o2}/4kT)$.

For two dissimilar spheres of radii a_1 and a_2 carrying scaled unperturbed surface potentials y_{o1} and y_{o2} in a 2-1 electrolyte solution of concentration n, we obtain from Eq. (11.43)

$$V(H) = \frac{192\pi\gamma'_1\gamma'_2 nkT}{\kappa^2}\left(\frac{2a_1a_2}{a_1+a_2}\right)e^{-\kappa H} \tag{12.19}$$

or

$$V(H) = 64\pi\varepsilon_r\varepsilon_o\kappa\gamma'_1\gamma'_2\left(\frac{a_1a_2}{a_1+a_2}\right)\left(\frac{kT}{ze}\right)^2 e^{-\kappa H} \tag{12.19a}$$

with

$$\gamma'_i = \frac{3}{2}\left\{\frac{\left(\frac{2}{3}e^{y_{oi}}+\frac{1}{3}\right)^{1/2}-1}{\left(\frac{2}{3}e^{y_{oi}}+\frac{1}{3}\right)^{1/2}+1}\right\}, \quad (i=1,2) \tag{12.20}$$

where κ is given by Eq. (1.313). For the case of a mixed solution of 1-1 electrolyte of concentration n_1 and 2-1 electrolyte of concentration n_2, we obtain

$$V(H) = \frac{128\pi\gamma''_1\gamma''_2(n_1+3n_2)kT}{\kappa^2}\left(\frac{a_1a_2}{a_1+a_2}\right)e^{-\kappa H} \tag{12.21}$$

or

$$V(H) = 64\pi\varepsilon_r\varepsilon_o\kappa\gamma''_1\gamma''_2\left(\frac{a_1a_2}{a_1+a_2}\right)\left(\frac{kT}{ze}\right)^2 e^{-\kappa H} \tag{12.21a}$$

with

$$\gamma''_i = \left(\frac{1}{1-\eta/3}\right)\left[\frac{\left\{\left(1-\frac{\eta}{3}\right)e^{y_{oi}}+\frac{\eta}{3}\right\}^{1/2}-1}{\left\{\left(1-\frac{\eta}{3}\right)e^{y_{oi}}+\frac{\eta}{3}\right\}^{1/2}+1}\right] \tag{12.22}$$

$$\eta = \frac{3n_2}{n_1+3n_2} \tag{12.23}$$

where κ is given by Eq. (1.16).

12.2.4 Curvature Correction to Derjaguin' Approximation

The next-order correction terms to Derjaguin's formula and HHF formula can be derived as follows [13] Consider two spherical particles 1 and 2 in an electrolyte solution, having radii a_1 and a_2 and surface potentials ψ_{o1} and ψ_{o2}, respectively, at a closest distance, H, between their surfaces (Fig. 12.2). We assume that ψ_{o1} and ψ_{o2} are constant, independent of H, and are small enough to apply the linear Debye–Hückel linearization approximation. The electrostatic interaction free energy $V^{\psi}(H)$ of two spheres at constant surface potential in the Debye–Hückel approximation is given by

$$V^{\psi}(H) = F^{\psi}(H) - F^{\psi}(\infty) \tag{12.24}$$

with

$$F^{\psi}(H) = -\frac{1}{2}\psi_{o1}\int_{S_1}\sigma_1(H)dS_1 - \frac{1}{2}\psi_{o2}\int_{S_2}\sigma_2(H)dS_2 \tag{12.25}$$

where $\sigma_i(H)$ is the surface charge density of sphere i ($i = 1, 2$) when the spheres are at separation H; $\sigma_i(\infty)$ is the surface charge density of sphere i when it is isolated (i.e., $H = \infty$); and the integration is taken over the surface S_i of sphere i. In order to calculate $V(H)$, we need expressions for $\sigma_i(H)$. Below we consider only $\sigma_1(H)$, since the expression for $\sigma_2(H)$ can be obtained by the interchange of indices 1 and 2.

We choose a spherical polar coordinate system (r, θ, φ) in which the origin O is located at the center of sphere 1, the $\theta = 0$ line coincides with the line joining the centers of the two spheres, and φ is the azimuthal angle about the $\theta = 0$ line. By symmetry, the electric potential ψ in the electrolyte solution does not depend on the angle φ. In the Debye–Hückel approximation, $\psi(r,\theta)$ satisfies

$$\frac{\partial^2\psi(r,\theta)}{\partial r^2} + \frac{2}{r}\frac{\partial\psi(r,\theta)}{\partial r} + \frac{1}{r^2\sin\theta}\left(\sin\theta\,\frac{\partial\psi(r,\theta)}{\partial\theta}\right) = \kappa^2\psi(r,\theta) \tag{12.26}$$

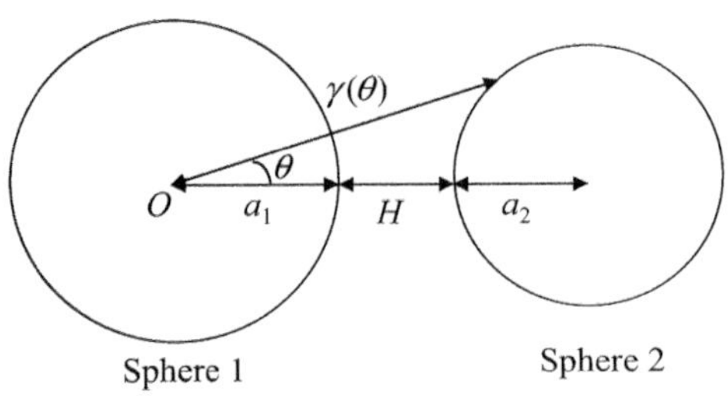

FIGURE 12.2 Interaction between two spheres 1 and 2 at a closest separation H, each having radii a_1 and a_2, respectively.

where κ is the Debye–Hückel parameter of the electrolyte solution. In the limit of large radii and in the region very near the particle surface, namely,

$$\kappa a_i \gg 1, \quad H \ll a_i, \quad (i = 1, 2) \tag{12.27}$$

the potential distribution should tend to that for the planar case and thus the contribution of the second and third terms on the left hand side of Eq. (12.26) becomes negligible. Following the method of successive approximations, we seek the solution of Eq. (12.26) in the form

$$\psi(r,\theta) = \psi^{(0)}(r,\theta) + \psi^{(1)}(r,\theta) + \cdots \tag{12.28}$$

where $\psi^{(0)}(r,\theta)$ is the zeroth-order solution, satisfying a ''plate-like'' equation obtained by neglecting the second and third terms on the left side of Eq. (12.26):

$$\frac{\partial^2 \psi^{(0)}(r,\theta)}{\partial r^2} = \kappa^2 \psi^{(0)}(r,\theta) \tag{12.29}$$

With the help of the above method, we finally obtain from Eq. (12.24) the following expression for the interaction energy $V_\psi(H)$

$$\begin{aligned} V^{\psi}(H) = {} & \pi\varepsilon_r\varepsilon_o \frac{a_1 a_2}{a_1 + a_2} \Big\{ (\psi_{o1} + \psi_{o2})^2 \ln(1 + e^{-\kappa H}) \\ & + (\psi_{o1} - \psi_{o2})^2 \ln(1 - e^{-\kappa H}) \Big\} - \frac{1}{2}\left(\frac{1}{\kappa a_1} + \frac{1}{\kappa a_2}\right) V_1(H) \end{aligned} \tag{12.30}$$

with

$$\begin{aligned} V_1(H) = {} & \pi\varepsilon_r\varepsilon_o \frac{a_1 a_2}{a_1 + a_2} \Bigg[(\psi_{o1} + \psi_{o2})^2 \Bigg\{ \frac{1}{2}\left(\frac{1}{3} - \left[\frac{a_1 - a_2}{a_1 + a_2}\right]^2\right) \kappa H \ln\left(1 + e^{-\kappa H}\right) \\ & + \frac{\kappa H e^{-\kappa H}}{3(1 + e^{-\kappa H})^2} - \frac{e^{-\kappa H}}{3(1 + e^{-\kappa H})} - \frac{1}{3} Li_2\left(-e^{-\kappa H}\right) \Bigg\} \\ & + (\psi_{o1} - \psi_{o2})^2 \Bigg\{ \frac{1}{2}\left(\frac{1}{3} - \left[\frac{a_1 - a_2}{a_1 + a_2}\right]^2\right) \kappa H \ln\left(1 - e^{-\kappa H}\right) \\ & - \frac{\kappa H e^{-\kappa H}}{3(1 - e^{-\kappa H})^2} + \frac{e^{-\kappa H}}{3(1 - e^{-\kappa H})} - \frac{1}{3} Li_2\left(e^{-\kappa H}\right) \Bigg\} \\ & - (\psi_{o1}^2 - \psi_{o2}^2) \left(\frac{a_1 - a_2}{a_1 + a_2}\right) \left\{ \kappa H \ln\left(1 - e^{-2\kappa H}\right) - \frac{1}{2} Li_2\left(e^{-2\kappa H}\right) \right\} \Bigg] \end{aligned} \tag{12.31}$$

where

$$\mathrm{Li}_s(z) = \sum_{k=1}^{\infty} \frac{z^k}{k^s} \tag{12.32}$$

is the polylogarithm function. The first term on the right-hand side of Eq. (12.30) agrees with the HHF formula (Eq. (12.10)), which is correct to order $1/(\kappa a_i)^0$, and the second term is the next-order correction of order $1/\kappa a_i$.

12.3 TWO PARALLEL CYLINDERS

We derive Derjaguin's approximation for obtaining the interaction energy between two parallel or crossed cylinders 1 and 2 of radii a_1 and a_2 at separation H from the corresponding interaction energy between two parallel plates [16, 17]. This method is applicable when conditions [12.11] hold.

Consider first the case of two parallel cylinders of radii a_1 and a_2 at separation H (Fig. 12.3). The interaction energy $V_{cy//}(H)$ is considered to be formed by the contributions of parallel thin plates of thickness dx, each pair of plates (having a planar intersection area of $1dx$) contributing to the interaction energy an amount equal to $V(h)dx$, where $V_{pl}(h)$ is the interaction energy per unit area between two parallel plates at a distance of h. The interaction energy $V_{cy//}(H)$ per unit length between

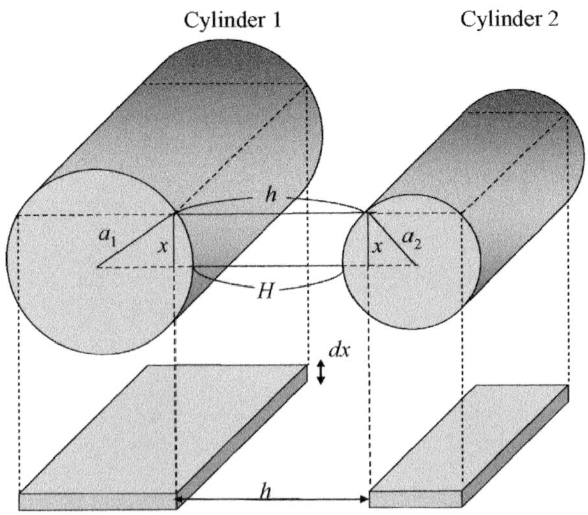

FIGURE 12.3 Derjaguin's approximation for the two interacting parallel cylinders 1 and 2 at separation H, having radii a_1 and a_2, respectively.

two cylinders 1 and 2, having radii a_1 and a_2, at separation H can be approximated by

$$V_{\mathrm{cy}//}(H) = \int_{x=-\infty}^{x=\infty} V_{\mathrm{pl}}(h)dx \tag{12.33}$$

where

$$h = H + \left(a_1 - \sqrt{a_1^2 - x^2}\right) + \left(a_2 - \sqrt{a_2^2 - x^2}\right) \tag{12.34}$$

For $x \ll a_1, a_2$, we obtain

$$h = H + \frac{a_1 + a_2}{2a_1 a_2} x^2 \tag{12.35}$$

from which

$$x\,dx = \frac{a_1 a_2}{a_1 + a_2} dh \tag{12.36}$$

and

$$x = \sqrt{\frac{2a_1 a_2}{a_1 + a_2}} \sqrt{h - H} \tag{12.37}$$

Substituting Eqs. (12.36) and (12.37) into Eq. (12.33) gives

$$V_{\mathrm{cy}//}(H) = \sqrt{\frac{2a_1 a_2}{a_1 + a_2}} \int_H^{\infty} V_{\mathrm{pl}}(h) \frac{dh}{\sqrt{h - H}} \tag{12.38}$$

The interaction energy $V^{\psi}_{\mathrm{cy}//}(H)$ per unit length between two parallel cylinders 1 and 2 with constant surface potentials $\psi_{\mathrm{o}1}$ and $\psi_{\mathrm{o}2}$ at separation H can be obtained by introducing Eq. (10.58) into Eq. (12.38), namely,

$$\begin{aligned} V^{\psi}_{\mathrm{cy}//}(H) &= -2\varepsilon_{\mathrm{r}}\varepsilon_{\mathrm{o}}\sqrt{\kappa}\sqrt{\frac{2\pi a_1 a_2}{a_1 + a_2}} \left[\left(\frac{\psi_{\mathrm{o}1} + \psi_{\mathrm{o}2}}{2}\right)^2 \mathrm{Li}_{1/2}(-e^{-\kappa H}) \right. \\ &\quad \left. + \left(\frac{\psi_{\mathrm{o}1} - \psi_{\mathrm{o}2}}{2}\right)^2 Li_{1/2}(e^{-\kappa H}) \right] \end{aligned} \tag{12.39}$$

The interaction energy $V^{\sigma}_{\mathrm{cy}//}(H)$ per unit length between two parallel cylinders 1 and 2 with constant surface charge densitiesσ_1 and σ_2 at separation H can be obtained by introducing Eq. (10.55) into Eq. (12.38), namely,

$$V^{\sigma}_{\mathrm{cy}//}(H) = 2\varepsilon_r\varepsilon_o\sqrt{\kappa}\sqrt{\frac{2\pi a_1 a_2}{a_1+a_2}}\left[\left(\frac{\psi_{o1}+\psi_{o2}}{2}\right)^2 \mathrm{Li}_{1/2}(e^{-\kappa H}) - \left(\frac{\psi_{o1}-\psi_{o2}}{2}\right)^2 \mathrm{Li}_{1/2}(-e^{-\kappa H})\right] \tag{12.40}$$

where $\psi_{o1}=\sigma_1/\varepsilon_r\varepsilon_o\kappa$ and $\psi_{o2}=\sigma_2/\varepsilon_r\varepsilon_o\kappa$ are, respectively, the uncharged surface potentials of plates 1 and 2.

If cylinder 1 has a constant surface potential ψ_{o1} and cylinder 2 has a constant surface charge density σ_2 (or, the unperturbed surface potential $\psi_{o2}=\sigma_2/\varepsilon_r\varepsilon_o\kappa$), then the interaction energy per unit length between two parallel cylinders 1 and 2 at separation H is given by

$$V^{\psi-\sigma}_{\mathrm{cy}//}(H) = \sqrt{2}\varepsilon_r\varepsilon_o\sqrt{\kappa}\sqrt{\frac{2\pi a_1 a_2}{a_1+a_2}}\left[\psi_{o1}\psi_{o2}\, e^{-\kappa H}\Phi\left(-e^{-\kappa H},\frac{1}{2},\frac{1}{2}\right) - \frac{1}{2}(\psi_{o1}^2-\psi_{o2}^2)\mathrm{Li}_{1/2}(-e^{-2\kappa H})\right] \tag{12.41}$$

where

$$\Phi(z,s,a) = \sum_{k=0}^{\infty}\frac{z^k}{(k+a)^s} \tag{12.42}$$

is a Lerch transcendent.

12.4 TWO CROSSED CYLINDERS

Similarly, for two crossed cylinders of radii a_1 and a_2, respectively, at separation H, the interaction energy $V_{\mathrm{cy}\perp}(H)$ is considered to be formed by the contributions of laths, each pair of laths (having a planar intersection area of $dx\,dy$) contributing to the interaction energy an amount equal to $V_{\mathrm{pl}}(h)dx\,dy$[16, 17] (Fig. 12.4)

$$V_{\mathrm{cy}\perp}(H) = \int_{x=-\infty}^{x=\infty} V_{\mathrm{pl}}(h)dx\,dy \tag{12.43}$$

where

$$h = H + \left(a_1 - \sqrt{a_1^2 - x^2}\right) + \left(a_2 - \sqrt{a_2^2 - y^2}\right) \tag{12.44}$$

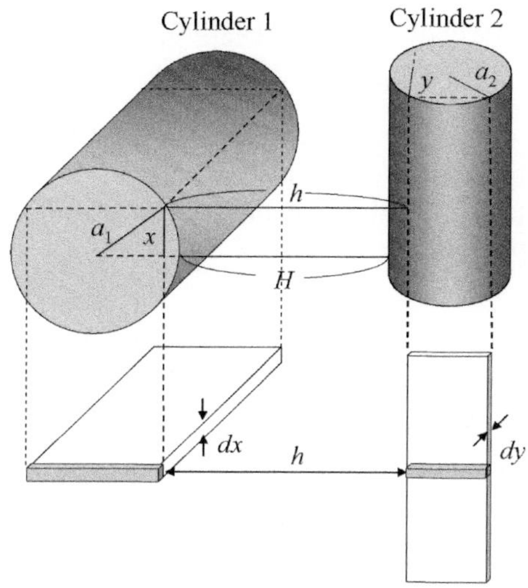

FIGURE 12.4 Derjaguin's approximation for the two interacting crossed cylinders 1 and 2 at separation H, having radii a_1 and a_2, respectively.

from which

$$h = H + \frac{x^2}{2a_1} + \frac{y^2}{2a_2} \tag{12.45}$$

If we put

$$\frac{x^2}{a_1^2} + \frac{y^2}{a_2^2} = r^2 \tag{12.46}$$

then we obtain

$$V_{cy\perp}(H) = \sqrt{2a_1}\sqrt{2a_2}\int_0^\infty V_{pl}\left(H + r^2\right)2\pi r\, dr \tag{12.47}$$

By putting $H + r^2 = h$ in Eq. (12.47), Eq. (12.47) can be rewritten as

$$V_{cy\perp}(H) = 2\pi\sqrt{a_1a_2}\int_H^\infty V_{pl}\left(h\right)dh \tag{12.48}$$

Comparison of Eqs. (12.2) and (12.48) gives

$$\frac{V_{cy\perp}(H)}{V_{sp}(H)} = \frac{a_1 + a_2}{\sqrt{a_1a_2}} \tag{12.49}$$

For the special case of two identical crossed cylinders ($a_1 = a_2$),

$$V_{cy\perp}(H) = 2V_{sp}(H) \tag{12.50}$$

That is, the interaction energy between two crossed identical cylinders equals twice the interaction energy between two identical spheres.

It follows from Eq. (12.25) that the interaction force $P_{cy\perp}(H)$ between two crossed cylinders at separation H is given by

$$P_{cy\perp}(H) = -\frac{dV_{cy\perp}(H)}{dH} = 2\pi\sqrt{a_1 a_2}V_{pl}(H) \tag{12.51}$$

The interaction energy $V^{\psi}_{cy\perp}(H)$ per unit area between two crossed cylinders with constant surface potentials ψ_{o1} and ψ_{o2} can be obtained by introducing Eq. (10.58) into Eq. (12.47), namely,

$$V^{\psi}_{cy\perp}(H) = 4\pi\varepsilon_r\varepsilon_o\sqrt{a_1 a_2}\left[\left(\frac{\psi_{o1}+\psi_{o2}}{2}\right)^2 \ln\left(1+e^{-\kappa H}\right) + \left(\frac{\psi_{o1}-\psi_{o2}}{2}\right)^2 \ln\left(\frac{1}{1-e^{-\kappa H}}\right)\right] \tag{12.52}$$

The interaction energy $V^{\sigma}_{cy\perp}(H)$ between two crossed cylinders with constant surface charge densities σ_1 and σ_2 can be obtained by introducing Eq. (10.55) into Eq. (12.47), namely,

$$V^{\sigma}_{cy\perp}(H) = 4\pi\varepsilon_r\varepsilon_o\sqrt{a_1 a_2}\left[\left(\frac{\psi_{o1}+\psi_{o2}}{2}\right)^2 \ln\left(\frac{1}{1-e^{-\kappa H}}\right) - \left(\frac{\psi_{o1}-\psi_{o2}}{2}\right)^2 \ln\left(1+e^{-\kappa H}\right)\right] \tag{12.53}$$

where $\psi_{o1} = \sigma_1/\varepsilon_r\varepsilon_o\kappa$ and $\psi_{o2} = \sigma_2/\varepsilon_r\varepsilon_o\kappa$ are, respectively, the uncharged surface potentials of cylinders 1 and 2.

The interaction energy $V^{\sigma}_{cy\perp}(H)$ two crossed cylinders 1 and 2 with constant surface potential ψ_{o1} and surface charge density σ_2, respectively, can be obtained by introducing Eq. (10.63) into Eq. (12.38), namely,

$$V^{\psi-\sigma}_{cy\perp}(H) = 4\pi\varepsilon_r\varepsilon_o\sqrt{a_1 a_2}\left\{\psi_{o1}\psi_{o2}\arctan\left(e^{-\kappa H}\right) + \frac{1}{4}\left(\psi_{o1}^2 - \psi_{o2}^2\right)\ln\left(1+e^{-2\kappa H}\right)\right\} \tag{12.54}$$

where $\psi_{o2} = \sigma_2/\varepsilon_r\varepsilon_o\kappa$ is the uncharged surface potentials of cylinder 2.

REFERENCES

1. B. V. Derjaguin, *Kolloid Z.* 69 (1934) 155.
2. B. V. Derjaguin and L. D. Landau, *Acta Physicochim.* 14 (1941) 633.
3. E. J. W. Verwey and J. Th. G. Overbeek, *Theory of the Stability of Lyophobic Colloids*, Elsevier, Amsterdam, 1948.
4. B. V. Derjaguin, *Theory of Stability of Colloids and Thin Films*, Consultants Bureau, New York, London, 1989.
5. J. N. Israelachvili, *Intermolecular and Surface Forces, 2nd edition*, Academic Press, New York, 1992.
6. J. Lyklema, *Fundamentals of Interface and Colloid Science, Solid–Liquid Interfaces*, Vol. 2, Academic Press, New York, 1995.
7. H. Ohshima, in H. Ohshima and K. Furusawa (Eds.), *Electrical Phenomena at Interfaces, Fundamentals, Measurements, and Applications, 2nd edition*, Revised and Expanded, Dekker, New York, 1998, Chapter 3.
8. H. Ohshima, *Theory of Colloid and Interfacial Electric Phenomena*, Elsevier/Academic Press, 1968.
9. T. F. Tadros (Ed.), *Colloid Stability, The Role of Surface Forces—Part 1*, Wiley-VCH, Weinheim, 2007.
10. R. Hogg, T. W. Healy, and D. W. Fuerstenau, *Trans. Faraday Soc.* 62 (1966) 1638.
11. G. R. Wiese and T. W. Healy, *Trans. Faraday Soc.*, 66 (1970) 490.
12. G. Kar, S. Chander, and T. S. Mika, *J. Colloid Interface Sci.* 44 (1973) 347.
13. H. Ohshima, T. W. Healy, and L. R. White, *J. Colloid Interface Sci.* 89 (1982) 494.
14. E. P. Honig and P. M. Mul, *J. Colloid Interface Sci.* 36 (1971) 258.
15. H. Ohshima and T. Kondo, *J. Colloid Interface Sci.* 122 (1988) 591.
16. M. J. Sparnaay, *Recueil*, 712 (1959) 6120.
17. H. Ohshima and A. Hyono, *J. Colloid Interface Sci.* 333 (2009) 202.

13 Donnan Potential-Regulated Interaction Between Porous Particles

13.1 INTRODUCTION

The solution to the Poisson–Boltzmann equation for the system of two interacting particles immersed in an electrolyte solution must satisfy the boundary conditions on the surface of the respective particles. For this reason, the sum of the unperturbed potentials of the interacting particles in general cannot be the solution to the corresponding Poisson–Boltzmann equation, even for the case where the linearized Poisson–Boltzmann is employed. Only one exception is the linearized Poisson–Boltzmann equation for the interaction between porous particles (or, soft particles without the particle core) (Fig. 13.1) provided that the relative permittivity takes the same value in the solutions both outside and inside the particles. One can thus calculate the interaction energy between two porous spheres or two parallel cylinders without recourse to Derjaguin's approximation. In such cases, the linear superposition of unperturbed potentials of interacting particles always gives the exact solution to the corresponding linearized Poisson–Boltzmann equation for all particle separations. For interactions involving hard particles, the linear superposition holds approximately good only for large particle separations. In this sense, porous particles can be regarded as a prototype of hard particles. It is also shown that if the particle size is much larger than the Debye length $1/\kappa$, then the potential in the region deep inside the particle remains unchanged at the Donnan potential in the interior region of the interacting particles. We call this type of interaction the Donnan potential-regulated interaction (the Donnan potential regulation model).

13.2 TWO PARALLEL SEMI-INFINITE ION-PENETRABLE MEMBRANES (POROUS PLATES)

Consider two parallel semi-infinite membranes 1 and 2 carrying constant densities $\rho_{\text{fix}1}$ and $\rho_{\text{fix}2}$ of fixed charges separation h between their surfaces in an electrolyte solution containing N ionic species with valence z_i and bulk concentration (number

Biophysical Chemistry of Biointerfaces By Hiroyuki Ohshima

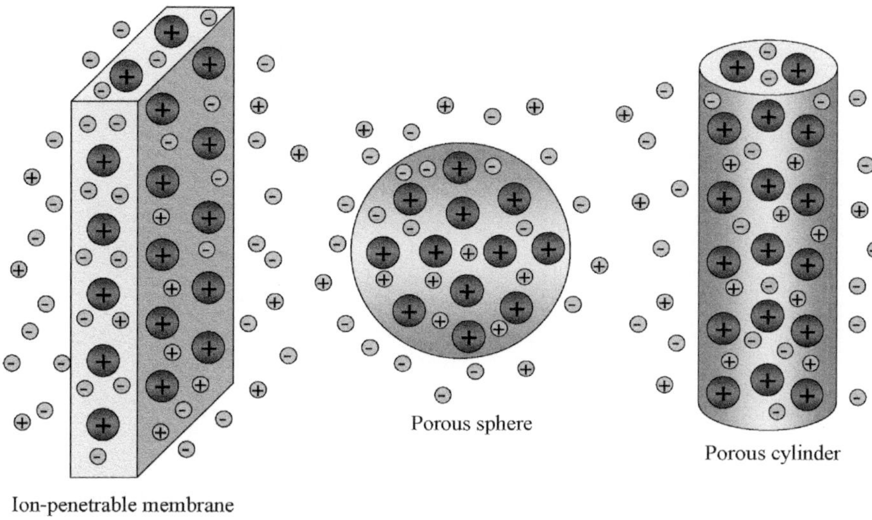

FIGURE 13.1 Various types of positively charged ion-penetrable porous particles with fixed charges (large circles with plus sign). Electrolyte ions (small circles with plus or minus signs) can penetrate the particle interior.

density) n_i^∞ $(i=1, 2, \ldots, N)$ (in units of m^{-3}) [1]. The fixed-charge density $\rho_{\mathrm{fix}i}$ in membrane i $(i=1, 2)$ is related to the density N_i of ionized groups of valence Z_i distributed in membrane i by $\rho_{\mathrm{fix}i}=Z_ieN_i$ $(i=1, 2)$. Without loss of generality, we may assume that membrane 1 is positively charged $(Z_1>0)$ and membrane 2 may be either positively or negatively charged and that

$$Z_1N_1 \geq |Z_2|N_2 > 0 \tag{13.1}$$

We also assume that the relative permittivity in membranes 1 and 2 tale the same value ε_r as that of the electrolyte solution. Suppose that membrane 1 is placed in the region $-\infty<x<0$ and membrane 2 is placed in the region $x>h$ (Fig. 13.2). The linearized Poisson–Boltzmann equations in the respective regions are

$$\frac{d^2\psi}{dx^2} = \kappa^2\psi - \frac{\rho_{\mathrm{fix1}}}{\varepsilon_r\varepsilon_o}, \quad -\infty < x < 0 \tag{13.2}$$

$$\frac{d^2\psi}{dx^2} = \kappa^2\psi, \quad 0 < x < h \tag{13.3}$$

$$\frac{d^2\psi}{dx^2} = \kappa^2\psi - \frac{\rho_{\mathrm{fix2}}}{\varepsilon_r\varepsilon_o}, \quad x > h \tag{13.4}$$

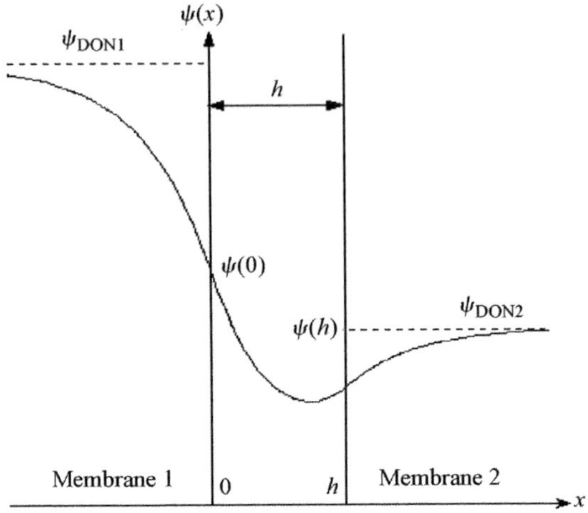

FIGURE 13.2 Schematic representation of the potential distribution $\psi(x)$ across two parallel interacting ion-penetrable semi-infinite membranes (soft plates) 1 and 2 at separation h. The potentials in the region far inside the membrane interior is practically equal to the Donnan potential $\psi_{\rm DON1}$ or $\psi_{\rm DON2}$.

with

$$\kappa = \left(\frac{1}{\varepsilon_r \varepsilon_o kT} \sum_{i=1}^{N} z_i^2 e^2 n_i^\infty \right)^{1/2} \tag{13.5}$$

where κ is the Debye–Hückel parameter. The boundary conditions are

$$\psi \text{ and } d\psi/dx \text{ are continuous at } x = 0 \text{ and } x = h \tag{13.6}$$

$$\frac{d\psi}{dx} \to 0 \quad \text{as } x \to \pm\infty \tag{13.7}$$

It can be shown that the solution $\psi(x)$ to Eqs. (13.2)–(13.4) subject to Eqs. (13.6) and (13.7) is given by a linear superposition of the unperturbed potential $\psi_1^{(0)}(x)$ produced by membrane 1 in the absence of membrane 2 and the corresponding unperturbed potential $\psi_2^{(0)}(x)$ for membrane 2, which are obtained by solving the linearized Poisson–Boltzmann equations for a single membrane, namely,

$$\psi(x) = \psi_1^{(0)}(x) + \psi_2^{(0)}(x) \tag{13.8}$$

with

$$\psi_1^{(0)}(x) = \begin{cases} 2\psi_{o1}\left(1 - \dfrac{1}{2}e^{\kappa x}\right), & x < 0 \\ \psi_{o1}\, e^{-\kappa x}, & x > 0 \end{cases} \tag{13.9}$$

$$\psi_1^{(0)}(x) = \begin{cases} \psi_{o2}\, e^{-\kappa(h-x)}, & x < h \\ 2\psi_{o2}\left(1 - \dfrac{1}{2}e^{\kappa(h-x)}\right), & x > h \end{cases} \tag{13.10}$$

$$\psi_{o1} = \frac{\rho_{\text{fix1}}}{2\varepsilon_r \varepsilon_o \kappa^2} \tag{13.11}$$

$$\psi_{o2} = \frac{\rho_{\text{fix2}}}{2\varepsilon_r \varepsilon_o \kappa^2} \tag{13.12}$$

where ψ_{o1} and ψ_{o2} are, respectively, the unperturbed surface potentials of membranes 1 and 2, which are half the Donnan potentials ψ_{DON1} and ψ_{DON2} for the low potential case, namely,

$$\psi_{\text{DON1}} = 2\psi_{o1} = \frac{\rho_{\text{fix1}}}{\varepsilon_r \varepsilon_o \kappa^2} \tag{13.13}$$

$$\psi_{\text{DON2}} = 2\psi_{o2} = \frac{\rho_{\text{fix2}}}{\varepsilon_r \varepsilon_o \kappa^2} \tag{13.14}$$

The solution to Eqs. (13.2)–(13.4) can thus be written as

$$\psi(x) = 2\psi_{o1}\left(1 - \frac{1}{2}e^{\kappa x}\right) + \psi_{o2}\, e^{-\kappa(h-x)}, \quad -\infty < x < 0 \tag{13.15}$$

$$\psi(x) = \psi_{o1}\, e^{-\kappa x} + \psi_{o2}\, e^{-\kappa(h-x)}, \quad 0 < x < h \tag{13.16}$$

$$\psi(x) = \psi_{o1}\, e^{-\kappa x} + 2\psi_{o2}\left(1 - \frac{1}{2}e^{\kappa(h-x)}\right), \quad h < x < -\infty \tag{13.17}$$

Note that the boundary condition (13.6) states that membranes 1 and 2 are ''electrically transparent'' to each other. As a result, the solution to Eqs. (13.2)–(13.4) is obtained by a linear superposition approximation (LSA), which, for the case of rigid membranes, holds only at large membrane separations.

In Fig. 13.3, we plot the potential distribution $\psi(x)$ between two parallel similar ion-penetrable membranes with $\psi_{\text{DON1}} = \psi_{\text{DON2}} = \psi_{\text{DON}}$ (or $\psi_{o1} = \psi_{o2} = \psi_o$) for $\kappa h = 0$, 1, 2, and ∞. In Fig. 13.3, we have introduced the following scaled potential y, scaled unperturbed surface potential y_o, and scaled Donnan potential y_{DON}:

$$y(x) = \frac{e\psi(x)}{kT} \tag{13.18}$$

$$y_o = \frac{e\psi_o}{kT} \tag{13.19}$$

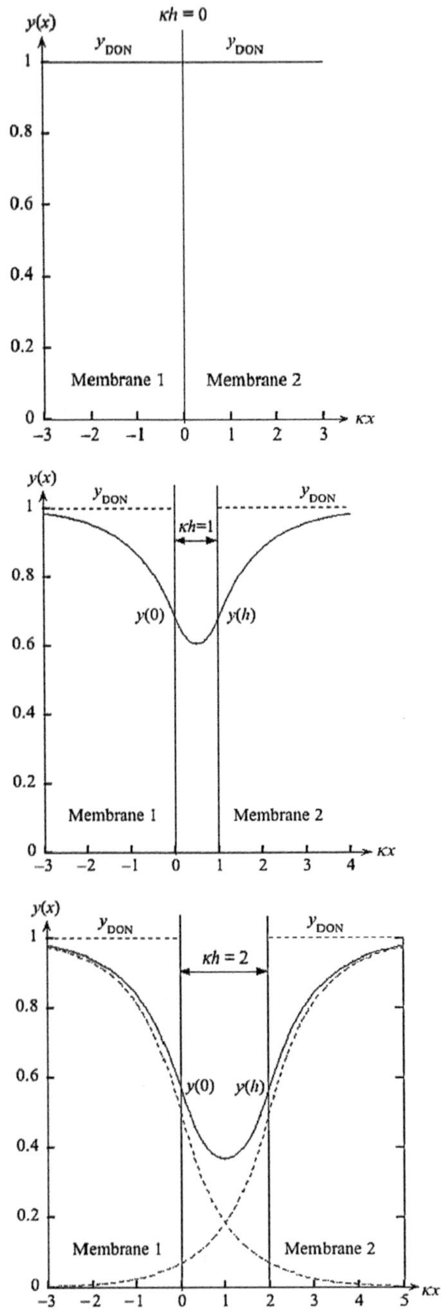

FIGURE 13.3 Scaled potential distribution $y(x) = e\psi(x)/kT$ between two parallel similar membranes 1 and 2 with $y_{DON} = e\psi_{DON}/kT = 1$ for $\kappa h = 0$, 1, and 2. The dotted curves are the scaled unperturbed potential distribution at $\kappa h = \infty$.

$$y_{\mathrm{DON}} = \frac{e\psi_{\mathrm{DON}}}{kT} \tag{13.20}$$

It follows from Fig. 13.3 and Eqs. (13.15) and (13.17) that the potentials in the region deep inside the membranes are equal to the Donnan potential ψ_{DON} in the membranes, which is independent of the membrane separation h. That is, the potential in the region deep inside the membrane remains unchanged at the Donnan potential during interaction. Even for the interaction between membranes of finite thickness, if the membrane thickness is much larger than the Debye length $1/\kappa$, then the potential in the region deep inside the membrane is always equal to the Donnan potential. Note also that the values of the surface potentials $\psi(0)$ and $\psi(h)$ increase in magnitude from $\psi_{\mathrm{DON}}/2$ at infinite separation ($h = \infty$) to ψ_{DON} at $h = 0$ and that when the membranes are in contact with each other, the potential inside the membranes is equal to the Donnan potential ψ_{DON} everywhere in the membranes.

Figures 13.4 and 13.5 give the results for the interaction between two dissimilar membranes, showing changes in the potential distribution $y(x) = e\psi(x)/kT$ due to the approach of two membranes for the cases when the two membranes are likely charged, that is, $Z_1 > 0$ and $Z_2 > 0$ (Fig. 13.4) and when they are oppositely charged, that is, $Z_1 > 0$ and $Z_2 < 0$ (Fig. 13.5). Here we have introduced the following scaled unperturbed surface potentials y_o, and scaled Donnan potentials y_{DON1} and y_{DON2}:

$$y_{\mathrm{DON1}} = \frac{e\psi_{\mathrm{DON1}}}{kT} \tag{13.21}$$

$$y_{\mathrm{DON2}} = \frac{e\psi_{\mathrm{DON2}}}{kT} \tag{13.22}$$

Figure 13.4 shows that when Z_1, $Z_1 > 0$, there is a potential minimum for large h and the surface potentials $y(0)$ and $y(h)$ increase with decreasing membrane separation h. This potential minimum disappears at $h = h_c$, where the surface potential of the weakly charged membrane, $y(h)$, coincides with its Donnan potential y_{DON2}. The value of h_c can be obtained from Eq. (13.16), namely,

$$\kappa h_c = \ln\left(\frac{\psi_{01}}{\psi_{02}}\right) = \ln\left(\frac{Z_1 N_1}{Z_2 N_2}\right) \tag{13.23}$$

At $h < h_c$, $y(h)$ exceeds y_{DON2}. Note that both $y(0)$ and $y(h)$ are always less than y_{DON1}. Figure 13.5 shows that when $Z_1 > 0$ and $Z_2 < 0$, there is a zero point of the potential $y(x)$, $0 < x < h$, for large h. As h decreases, both $y(0)$ and $y(h)$ decrease in magnitude. Then, at some point $h = h_o$, the surface potential $y(h)$ of the negatively charged membrane becomes zero. The value of h_c can be obtained from Eq. (13.16), namely,

$$\kappa h_0 = \ln\left(\frac{\psi_{01}}{|\psi_{02}|}\right) = \ln\left(\frac{Z_1 N_1}{|Z_2| N_2}\right) \tag{13.24}$$

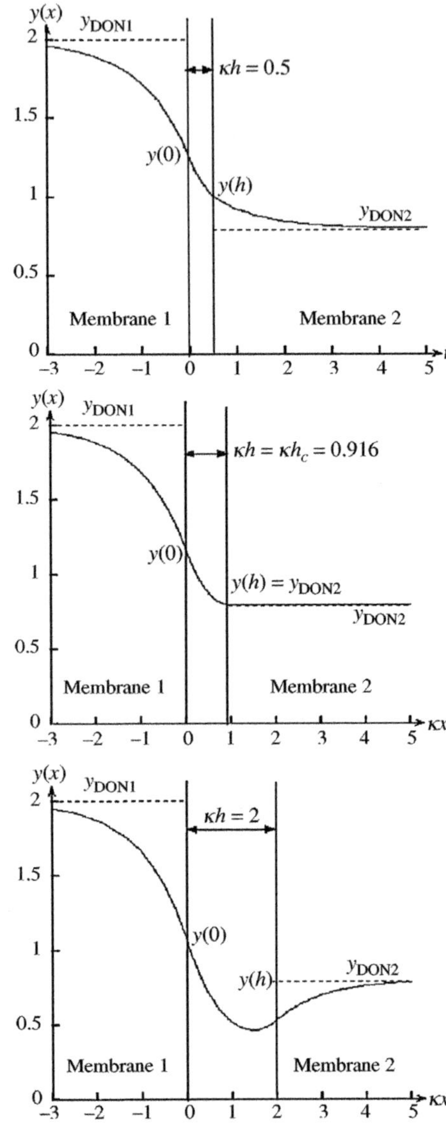

FIGURE 13.4 Scaled potential distribution $y(x) = e\psi(x)/kT$ between two parallel likely charged dissimilar membranes 1 and 2 with $y_{DON1} = e\psi_{DON}/kT = 2$ and $y_{DON2} = e\psi_{DON2}/kT = 0.8$ (or $y_{o1} = e\psi_{o1}/kT = 1$ and $y_{o2} = e\psi_{o2}/kT = 0.4$) for $\kappa h = 0.5$, 0.916 ($= \kappa h_c$), and 2.

At $h < h_o$, $y(h)$ reverses its sign and both $y(0)$ and $y(h)$ become positive. Note that as in the case where $Z_1 > 0$ and $Z_2 > 0$, both $y(0)$ and $y(h)$ never exceed y_{DON1}.

The interaction energy $V(h)$ per unit area between two parallel membranes 1 and 2 is obtained from the free energy of the system, namely,

$$V(h) = F(h) - F(\infty) \tag{13.25}$$

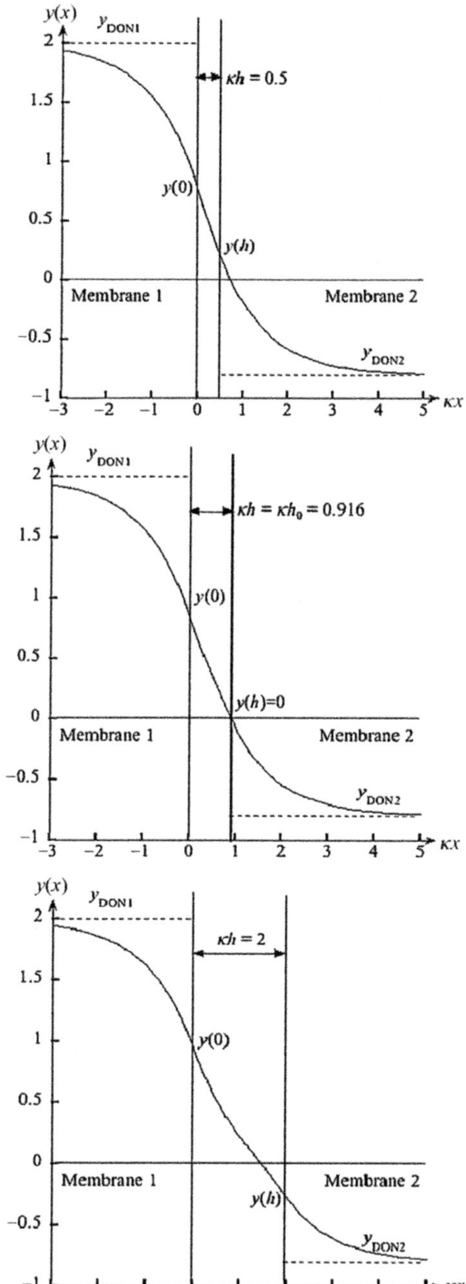

FIGURE 13.5 Scaled potential distribution $y(x) = e\psi(x)/kT$ between two parallel oppositely charged dissimilar membranes 1 and 2 with $y_{DON1} = e\psi_{DON}/kT = 2$ and $y_{DON2} = e\psi_{DON2}/kT = -0.8$ (or $y_{o1} = e\psi_{o1}/kT = 1$ and $y_{o2} = e\psi_{o2}/kT = -0.4$) for $\kappa h = 0.5$, 0.916 $(= \kappa h_o)$, and 2.

with

$$F(h) = \frac{1}{2}\int_{-\infty}^{0} \rho_{\text{fix1}}\psi\, dx + \frac{1}{2}\int_{h}^{\infty} \rho_{\text{fix2}}\psi\, dx \tag{13.26}$$

Substitution of Eqs. (13.15) and (13.17) into Eq. (13.26) gives the following expression for the interaction energy $V(h)$ per unit area between two parallel dissimilar membranes

$$V(h) = 2\varepsilon_r\varepsilon_o\kappa\psi_{o1}\psi_{o2}\, e^{-\kappa h} = \frac{\rho_{\text{fix1}}\rho_{\text{fix2}}}{2\varepsilon_r\varepsilon_o\kappa^3} e^{-\kappa h} \tag{13.27}$$

The interaction force $P(h)$ per unit area between membranes 1 and 2 is thus given by

$$P(h) = 2\varepsilon_r\varepsilon_o\kappa^2\psi_{o1}\psi_{o2}\, e^{-\kappa h} = \frac{\rho_{\text{fix1}}\rho_{\text{fix2}}}{2\varepsilon_r\varepsilon_o\kappa^2} e^{-\kappa h} \tag{13.28}$$

We thus see that the interaction force $P(h)$ is positive (repulsive) for the interaction between likely charged membranes and negative (attractive) for oppositely charged membranes. It must be stressed that the sign of the interaction force $P(h)$ remains unchanged even when the potential minimum disappears (for the case Z_1, $Z_2 > 0$) or even when the surface potential of one of the membranes reverses its sign (for the case $Z_1 > 0$ and $Z_2 < 0$).

13.3 TWO POROUS SPHERES

Consider two interacting ion-penetrable porous spheres 1 and 2 having radii a_1 and a_2, respectively, at separation R between their centers O_1 and O_2 (or, at separation $H = R - a_1 - a_2$ between their closest distances (Fig. 13.6) [1–3].

The linearized Poisson–Boltzmann equations in the respective regions are

$$\Delta\psi = \kappa^2\psi, \quad \text{outside spheres 1 and 2} \tag{13.29}$$

$$\Delta\psi = \kappa^2\psi - \frac{\rho_{\text{fix}i}}{\varepsilon_r\varepsilon_o}, \quad \text{inside sphere } i\ (i = 1, 2) \tag{13.30}$$

The boundary conditions are

$$\psi \to 0 \quad \text{as } r \to \infty \tag{13.31}$$

$$\psi \text{ and } d\psi/dn \text{ are continuous at the surfaces of spheres 1 and 2} \tag{13.32}$$

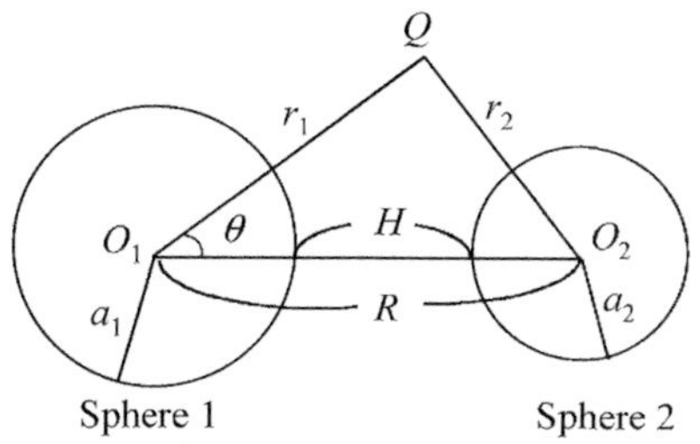

FIGURE 13.6 Interaction between two porous spheres 1 and 2 having radii a_1 and a_2, respectively, at separation R between their centers O_1 and O_2. $H = R - a_1 - a_2$. Q is a field point.

The derivative of ψ being taken along the outward normal to the surface of each sphere.

The solution to Eqs. (13.29) and (13.30) can be expressed as the sum

$$\psi = \psi_1^{(0)} + \psi_2^{(0)} \tag{13.33}$$

where $\psi_1^{(0)}$ and $\psi_2^{(0)}$ are, respectively, the unperturbed potentials for spheres 1 and 2. This is because when the boundary conditions are given by Eq. (13.32), the unperturbed potential of one sphere automatically satisfies the boundary conditions at the surface of the other sphere. The potential distribution for two interacting ion-penetrable spheres are thus simply given by linear superposition of the unperturbed potentials produced by the respective spheres, as in the case of ion-penetrable membranes. Thus, one needs to solve only the potential distribution for a single isolated sphere. Consider the unperturbed potential $\psi_1^{(0)}$ produced by sphere 1, for which Eqs. (13.29) and (13.30) reduce to

$$\frac{d^2\psi_1^{(0)}}{dr_1^2} + \frac{2}{r}\frac{d\psi_1^{(0)}}{dr_1} = \kappa^2\psi_1^{(0)}, \quad r_1 > a_1 \tag{13.34}$$

$$\frac{d^2\psi_1^{(0)}}{dr_1^2} + \frac{2}{r}\frac{d\psi_1^{(0)}}{dr_1} = \kappa^2\psi_1^{(0)} - \frac{\rho_{\text{fix1}}}{\varepsilon_r\varepsilon_o}, \quad 0 \le r < a_1 \tag{13.35}$$

and the boundary conditions to

$$\psi \to 0 \quad \text{as } r_1 \to \infty \tag{13.36}$$

$$\psi_1^{(0)}(a_1^-) = \psi_1^{(0)}(a_1^+) \tag{13.37}$$

$$\left.\frac{d\psi_1^{(0)}}{dr_1}\right|_{r_1=a_1^-} = \left.\frac{d\psi_1^{(0)}}{dr_1}\right|_{r_1=a_1^+} \tag{13.38}$$

where r_1 is the distance measured from the center O_1 of sphere 1 (Fig. 13.6). The solution to Eqs. (13.34) and (13.35) subject to Eqs. (13.36)–(13.38) are

$$\psi_1^{(0)}(r_1) = \begin{cases} \dfrac{\rho_{\text{fix1}}}{\varepsilon_r \varepsilon_o \kappa^2} \left\{ \cosh(\kappa a_1) - \dfrac{\sinh(\kappa a_1)}{\kappa a_1} \right\} a_1 \dfrac{e^{-\kappa r_1}}{r_1}, & r_1 \geq a_1 \\ \dfrac{\rho_{\text{fix1}}}{\varepsilon_r \varepsilon_o \kappa^2} \left\{ 1 - \left(1 + \dfrac{1}{\kappa a_1}\right) a_1 \, e^{-\kappa a_1} \dfrac{\sinh(\kappa r_1)}{r_1} \right\}, & 0 \leq r_1 \leq a_1 \end{cases} \tag{13.39}$$

Similarly, we can derive the potential ψ_2 produced by sphere 2 in the absence of sphere 1, which ψ_2 is obtained by replacing r_1 with r_2 and a_1 with a_2 in Eq. (13.39). Here r_2 is the radial coordinate measured from the center O_2 of sphere 2, which is related to r_1 via (Fig. 13.6)

$$r_2 = (R^2 + r_1^2 - 2Rr_1 \cos\theta)^{1/2} \tag{13.40}$$

The potential distribution for two interacting spheres 1 and 2 is given by the sum of ψ_1 and ψ_2, namely,

$$\begin{aligned} \psi(r_1, \theta) &= \frac{\rho_{\text{fix1}}}{\varepsilon_r \varepsilon_o \kappa^2} \left\{ 1 - \left(1 + \frac{1}{\kappa a_1}\right) a_1 \, e^{-\kappa a_1} \frac{\sinh(\kappa r_1)}{r_1} \right\} \\ &\quad + \frac{\rho_{\text{fix2}}}{\varepsilon_r \varepsilon_o \kappa^2} \left\{ \cosh(\kappa a_2) - \frac{\sinh(\kappa a_2)}{\kappa a_2} \right\} a_2 \frac{e^{-\kappa r_2}}{r_2}, \\ &\quad 0 \leq r_1 \leq a_1 (\text{inside sphere 1}) \end{aligned} \tag{13.41}$$

$$\begin{aligned} \psi(r_1, \theta) &= \frac{\rho_{\text{fix1}}}{\varepsilon_r \varepsilon_o \kappa^2} \left\{ \cosh(\kappa a_1) - \frac{\sinh(\kappa a_1)}{\kappa a_1} \right\} a_1 \frac{e^{-\kappa r_1}}{r_1} \\ &\quad + \frac{\rho_{\text{fix2}}}{\varepsilon_r \varepsilon_o \kappa^2} \left\{ \cosh(\kappa a_2) - \frac{\sinh(\kappa a_2)}{\kappa a_2} \right\} a_2 \frac{e^{-\kappa r_2}}{r_2}, \\ &\quad r_1 \geq a_1, \; r_2 \geq a_2 \end{aligned} \tag{13.42}$$

$$\begin{aligned} \psi(r_1, \theta) &= \frac{\rho_{\text{fix2}}}{\varepsilon_r \varepsilon_o \kappa^2} \left\{ 1 - \left(1 + \frac{1}{\kappa a_2}\right) a_2 \, e^{-\kappa a_2} \frac{\sinh(\kappa r_2)}{r_2} \right\} \\ &\quad + \frac{\rho_{\text{fix1}}}{\varepsilon_r \varepsilon_o \kappa^2} \left\{ \cosh(\kappa a_1) - \frac{\sinh(\kappa a_1)}{\kappa a_1} \right\} a_1 \frac{e^{-\kappa r_1}}{r_1}, \\ &\quad 0 \leq r_1 \leq a_1 (\text{inside sphere 2}) \end{aligned} \tag{13.43}$$

We see that when $\kappa a_i \gg 1$, the potential in the region deep inside the spheres, that is, around the sphere center, becomes the Donnan potential (Eqs. (13.13) and (13.14)).

The interaction energy $V(R)$ is most easily obtained from the free energy of the system, namely, $V(R)=F(R)-F(\infty)$, where

$$F(R)=\frac{1}{2}\int_{V_1}\rho_{\text{fix}1}\psi\,dV_1+\frac{1}{2}\int_{V_2}\rho_{\text{fix}2}\psi\,dV_2 \tag{13.44}$$

Thus, we need only to obtain the unperturbed potential distribution produced within sphere 1 by sphere 2. By using the following relation

$$\begin{aligned}\int_{V_1}\frac{e^{-\kappa r_2}}{r_2}dV_1 &= 2\pi\int_0^{a_1}\int_0^{\pi}\frac{\exp\left(-\kappa\sqrt{R^2+r_1^2-2Rr_1\cos\theta}\right)}{\sqrt{R^2+r_1^2-2Rr_1\cos\theta}}r_1^2\sin\theta\,d\theta\,dr_1\\ &=\frac{4\pi}{\kappa^2}\left\{\cosh(\kappa a_1)-\frac{\sinh(\kappa a_1)}{\kappa a_1}\right\}\frac{a_1}{R}e^{-\kappa R}\end{aligned} \tag{13.45}$$

one can calculate the first term on the right-hand side of Eq. (13.44). Using this result and adding the corresponding contribution from the integration over the volume V_2 of sphere 2, we finally obtain

$$V(R)=4\pi\varepsilon_r\varepsilon_o a_1a_2\psi_{o1}\psi_{o2}\frac{e^{-\kappa(R-a_1-a_2)}}{R} \tag{13.46}$$

where

$$\begin{aligned}\psi_{oi}&=\frac{\rho_{\text{fix}i}}{2\varepsilon_r\varepsilon_o\kappa^2}\left(1+e^{-2\kappa a_i}-\frac{1-e^{-2\kappa a_i}}{\kappa a_i}\right)\\ &=\frac{\rho_{\text{fix}i}}{\varepsilon_r\varepsilon_o\kappa^2}\left\{\cosh(\kappa a_i)-\frac{\sinh(\kappa a_i)}{\kappa a_i}\right\}\end{aligned} \tag{13.47}$$

is the unperturbed surface potential of sphere i ($i=1, 2$). Equation (13.46) coincides with the interaction energy obtained by the linear superposition approximation (see Chapter .

We consider the limiting case of $\kappa a_i \to 0$ ($i=1, 2$). In this case, Eq. (13.47) becomes

$$\psi_{oi}=\frac{\rho_{\text{fix}i}a_i^2}{3\varepsilon_r\varepsilon_o\kappa^2}=\frac{Q_i}{4\pi\varepsilon_r\varepsilon_o a_i} \tag{13.48}$$

where

$$Q_i=\frac{4\pi a^3}{3}\rho_{\text{fix}i} \tag{13.49}$$

is the total charge amount of sphere i. Thus, Eq. (13.46) tends to

$$V(R) = \frac{Q_1 Q_2}{4\pi\varepsilon_r\varepsilon_o R} \tag{13.50}$$

which agrees with a Coulomb interaction potential between two point charges Q_1 and Q_2, as should be expected.

Consider the validity of Derjaguin's approximation. In this approximation, the interaction energy between two spheres of radii a_1 and a_2 at separation H between their surfaces is obtained by integrating the corresponding interaction energy between two parallel membranes at separation h via Eq. (13.28). We thus obtain

$$V(H) = 2\pi\varepsilon_r\varepsilon_o\left(\frac{2a_1a_2}{a_1 + a_2}\right)\psi_{o1}\psi_{o2}\, e^{-\kappa H} \tag{13.51}$$

This is the result obtained via Derjaguin's approximation [4] (Chapter 12). On the other hand, by expanding the exact expression (13.46), we obtain

$$V(H) = 2\pi\varepsilon_r\varepsilon_o\left(\frac{2a_1a_2}{a_1 + a_2}\right)\psi_{o1}\psi_{o2}\, e^{-\kappa H}\left(1 - \frac{1}{\kappa a_1} - \frac{1}{\kappa a_2} - \frac{H}{a_1 + a_2} + \cdots\right) \tag{13.52}$$

We see that the first term of the expansion of Eq. (13.52) indeed agrees with Derjaguin's approximation (Eq. (13.51)). That is, Derjaguin's approximation yields the correct leading order expression for the interaction energy and the next-order correction terms are of the order of $1/\kappa a_1$, $1/\kappa a_2$, and $H/(a_1 + a_2)$.

13.4 TWO PARALLEL POROUS CYLINDERS

Consider the double-layer interaction between two parallel porous cylinders 1 and 2 of radii a_1 and a_2, respectively, separated by a distance R between their axes in an electrolyte solution (or, at separation $H = R - a_1 - a_2$ between their closest distances) [5]. Let the fixed-charge densities of cylinders 1 and 2 be ρ_{fix1} and ρ_{fix2}, respectively. As in the case of ion-penetrable membranes and porous spheres, the potential distribution for the system of two interacting parallel porous cylinders is given by the sum of the two unperturbed potentials

$$\psi = \psi_1^{(0)} + \psi_2^{(0)} \tag{13.53}$$

with

$$\psi_i^{(0)}(r_i) = \begin{cases} \psi_{oi}\dfrac{K_0(\kappa r_i)}{K_0(\kappa a_i)}, & r_i \geq a_i \\[2ex] \dfrac{\rho_{\text{fix}i}}{\varepsilon_r\varepsilon_o\kappa^2} - \psi_{oi}\dfrac{K_1(\kappa a_i)I_0(\kappa r_i)}{K_0(\kappa a_i)I_1(\kappa a_i)}, & 0 \leq r_i \leq a_i \end{cases} \tag{13.54}$$

where

$$\psi_{oi} = \frac{\rho_{\text{fix}i}}{\varepsilon_r \varepsilon_o \kappa} a_i K_0(\kappa a_i) I_1(\kappa a_i), \quad (i = 1, 2) \tag{13.55}$$

is the unperturbed surface potential of cylinder i and r_i is the distance measured from the axis of cylinder i. We calculate the interaction energy per unit length between soft cylinders 1 and 2 from $V(R) = F(R) - F(\infty)$, with the result that [5].

$$V(R) = 2\pi\varepsilon_r\varepsilon_o\psi_{o1}\psi_{o2}\frac{K_0(\kappa R)}{K_0(\kappa a_1)K_0(\kappa a_2)} \tag{13.56}$$

For large κa_1 and κa_2, Eq. (13.56) reduces to

$$V(R) = 2\sqrt{2\pi}\varepsilon_r\varepsilon_o\psi_{o1}\psi_{o2}\sqrt{\kappa a_1 a_2}\frac{e^{-\kappa(R-a_1-a_2)}}{R} \tag{13.57}$$

If, further, $H \ll a_1$ and $H \ll a_1$, then Eq. (13.61) becomes

$$V(R) = 2\sqrt{2\pi}\varepsilon_r\varepsilon_o\psi_{o1}\psi_{o2}\sqrt{\frac{\kappa a_1 a_2}{a_1 + a_2}}e^{-\kappa(R-a_1-a_2)} \tag{13.58}$$

which agrees with the result obtained via Derjaguin's approximation [6] (Chapter 12).

13.5 TWO PARALLEL MEMBRANES WITH ARBITRARY POTENTIALS

13.5.1 Interaction Force and Isodynamic Curves

Consider two parallel planar dissimilar ion-penetrable membranes 1 and 2 at separation h immersed in a solution containing a symmetrical electrolyte of valence z and bulk concentration n. We take an x-axis as shown in Fig. 13.2 [7–9]. We denote by N_1 and Z_1, respectively, the density and valence of charged groups in membrane 1 and by N_2 and Z_2 the corresponding quantities of membrane 2. Without loss of generality we may assume that $Z_1 > 0$ and Z_2 may be either positive or negative and that Eq. (13.1) holds. The Poisson–Boltzmann equations (13.2)–(13.4) for the potential distribution $\psi(x)$ are rewritten in terms of the scaled potential $y = ze\psi/kT$ as

$$\frac{d^2y}{dx^2} = \kappa^2\left(\sinh y - \frac{Z_1N_1}{2zn}\right), \quad x < 0 \tag{13.59}$$

$$\frac{d^2y}{dx^2} = \kappa^2 \sinh y, \quad 0 < x < h \tag{13.60}$$

$$\frac{d^2y}{dx^2} = \kappa^2\left(\sinh y - \frac{Z_2N_2}{2zn}\right), \quad x > h \tag{13.61}$$

where

$$\kappa = \sqrt{\frac{2nz^2e^2}{\varepsilon_r\varepsilon_o kT}} \tag{13.62}$$

is the Debye–Hückel parameter for a symmetrical electrolyte solution. Equations (13.59)–(13.61) are subject to the boundary conditions

$$y(0^-) = y(0^+) \tag{13.63}$$

$$y(h^-) = y(h^+) \tag{13.64}$$

$$\left.\frac{dy}{dx}\right|_{x=0^-} = \left.\frac{dy}{dx}\right|_{x=0^+} \tag{13.65}$$

$$\left.\frac{dy}{dx}\right|_{x=h^-} = \left.\frac{dy}{dx}\right|_{x=h^+} \tag{13.66}$$

$$y \rightarrow y_{\mathrm{DON1}} \quad \text{as } x \rightarrow -\infty \tag{13.67}$$

$$y \rightarrow y_{\mathrm{DON2}} \quad \text{as } x \rightarrow +\infty \tag{13.68}$$

Equations (13.67) and (13.68) correspond to the assumption that the potential far inside the membrane is always equal to the Donnan potential. Expressions for y_{DON1} and y_{DON2} can be derived by setting the right-hand sides of Eqs. (13.59) and (13.61) equal to zero, namely,

$$y_{\mathrm{DON1}} = \mathrm{arcsinh}\left(\frac{Z_1N_1}{2zn}\right) \tag{13.69}$$

$$y_{\mathrm{DON2}} = \mathrm{arcsinh}\left(\frac{Z_2N_2}{2zn}\right) \tag{13.70}$$

which, by Eq. (13.1), satisfy

$$y_{\mathrm{DON1}} \geq |y_{\mathrm{DON2}}| > 0 \tag{13.71}$$

Equations (13.59)–(13.61) can be integrated once to give

$$\left(\frac{dy}{dx}\right)^2 = \kappa^2\left\{2(\cosh y - \cosh y_{\mathrm{DON1}}) - \frac{Z_1N_1}{zn}(y - y_{\mathrm{DON1}})\right\}, \quad x < 0 \qquad (13.72)$$

$$\left(\frac{dy}{dx}\right)^2 = \kappa^2(2\cosh y + C), \quad 0 < x < h \qquad (13.73)$$

$$\left(\frac{dy}{dx}\right)^2 = \kappa^2\left\{2(\cosh y - \cosh y_{\mathrm{DON2}}) - \frac{Z_2N_2}{zn}(y - y_{\mathrm{DON2}})\right\}, \quad x > h \qquad (13.74)$$

where in deriving Eqs. (13.72) and (13.74), boundary conditions (13.67) and (13.68) have been used and C is an integration constant. As will be shown later (Eq. (13.88)), the interaction force is related to C.

To obtain a relationship between C and h, one must further integrate Eq. (13.73). If $y(x)$ passes through a minimum y_{m} at some point $x = x_{\mathrm{m}}$ $(0 < x_{\mathrm{m}} < h)$, then further integration of Eq. (13.73) for $0 < x < x_{\mathrm{m}}$ yields

$$\kappa x = \int_y^{y(0)} \frac{dy}{\sqrt{2\cosh y + C}} \qquad (13.75)$$

and for $x_{\mathrm{m}} < x < h$

$$\kappa(x - x_{\mathrm{m}}) = \int_{y_{\mathrm{m}}}^{y} \frac{dy}{\sqrt{2\cosh y + C}} \qquad (13.76)$$

where x_{m} must satisfy

$$\kappa x_{\mathrm{m}} = \int_{y_{\mathrm{m}}}^{y(0)} \frac{dy}{\sqrt{2\cosh y + C}} \qquad (13.77)$$

It follows from Eq. (13.73) that

$$C = -2\cosh y_{\mathrm{m}} \qquad (13.78)$$

If there is no potential minimum, Eq. (13.75) is valid for any point in the region $0 < x < h$. In this case Eq. (13.78) does not hold.

By evaluating integrals (13.75) and (13.76) at $x = x_{\mathrm{m}}$ and $x = h$, we can obtain relationships between C and h, as follows. If there exists a potential minimum, then

$$\kappa h = \int_{y_{\mathrm{m}}}^{y(0)} \frac{dy}{\sqrt{2\cosh y + C}} + \int_{y_{\mathrm{m}}}^{y(h)} \frac{dy}{\sqrt{2\cosh y + C}} \qquad (13.79)$$

If there is no potential minimum, then

$$\kappa h = \int_{y(h)}^{y(0)} \frac{dy}{\sqrt{2\cosh y + C}} \tag{13.80}$$

Note that $y(0)$ and $y(h)$, which are the membrane surface potentials, are not constant but depend on C. Therefore one needs another relationship connecting $y(0)$ and $y(h)$ with C, which can be derived as follows. Substituting Eqs. (13.72)–(13.74) into Eqs. (13.65) and (13.66), we obtain

$$y(0) = y_{\rm DON1} - \tanh\left(\frac{y_{\rm DON1}}{2}\right) - \frac{C+2}{2\sinh y_{\rm DON1}} \tag{13.81}$$

$$y(h) = y_{\rm DON2} - \tanh\left(\frac{y_{\rm DON2}}{2}\right) - \frac{C+2}{2\sinh y_{\rm DON2}} \tag{13.82}$$

where Z_1N_1/zn and Z_2N_2/zn have been replaced by $2\sinh y_{\rm DON1}$ and $2\sinh y_{\rm DON2}$, respectively (via Eqs. (13.69) and (13.70). If we put $C = -2$ (or $P = 0$) in Eqs. (13.81) and (13.82), then these equations give the unperturbed surface potentials ψ_{o1} and ψ_{o2} of membranes 1 and 2, respectively, in the absence of interaction (or, the membranes are at infinite separation $h = \infty$), namely,

$$y(0) = y_{o1} = y_{\rm DON1} - \tanh(y_{\rm DON1}/2) \tag{13.83}$$

$$y(h) = y_{o2} = y_{\rm DON2} - \tanh(y_{\rm DON2}/2) \tag{13.84}$$

with

$$y_{o1} = \frac{ze\psi_{o1}}{kT} \tag{13.85}$$

$$y_{o2} = \frac{ze\psi_{o2}}{kT} \tag{13.86}$$

where y_{o1} and y_{o2} are the scaled unperturbed surface potentials of membranes 1 and 2, respectively. Coupled equations (13.79) (or (13.80)), (13.81), and (13.82) can provide $y(0)$, $y(h)$, and C as functions of h for given values of Z_1N_1/zn and Z_2N_2/zn (or $y_{\rm DON1}$ and $y_{\rm DON2}$).

The electrostatic force acting between the two membranes can be obtained by integrating the Maxwell stress and the osmotic pressure over an arbitrary surface enclosing any one of the membranes. We may choose two planes $x = -\infty$ (in the solution) and $x = x$ at an arbitrary point in the region $0 < x < h$ as a surface enclosing membrane 1. On the plane $x = -\infty$, Maxwell's stress is zero. The force driving the two membranes apart per unit area is thus given by (see Eq. (8.20))

$$P(h) = 2nkT\{\cosh y(x) - 1\} - \frac{1}{2}\varepsilon_r\varepsilon_o\left(\frac{d\psi}{dx}\right)^2, \quad 0 < x < h \tag{13.87}$$

which can be rewritten using Eq. (13.73) as

$$P(h) = -nkT(C + 2) \tag{13.88}$$

where $P > 0\,(C < -2)$ corresponds to repulsion and $P < 0$ $(C > -2)$ to attraction. As shown before, $C(h)$ and thus $P(h)$ are determined by y_{DON1} and y_{DON2}.

Consider the sign of $P(h)$ when the values of y_{DON1} and y_{DON2} are given. According to Eq. (13.73), we plot isodynamic curves (i.e., y as a function of x for a given value of C) for the case $C < -2$ in Fig. 13.7 and for the case $C > -2$ in Fig. 13.8. The idea of isodynamic curves has been introduced first by Derjaguin [10]. The distance between two intersections of one of these isodynamic curves with two straight lines $y = y(0)$ (solid lines) and $y = y(h)$ (dotted lines), both of which must correspond to the same value of C, gives possible values for h. Three isodynamic curves ($C = C_1$, C_2, and C_3) and the corresponding three values of h (h_1, h_2, and h_3) are illustrated in Figs 13.7 and 13.8. It must be stressed that both $y(0)$ and $y(h)$ are not constant but change with C according to Eqs. (13.81) and (13.82). (In the system treated by Derjaguin [10], $y(0)$ and $y(h)$ are constant.) Consequently, the pair of straight lines $y = y(0)$ and $y = y(h)$ moves with changing C. In Fig. 13.7 (where $C_3 < C_2 < C_1 < -2$), both lines $y = y(0)$ and $y = y(h)$ shift upward with decreasing C, while in Fig. 13.8 (where $C_3 > C_2 > C_1 > -2$), $y = y(0)$ moves downward and $y = y(h)$ moves upward with increasing C.

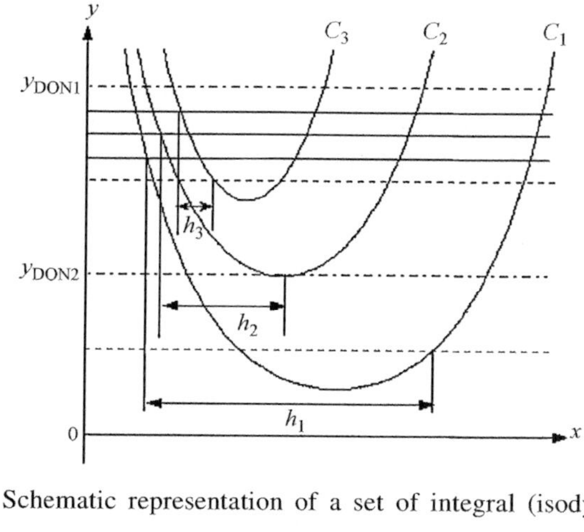

FIGURE 13.7 Schematic representation of a set of integral (isodynamic) curves for $C < -2$ (see text).

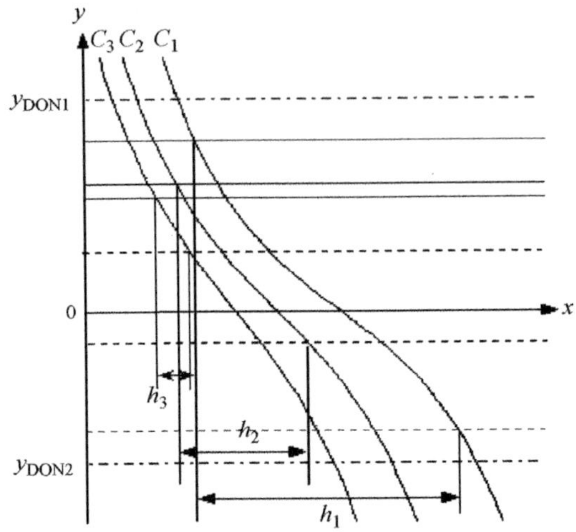

FIGURE 13.8 Schematic representation of a set of integral (isodynamic) curves for $C > -2$.

Figure 13.7 shows that for the case $C < -2$ there exist intersections of the isodynamic curve with $y = y(0)$ and $y = y(h)$ only when $y_{\mathrm{DON1}} > 0$ and $y_{\mathrm{DON2}} > 0$ (or $Z_1 > 0$ and $Z_2 > 0$). In other words, if $Z_1 > 0$ and $Z_2 > 0$, then $C < -2$ (or $P > 0$). Similarly, it can be proven by Fig. 13.8 that if $Z_1 > 0$ and $Z_2 < 0$, then $C > -2$ (or $P < 0$). Figure 13.7 also shows that for the curve with $C = C_2$, $y(h)$ coincides with y_{DON2}. We denote the corresponding value of h (i.e., h_2) by h_c. In this situation, from Eq. (13.82) we have $C = C_2 = -2\cosh y_{\mathrm{DON2}}$ so that Eq. (13.80) yields

$$\kappa h_c = \int_{y_{\mathrm{DON2}}}^{y(0)} \frac{dy}{\sqrt{2(\cosh y - \cosh y_{\mathrm{DON2}})}} \tag{13.89}$$

which, for low potentials, reduces to Eq. (13.23). Clearly, if $h > h_c$, there exists a potential minimum in the region $0 < x < h$, while if $h \leq h_c$, there is no potential minimum. Therefore, one must use Eq. (13.79) if $Z_1 > 0$, $Z_2 > 0$, and $h > h_c$, and Eq. (13.80) if $Z_1 > 0$, $Z_2 > 0$, and $h \leq h_c$, or if $Z_1 > 0$ and $Z_2 < 0$. When $Z_1 > 0$ and $Z_2 < 0$, at some point $h = h_o$, the surface potential $y(h)$ of the negatively charged membrane becomes zero. It follows from Eqs. (13.80) and (13.82) that the value of h_o is given by

$$\kappa h_o = \int_{0}^{y(0)} \frac{dy}{\sqrt{2\cosh y + C}} \tag{13.90}$$

and

$$C = -2(\cosh y_{\mathrm{DON2}} - y_{\mathrm{DON2}} \sinh y_{\mathrm{DON2}}) \tag{13.91}$$

The magnitude of $P(h)$ has its largest value at $h=0$. The value of $P(0)$ can be derived by equating $y(0)$ and $y(h)$ in Eqs. (13.81) and (13.82), namely,

$$P(0) = -\frac{2[y_{\rm DON1} - y_{\rm DON2} - \{\tanh(y_{\rm DON1}/2) - \tanh(y_{\rm DON2}/2)\}]}{\operatorname{cosech} y_{\rm DON1} - \operatorname{cosech} y_{\rm DON2}} nkT \tag{13.92}$$

For the special case of two similar membranes, that is, $y_{\rm DON1}=y_{\rm DON2}=y_{\rm DON}$, $y(x)$ reaches a minimum $y_{\rm m}=y(h/2)$ at the midpoint between the two membranes, that is, $x=x_{\rm m}=h/2$ so that Eqs. (13.77) and (13.78) give

$$\frac{\kappa h}{2} = \int_{y_{\rm m}}^{y_{\rm o}} \frac{dy}{\sqrt{2(\cosh y - \cosh y_{\rm m})}} \tag{13.93}$$

or equivalently

$$\cosh\left(\frac{y_{\rm o}}{2}\right) = \cosh\left(\frac{y_{\rm m}}{2}\right) dc\left(\frac{\kappa h}{2}\cosh\left(\frac{y_{\rm m}}{2}\right)\frac{1}{\cosh(y_{\rm m}/2)}\right) \tag{13.94}$$

where dc is a Jacobian elliptic function with modulus $1/\cosh(y_{\rm m}/2)$. Also, Eqs. (13.81) and (13.82) become

$$y_{\rm o} = y_{\rm DON} - \tanh(y_{\rm DON}/2) + \frac{2\sinh^2(y_{\rm m}/2)}{\sinh y_{\rm DON}} \tag{13.95}$$

where $y_{\rm o}=y(0)$ is the scaled membrane surface potential. Equations (13.94) and (13.95) form coupled equation that determine $y_{\rm o}$ and $y_{\rm m}$ for given values of κh and $y_{\rm DON}$. These equations can be numerically solved. The interaction force $P(h)$ per unit area between membranes 1 and 2 is calculated from Eqs. (13.88) and (13.88), namely,

$$P(h) = 4nkT\sinh^2\left(\frac{y_{\rm m}}{2}\right) = 4nkT\sinh^2\left(\frac{y(h/2)}{2}\right) \tag{13.96}$$

which reaches its largest value $P(0)$ at $h=0$, that is,

$$P(0) = 4nkT\sinh^2\left(\frac{y_{\rm DON}}{2}\right) \tag{13.97}$$

13.5.2 Interaction Energy

We calculate the potential energy of the double-layer interaction between two parallel ion-penetrable dissimilar membranes. We imagine a charging process in which all the membrane-fixed charges are increased at the same rate. Let v be a parameter that expresses a stage of the charging process and varies from 0 to 1. Then, at stage

v, the fixed charges of membranes 1 and 2 are vZ_1eN_1 and vZ_2eN_2, respectively. We denote by $\psi(x, v)$ the potential at stage v. The double-layer free energy per unit area of the present system is thus (see Eq. (5.90))

$$F = Z_1eN_1 \int_0^1 dv \int_{-\infty}^0 \psi(x, v)dx + Z_2eN_2 \int_0^1 dv \int_h^{\infty} \psi(x, v)dx \tag{13.98}$$

An alternative expression for the double-layer free energy can be derived by considering a discharging process in which all the charges in the system (both the membrane-fixed charges and the charges of electrolyte ions) are decreased at the same rate. Let λ be a parameter that expresses a stage of the discharging process, varying from 1 to 0, and $\psi(x, \lambda)$ be the potential at stage X. Then the free energy per unit area is given by Eq. (5.86), namely,

$$F = \int_0^1 d\lambda \frac{2}{\lambda} E(\lambda) \tag{13.99}$$

with

$$E(\lambda) = \frac{1}{2} \int_{-\infty}^{\infty} \psi(x, \lambda)\rho(x, \lambda)dX \tag{13.100}$$

where $E(\lambda)$ and $\rho(x, \lambda)$ are, respectively, the internal energy per unit area and the volume charge density at stage λ. The Poisson–Boltzmann equation at stage λ is

$$\frac{\partial^2 \psi(x, \lambda)}{\partial x^2} = -\frac{\rho(x, \lambda)}{\varepsilon_r \varepsilon_o} \tag{13.101}$$

where

$$\rho(x, \lambda) = -2zn\lambda e\left\{\sinh\left(\frac{z\lambda e\psi(x, \lambda)}{kT}\right) - \frac{Z_1N_1}{2zn}\right\}, \quad x < 0 \tag{13.102}$$

$$\rho(x, \lambda) = -2zn\lambda e \sinh\left(\frac{z\lambda e\psi(x, \lambda)}{kT}\right), \quad 0 < x < h \tag{13.103}$$

$$\rho(x, \lambda) = -2zn\lambda e\left\{\sinh\left(\frac{z\lambda e\psi(x, \lambda)}{kT}\right) - \frac{Z_2N_2}{2zn}\right\}, \quad x > h \tag{13.104}$$

Substituting Eqs. (13.102)–(13.104) into (13.99), we obtain after lengthy algebra

$$F = nkT(C + 2)h - \varepsilon_r\varepsilon_o \int_{-\infty}^{\infty} \left(\frac{d\psi}{dx}\right)^2 dx + F_c \tag{13.105}$$

with

$$F_c = 2nkT\left[\sinh y_{DON1}\left\{y_{DON1} - \tanh\left(\frac{y_{DON1}}{2}\right)\right\}\int_{-\infty}^{0} dx\right.$$
$$= \left. +\sinh y_{DON2}\left\{y_{DON2} - \tanh\left(\frac{y_{DON2}}{2}\right)\right\}\int_{h}^{\infty} dx\right] \tag{13.106}$$

Note that although F_c is infinitely large, it does not contribute to the potential energy of interaction since it is independent of h. To calculate the integral in Eq. (13.105), one must substitute Eqs. (13.72)–(13.74) separately in the three regions $x < 0$, $0 < x < h$, and $x > h$. The potential energy $V(h)$ of double-layer interaction per unit area is given by

$$V(h) = F(h) - F(\infty) \tag{13.107}$$

In Fig. 13.9, we give the reduced potential energy $V^* = (\kappa/64nkT)V$ of the double--layer interaction between two ion-penetrable semi-infinite similar membranes as a

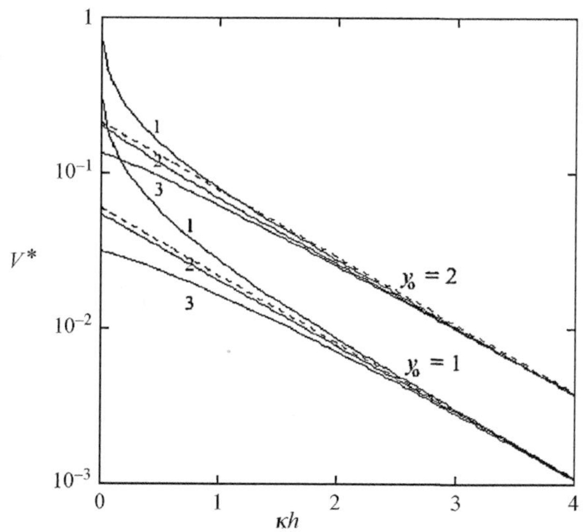

FIGURE 13.9 The reduced potential energy $V^* = (\kappa/64nkT)V$ of the double-layer interaction between two ion-penetrable semi-infinite similar membranes as a function of the scaled membrane separation κh for $y_{o1} = y_{o2} = y_o = 1$ and 2. Comparison is made with the results for the two conventional models for hard plates given by Honig and Mul [11]. Curve 1, constant surface charge density model; curve 2, Donnan potential regulation model; curve 3, constant surface potential model. The dotted line shows the result for the linear superposition approximation. From Ref. [9].

function of the scaled membrane separation κh for $y_{o1}=y_{o2}=y_o=1$ and 2 calculated via Eqs. (13.105) and (13.106), which become

$$F = -4nkT \sinh^2\left(\frac{y(h/2)}{2}\right) \cdot h - \varepsilon_r\varepsilon_o \int_{-\infty}^{h/2} \left(\frac{d\psi}{dx}\right)^2 dx + F_c \tag{13.108}$$

with

$$F_c = 4nkT\left[\sinh y_{DON}\left\{y_{DON} - \tanh\left(\frac{y_{DON}}{2}\right)\right\} \int_{-\infty}^{0} dx\right. \tag{13.109}$$

Comparison is made with the results for the two conventional models for hard plates given by Honig and Mul [11]. We see that the values of the interaction energy calculated on the basis of the Donnan potential regulation model lie between those calculated from the conventional interaction models (i.e., the constant surface potential model and the constant surface charge density model) and are close to the results obtained the linear superposition approximation.

In the linear superposition approximation (see Eqs. (11.11) and (11.14)), the interaction force $P(h)$ and potential energy $V(h)$ per unit area between membranes 1 and 2 are, respectively, given by

$$P(h) = 64 \tanh\left(\frac{y_o}{4}\right)^2 nkT \exp(-\kappa h) \tag{13.110}$$

$$V(h) = \frac{64}{\kappa} \tanh\left(\frac{y_o}{4}\right)^2 nkT \exp(-\kappa h) \tag{13.111}$$

Note that for the low potential case, the Donnan potential regulation model agrees exactly with the LSA results.

13.6 pH DEPENDENCE OF ELECTROSTATIC INTERACTION BETWEEN ION-PENETRABLE MEMBRANES

Consider two parallel planar similar ion-penetrable membranes 1 and 2 at separation h immersed in a solution containing a symmetrical electrolyte of valence z and bulk concentration n [12]. We take an x-axis as shown in Fig. 13.2. The surface is in equilibrium with a monovalent electrolyte solution of bulk concentration n. Note here that n represents the total concentration of monovalent cations including H^+ ions and that of monovalent anions including OH^- ions. Let n_H be the H^+ concentration in the bulk solution phase. In the membrane phase, monovalent acidic groups of dissociation constant K_a are distributed at a density N_{max}. The mass action law for the dissociation of acidic groups AH (AH $\Longleftrightarrow$ $A^- + H^+$) gives the number density $N(x)$ of dissociated groups at position x in membrane 1, namely,

$$N(x) = \frac{N_{\max}}{1 + n_{\mathrm{H}}/K_{\mathrm{a}} \exp(-e\psi(x)/kT)} \tag{13.112}$$

Here $n_{\mathrm{H}} \exp(-e\psi(x)/kT)$ is the H^+ concentration at position x. The charge density $\rho_{\mathrm{fix}}(x)$ resulting from dissociated acidic groups at position x in the membrane is thus given by

$$\rho_{\mathrm{fix}}(x) = -eN(x) = -\frac{eN_{\max}}{1 + n_{\mathrm{H}}/K_{\mathrm{a}} \exp(-e\psi(x)/kT)} \tag{13.113}$$

The Poisson–Boltzmann equations for $\psi(x)$ to be solved are then given by

$$\frac{d^2\psi}{dx^2} = \frac{2en}{\varepsilon_{\mathrm{r}}\varepsilon_{\mathrm{o}}} \sinh\left(\frac{e\psi}{kT}\right) - \frac{\rho_{\mathrm{fix}}(x)}{\varepsilon_{\mathrm{r}}\varepsilon_{\mathrm{o}}} \tag{13.114}$$

which is

$$\frac{d^2\psi}{dx^2} = \frac{2en}{\varepsilon_{\mathrm{r}}\varepsilon_{\mathrm{o}}} \sinh\left(\frac{e\psi}{kT}\right) + \frac{e}{\varepsilon_{\mathrm{r}}\varepsilon_{\mathrm{o}}} \frac{N_{\max}}{1 + n_{\mathrm{H}}/K_{\mathrm{a}} \exp(-e\psi(x)/kT)}, \quad x < 0 \tag{13.115}$$

For the solution phase ($x > 0$) we have

$$\frac{d^2\psi}{dx^2} = \frac{2en}{\varepsilon_{\mathrm{r}}\varepsilon_{\mathrm{o}}} \sinh\left(\frac{e\psi}{kT}\right), \quad 0 < x \leq h/2 \tag{13.116}$$

The Donnan potential in the membrane 2, which we denote by ψ_{DON} is obtained by setting the right-hand side of the Poisson–Boltzmann equation (13.115) equal to zero. That is, ψ_{DON} is the solution to the following transcendental equation:

$$\sinh\left(\frac{e\psi_{\mathrm{DON}}}{kT}\right) + \frac{N_{\max}}{2n} \frac{1}{1 + n_{\mathrm{H}}/K_{\mathrm{a}} \exp(-e\psi_{\mathrm{DON}}/kT)} = 0 \tag{13.117}$$

The double-layer free energy per unit area of the present system is given by

$$F(h) = 2\int_{-\infty}^{0} f(x)dx \tag{13.118}$$

where the factor 2 corresponds to two membranes 1 and 2, and $f(x)$ is the free energy density at position x in membrane 1, given by

$$\begin{aligned} f(x) = & \int_0^{\rho_{\mathrm{fix}}(x)} \psi(x)dxe - \rho_{\mathrm{fix}}(x)\psi(x) \\ & -NkT \ln\left(1 + \frac{n_{\mathrm{H}}\, e^{-y_{\mathrm{o}}}}{K_{\mathrm{d}}} \exp\left(-\frac{e\psi(x)}{kT}\right)\right) \end{aligned} \tag{13.119}$$

The potential energy $V(h)$ of the double-layer interaction per unit area between two membranes 1 and 2 is given by

$$V(h) = F(h) - F(\infty) \tag{13.120}$$

REFERENCES

1. H. Ohshima, *Theory of Colloid and Interfacial Electric Phenomena*, Elsevier/Academic Press, 2006, Chapter 16.
2. H. Ohshima and T. Kondo, *J. Colloid Interface Sci.* 155 (1993) 499.
3. H. Ohshima, *Adv. Colloid Interface Sci.* 53 (1994) 77.
4. B. V. Derjaguin, *Kolloid Z.* 69 (1934) 155.
5. H. Ohshima, *Colloid Polym. Sci.* 274 (1998) 1176.
6. M. J. Sparnaay, *Recueil* 78 (1959) 680.
7. H. Ohshima and T. Kondo, *J. Theor. Biol.* 128 (1987) 187.
8. H. Ohshima and T. Kondo, *J. Colloid Interface Sci.* 123 (1988) 136.
9. H. Ohshima and T. Kondo, *J. Colloid Interface Sci.* 123 (1989) 523.
10. B. V. Derjaguin, *Discuss. Faraday Soc.* 18 (1954) 85.
11. E. P. Honig and P. M. Mul, *J. Colloid Interface Sci.* 36 (1973) 258.
12. H. Ohshima and T. Kondo, *Biophys. Chem.* 32 (1988) 131.

14 Series Expansion Representations for the Double-Layer Interaction Between Two Particles

14.1 INTRODUCTION

This chapter deals with a method for obtaining the exact solution to the linearized Poisson–Boltzmann equation on the basis of Schwartz's method [1] without recourse to Derjaguin's approximation [2]. Then we apply this method to derive series expansion representations for the double-layer interaction between spheres [3–13] and those between two parallel cylinders [14, 15].

14.2 SCHWARTZ'S METHOD

We start with the simplest problem of the plate–plate interaction. Consider two parallel plates 1 and 2 in an electrolyte solution, having constant surface potentials ψ_{o1} and ψ_{o2}, separated at a distance H between their surfaces (Fig. 14.1). We take an x-axis perpendicular to the plates with its origin 0 at the surface of one plate so that the region $0 < x < h$ corresponds to the solution phase. We derive the potential distribution for the region between the plates ($0 < x < h$) on the basis of Schwartz's method [1]. The linearized Poisson–Boltzmann equation in the one-dimensional case is

$$\frac{d^2\psi}{dx^2} = \kappa^2\psi, \quad 0 < x < h \tag{14.1}$$

with the following boundary conditions at plate surfaces $x = 0$ and $x = h$:

$$\psi(0) = \psi_{o1} \tag{14.2}$$

$$\psi(h) = \psi_{o2} \tag{14.3}$$

Biophysical Chemistry of Biointerfaces By Hiroyuki Ohshima

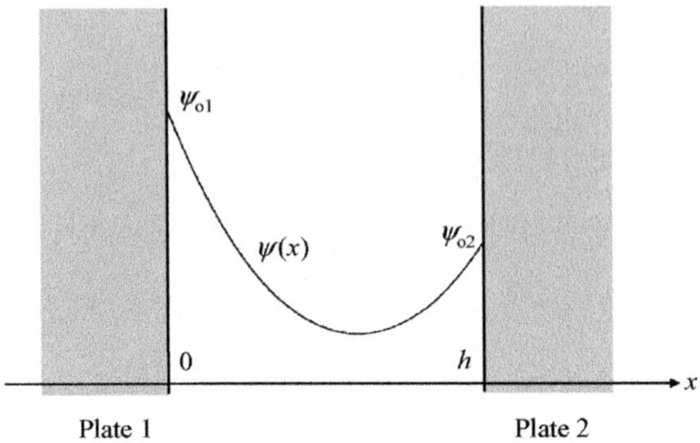

FIGURE 14.1 Interaction between two parallel dissimilar hard plates 1 and 2 at separation h.

We write the solution to Eq. (14.1) as

$$\psi(x) = \psi_1(x) + \psi_2(x), \quad 0 < x < h \tag{14.4}$$

with

$$\psi_1(x) = \psi_1^{(0)}(x) + \psi_1^{(1)}(x) + \psi_1^{(2)}(x) + \cdots \tag{14.5}$$

$$\psi_2(x) = \psi_2^{(0)}(x) + \psi_2^{(1)}(x) + \psi_2^{(2)}(x) + \cdots \tag{14.6}$$

As the zeroth-order approximate solutions $\psi_1^{(0)}(x)$ and $\psi_2^{(0)}(x)$, we choose the unperturbed potentials produced by plates 1 and 2 in the absence of interaction (i.e., when they are isolated), which are (see Chapter 1)

$$\psi_1^{(0)}(x) = \psi_{o1}\, e^{-\kappa x} \tag{14.7}$$

$$\psi_2^{(0)}(x) = \psi_{o2}\, e^{-\kappa(H-x)} \tag{14.8}$$

Note that $\psi_1^{(0)}(x)$ and $\psi_2^{(0)}(x)$, respectively, satisfy the boundary conditions (14.2) and (14.3). We construct the functions $\psi_1^{(k)}(x)$ and $\psi_2^{(k)}(x)$ ($k = 1, 2, \ldots$) as follows. The unperturbed potential $\psi_1^{(0)}(x)$ satisfies the boundary condition (14.2) on plate 1. The boundary condition (14.3) on plate 2, on the other hand, which has been satisfied by $\psi_2^{(0)}(x)$, is now violated, since $\psi_1^{(0)}(x)$ gives rise to the following nonzero value on plate 2:

$$\psi_1^{(0)}(h) = \psi_{o1} e^{-\kappa h} \tag{14.9}$$

which is obtained from Eq. (14.7). We thus construct the first-order approximate solution $\psi_1^{(1)}(x)$ so as to cancel $\psi_1^{(0)}(h)$ on plate 2, namely,

$$\psi_1^{(0)}(h) + \psi_1^{(1)}(h) = 0 \quad \text{(on plate 2)} \tag{14.10}$$

or

$$\psi_1^{(1)}(h) = -\psi_1^{(0)}(h) = -\psi_{o1}e^{-\kappa h} \tag{14.11}$$

Therefore, the function $\psi_1^{(1)}(x)$ must take the form

$$\psi_1^{(1)}(x) = (-\psi_{o1}\, e^{-\kappa h})e^{-\kappa(h-x)} \tag{14.12}$$

The function $\psi_1^{(1)}(x)$ can be interpreted as the "image" potential of $\psi_1^{(0)}(x)$ with respect to plate 2 by analogy with "the method of images" in electrostatics, as schematically shown in Fig. 14.2.

The function $\psi_1^{(1)}(x)$ in turn gives rise to the following nonzero value on plate 1,

$$\psi_1^{(1)}(0) = -\psi_{o1}e^{-2\kappa h} \tag{14.13}$$

violating the boundary condition (Eq. (14.2)), which has already been satisfied by $\psi_1^{(0)}(x)$. Therefore, the second-order approximate solution $\psi_1^{(2)}(x)$ must be constructed so as to cancel $\psi_1^{(1)}(0)$ on plate 1, namely,

$$\psi_1^{(1)}(0) + \psi_1^{(2)}(0) = 0 \quad \text{(on plate 1)} \tag{14.14}$$

or

$$\psi_1^{(2)}(0) = -\psi_1^{(1)}(0) = +\psi_{o1}\, e^{-2\kappa h} \tag{14.15}$$

Thus, $\psi_1^{(2)}(x)$, which can be interpreted as the "image" potential of $\psi_1^{(1)}(x)$ with respect to plate 1, is given by

$$\psi_1^{(1)}(x) = (+\psi_{o1}e^{-2\kappa h})e^{-\kappa x} \tag{14.16}$$

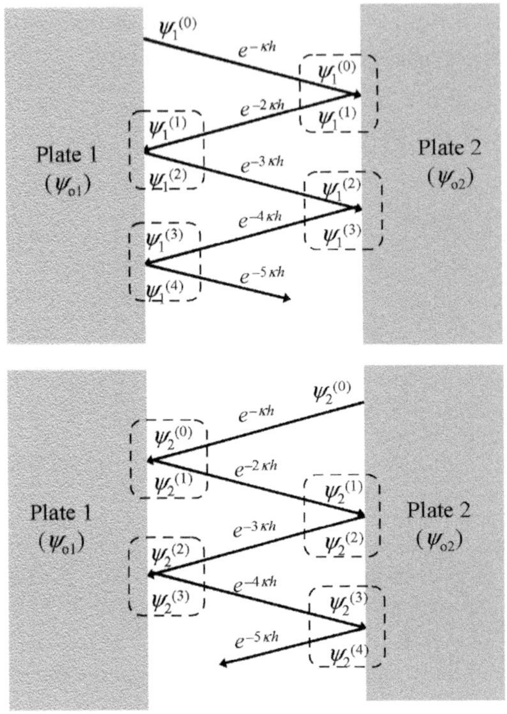

FIGURE 14.2 The unperturbed potentials $\psi_1^{(0)}$ and $\psi_2^{(0)}$, and the correction terms $\psi_1^{(k)}$ and $\psi_2^{(k)}$ ($k = 1, 2, \ldots$). Squares (dotted lines) mean that $\psi_i^{(2n)}$ is the image potential of $\psi_i^{(2n-1)}$ with respect to plate i, while $\psi_i^{(2n-1)}$ is the image potential of $\psi_i^{(2n-2)}$ with respect to plate j ($n = 1, 2, \ldots$; $i, j = 1, 2$; $i \neq j$). $\exp[-(k+1)\kappa H)$ indicates that $\psi_1^{(k)}$ and $\psi_2^{(k)}$ ($k = 0, 1, 2, \ldots$) are proportional to $\exp(-(k+1)kH)$.

The function $\psi_1^{(3)}(x)$ then must be constructed so as to satisfy

$$\psi_1^{(2)}(h) + \psi_1^{(3)}(h) = 0 \quad \text{(on plate 2)} \tag{14.17}$$

In this way one can construct $\psi_1^{(k)}(x)$ ($k = 1, 2, \ldots$) that satisfy the boundary condition (14.2) namely,

$$\psi_1(0) = \psi_1^{(0)}(x) + \sum_{n=1}^{\infty} \left\{ \psi_1^{(2n-1)}(0) + \psi_1^{(2n)}(0) \right\} = \psi_{o1} \quad \text{(on plate 1)} \tag{14.18}$$

$$\psi_1(h) = \sum_{n=0}^{\infty} \left\{ \psi_1^{(2n)}(h) + \psi_1^{(2n+1)}(h) \right\} = 0 \quad \text{(on plate 2)} \tag{14.19}$$

Thus, we find that

$$\begin{aligned}
\psi_1(x) &= \psi_{o1}\, e^{-\kappa x} + (-\psi_{o1}\, e^{-\kappa h})e^{-\kappa(h-x)} + (+\psi_{o1}\, e^{-2\kappa h})e^{-\kappa x} \\
&\quad + (-\psi_{o1}\, e^{-3\kappa h})e^{-\kappa(h-x)} + \cdots \\
&= \psi_{o1}\, e^{-\kappa x}(1 + e^{-2\kappa h} + e^{-4\kappa h} + \cdots) - \psi_{o1}\, e^{-\kappa(h-x)}(e^{-\kappa h} + e^{-3\kappa h} + \cdots) \\
&= \psi_{o1}\left\{ e^{-\kappa x}\frac{1}{1 - e^{-2\kappa h}} - e^{-\kappa(h-x)}\frac{e^{-\kappa h}}{1 - e^{-2\kappa h}} \right\} \\
&= \psi_{o1}\frac{\sinh(\kappa(h-x))}{\sinh(\kappa h)}
\end{aligned} \tag{14.20}$$

$$\begin{aligned}
\psi_2(x) &= \psi_{o2}\, e^{-\kappa(h-x)} + (-\psi_{o2}\, e^{-\kappa h})e^{-\kappa x} \\
&\quad + (+\psi_{o2}\, e^{-2\kappa h})e^{-\kappa(h-x)}(-\psi_{o2}\, e^{-3\kappa h})e^{-\kappa x} + \cdots \\
&= \psi_{o2}\frac{\sinh(\kappa x)}{\sinh(\kappa h)}
\end{aligned} \tag{14.21}$$

From Eqs. (14.20) and (14.21) we obtain

$$\begin{aligned}
\psi(x) &= \psi_1(x) + \psi_2(x) \\
&= \frac{\psi_{o1}\sinh(\kappa(h-x)) + \psi_{o1}\sinh(\kappa x)}{\sinh(\kappa h)}, \quad 0 < x < h
\end{aligned} \tag{14.22}$$

which agrees with an expression for the potential derived by directly solving Eq. (14.1) to the boundary conditions (14.2) and (14.3).

14.3 TWO SPHERES

By applying the above method to two interacting spheres on the basis of the linearized Poisson–Boltzmann equations, we can derive series expansion representations for the double-layer interaction between two spheres 1 and 2 (Fig. 14.3).

Case (a): spheres 1 and 2 both at constant surface potential

For the interaction between two hard spheres 1 and 2 having radii a_1 and a_2, and constant surface potentials ψ_{o1} and ψ_{o2}, respectively, at separation H between the two spheres, the interaction energy $V^{\psi}(R)$ is given by [3, 6, 8, 10]

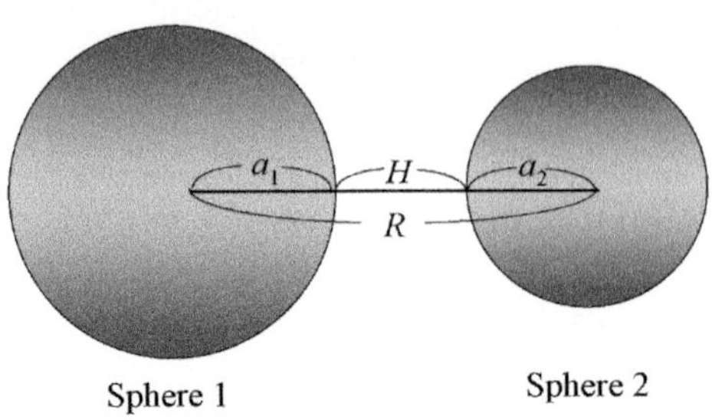

FIGURE 14.3 Interaction between two charged hard spheres 1 and 2 of radii a_1 and a_2 at a separation R between their centers. H $(= R - a_1 - a_2)$ is the closest distance between their surfaces. From Ref. 6.

$$
\begin{aligned}
V^{\psi}(R) &= 4\pi\varepsilon_r\varepsilon_o\psi_{o1}\psi_{o2}a_1a_2\frac{e^{-\kappa(R-a_1-a_2)}}{R} \\
&\quad +2\pi\varepsilon_r\varepsilon_o\psi_{o1}^2a_1^2\frac{e^{2\kappa a_1}}{R}\sum_{n=0}^{\infty}(2n+1)G_n(2)K_{n+1/2}^2(\kappa R) \\
&\quad +2\pi\varepsilon_r\varepsilon_o\psi_{o2}^2a_2^2\frac{e^{2\kappa a_2}}{R}\sum_{n=0}^{\infty}(2n+1)G_n(1)K_{n+1/2}^2(\kappa R) \\
&\quad +4\pi\varepsilon\varepsilon_r\varepsilon_o\psi_{o1}\psi_{o2}a_1a_2\frac{e^{\kappa(a_1+a_2)}}{R} \\
&\quad \times\sum_{n=0}^{\infty}\sum_{m=0}^{\infty}(2n+1)(2m+1)B_{nm}G_n(2)G_m(1)K_{n+1/2}(\kappa R)K_{m+1/2}(\kappa R)+\cdots \\
&\quad +2\pi\varepsilon\varepsilon_r\varepsilon_o\psi_{o1}\psi_{o2}a_1a_2\frac{e^{\kappa(a_1+a_2)}}{R} \\
&\quad \times\sum_{n_1=0}^{\infty}\sum_{n2=0}^{\infty}\cdots\sum_{n_{2\nu}=0}^{\infty}[L_{12}(n_1,n_2,\ldots,n_{2\nu})+L_{21}(n_1,n_2,\ldots,n_{2\nu})] \\
&\quad \times K_{n_1+1/2}(\kappa R)K_{n_{2\nu}+1/2}(\kappa R)+2\pi\varepsilon_r\varepsilon_o\sum_{n_1=0}^{\infty}\sum_{n_2=0}^{\infty}\cdots\sum_{n_{2\nu-1}=0}^{\infty}(2n_{2\nu-1}+1)B_{n_{2\nu-2}n_{2\nu-1}} \\
&\quad \times\left[\psi_{o1}^2a_1^2\frac{e^{2\kappa a_1}}{R}L_{21}(n_1,n_2,\ldots,n_{2\nu-2})G_{n_{2\nu-1}}(2)\right. \\
&\quad \left.+\psi_{o2}^2a_2^2\frac{e^{2\kappa a_2}}{R}L_{12}(n_1,n_2,\ldots,n_{2\nu-2})G_{n_{2\nu-1}}(1)\right] \\
&\quad \times K_{n_1+1/2}(\kappa R)K_{n_{2\nu-1}+1/2}(\kappa R)+\cdots
\end{aligned}
\tag{14.23}
$$

where

$$L_{21}(n_1,n_2,\ldots,n_{2v-2}) = (2n_1+1)(2n_2+1)\cdots(2n_{2v}+1) \times B_{n_1n_2}B_{n_2n_3}\cdots B_{n_{2v-1}n_{2v}}G_{n_1}(2)G_{n_2}(1)\times\cdots \times G_{n_{2v-1}}(2)G_{n_{2v}}(1), \quad (v=1,2,\ldots) \tag{14.24}$$

$$L_{12}(n_1,n_2,\ldots,n_{2v}) = (2n_1+1)(2n_2+1)\cdots(2n_{2v}+1) \times B_{n_1n_2}B_{n_2n_3}\cdots B_{n_{2v-1}n_{2v}}G_{n_1}(1)G_{n_2}(2)\times\cdots \times G_{n_{2v-1}}(1)G_{n_{2v}}(2), \quad (v=1,2,\ldots) \tag{14.25}$$

$$G_n(i) = -\frac{I_{n+1/2}(\kappa a_i)}{K_{n+1/2}(\kappa a_i)}, \quad (i=1,2) \tag{14.26}$$

$$B_{nm} = \sum_{r=0}^{\min\{n,m\}} A_{nmr}\left(\frac{\pi}{2\kappa R}\right)^{1/2} K_{n+m-2r+1/2}(\kappa R) \tag{14.27}$$

$$A_{nmr} = \frac{\Gamma(n-r+\frac{1}{2})\Gamma(m-r+\frac{1}{2})\Gamma(r+\frac{1}{2})(n+m-r)!(n+m-2r+\frac{1}{2})}{\pi\Gamma(m+n-r+\frac{3}{2})(n-r)!(m-r)!r!} \tag{14.28}$$

Case (b): spheres 1 and 2 both at constant surface charge density

For the interaction between two hard spheres 1 and 2 having radii a_1 and a_2, and constant surface charge densities σ_1 and σ_2, respectively, at separation H between the two spheres, the interaction energy $V^{\sigma}(R)$ is given by the following equation [3, 8, 9, 10]:

$$\begin{aligned} V^{\sigma}(R) &= 4\pi\varepsilon_r\varepsilon_o\psi_{o1}\psi_{o2}a_1a_2\frac{e^{-\kappa(R-a_1-a_2)}}{R} \\ &+ 2\pi\varepsilon_r\varepsilon_o\psi_{o1}^2a_1^2\frac{e^{2\kappa a_1}}{R}\sum_{n=0}^{\infty}(2n+1)H_n(2)K_{n+1/2}^2(\kappa R) \\ &+ 2\pi\varepsilon_r\varepsilon_o\psi_{o2}^2a_2^2\frac{e^{2\kappa a_2}}{R}\sum_{n=0}^{\infty}(2n+1)H_n(1)K_{n+1/2}^2(\kappa R) \\ &+ 4\pi\varepsilon\varepsilon_r\varepsilon_o\psi_{o1}\psi_{o2}a_1a_2\frac{e^{\kappa(a_1+a_2)}}{R} \\ &\times\sum_{n=0}^{\infty}\sum_{m=0}^{\infty}(2n+1)(2m+1)B_{nm}H_n(2)H_m(1)K_{n+1/2}(\kappa R)K_{m+1/2}(\kappa R)+\cdots \end{aligned}$$

$$
\begin{aligned}
&+2\pi\varepsilon\varepsilon_r\varepsilon_o\psi_{o1}\psi_{o2}a_1a_2\frac{e^{\kappa(a_1+a_2)}}{R}\\
&\times\sum_{n_1=0}^{\infty}\sum_{n_2=0}^{\infty}\cdots\sum_{n_{2v}=0}^{\infty}[M_{12}(n_1,n_2,\ldots,n_{2v})+M_{21}(n_1,n_2,\ldots,n_{2v})]\\
&\times K_{n_1+1/2}(\kappa R)K_{n_{2v}+1/2}(\kappa R)+2\pi\varepsilon_r\varepsilon_o\sum_{n_1=0}^{\infty}\sum_{n_2=0}^{\infty}\cdots\sum_{n_{2v-1}=0}^{\infty}(2n_{2v-1}+1)B_{n_{2v-2}n_{2v-1}}\\
&\times\left[\psi_{o1}^2a_1^2\frac{e^{2\kappa a_1}}{R}M_{21}(n_1,n_2,\ldots,n_{2v-2})H_{n_{2v-1}}(2)\right.\\
&\left.+\psi_{o2}^2a_2^2\frac{e^{2\kappa a_2}}{R}M_{12}(n_1,n_2,\ldots,n_{2v-2})H_{n_{2v-1}}(1)\right]\\
&\times K_{n_1+1/2}(\kappa R)K_{n_{2v-1}+1/2}(\kappa R)+\cdots
\end{aligned}
\tag{14.29}
$$

with

$$
\begin{aligned}
M_{21}(n_1,n_2,\ldots,n_{2v-2}) &= (2n_1+1)(2n_2+1)\cdots(2n_{2v}+1)\\
&\times B_{n_1n_2}B_{n_2n_3}\cdots B_{n_{2v-1}n_{2v}}H_{n_1}(2)H_{n_2}(1)\times\cdots\\
&\times H_{n_{2v-1}}(2)H_{n_{2v}}(1),\quad (v=1,2,\ldots)
\end{aligned}
\tag{14.30}
$$

$$
\begin{aligned}
M_{12}(n_1,n_2,\ldots,n_{2v}) &= (2n_1+1)(2n_2+1)\cdots(2n_{2v}+1)\\
&\times B_{n_1n_2}B_{n_2n_3}\cdots B_{n_{2v-1}n_{2v}}H_{n_1}(1)H_{n_2}(2)\times\cdots\\
&\times H_{n_{2v-1}}(1)H_{n_{2v}}(2),\quad (v=1,2,\ldots)
\end{aligned}
\tag{14.31}
$$

$$
H_n(i)=\frac{I_{n-1/2}(\kappa a_i)-(n+1+n\varepsilon_{pi}/\varepsilon_r)I_{n+1/2}(\kappa a_i)/\kappa a_i}{K_{n-1/2}(\kappa a_i)+(n+1+n\varepsilon_{pi}/\varepsilon_r)K_{n+1/2}(\kappa a_i)/\kappa a_i},\quad (i=1,2)
\tag{14.32}
$$

where ε_{pi} is the relative permittivity of sphere i and ψ_{oi} is the unperturbed surface potential of sphere i and is related to the surface charge density σ_i of sphere i as

$$
\psi_{oi}=\frac{\sigma_i}{\varepsilon_r\varepsilon_o\kappa(1+\kappa a_i)},\quad (i=1,2)
\tag{14.33}
$$

Case (c): sphere1 maintained at constant surface potential and sphere 2 at constant surface charge density

For the interaction between two hard spheres 1 and 2 having radii a_1 and a_2 for the case where sphere 1 is maintained at constant surface potential ψ_1 and sphere 2 at constant surface charge density σ_2, at separation H between the two spheres, the interaction energy $V^{\psi-\sigma}(R)$ is given by the following equation [3, 10]:

$$
\begin{aligned}
V^{\psi-\sigma}(R) = {} & 4\pi\varepsilon_r\varepsilon_o\psi_{o1}\psi_{o2}a_1a_2\frac{e^{-\kappa(R-a_1-a_2)}}{R} \\
& +2\pi\varepsilon_r\varepsilon_o\psi_{o1}^2a_1^2\frac{e^{2\kappa a_1}}{R}\sum_{n=0}^{\infty}(2n+1)H_n(2)K_{n+1/2}^2(\kappa R) \\
& +2\pi\varepsilon_r\varepsilon_o\psi_{o2}^2a_2^2\frac{e^{2\kappa a_2}}{R}\sum_{n=0}^{\infty}(2n+1)G_n(1)K_{n+1/2}^2(\kappa R) \\
& +4\pi\varepsilon\varepsilon_r\varepsilon_o\psi_{o1}\psi_{o2}a_1a_2\frac{e^{\kappa(a_1+a_2)}}{R} \\
& \times\sum_{n=0}^{\infty}\sum_{m=0}^{\infty}(2n+1)(2m+1)B_{nm}H_n(2)G_m(1)K_{n+1/2}(\kappa R)K_{m+1/2}(\kappa R)+\cdots \\
& +2\pi\varepsilon\varepsilon_r\varepsilon_o\psi_{o1}\psi_{o2}a_1a_2\frac{e^{\kappa(a_1+a_2)}}{R} \\
& \times\sum_{n_1=0}^{\infty}\sum_{n_2=0}^{\infty}\cdots\sum_{n_{2v}=0}^{\infty}[N_{12}(n_1,n_2,\ldots,n_{2v})+N_{21}(n_1,n_2,\ldots,n_{2v})] \\
& \times K_{n_1+1/2}(\kappa R)K_{n_{2v}+1/2}(\kappa R)+2\pi\varepsilon_r\varepsilon_o\sum_{n_1=0}^{\infty}\sum_{n_2=0}^{\infty}\cdots \\
& \times\sum_{n_{2v-1}=0}^{\infty}(2n_{2v-1}+1)B_{n_{2v-2}n_{2v-1}} \\
& \times\left[\psi_{o1}^2a_1^2\frac{e^{2\kappa a_1}}{R}N_{21}(n_1,n_2,\ldots,n_{2v-2})H_{n_{2v-1}}(2)\right. \\
& \left.+\psi_{o2}^2a_2^2\frac{e^{2\kappa a_2}}{R}N_{12}(n_1,n_2,\ldots,n_{2v-2})G_{n_{2v-1}}(1)\right] \\
& \times K_{n_1+1/2}(\kappa R)K_{n_{2v-1}+1/2}(\kappa R)+\cdots
\end{aligned}
\tag{14.34}
$$

where

$$
\begin{aligned}
N_{21}(n_1,n_2,\ldots,n_{2v-2}) = {} & (2n_1+1)(2n_2+1)\cdots(2n_{2v}+1) \\
& \times B_{n_1n_2}B_{n_2n_3}\cdots B_{n_{2v-1}n_{2v}}H_{n_1}(2)G_{n_2}(1)\times\cdots \\
& \times H_{n_{2v-1}}(2)G_{n_{2v}}(1),\quad (v=1,2,\ldots)
\end{aligned}
\tag{14.35}
$$

$$N_{12}(n_1, n_2, \ldots, n_{2v}) = (2n_1 + 1)(2n_2 + 1) \cdots (2n_{2v} + 1) \times B_{n_1 n_2} B_{n_2 n_3} \cdots B_{n_{2v-1} n_{2v}} G_{n_1}(1) H_{n_2}(2) \times \cdots \times G_{n_{2v-1}}(1) H_{n_{2v}}(2), \quad (v = 1, 2, \ldots) \tag{14.36}$$

where ψ_{o2} is the unperturbed surface potential of sphere 2 and is related to the surface charge density σ_2 of sphere 2 as

$$\psi_{o2} = \frac{\sigma_2}{\varepsilon_r \varepsilon_o \kappa (1 + \kappa a_2)} \tag{14.37}$$

The interaction energy between spheres at constant surface potential involves only the function $G_n(i)$ (which depends only on the sphere radius a_i), while the interaction at constant surface charge density is characterized by the function $H_n(i)$ (which depends on both sphere radius a_i and relative permittivity ε_{pi}). The interaction energy in the mixed case involves both $G_n(i)$ and $H_n(i)$.

The first term (the leading term) of each of Eqs. (14.23), (14.29), and (14.34) is independent of the type of the boundary conditions at the sphere surface and takes the same form, namely,

$$V_{\mathrm{LSA}} = V^{(0)}(R) = 4\pi \varepsilon_r \varepsilon_o \psi_{o1} \psi_{o2} a_1 a_2 \frac{e^{-\kappa(R - a_1 - a_2)}}{R} \tag{14.38}$$

This expression coincides with that obtained by the linear superposition of the unperturbed potentials $\psi_1^{(0)}(r)$ and $\psi_2^{(0)}(r)$ (Eq. (11.26)). Note that if spheres 1 and 2 were both ion-penetrable porous spheres (we here call this type of particles "soft" particles), the interaction energy would be given by only this term (see Chapter 13), namely,

$$V_{\mathrm{soft/soft}}(R) = V^{(0)}(R) \tag{14.39}$$

The first-order correction to the linear superposition approximation V_{LSA} is given by the sum of the second and third terms on the right-hand side of Eqs. (14.23), (14.29) and (14.34), each corresponding to the image interaction of one sphere with respect to the other sphere. We denote this image interaction by V_{image} and expressed it as

$$V_{\mathrm{LSA}}(R) = V_1^{(1)}(R) + V_2^{(1)}(R) \tag{14.40}$$

Here if sphere i is maintained at constant surface potential, then $V_i^{(0)}(R)$ $(i, j = 1, 2; i \neq j)$ is given by

$$V_i^{\psi(1)}(R) = 2\pi\varepsilon_r\varepsilon_o\psi_{oi}^2 a_i^2 \frac{e^{2\kappa a_i}}{R}\sum_{n=0}^{\infty}(2n+1)G_n(i)K_{n+1/2}^2(\kappa R) \tag{14.41}$$

while if sphere J is maintained at constant surface charge density, then it is given by

$$V_i^{\sigma(1)}(R) = 2\pi\varepsilon_r\varepsilon_o\psi_{oi}^2 a_i^2 \frac{e^{2\kappa a_i}}{R}\sum_{n=0}^{\infty}(2n+1)H_n(i)K_{n+1/2}^2(\kappa R) \tag{14.42}$$

If sphere i were not hard but ion-penetrable (a soft sphere), with sphere j ($i, j = 1$, 2; $i \neq j$) being a hard sphere with constant surface charge density, then the interaction energy would be equal to the sum of only $V^{(0)}(R)$ and $V_i^{(1)}(R)$, namely,

$$V_{\text{soft/hard}}(R) = V^{(0)}(R) + V_i^{(1)}(R) \quad (i = 1, 2) \tag{14.43}$$

The interaction $V_i^{(1)}(R)$ depends only on the unperturbed surface potential ψ_{oi} of sphere i ($i = 1$, 2) and can be interpreted as the interaction between sphere i and its "image" with respect to sphere j ($j = 1$, 2; $j \neq i$). When both spheres are hard, the interaction energy is given by

$$\begin{aligned} V_{\text{hardt/hard}}(R) &= V^{(0)}(R) + V_1^{(1)}(R) + V_2^{(1)}(R) + \cdots \\ &= V_{\text{LSA}}(R) + V_{\text{image}}(R) + \cdots \end{aligned} \tag{14.44}$$

Figure 14.4 shows the linear superposition approximation (V_{LSA}) and the image interaction correction (V_{image}) to V_{LSA} as well as the full solutions, Eqs. (14.23) and (14.29), (given as circles) for the interaction energy between two identical spheres at $\kappa a_1 = \kappa a_2 = 5$ and $\psi_{o1} = \psi_{o2}$ as functions of the scaled separation κH for two cases where both spheres are maintained at constant surface potential or at constant surface charge density. In the latter case the relative permittivity of spheres at constant surface charge density is assumed to be zero ($\varepsilon_{p1} = \varepsilon_{p2} = 0$). We see that the interaction energy is well approximated by the leading term plus the first-order correction especially for $\kappa H > 0.5$. This is because the leading term and the first-order correction are, respectively, of the order of $e^{-\kappa H}$ and $e^{-2\kappa H}$ and the higher correction terms are of $O(e^{-3\kappa H})$.

As is seen in Fig. 14.4, the first-order correction (the image interaction) is negative (attractive) for the constant surface potential case (a) and positive (repulsive) for the constant surface charge density case (b) at $\varepsilon_{p1} = \varepsilon_{p2} = 0$. This explains the following well-known behavior observed in the interaction between two unlike spheres (see Fig. 14.5). Namely, in case (a) if the surface potentials of the two interacting spheres are of the different magnitude and of the same sign ($\psi_{o1} = 4\psi_{o2}$ and $\kappa a_1 = \kappa a_2 = 5$ in Fig. 14.5), then the interaction energy, when plotted as a function

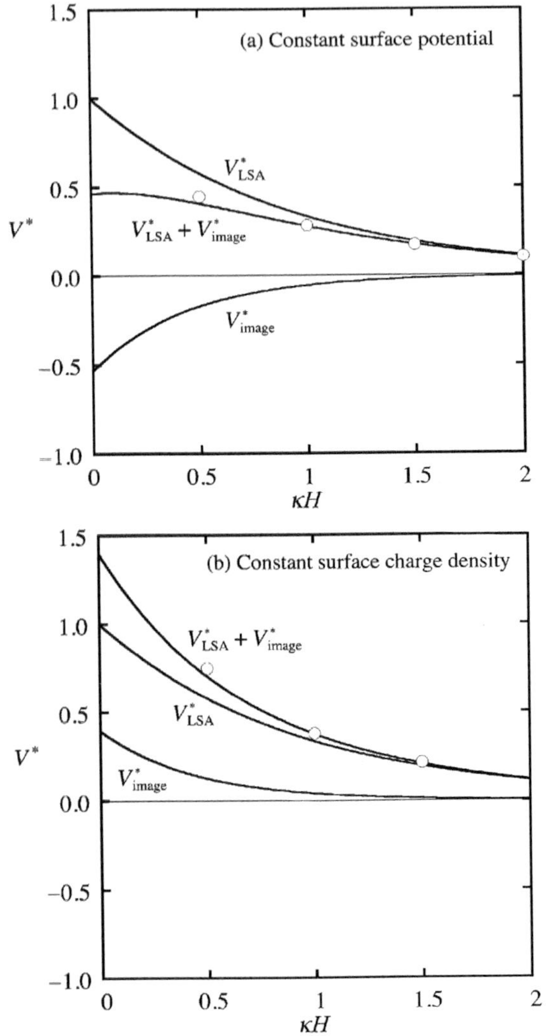

FIGURE 14.4 Comparison of the linear superposition approximation $V_{LSA} = V^{(0)}$, the image interaction correction $V_{image} = V^{(1)} + V^{(2)}$, their sum $V_{LSA} + V_{image}$, and the full solutions, Eqs. (14.23) and (14.29), (represented by circles) as functions of the scaled separation κH at $\kappa a_1 = \kappa a_2 = \kappa a = 5$ and $\psi_{01} = \psi_{02} = \psi_0$ for the interaction of two similar spheres at constant surface potential (a) and at constant surface charge density (b). The reduced quantities are given in the figure: $V^*_{LSA} = V_{LSA}/2\pi\varepsilon_r\varepsilon_o\psi_o{}^2 a$ and $V^*_{image} = V_{image}/2\pi\varepsilon_r\varepsilon_o\psi_o{}^2 a$. From Ref. [10].

of κH, exhibits a maximum. In other words, the interaction force, which is repulsive at large separations, may become attractive at small separations. In case (b), on the other hand, if the surface potentials of the two interacting spheres are of the different magnitude and of the opposite sign ($\psi_{o1} = -4\psi_{o2}$ and $\kappa a_1 = \kappa a_2 = 5$ in

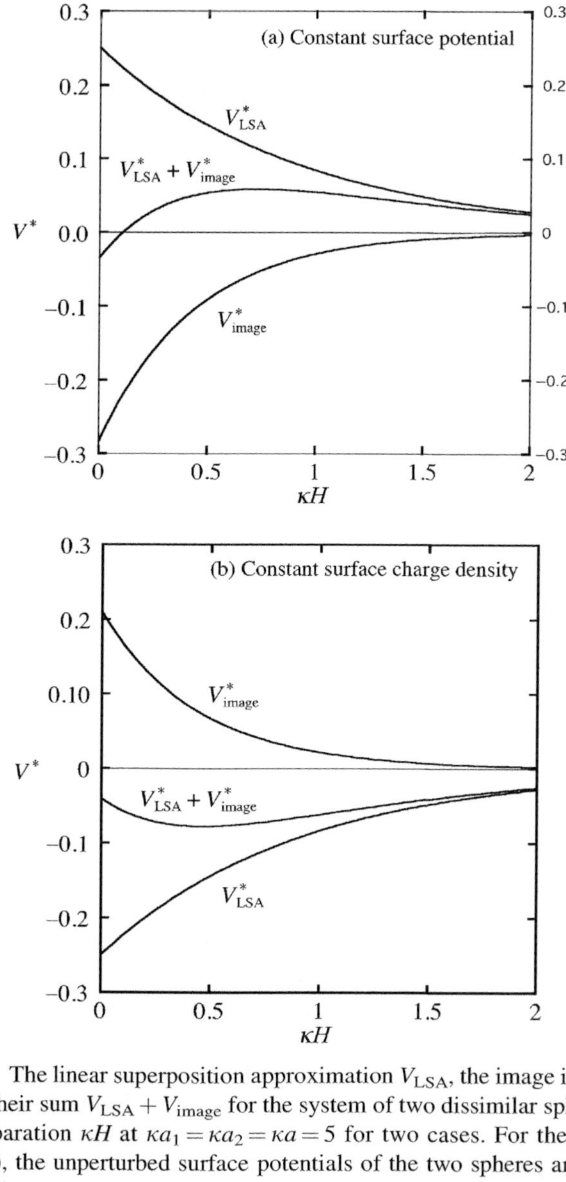

FIGURE 14.5 The linear superposition approximation V_{LSA}, the image interaction correction V_{image}, and their sum $V_{LSA} + V_{image}$ for the system of two dissimilar spheres as functions of the scaled separation κH at $\kappa a_1 = \kappa a_2 = \kappa a = 5$ for two cases. For the constant surface potential case (a), the unperturbed surface potentials of the two spheres are of the different magnitude and of the same sign ($\psi_{o1} = 4\psi_{o2} = \psi_o$). For the constant surface charge density case (b), the unperturbed surface potentials are of the different magnitude and of opposite sign ($\psi_{o1} = -4\psi_{o2} = -\psi_o$). The spheres at constant surface charge density have $\varepsilon_p = 0$. The reduced quantities are given in the figure: $V^*_{LSA} = V_{LSA}/2\pi\varepsilon_r\varepsilon_o\psi_o{}^2 a$ and $V^*_{image} = V_{image}/2\pi\varepsilon_r\varepsilon_o\psi_o{}^2 a$. From Ref. [10].

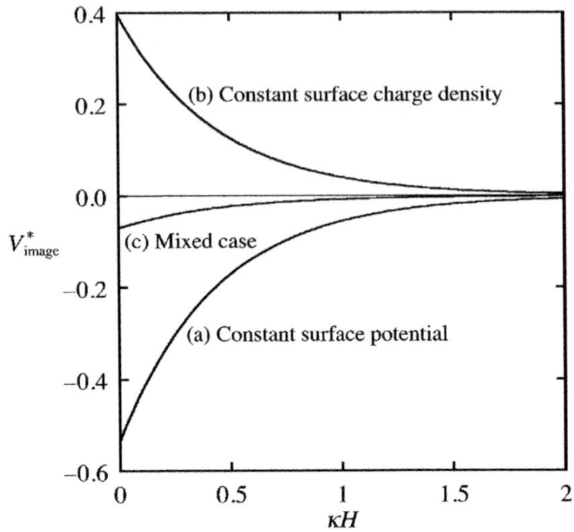

FIGURE 14.6 The reduced image interaction energy $V^*_{\text{image}} = V_{\text{image}}/2\pi\varepsilon_r\varepsilon_o\psi_o{}^2 a$ as a function of the scaled separation κH at $\kappa a_1 = \kappa a_2 = \kappa a = 5$ and $\psi_{o1} = \psi_{o2} = \psi_o$ for the constant surface potential case (a), the constant surface charge density (b), and the mixed cases (c). The sphere at constant surface charge density has $\varepsilon_{pi} = 0$. From Ref. [10].

Fig. 14.5), then the interaction energy shows a minimum. That is, the interaction force, which is attractive at large separations, may become repulsive at small separations. As Fig. 14.5 shows, the change in sign of the interaction force or the appearance of the extremum in the interaction energy occurs when the contribution of the image interaction correction exceeds that of the leading term (or the linear superposition approximation term).

In case (c), the image interaction energy carries both characters of cases (a) and (b). When the unperturbed surface potentials and the radii of the two spheres become similar, these two contributions from cases (a) and (b) tend to cancel each other so that the total image interaction for case (c) becomes small, as shown in Fig. 14.6, in which the interacting spheres are identical ($\kappa a_1 = \kappa a_2 = 5$ and $\psi_{o1} = \psi_{o2}$). In the opposite case where the difference in the two unperturbed potentials is large, the image interaction for case (c) is determined almost only by the larger unperturbed surface potential.

Figure 14.7 shows the dependence of the image interaction between two similar spheres ($\kappa a_1 = \kappa a_2 = \kappa a$ and $\psi_{o1} = \psi_{o2}$) on the value of κa at $\kappa H = 0.5$ for three cases: the constant surface potential case (a), the constant surface charge density case (b), and the mixed case (c). In case (b), the relative permittivity values of spheres at constant surface charge density are assumed to be zero ($\varepsilon_{p1} = \varepsilon_{p2} = 0$). We see that in the limit of large κa, the image interaction energies for cases (a) and (b) tend to have the same magnitude but the opposite sign so that the image interaction becomes zero for case (c).

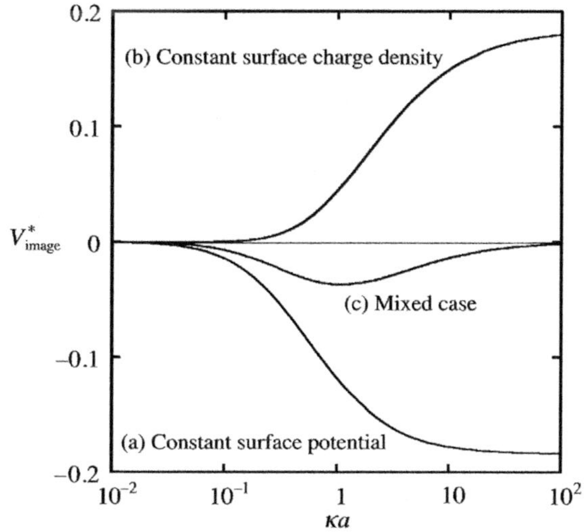

FIGURE 14.7 The reduced image interaction energy $V^*_{\text{image}} = V_{\text{image}}/2\pi\varepsilon_r\varepsilon_o\psi_o{}^2a$ as a function of the scaled sphere radius $\kappa a_1 = \kappa a_2 = \kappa a$ for $\psi_{o1} = \psi_{o2} = \psi_o$ at $\kappa H = 0.5$ for the constant surface potential (a), the constant surface charge density (b), and the mixed cases (c). The sphere at constant surface charge density has $\varepsilon_{pi} = 0$. From Ref. [10].

In the limit of large κa_1 and κa_2, one can derive analytic approximate expressions for the image interaction V_{image} on the basis of the following approximations:

$$\sum_{n=0}^{\infty}(2n+1)G_n(i)K^2_{n+1/2}(\kappa R) \approx -\frac{a_i\, e^{-2\kappa(H+a_j)}}{2(H+a_j)}, \quad (i=1,2) \tag{14.45}$$

$$\sum_{n=0}^{\infty}(2n+1)H_n(i)K^2_{n+1/2}(\kappa R) \approx \frac{a_i\, e^{-2\kappa(H+a_j)}}{2(H+a_j)} \times \left\{1 - \frac{\varepsilon_{pi}}{\varepsilon_r}\sqrt{\frac{\pi(H+a_1+a_2)}{\kappa a_i(H+a_j)}}\right\}, \quad (i=1,2) \tag{14.46}$$

The results are given as follows.

Case (a): spheres 1 and 2 both at constant surface potential

$$V^{\psi}_{\text{image}} = -\pi\varepsilon_r\varepsilon_o\psi^2_{o1}a_1^2a_2\frac{e^{-2\kappa H}}{R(a_1+H)} - \pi\varepsilon_r\varepsilon_o\psi^2_{o2}a_1a_2^2\frac{e^{-2\kappa H}}{R(a_2+H)} \tag{14.47}$$

Case (b): spheres 1 and 2 both at constant surface charge density

$$V^{\sigma}_{\text{image}} = \pi\varepsilon_r\varepsilon_o\psi_{o1}^2 a_1^2 a_2 \frac{e^{-2\kappa H}}{R(a_1+H)}\left\{1-\frac{\varepsilon_{p2}}{\varepsilon_r}\sqrt{\frac{\pi(H+a_1+a_2)}{\kappa a_2(H+a_1)}}\right\}$$
$$+\pi\varepsilon_r\varepsilon_o\psi_{o2}^2 a_2^2 a_1 \frac{e^{-2\kappa H}}{R(a_1+H)}\left\{1-\frac{\varepsilon_{p1}}{\varepsilon_r}\sqrt{\frac{\pi(H+a_1+a_2)}{\kappa a_1(H+a_2)}}\right\} \quad (14.48)$$

Case (c): sphere 1 at constant surface potential and sphere 2 at constant surface charge density

$$V^{\psi\text{-}\sigma}_{\text{image}} = \pi\varepsilon_r\varepsilon_o\psi_{o1}^2 a_1^2 a_2 \frac{e^{-2\kappa H}}{R(a_1+H)}\left\{1-\frac{\varepsilon_{p2}}{\varepsilon_r}\sqrt{\frac{\pi(H+a_1+a_2)}{\kappa a_2(H+a_1)}}\right\}$$
$$-\pi\varepsilon_r\varepsilon_o\psi_{o2}^2 a_1 a_2^2 \frac{e^{-2\kappa H}}{R(a_2+H)} \quad (14.49)$$

We thus again find that in the limit of large κa_1 and κa_2 (at finite $\varepsilon_{p1}$and ε_{p2}), the image interaction energies for cases (a) and (b) tend to have the same magnitude but the opposite sign so that for case (c) $V^{\psi-\sigma}_{\text{image}}$ becomes very small if $\psi_{o1} \approx \psi_{o2}$.

In Figs 14.5–14.7, the image interaction for the constant surface charge density case (b) is positive. Note, however, that this holds only when the relative permittivity is small. The image interaction in case (b) may become negative as the magnitude of the ratio of the relative permittivity of the sphere to that of the solution $\varepsilon_{pi}/\varepsilon_r$ increases. In the limit $\varepsilon_{pi} \to \infty$ (this case corresponds to metallic particles), in particular, the image interaction is always negative. Figure 14.8 shows this dependence together with those in cases (a) and (c) for $\varepsilon_{p1} = \varepsilon_{p2}$, $\kappa a_1 = \kappa a_2 = 5$, and $\psi_{o1} = \psi_{o2}$ at $\kappa H = 0.5$. For two similar spheres with constant surface charge density ($\kappa a_1 = \kappa a_2 = \kappa a$, $\varepsilon_{p1} = \varepsilon_{p2} = \varepsilon_p$, and $\psi_{o1} = \psi_{o2} = \psi_o$) at $\kappa a \gg 1$ and $H \ll a$, the image interaction energy (Eq. (14.48)) becomes

$$V^{\sigma}_{\text{image}} = \pi\varepsilon_r\varepsilon_o\psi_o^2 a e^{-2\kappa H}\left(1-\frac{\varepsilon_{p2}}{\varepsilon_r}\sqrt{\frac{2\pi}{\kappa a}}\right) \quad (14.50)$$

Thus, we see that for $\kappa a \gg 1$ and $H \ll a$,

$$V^{\sigma}_{\text{image}} \geq 0 \quad \text{for } \frac{\varepsilon_{p2}}{\varepsilon_r} \leq \sqrt{\frac{\kappa a}{2\pi}} \quad (14.51)$$

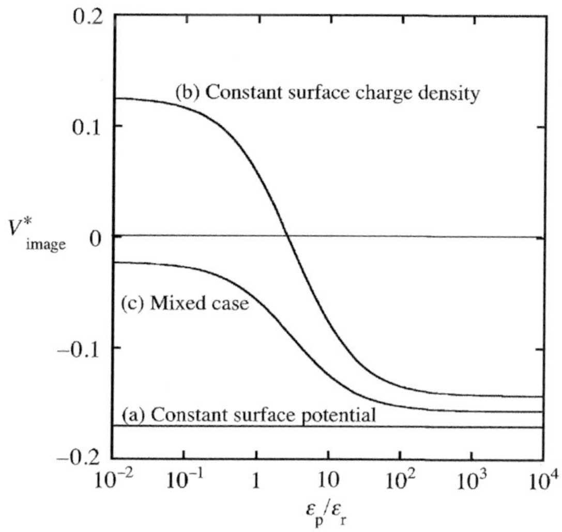

FIGURE 14.8 The reduced image interaction energy $V^*_{\text{image}} = V_{\text{image}}/2\pi\varepsilon_r\varepsilon_o\psi_0{}^2a$ as a function of the ratio of the relative permittivity of the sphere to that of the solution $\varepsilon_p/\varepsilon_r$ for $\varepsilon_{p1} = \varepsilon_{p2} = \varepsilon_p$, $\kappa a_1 = \kappa a_2 = \kappa a = 5$, and $\psi_{o1} = \psi_{o2} = \psi_o$ at $\kappa H = 0.5$. The image interaction energy at constant surface charge density changes its sign as the value of $\varepsilon_p/\varepsilon_r$ increases, while it does not depend on $\varepsilon_p/\varepsilon_r$ for the constant surface potential case. From Ref. [10].

$$V^{\sigma}_{\text{images}} < 0 \quad \text{for} \ \frac{\varepsilon_{p2}}{\varepsilon_r} > \sqrt{\frac{\kappa a}{2\pi}} \tag{14.52}$$

For case (a), if, further,

$$H \ll a_1 \quad \text{and} \quad H \ll a_2 \tag{14.53}$$

that is, if $\kappa a_1 \gg 1$ and $\kappa a_2 \gg 1$, $H \ll a_1$ and $H \ll a_2$, then Eq. (14.47) tends to

$$V^{\psi}_{\text{image}} = -\pi\varepsilon_r\varepsilon_o \frac{a_1 a_2}{a_1 + a_2} (\psi_{o1}^2 + \psi_{o2}^2) e^{-2\kappa H} \tag{14.54}$$

so that Eq. (14.44) gives

$$V^{\psi}(H) = 4\pi\varepsilon_r\varepsilon_o \frac{a_1 a_2}{a_1 + a_2} \left\{ \psi_{o1}\psi_{o2}\, e^{-\kappa H} - \frac{1}{4}(\psi_{o1}^2 + \psi_{o2}^2) e^{-2\kappa H} \right\} + O(e^{-3\kappa H}) \tag{14.55}$$

Equation (14.55) agrees with the HHF formula (Eq. (12.10)) [18], namely,

$$V^{\psi}(H) = \pi\varepsilon_r\varepsilon_o \frac{a_1 a_2}{a_1 + a_2}\left\{(\psi_{o1} + \psi_{o2})^2 \ln(1 + e^{-\kappa H}) + (\psi_{o1} - \psi_{o2})^2 \ln(1 - e^{-\kappa H})\right\} \tag{14.56}$$

which can be rewritten as

$$\begin{aligned} V^{\psi}(H) &= 4\pi\varepsilon_r\varepsilon_o \frac{a_1 a_2}{a_1 + a_2}\left\{\psi_{o1}\psi_{o2}\sum_{n=0}^{\infty}\frac{e^{-(2n+1)\kappa H}}{2n+1} - \frac{1}{2}(\psi_{o1}^2 + \psi_{o2}^2)\sum_{n=1}^{\infty}\frac{e^{-2n\kappa H}}{2n}\right\} \\ &= 4\pi\varepsilon_r\varepsilon_o \frac{a_1 a_2}{a_1 + a_2}\left\{\psi_{o1}\psi_{o2}\, e^{-\kappa H} - \frac{1}{4}(\psi_{o1}^2 + \psi_{o2}^2)e^{-2\kappa H}\right\} + O(e^{-3\kappa H}) \end{aligned} \tag{14.57}$$

For case (b), if $\kappa a_1 \gg 1$ and $\kappa a_2 \gg 1$, $H \ll a_1$ and $H \ll a_2$, then Eq. (14.48) becomes

$$\begin{aligned} V^{\sigma}_{\text{image}} = \pi\varepsilon_r\varepsilon_o \frac{a_1 a_2}{a_1 + a_2}&\left[\psi_{o1}^2\, e^{-2\kappa H}\left\{1 - \sqrt{\pi}\frac{\varepsilon_{rp2}}{\varepsilon_r}\sqrt{\frac{1}{\kappa a_1} + \frac{1}{\kappa a_2}}\right\}\right. \\ &\left.+\psi_{o2}^2\, e^{-2\kappa H}\left\{1 - \sqrt{\pi}\frac{\varepsilon_{rp1}}{\varepsilon_r}\sqrt{\frac{1}{\kappa a_1} + \frac{1}{\kappa a_2}}\right\}\right] \end{aligned} \tag{14.58}$$

so that Eq. (14.44) gives

$$\begin{aligned} V^{\sigma}(H) = 4\pi\varepsilon_r\varepsilon_o \frac{a_1 a_2}{a_1 + a_2}&\left[\psi_{o1}\psi_{o2}\, e^{-\kappa H}\right. \\ &+\frac{1}{4}\psi_{o1}^2\, e^{-2\kappa H}\left\{1 - \sqrt{\pi}\frac{\varepsilon_{p2}}{\varepsilon_r}\sqrt{\frac{1}{\kappa a_1} + \frac{1}{\kappa a_2}}\right\} \\ &\left.+\frac{1}{4}\psi_{o2}^2\, e^{-2\kappa H}\left\{1 - \sqrt{\pi}\frac{\varepsilon_{p1}}{\varepsilon_r}\sqrt{\frac{1}{\kappa a_1} + \frac{1}{\kappa a_2}}\right\}\right] + O(e^{-3\kappa H}) \end{aligned} \tag{14.59}$$

If, further, the terms of the order of $1/\sqrt{\kappa a_i}$ $(i = 1, 2)$ are neglected, Eq. (14.59) tends to

$$V^{\psi-\sigma}_{\text{image}} = \pi\varepsilon_r\varepsilon_o \frac{a_1 a_2}{a_1 + a_2}(\psi_{o1}^2 + \psi_{o2}^2)e^{-2\kappa H} \tag{14.60}$$

so that Eq. (14.44) gives

$$V^{\psi}(H) = 4\pi\varepsilon_r\varepsilon_o \frac{a_1 a_2}{a_1 + a_2}\left\{\psi_{o1}\psi_{o2}\, e^{-\kappa H} + \frac{1}{4}(\psi_{o1}^2 + \psi_{o2}^2)e^{-2\kappa H}\right\} + O(e^{-3\kappa H}) \tag{14.61}$$

which agrees with Wiese and Healy's formula (Eq. (12.11)) [19], namely,

$$\begin{aligned} V^{\sigma}(H) &= \pi\varepsilon_r\varepsilon_o \frac{a_1 a_2}{a_1 + a_2}\{-(\psi_{o1} + \psi_{o2})^2 \ln(1 - e^{-\kappa H}) - (\psi_{o1} - \psi_{o2})^2 \ln(1 + e^{-\kappa H})\} \\ &= 4\pi\varepsilon_r\varepsilon_o \frac{a_1 a_2}{a_1 + a_2}\left\{\psi_{o1}\psi_{o2}\sum_{n=0}^{\infty}\frac{e^{-(2n+1)\kappa H}}{2n+1} + \frac{1}{2}(\psi_{o1}^2 + \psi_{o2}^2)\sum_{n=1}^{\infty}\frac{e^{-2n\kappa H}}{2n}\right\} \\ &= 4\pi\varepsilon_r\varepsilon_o \frac{a_1 a_2}{a_1 + a_2}\left\{\psi_{o1}\psi_{o2}\, e^{-\kappa H} + \frac{1}{4}(\psi_{o1}^2 + \psi_{o2}^2)e^{-2\kappa H}\right\} + O(e^{-3\kappa H}) \end{aligned} \tag{14.62}$$

Also we see from Eq. (14.59) that the next-order curvature correction to Derjaguin's approximation [2] is of the order of $1/\sqrt{\kappa a_i}$ ($i = 1, 2$). This has been suggested first by Dukhin and Lyklema [20] and discussed by Kijlstra [21]. In the case of the interaction between particles with constant surface potential, on the other hand, the next-order curvature correction to Derjaguin's approximation is of the order of $1/\kappa a_i$ ($i = 1, 2$), since in this case no electric fields are induced within the interacting particles.

For the mixed case where sphere 1 has a constant surface potential and sphere 2 has a constant surface charge density (case (c)), if $\kappa a_1 \gg 1$ and $\kappa a_2 \gg 1$, $H \ll a_1$ and $H \ll a_2$, then Eq. (14.48) becomes

$$V_{\text{image}}^{\psi-\sigma} = \pi\varepsilon_r\varepsilon_o \frac{a_1 a_2}{a_1 + a_2}\left[\psi_{o1}^2\, e^{-2\kappa H}\left\{1 - \sqrt{\pi}\frac{\varepsilon_{rp2}}{\varepsilon_r}\sqrt{\frac{1}{\kappa a_1} + \frac{1}{\kappa a_2}}\right\} - \psi_{o2}^2\, e^{-2\kappa H}\right] \tag{14.63}$$

so that Eq. (14.44) gives

$$\begin{aligned} V^{\psi-\sigma}(H) &= 4\pi\varepsilon_r\varepsilon_o \frac{a_1 a_2}{a_1 + a_2}\Bigg[\psi_{o1}\psi_{o2}\, e^{-\kappa H} \\ &+ \frac{1}{4}\psi_{o1}^2\, e^{-2\kappa H}\left\{1 - \sqrt{\pi}\frac{\varepsilon_{p2}}{\varepsilon_r}\sqrt{\frac{1}{\kappa a_1} + \frac{1}{\kappa a_2}}\right\} - \frac{1}{4}\psi_{o2}^2\, e^{-2\kappa H}\Bigg] + O(e^{-3\kappa H}) \end{aligned} \tag{14.64}$$

If, further, the terms of the order of $1/\sqrt{\kappa a_i}$ ($i = 1, 2$) are neglected, Eq. (14.64) tends to

$$V^{\psi-\sigma}(H) = 4\pi\varepsilon_r\varepsilon_o \frac{a_1 a_2}{a_1 + a_2} \left\{ \psi_{o1}\psi_{o2}\, e^{-\kappa H} + \frac{1}{4}(\psi_{o1}^2 - \psi_{o2}^2)\, e^{-2\kappa H} \right\} + O(e^{-3\kappa H}) \tag{14.65}$$

Equation (14.65) agrees with Kar et al's equation (Eq. (12.12)) [22]

$$V^{\psi-\sigma}(H) = 4\pi\varepsilon_r\varepsilon_o \frac{a_1 a_2}{a_1 + a_2} \left\{ \psi_{o1}\psi_{o2} \arctan(e^{-\kappa H}) + \frac{1}{4}(\psi_{o1}^2 - \psi_{o2}^2) \ln(1 + e^{-2\kappa H}) \right\} \tag{14.66}$$

which is rewritten as

$$\begin{aligned} V^{\psi-\sigma}(H) &= 4\pi\varepsilon_r\varepsilon_o \frac{a_1 a_2}{a_1 + a_2} \left\{ \psi_{o1}\psi_{o2} \sum_{n=0}^{\infty} (-1)^n \frac{e^{-(2n+1)\kappa H}}{2n+1} \right. \\ &\quad \left. - \frac{1}{2}(\psi_{o1}^2 - \psi_{o2}^2) \sum_{n=1}^{\infty} (-1)^n \frac{e^{-2n\kappa H}}{2n} \right\} \\ &= 4\pi\varepsilon_r\varepsilon_o \frac{a_1 a_2}{a_1 + a_2} \left\{ \psi_{o1}\psi_{o2}\, e^{-\kappa H} + \frac{1}{4}(\psi_{o1}^2 - \psi_{o2}^2) e^{-2\kappa H} \right\} + O(e^{-3\kappa H}) \end{aligned} \tag{14.67}$$

14.4 PLATE AND SPHERE

Although when the radius of either of the two interacting spheres is very large, the interaction energy between two spheres gives the interaction energy between a sphere and a plate, the following alternative expression is more convenient for practical calculations of the interaction energy.

Case (a): plate 1 and sphere 2 both at constant surface potential
The result for the interaction energy V^{ψ} between a plate 1 carrying a constant surface potential ψ_{o1} and a sphere 2 of radius a_2 carrying a constant surface potential ψ_{o2}, separated by a distance H between their surfaces, immersed in an electrolyte solution is (Fig. 14.9)

$$\begin{aligned} V^{\psi}(H) &= 4\pi\varepsilon_r\varepsilon_o\psi_{o1}\psi_{o2}a_2\, e^{-\kappa H} + \pi^2\varepsilon_r\varepsilon_o\psi_{o1}^2 \frac{1}{\kappa} e^{-2\kappa(H+a_2)} \sum_{n=0}^{\infty}(2n+1)G_n(2) \\ &\quad - 4\varepsilon_r\varepsilon_o\psi_{o2}^2\kappa a_2^2\, e^{2\kappa a_2}\beta_{00} - 2\pi\varepsilon_r\varepsilon_o\psi_{o1}\psi_{o2}a_2\, e^{-\kappa H}\sum_{n=0}^{\infty}(2n+1)(\beta_{0n}+\beta_{n0})G_n(2) \\ &\quad - \pi^2\varepsilon_r\varepsilon_o\psi_{o1}^2 \frac{1}{\kappa} e^{-2\kappa(H+a_2)} \sum_{n=0}^{\infty}\sum_{m=0}^{\infty}(2n+1)(2m+1)\beta_{nm}G_n(2)G_m(2) \\ &\quad + 4\varepsilon_r\varepsilon_o\psi_{o2}^2\kappa a_2^2\, e^{2\kappa a_2}\sum_{n=0}^{\infty}(2n+1)\beta_{0n}\beta_{n0}G_n(2) \end{aligned}$$

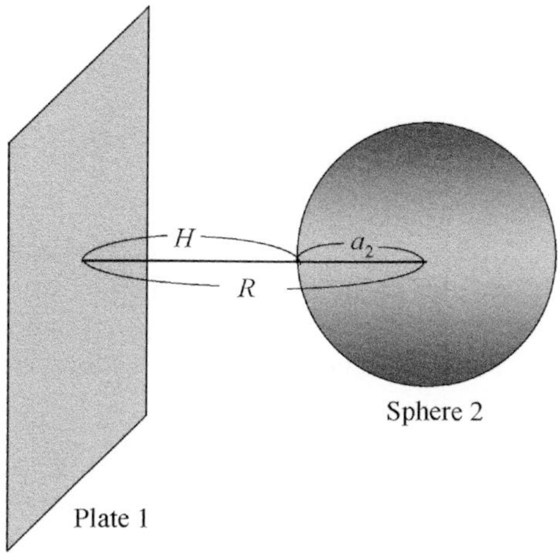

FIGURE 14.9 Interaction between a plate 1 and a sphere 2 of radius a_2 at a separation R. $H\,(=R-a_2)$ is the closest distance between their surfaces.

$$+2\pi\varepsilon_r\varepsilon_o\psi_{o1}\psi_{o2}a_2\,e^{-\kappa H}\sum_{n=0}^{\infty}\sum_{m=0}^{\infty}(2n+1)(2m+1)\beta_{nm}(\beta_{0n}+\beta_{m0})G_n(2)G_m(2)+\cdots$$

$$-2\pi\varepsilon_r\varepsilon_o\psi_{o1}\psi_{o2}a_2\,e^{-\kappa H}\sum_{n_1=0}^{\infty}\sum_{n_2=0}^{\infty}\cdots\sum_{n_v=0}^{\infty}(-1)^{v-1}(2n_1+1)(2n_2+1)\cdots(2n_v+1)$$

$$\beta_{n_1n_2}\beta_{n_2n_3}\cdots\beta_{n_{v-1}n_v}(\beta_{0n_1}+\beta_{n_v0})G_{n_1}(2)G_{n_2}(2)\cdots G_{n_v}(2)$$

$$+\pi^2\varepsilon_r\varepsilon_o\psi_{o1}^2\frac{1}{\kappa}e^{-2\kappa(H+a_2)}\sum_{n_1=0}^{\infty}\sum_{n_2=0}^{\infty}\cdots\sum_{n_v=0}^{\infty}(-1)^{v-1}(2n_1+1)(2n_2+1)\cdots(2n_v+1)$$

$$\times\beta_{n_1n_2}\beta_{n_2n_3}\cdots\beta_{n_{v-1}n_v}G_{n_1}(2)G_{n_2}(2)\cdots G_{n_v}(2)$$

$$+4\varepsilon_r\varepsilon_o\psi_{o2}^2\kappa a_2^2\,e^{2\kappa a_2}\sum_{n_1=0}^{\infty}\sum_{n_2=0}^{\infty}\cdots\sum_{n_v=0}^{\infty}(-1)^{v-1}(2n_1+1)(2n_2+1)\cdots(2n_v+1)$$

$$\times\beta_{n_1n_2}\beta_{n_2n_3}\cdots\beta_{n_{v-1}n_v}\beta_{0n_1}\beta_{n_v0}G_{n_1}(2)G_{n_2}(2)\cdots G_{n_v}(2)+\cdots \tag{14.68}$$

where

$$\begin{aligned}\beta_{nm}&=\frac{\pi}{2\kappa}\int_0^{\infty}\frac{\exp(-2\sqrt{k^2+\kappa^2}(H+a_2))}{\sqrt{k^2+\kappa^2}}P_n\left(\frac{\sqrt{k^2+\kappa^2}}{\kappa}\right)P_m\left(\frac{\sqrt{k^2+\kappa^2}}{\kappa}\right)k\,dk\\&=\frac{\pi}{2}\int_1^{\infty}e^{-2\kappa t(H+a_2)}P_n(t)P_m(t)dt\end{aligned} \tag{14.69}$$

where $P_n(t)$ are Legendre polynomials. Or, alternatively

$$\beta_{nm} = \sum_{r=0}^{\min\{m,n\}} A_{nmr} \sqrt{\frac{\pi}{4\kappa(H + a_2)}} K_{n+m-2r+1/2}(2\kappa(H + a_2)) \quad (14.70)$$

Case (b): plate 1 and sphere 2 both at constant surface potential

For the case where the surface charge densities (instead of the surface potentials) of plate 1 (of relative permittivity ε_{p1}) and sphere 2 (of relative permittivity ε_{p2}) remain constant, the corresponding expression for the interaction energy V^{σ} is given by

$$\begin{aligned}
V^{\sigma}(H) &= 4\pi\varepsilon_r\varepsilon_o\psi_{o1}\psi_{o2}a_2\, e^{-\kappa H} + \pi^2\varepsilon_r\varepsilon_o\psi_{o1}^2 \frac{1}{\kappa} e^{-2\kappa(H+a_2)} \sum_{n=0}^{\infty}(2n+1)H_n(2) \\
&+ 4\varepsilon_r\varepsilon_o\psi_{o2}^2\kappa a_2^2\, e^{2\kappa a_2}\gamma_{00} + 2\pi\varepsilon_r\varepsilon_o\psi_{o1}\psi_{o2}a_2\, e^{-\kappa H} \\
&\times \sum_{n=0}^{\infty}(2n+1)(\gamma_{0n} + \gamma_{n0})H_n(2) \\
&+ \pi^2\varepsilon_r\varepsilon_o\psi_{o1}^2 \frac{1}{\kappa} e^{-2\kappa(H+a_2)} \sum_{n=0}^{\infty}\sum_{m=0}^{\infty}(2n+1)(2m+1)\gamma_{nm}H_n(2)H_m(2) \\
&+ 4\varepsilon_r\varepsilon_o\psi_{o2}^2\kappa a_2^2\, e^{2\kappa a_2} \sum_{n=0}^{\infty}(2n+1)\gamma_{0n}\gamma_{n0}H_n(2) \\
&+ 2\pi\varepsilon_r\varepsilon_o\psi_{o1}\psi_{o2}a_2\, e^{-\kappa H} \sum_{n=0}^{\infty}\sum_{m=0}^{\infty}(2n+1)(2m+1)\gamma_{nm} \\
&\times (\gamma_{0n} + \gamma_{m0})H_n(2)H_m(2) + \cdots \\
&+ 2\pi\varepsilon_r\varepsilon_o\psi_{o1}\psi_{o2}a_2\, e^{-\kappa H} \sum_{n_1=0}^{\infty}\sum_{n_2=0}^{\infty}\cdots\sum_{n_\nu=0}^{\infty}(2n_1+1)(2n_2+1)\cdots(2n_\nu+1) \\
&\times \gamma_{n_1n_2}\gamma_{n_2n_3}\cdots\gamma_{n_{\nu-1}n_\nu}(\gamma_{0n_1} + \gamma_{n_\nu 0})H_{n_1}(2)H_{n_2}(2)\cdots H_{n_\nu}(2) \\
&+ \pi^2\varepsilon_r\varepsilon_o\psi_{o1}^2 \frac{1}{\kappa} e^{-2\kappa(H+a_2)} \sum_{n_1=0}^{\infty}\sum_{n_2=0}^{\infty}\cdots\sum_{n_\nu=0}^{\infty}(2n_1+1)(2n_2+1)\cdots(2n_\nu+1) \\
&\times \gamma_{n_1n_2}\gamma_{n_2n_3}\cdots\gamma_{n_{\nu-1}n_\nu}H_{n_1}(2)H_{n_2}(2)\cdots H_{n_\nu}(2) \\
&+ 4\varepsilon_r\varepsilon_o\psi_{o2}^2\kappa a_2^2\, e^{2\kappa a_2} \sum_{n_1=0}^{\infty}\sum_{n_2=0}^{\infty}\cdots\sum_{n_\nu=0}^{\infty}(2n_1+1)(2n_2+1)\cdots(2n_\nu+1) \\
&\times \gamma_{n_1n_2}\gamma_{n_2n_3}\cdots\gamma_{n_{\nu-1}n_\nu}\gamma_{0n_1}\gamma_{n_\nu 0}H_{n_1}(2)H_{n_2}(2)\cdots H_{n_\nu}(2) + \cdots
\end{aligned} \quad (14.71)$$

where γ_{nm} is defined by

$$\gamma_{nm} = -\frac{\pi}{2\kappa}\left(\frac{\varepsilon_{p1}k - \varepsilon_r\sqrt{k^2+\kappa^2}}{\varepsilon_{p1}k + \varepsilon_r\sqrt{k^2+\kappa^2}}\right)$$
$$\times\int_0^\infty \frac{\exp(-2\sqrt{k^2+\kappa^2}(H+a_2))}{\sqrt{k^2+\kappa^2}} P_n\left(\frac{\sqrt{k^2+\kappa^2}}{\kappa}\right) P_m\left(\frac{\sqrt{k^2+\kappa^2}}{\kappa}\right) k\,dk$$
$$= \frac{\pi}{2}\int_1^\infty \left(\frac{\varepsilon_r t - \varepsilon_{p1}\sqrt{t^2-1}}{\varepsilon_r t + \varepsilon_{p1}\sqrt{t^2-1}}\right) e^{-2\kappa t(H+a_2)} P_n(t)P_m(t)dt \tag{14.72}$$

Case (c): plate 1 at constant surface potential and sphere 2 at constant surface charge density

When plate 1 has a constant surface potential and sphere 2 has a constant surface charge density, the interaction energy $V^{\psi-\sigma}$ is given by

$$V^{\psi-\sigma}(H) = 4\pi\varepsilon_r\varepsilon_o\psi_{o1}\psi_{o2}a_2\, e^{-\kappa H} + \pi^2\varepsilon_r\varepsilon_o\psi_{o1}^2\frac{1}{\kappa}e^{-2\kappa(H+a_2)}\sum_{n=0}^{\infty}(2n+1)H_n(2)$$
$$- 4\varepsilon_r\varepsilon_o\psi_{o2}^2\kappa a_2^2\, e^{2\kappa a_2}\beta_{00} - 2\pi\varepsilon_r\varepsilon_o\psi_{o1}\psi_{o2}a_2\, e^{-\kappa H}$$
$$\times\sum_{n=0}^{\infty}(2n+1)(\beta_{0n}+\beta_{n0})H_n(2)$$
$$- \pi^2\varepsilon_r\varepsilon_o\psi_{o1}^2\frac{1}{\kappa}e^{-2\kappa(H+a_2)}\sum_{n=0}^{\infty}\sum_{m=0}^{\infty}(2n+1)(2m+1)\beta_{nm}H_n(2)H_m(2)$$
$$+ 4\varepsilon_r\varepsilon_o\psi_{o2}^2\kappa a_2^2\, e^{2\kappa a_2}\sum_{n=0}^{\infty}(2n+1)\beta_{0n}\beta_{n0}H_n(2)$$
$$+ 2\pi\varepsilon_r\varepsilon_o\psi_{o1}\psi_{o2}a_2\, e^{-\kappa H}\sum_{n=0}^{\infty}\sum_{m=0}^{\infty}(2n+1)(2m+1)\beta_{nm}$$
$$\times(\beta_{0n}+\beta_{m0})H_n(2)H_m(2) + \cdots$$
$$- 2\pi\varepsilon_r\varepsilon_o\psi_{o1}\psi_{o2}a_2\, e^{-\kappa H}\sum_{n_1=0}^{\infty}\sum_{n_2=0}^{\infty}\cdots\sum_{n_\nu=0}^{\infty}(-1)^{\nu-1}(2n_1+1)$$
$$\times(2n_2+1)\cdots(2n_\nu+1)$$
$$\beta_{n_1n_2}\beta_{n_2n_3}\cdots\beta_{n_{\nu-1}n_\nu}(\beta_{0n_1}+\beta_{n_\nu 0})H_{n_1}(2)H_{n_2}(2)\cdots H_{n_\nu}(2)$$
$$+ \pi^2\varepsilon_r\varepsilon_o\psi_{o1}^2\frac{1}{\kappa}e^{-2\kappa(H+a_2)}\sum_{n_1=0}^{\infty}\sum_{n_2=0}^{\infty}\cdots\sum_{n_\nu=0}^{\infty}(-1)^{\nu-1}(2n_1+1)$$
$$\times(2n_2+1)\cdots(2n_\nu+1)$$

$$\times \beta_{n_1 n_2}\beta_{n_2 n_3}\cdots\beta_{n_{\nu-1}n_\nu}H_{n_1}(2)H_{n_2}(2)\cdots H_{n_\nu}(2)$$
$$+ 4\varepsilon_r\varepsilon_o\psi_{o2}^2\kappa a_2^2\, e^{2\kappa a_2}\sum_{n_1=0}^{\infty}\sum_{n_2=0}^{\infty}\cdots\sum_{n_\nu=0}^{\infty}(-1)^{\nu-1}(2n_1+1)$$
$$\times(2n_2+1)\cdots(2n_\nu+1)$$
$$\times\beta_{n_1 n_2}\beta_{n_2 n_3}\cdots\beta_{n_{\nu-1}n_\nu}\beta_{0n_1}\beta_{n_\nu 0}H_{n_1}(2)H_{n_2}(2)\cdots H_{n_\nu}(2)+\cdots \tag{14.73}$$

Case (d): plate 1 at constant surface charge density and sphere 2 at constant surface potential

When plate 1 has a constant surface charge density and sphere 2 has a constant surface potential, the interaction energy $V^{\sigma-\psi}$ is given by

$$V^{\sigma-\phi}(H) = 4\pi\varepsilon_r\varepsilon_o\psi_{o1}\psi_{o2}a_2\, e^{-\kappa H} + \pi^2\varepsilon_r\varepsilon_o\psi_{o1}^2\frac{1}{\kappa}e^{-2\kappa(H+a_2)}\sum_{n=0}^{\infty}(2n+1)G_n(2)$$
$$+ 4\varepsilon_r\varepsilon_o\psi_{o2}^2\kappa a_2^2\, e^{2\kappa a_2}\gamma_{00} + 2\pi\varepsilon_r\varepsilon_o\psi_{o1}\psi_{o2}a_2\, e^{-\kappa H}$$
$$\times\sum_{n=0}^{\infty}(2n+1)(\gamma_{0n}+\gamma_{n0})G_n(2)$$
$$+ \pi^2\varepsilon_r\varepsilon_o\psi_{o1}^2\frac{1}{\kappa}e^{-2\kappa(H+a_2)}\sum_{n=0}^{\infty}\sum_{m=0}^{\infty}(2n+1)(2m+1)\gamma_{nm}G_n(2)G_m(2)$$
$$+ 4\varepsilon_r\varepsilon_o\psi_{o2}^2\kappa a_2^2\, e^{2\kappa a_2}\sum_{n=0}^{\infty}(2n+1)\gamma_{0n}\gamma_{n0}G_n(2)$$
$$+ 2\pi\varepsilon_r\varepsilon_o\psi_{o1}\psi_{o2}a_2\, e^{-\kappa H}\sum_{n=0}^{\infty}\sum_{m=0}^{\infty}(2n+1)(2m+1)\gamma_{nm}$$
$$\times(\gamma_{0n}+\gamma_{m0})G_n(2)G_m(2)+\cdots$$
$$+ 2\pi\varepsilon_r\varepsilon_o\psi_{o1}\psi_{o2}a_2\, e^{-\kappa H}\sum_{n_1=0}^{\infty}\sum_{n_2=0}^{\infty}\cdots\sum_{n_\nu=0}^{\infty}(2n_1+1)$$
$$\times(2n_2+1)\cdots(2n_\nu+1)$$
$$\times\gamma_{n_1 n_2}\gamma_{n_2 n_3}\cdots\gamma_{n_{\nu-1}n_\nu}(\gamma_{0n_1}+\gamma_{n_\nu 0})G_{n_1}(2)G_{n_2}(2)\cdots G_{n_\nu}(2)$$
$$+ \pi^2\varepsilon_r\varepsilon_o\psi_{o1}^2\frac{1}{\kappa}e^{-2\kappa(H+a_2)}\sum_{n_1=0}^{\infty}\sum_{n_2=0}^{\infty}\cdots\sum_{n_\nu=0}^{\infty}(2n_1+1)$$
$$\times(2n_2+1)\cdots(2n_\nu+1)$$
$$\times\gamma_{n_1 n_2}\gamma_{n_2 n_3}\cdots\gamma_{n_{\nu-1}n_\nu}G_{n_1}(2)G_{n_2}(2)\cdots G_{n_\nu}(2)$$
$$+ 4\varepsilon_r\varepsilon_o\psi_{o2}^2\kappa a_2^2\, e^{2\kappa a_2}\sum_{n_1=0}^{\infty}\sum_{n_2=0}^{\infty}\cdots\sum_{n_\nu=0}^{\infty}(2n_1+1)(2n_2+1)\cdots(2n_\nu+1)$$
$$\times\gamma_{n_1 n_2}\gamma_{n_2 n_3}\cdots\gamma_{n_{\nu-1}n_\nu}\gamma_{0n_1}\gamma_{n_\nu 0}G_{n_1}(2)G_{n_2}(2)\cdots G_{n_\nu}(2)+\cdots \tag{14.74}$$

We compare the image interactions appearing in the present theory with the usual image interaction. For this purpose we consider the case where the surface charge density of plate 1 is always zero $\psi_{o1}^{(0)}=0$ and $\kappa a \rightarrow 0$, since the usual image interaction refers to a point charge interacting with a uncharged plate. In the case of $\psi_{o1}^{(0)}=0$, the interaction energy (Eq. (14.71) becomes

$$
\begin{aligned}
V^{\sigma}(H) = {} & 4\varepsilon_r\varepsilon_o\psi_{o2}^2\kappa a_2^2\, e^{2\kappa a_2}\gamma_{00} \\
& +4\varepsilon_r\varepsilon_o\psi_{o2}^2\kappa a_2^2\, e^{2\kappa a_2}\sum_{n=0}^{\infty}(2n+1)\gamma_{0n}\gamma_{n0}H_n(2)+\cdots \\
& +4\varepsilon_r\varepsilon_o\psi_{o2}^2\kappa a_2^2\, e^{2\kappa a_2}\sum_{n_1=0}^{\infty}\sum_{n_2=0}^{\infty}\cdots\sum_{n_\nu=0}^{\infty}(2n_1+1)(2n_2+1)\cdots(2n_\nu+1) \\
& \times\gamma_{n_1n_2}\gamma_{n_2n_3}\cdots\gamma_{n_{\nu-1}n_\nu}\gamma_{0n_1}\gamma_{n_\nu 0}H_{n_1}(2)H_{n_2}(2)\cdots H_{n_\nu}(2)+\cdots
\end{aligned}
\tag{14.75}
$$

In the limit $\kappa a \rightarrow 0$, all the terms on the right-hand side except the first term vanish and γ_{00} tends to

$$
\gamma_{00} \rightarrow \frac{\pi}{4\kappa H}\frac{\varepsilon_r-\varepsilon_{p1}}{\varepsilon_r+\varepsilon_{p1}} \tag{14.76}
$$

We introduce the total charge $Q \equiv 4\pi a^2\sigma$ on sphere 2, which is related to the unperturbed surface potential ψ_{o2} of sphere 2 by

$$
\psi_{o2} = \frac{Q}{4\pi\varepsilon_r\varepsilon_o a_2(1+\kappa a_2)} \tag{14.77}
$$

Then Eq. (14.75) tends to

$$
V^{\sigma}(H) = \frac{Q^2 e^{-\kappa H}}{16\pi\varepsilon_r\varepsilon_o H}\frac{\varepsilon_r-\varepsilon_{p1}}{\varepsilon_r+\varepsilon_{p1}} \tag{14.78}
$$

This is the screened image interaction between a point charge and an uncharged plate, both immersed in an electrolyte solution of Debye–Hückel parameter κ. Further, in the absence of electrolytes ($\kappa \rightarrow 0$), Eq. (14.78) becomes

$$
V^{\sigma}(H) = \frac{Q^2}{16\pi\varepsilon_r\varepsilon_o H}\frac{\varepsilon_r-\varepsilon_{p1}}{\varepsilon_r+\varepsilon_{p1}} \tag{14.79}
$$

which is the usual image interaction energy [23].

We can thus conclude that Eq. (14.75) is a generalization of the usual image interaction of point charge to a colloidal particle of finite size (Fig. 14.9).

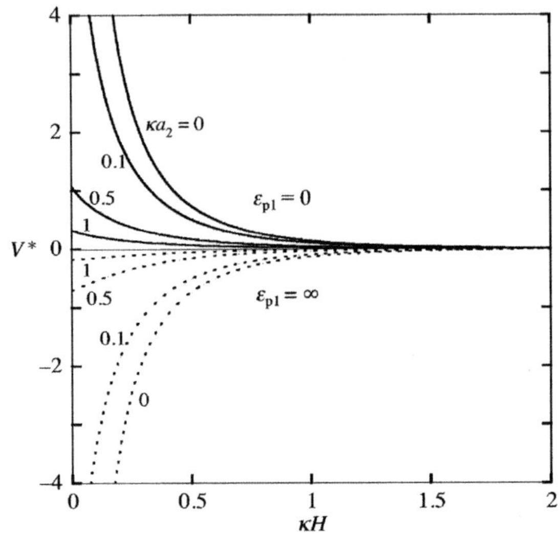

FIGURE 14.10 Reduced potential energy $V^* = 16\pi\varepsilon_r\varepsilon_o V/\kappa Q^2$ of the image interaction between a hard plate (plate 1) and a hard sphere (sphere 2) of radius a_2 with $\varepsilon_2 = 0$ as a function of κH for several values of the reduced radius κa_2 of sphere 2. Solid lines: $\varepsilon_1 = 0$; dashed lines: $\varepsilon_1 = \infty$ (plate 1 is a metal). From Ref. [14].

14.5 TWO PARALLEL CYLINDERS

Similarly, series expansion representations for the double-layer interaction between two parallel cylinders can be obtained [13, 14]. The results are given below.

Case (a): cylinders 1 and 2 both at constant surface potential
Consider a cylinder of radius a_1 carrying a constant surface potential ψ_{o1} (cylinder 1) and a cylinder of radius a_2 carrying a constant surface potential ψ_{o2} (cylinder 2), separated by a distance R between their axes, immersed in an electrolyte solution (Fig. 14.11). The interaction energy $V^{\psi}(R)$ between cylinders 1 and 2 per unit length is given by [13]

$$V^{\psi}(R) = 2\pi\varepsilon_r\varepsilon_o\psi_{o1}\psi_{o2}\frac{K_0(\kappa R)}{K_0(\kappa a_1)K_0(\kappa a_2)}$$

$$+\pi\varepsilon_r\varepsilon_o\psi_{o1}^2\frac{1}{K_0^2(\kappa a_1)}\sum_{n=-\infty}^{\infty}G_n(2)K_n^2(\kappa R) + \pi\varepsilon_r\varepsilon_o\psi_{o2}^2\frac{1}{K_0^2(\kappa a_2)}\sum_{n=-\infty}^{\infty}G_n(1)K_n^2(\kappa R)$$

$$+2\pi\varepsilon_r\varepsilon_o\psi_{o1}\psi_{o2}\frac{1}{K_0(\kappa a_1)K_0(\kappa a_2)}$$

$$
\begin{aligned}
&\times \sum_{n=-\infty}^{\infty} \sum_{m=-\infty}^{\infty} G_n(2)G_m(1)K_n(\kappa R)K_{n+m}(\kappa R)K_m(\kappa R) \\
&+\cdots+\pi\varepsilon_r\varepsilon_o\psi_{o1}\psi_{o2}\frac{1}{K_0(\kappa a_1)K_0(\kappa a_2)} \\
&\times \sum_{n_1=-\infty}^{\infty} \sum_{n_2=-\infty}^{\infty} \cdots \sum_{n_{2\nu}=-\infty}^{\infty} \{L_{21}(n_1,n_2,\ldots,n_{2\nu})+L_{12}(n_1,n_2,\ldots,n_{2\nu})\} \\
&\times K_{n_1}(\kappa R)K_{n_{2\nu}}(\kappa R) \\
&+\pi\varepsilon_r\varepsilon_o \sum_{n_1=-\infty}^{\infty} \sum_{n_2=-\infty}^{\infty} \cdots \sum_{n_{2\nu-1}=-\infty}^{\infty} \left[\frac{\psi_{o1}^2}{K_0^2(\kappa a_1)}L_{21}(n_1,n_2,\ldots,n_{2\nu-2})G_{2\nu-1}(2)\right. \\
&\left.+\frac{\psi_{o2}^2}{K_0^2(\kappa a_2)}L_{12}(n_1,n_2,\ldots,n_{2\nu-2})G_{2\nu-1}(1)\right] \\
&\times K_{n_1}(\kappa R)K_{n_{2\nu-2}+n_{2\nu-1}}(\kappa R)K_{n_{2\nu-1}}(\kappa R)+\cdots
\end{aligned}
\tag{14.80}
$$

with

$$
\begin{aligned}
L_{21}(n_1,n_2,\ldots,n_{2\nu-2}) &= K_{n_1+n_2}(\kappa R)K_{n_2+n_3}(\kappa R)\times\cdots \\
&\times K_{n_{2\nu-2}+n_{2\nu-1}}(\kappa R)K_{n_{2\nu-1}+n_{2\nu}}(\kappa R) \\
&\times G_{n_1}(2)G_{n_2}(1)\times\cdots\times G_{n_\nu-1}(2)G_{n_{2\nu}}(1)
\end{aligned}
\tag{14.81}
$$

$$
\begin{aligned}
L_{12}(n_1,n_2,\ldots,n_{2\nu-2}) &= K_{n_1+n_2}(\kappa R)K_{n_2+n_3}(\kappa R)\times\cdots \\
&\times K_{n_{2\nu-2}+n_{2\nu-1}}(\kappa R)K_{n_{2\nu-1}+n_{2\nu}}(\kappa R) \\
&\times G_{n_1}(1)G_{n_2}(2)\times\cdots\times G_{n_\nu-1}(1)G_{n_{2\nu}}(2)
\end{aligned}
\tag{14.82}
$$

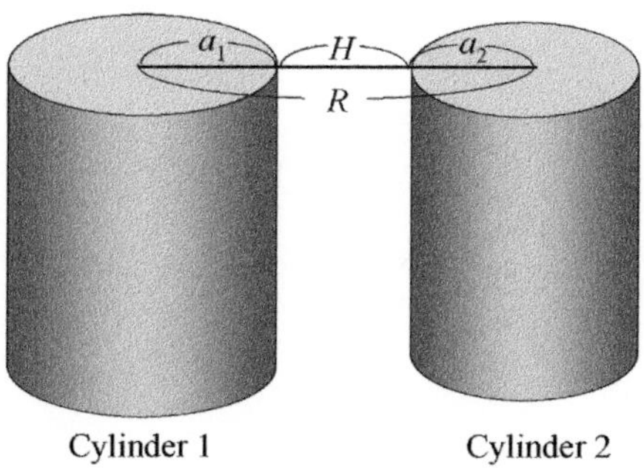

FIGURE 14.11 Interaction between two infinitely long charged hard cylinders 1 and 2 of radii a_1 and a_2 at a separation R between their axes. H $(=R-a_1-a_2)$ is the closest distance between their surfaces.

$$G_n(i) = -\frac{I_n(\kappa a_i)}{K_n(\kappa a_i)}, \quad (i = 1,2) \tag{14.83}$$

The leading term of $V^{\psi}(R)$ and $V^{\sigma}(R)$ agrees with Eq. (11.82) obtained by the linear superposition approximation.

Case (b): cylinders 1 and 2 both at constant surface charge density
It can be shown that the interaction energy $V^{\sigma}(R)$ per unit length between cylinder 1 (of relative permittivity ε_{p1}) and cylinder 2 (of relative permittivity ε_{p2}) at constant surface charge density is obtained by the interchange $G_n(i) \leftrightarrow H_n(i)$ with the result that

$$\begin{aligned}
V^{\sigma}(R) &= 2\pi\varepsilon_r\varepsilon_o\psi_{o1}\psi_{o2}\frac{K_0(\kappa R)}{K_0(\kappa a_1)K_0(\kappa a_2)} \\
&+ \pi\varepsilon_r\varepsilon_o\psi_{o1}^2\frac{1}{K_0^2(\kappa a_1)}\sum_{n=-\infty}^{\infty}H_n(2)K_n^2(\kappa R) + \pi\varepsilon_r\varepsilon_o\psi_{o2}^2\frac{1}{K_0^2(\kappa a_2)} \\
&\times \sum_{n=-\infty}^{\infty}H_n(1)K_n^2(\kappa R) \\
&+ 2\pi\varepsilon_r\varepsilon_o\psi_{o1}\psi_{o2}\frac{1}{K_0(\kappa a_1)K_0(\kappa a_2)} \\
&\times \sum_{n=-\infty}^{\infty}\sum_{m=-\infty}^{\infty}H_n(2)H_m(1)K_n(\kappa R)K_{n+m}(\kappa R)K_m(\kappa R) \\
&+ \cdots + \pi\varepsilon_r\varepsilon_o\psi_{o1}\psi_{o2}\frac{1}{K_0(\kappa a_1)K_0(\kappa a_2)} \\
&\times \sum_{n_1=-\infty}^{\infty}\sum_{n_2=-\infty}^{\infty}\cdots\sum_{n_{2v}=-\infty}^{\infty}\{M_{21}(n_1,n_2,\ldots,n_{2v}) + M_{12}(n_1,n_2,\ldots,n_{2v})\} \\
&\times K_{n_1}(\kappa R)K_{n_{2v}}(\kappa R) \\
&+ \pi\varepsilon_r\varepsilon_o\sum_{n_1=-\infty}^{\infty}\sum_{n_2=-\infty}^{\infty}\cdots\sum_{n_{2v-1}=-\infty}^{\infty}\left[\frac{\psi_{o1}^2}{K_0^2(\kappa a_1)}M_{21}(n_1,n_2,\ldots,n_{2v-2})H_{2v-1}(2)\right. \\
&\left.+ \frac{\psi_{o2}^2}{K_0^2(\kappa a_2)}M_{12}(n_1,n_2,\ldots,n_{2v-2})H_{2v-1}(1)\right] \\
&\times K_{n_1}(\kappa R)K_{n_{2v-2}+n_{2v-1}}(\kappa R)K_{n_{2v-1}}(\kappa R) + \cdots
\end{aligned} \tag{14.84}$$

with

$$\begin{aligned}
M_{21}(n_1,n_2,\ldots,n_{2v-2}) = {}& K_{n_1+n_2}(\kappa R)K_{n_2+n_3}(\kappa R) \times \cdots \\
&\times K_{n_{2v-2}+n_{2v-1}}(\kappa R)K_{n_{2v-1}+n_{2v}}(\kappa R) \\
&\times H_{n_1}(2)H_{n_2}(1) \times \cdots \times H_{n_v-1}(2)H_{n_{2v}}(1)
\end{aligned} \tag{14.85}$$

$$
\begin{aligned}
M_{12}(n_1, n_2, \ldots, n_{2\nu-2}) = {} & K_{n_1+n_2}(\kappa R)K_{n_2+n_3}(\kappa R) \times \cdots \\
& \times K_{n_{2\nu-2}+n_{2\nu-1}}(\kappa R)K_{n_{2\nu-1}+n_{2\nu}}(\kappa R) \\
& \times H_{n_1}(1)H_{n_2}(2) \times \cdots \times H_{n_\nu-1}(1)H_{n_{2\nu}}(2)
\end{aligned} \tag{14.86}
$$

where

$$
H_n(i) = -\frac{I'_n(\kappa a_i) - (\varepsilon_{\mathrm{p}i}|n|/\varepsilon_{\mathrm{r}}\kappa a_i)I_n(\kappa a_i)}{K'_n(\kappa a_i) - (\varepsilon_{\mathrm{p}i}|n|/\varepsilon_{\mathrm{r}}\kappa a_i)K_n(\kappa a_i)}, \quad (i = 1, 2) \tag{14.87}
$$

Case (c): cylinder 1 maintained at constant surface potential and cylinder 2 at constant surface charge density

It can also be shown that when cylinder 1 has a constant surface potential and cylinder 2 has a constant surface charge density, the interaction energy $V^{\psi-\sigma}$ between plates 1 and 2 per unit length s given by [13]

$$
\begin{aligned}
V^{\psi-\sigma}(R) = {} & 2\pi\varepsilon_{\mathrm{r}}\varepsilon_o\psi_{\mathrm{o}1}\psi_{\mathrm{o}2}\frac{K_0(\kappa R)}{K_0(\kappa a_1)K_0(\kappa a_2)} \\
& + \pi\varepsilon_{\mathrm{r}}\varepsilon_{\mathrm{o}}\psi_{\mathrm{o}1}^2\frac{1}{K_0^2(\kappa a_1)}\sum_{n=-\infty}^{\infty} H_n(2)K_n^2(\kappa R) + \pi\varepsilon_{\mathrm{r}}\varepsilon_{\mathrm{o}}\psi_{o2}^2\frac{1}{K_0^2(\kappa a_2)} \\
& \times \sum_{n=-\infty}^{\infty} G_n(1)K_n^2(\kappa R) \\
& + 2\pi\varepsilon_{\mathrm{r}}\varepsilon_{\mathrm{o}}\psi_{\mathrm{o}1}\psi_{\mathrm{o}2}\frac{1}{K_0(\kappa a_1)K_0(\kappa a_2)}\sum_{n=-\infty}^{\infty}\sum_{m=-\infty}^{\infty} H_n(2)G_m(1)K_n(\kappa R) \\
& \times K_{n+m}(\kappa R)K_m(\kappa R) \\
& + \cdots + \pi\varepsilon_{\mathrm{r}}\varepsilon_{\mathrm{o}}\psi_{\mathrm{o}1}\psi_{\mathrm{o}2}\frac{1}{K_0(\kappa a_1)K_0(\kappa a_2)} \\
& \times \sum_{n_1=-\infty}^{\infty}\sum_{n_2=-\infty}^{\infty}\cdots\sum_{n_{2\nu}=-\infty}^{\infty}\{N_{21}(n_1, n_2, \ldots, n_{2\nu}) + N_{12}(n_1, n_2, \ldots, n_{2\nu})\} \\
& \times K_{n_1}(\kappa R)K_{n_{2\nu}}(\kappa R) \\
& + \pi\varepsilon_{\mathrm{r}}\varepsilon_{\mathrm{o}}\sum_{n_1=-\infty}^{\infty}\sum_{n_2=-\infty}^{\infty}\cdots\sum_{n_{2\nu-1}=-\infty}^{\infty}\left[\frac{\psi_{\mathrm{o}1}^2}{K_0^2(\kappa a_1)}N_{21}(n_1, n_2, \ldots, n_{2\nu-2})H_{2\nu-1}(2)\right. \\
& \left. + \frac{\psi_{\mathrm{o}2}^2}{K_0^2(\kappa a_2)}N_{12}(n_1, n_2, \ldots, n_{2\nu-2})G_{2\nu-1}(1)\right] \\
& \times K_{n_1}(\kappa R)K_{n_{2\nu-2}+n_{2\nu-1}}(\kappa R)K_{n_{2\nu-1}}(\kappa R) + \cdots
\end{aligned} \tag{14.88}
$$

with

$$\begin{aligned} N_{21}(n_1, n_2, \ldots, n_{2\nu-2}) = & K_{n_1+n_2}(\kappa R) K_{n_2+n_3}(\kappa R) \times \cdots \\ & \times K_{n_{2\nu-2}+n_{2\nu-1}}(\kappa R) K_{n_{2\nu-1}+n_{2\nu}}(\kappa R) \\ & \times H_{n_1}(2) G_{n_2}(1) \times \cdots \times H_{n_\nu-1}(2) G_{n_{2\nu}}(1) \end{aligned} \tag{14.89}$$

$$\begin{aligned} N_{12}(n_1, n_2, \ldots, n_{2\nu-2}) = & K_{n_1+n_2}(\kappa R) K_{n_2+n_3}(\kappa R) \times \cdots \\ & \times K_{n_{2\nu-2}+n_{2\nu-1}}(\kappa R) K_{n_{2\nu-1}+n_{2\nu}}(\kappa R) \\ & \times G_{n_1}(1) H_{n_2}(2) \times \cdots \times G_{n_\nu-1}(1) H_{n_{2\nu}}(2) \end{aligned} \tag{14.90}$$

Consider the case where $\kappa a_1 \gg 1$ and $\kappa a_2 \gg 1$. For the constant surface potential case, V^{ψ} becomes

$$\begin{aligned} V^{\psi}(R) = & 2\sqrt{2\pi}\varepsilon_r\varepsilon_o\psi_{o1}\psi_{o2}\sqrt{\kappa a_1 a_2}\frac{e^{-\kappa H}}{\sqrt{R}} \\ & -\sqrt{\pi}\varepsilon_r\varepsilon_o\frac{e^{-2\kappa H}}{R}\left[\psi_{o1}^2 a_1\sqrt{\frac{\kappa a_2 R}{R-a_2}} + \psi_{o2}^2 a_2\sqrt{\frac{\kappa a_1 R}{R-a_1}}\right] + O(e^{-3\kappa H}) \end{aligned} \tag{14.91}$$

For the constant surface charge density case, if ε_{p1} and ε_{p2} are finite, we obtain

$$\begin{aligned} V^{\sigma}(R) = & 2\sqrt{2\pi}\varepsilon_r\varepsilon_o\psi_{o1}\psi_{o2}\sqrt{\kappa a_1 a_2}\frac{e^{-\kappa H}}{\sqrt{R}} \\ & +\sqrt{\pi}\varepsilon_r\varepsilon_o\frac{e^{-2\kappa H}}{R}\left[\psi_{o1}^2 a_1\sqrt{\frac{\kappa a_2 R}{R-a_2}}\left\{1-\frac{2\varepsilon_{p2}}{\varepsilon_r}\sqrt{\frac{R}{\pi\kappa a_2(R-a_2)}}\right\}\right. \\ & \left. +\psi_{o2}^2 a_2\sqrt{\frac{\kappa a_1 R}{R-a_1}}\left\{1-\frac{2\varepsilon_{p1}}{\varepsilon_r}\sqrt{\frac{R}{\pi\kappa a_1(R-a_1)}}\right\}\right] + O(e^{-3\kappa H}) \end{aligned} \tag{14.92}$$

For the mixed case,

$$\begin{aligned} V^{\psi-\sigma}(R) = & 2\sqrt{2\pi}\varepsilon_r\varepsilon_o\psi_{o1}\psi_{o2}\sqrt{\kappa a_1 a_2}\frac{e^{-\kappa H}}{\sqrt{R}} \\ & +\sqrt{\pi}\varepsilon_r\varepsilon_o\frac{e^{-2\kappa H}}{R}\left[\psi_{o1}^2 a_1\sqrt{\frac{\kappa a_2 R}{R-a_2}}\left\{1-\frac{2\varepsilon_{p2}}{\varepsilon_r}\sqrt{\frac{R}{\pi\kappa a_2(R-a_2)}}\right\}\right. \\ & \left. -\psi_{o2}^2 a_2\sqrt{\frac{\kappa a_1 R}{R-a_1}}\right] + O(e^{-3\kappa H}) \end{aligned} \tag{14.93}$$

Further, if $H \ll a_1$ and $H \ll a_2$, Eqs. (14.91)–(14.93) reduce to

$$V^{\psi}(R) = 2\sqrt{2\pi}\varepsilon_r\varepsilon_o\psi_{o1}\psi_{o2}\sqrt{\frac{\kappa a_1 a_2}{a_1 + a_2}}e^{-\kappa H} - \sqrt{\pi}\varepsilon_r\varepsilon_o\sqrt{\frac{\kappa a_1 a_2}{a_1 + a_2}}e^{-2\kappa H}(\psi_{o1}^2 + \psi_{o2}^2) \tag{14.94}$$

$$\begin{aligned} V^{\sigma}(R) = {} & 2\sqrt{2\pi}\varepsilon_r\varepsilon_o\psi_{o1}\psi_{o2}\sqrt{\frac{\kappa a_1 a_2}{a_1 + a_2}}e^{-\kappa H} \\ & + \sqrt{\pi}\varepsilon_r\varepsilon_o\sqrt{\frac{\kappa a_1 a_2}{a_1 + a_2}}e^{-2\kappa H}\left[\psi_{o1}^2\left(1 - \frac{2}{\sqrt{\pi}}\frac{\varepsilon_{p2}}{\varepsilon_r}\sqrt{\frac{1}{\kappa a_1} + \frac{1}{\kappa a_2}}\right)\right. \\ & \left. + \psi_{o2}^2\left(1 - \frac{2}{\sqrt{\pi}}\frac{\varepsilon_{p1}}{\varepsilon_r}\sqrt{\frac{1}{\kappa a_1} + \frac{1}{\kappa a_2}}\right)\right] \end{aligned} \tag{14.95}$$

$$\begin{aligned} V^{\phi-\sigma}(R) = {} & 2\sqrt{2\pi}\varepsilon_r\varepsilon_o\psi_{o1}\psi_{o2}\sqrt{\frac{\kappa a_1 a_2}{a_1 + a_2}}e^{-\kappa H} \\ & + \sqrt{\pi}\varepsilon_r\varepsilon_o\sqrt{\frac{\kappa a_1 a_2}{a_1 + a_2}}e^{-2\kappa H}\left[\psi_{o1}^2\left(1 - \frac{2}{\sqrt{\pi}}\frac{\varepsilon_{p2}}{\varepsilon_r}\sqrt{\frac{1}{\kappa a_1} + \frac{1}{\kappa a_2}}\right) - \psi_{o2}^2\right] \end{aligned} \tag{14.96}$$

Equations (14.94)–(14.96) are consistent with the results obtained via Derjaguin's approximation. Equation (14.94) shows that the next-order curvature correction to Derjaguin's approximation is of the order of $1/\sqrt{\kappa a_i}$ $(i = 1, 2)$, as in the case of sphere–sphere interaction.

14.6 PLATE AND CYLINDER

One can also obtain the interaction energy between a cylinder and a hard plate, both having constant surface potential for the case where the cylinder axis is perpendicular to the plate surface (Fig. 14.12). The result is [14]

$$\begin{aligned} V^{\psi}(H) = {} & 2\pi\varepsilon_r\varepsilon_o\psi_{o1}\psi_{o2}\frac{e^{-\kappa(H+a_2)}}{K_0(\kappa a_2)} + \pi\varepsilon_r\varepsilon_o\psi_{o1}^2\, e^{-2\kappa(H+a_2)}\sum_{n=-\infty}^{\infty}G_n(2) \\ & - \pi\varepsilon_r\varepsilon_o\psi_{o2}^2\frac{\beta_{00}}{\{K_0(\kappa a_2)\}^2} - \pi\varepsilon_r\varepsilon_o\psi_{o1}\psi_{o2}\frac{e^{-\kappa(H+a_2)}}{K_0(\kappa a_2)}\sum_{n=-\infty}^{\infty}(\beta_{0n} + \beta_{n0})G_n(2) \\ & - \pi\varepsilon_r\varepsilon_o\psi_{o1}^2\, e^{-2\kappa(H+a_2)}\sum_{n=-\infty}^{\infty}\sum_{n=-\infty}^{\infty}\beta_{nm}G_n(2)G_m(2) \end{aligned}$$

$$
\begin{aligned}
&+\pi\varepsilon_r\varepsilon_o\psi_{o2}^2\frac{1}{\{K_0(\kappa a_2)\}^2}\sum_{n=-\infty}^{\infty}\beta_{0n}\beta_{n0}G_n(2)\\
&+\pi\varepsilon_r\varepsilon_o\psi_{o1}\psi_{o2}\frac{e^{-\kappa(H+a_2)}}{K_0(\kappa a_2)}\sum_{n=-\infty}^{\infty}\sum_{n=-\infty}^{\infty}\beta_{nm}(\beta_{0n}+\beta_{n0})G_n(2)G_m(2)+\cdots\\
&-\pi\varepsilon_r\varepsilon_o\psi_{o1}\psi_{o2}\frac{e^{-\kappa(H+a_2)}}{K_0(\kappa a_2)}\sum_{n_1=-\infty}^{\infty}\sum_{n_2=-\infty}^{\infty}\cdots\sum_{n_\nu=-\infty}^{\infty}(-1)^{\nu-1}\\
&\times\beta_{n_1n_2}\beta_{n_2n_3}\cdots\beta_{n_{\nu-1}n_\nu}(\beta_{0n_1}+\beta_{n_\nu 0})G_{n_1}(2)G_{n_2}(2)\cdots G_{n_\nu}(2)\\
&+\pi\varepsilon_r\varepsilon_o\psi_{o1}^2\,e^{-2\kappa(H+a_2)}\sum_{n_1=-\infty}^{\infty}\sum_{n_2=-\infty}^{\infty}\cdots\sum_{n_\nu=-\infty}^{\infty}(-1)^{\nu-1}\\
&\times\beta_{n_1n_2}\beta_{n_2n_3}\times\cdots\times\beta_{n_{\nu-1}n_\nu}G_{n_1}(2)G_{n_2}(2)\cdots G_{n_\nu}(2)\\
&+\pi\varepsilon_r\varepsilon_o\psi_{o2}^2\frac{1}{\{K_0(\kappa a_2)\}^2}\sum_{n_1=-\infty}^{\infty}\sum_{n_2=-\infty}^{\infty}\cdots\sum_{n_\nu=-\infty}^{\infty}(-1)^{\nu-1}\\
&\times\beta_{n_1n_2}\beta_{n_2n_3}\cdots\beta_{n_{\nu-1}n_\nu}\beta_{0n_1}\beta_{n_\nu 0}G_{n_1}(2)G_{n_2}(2)\cdots G_{n_\nu}(2)+\cdots
\end{aligned}
\tag{14.97}
$$

with

$$
\begin{aligned}
\beta_{nm} &= \int_0^{\infty}\frac{\exp[-2\sqrt{k^2+\kappa^2}(H+a_2)]}{\sqrt{k^2+\kappa^2}}T_n\left(\frac{\sqrt{k^2+\kappa^2}}{\kappa}\right)T_m\left(\frac{\sqrt{k^2+\kappa^2}}{\kappa}\right)dk\\
&= \int_1^{\infty}\frac{\exp[-2\kappa t(H+a_2)]}{\sqrt{t^2-1}}T_n(t)T_m(t)dt
\end{aligned}
\tag{14.98}
$$

where $T_n(x)$ is the nth-order Tchebycheff's polynomial.

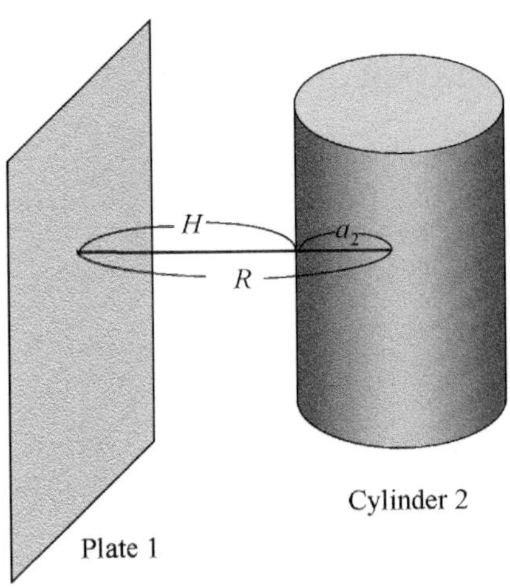

FIGURE 14.12 Interaction between a hard plate 1 and an infinitely long charged hard cylinder 2 of radius a_2 at a separation R. H $(=R-a_2)$ is the closest distance between their surfaces.

When plate 1 and cylinder 2 have constant surface charge densities, the interaction energy V^{σ} is obtained by Eq. (18.14) with $G_n(2)$ and β_{nm} replaced by $H_n(2)$ and $-\gamma_{nm}$, respectively, where γ_{nm} is defined by

$$\gamma_{nm} = -\int_0^{\infty} \frac{\varepsilon_1 k - \varepsilon\sqrt{k^2+\kappa^2}}{\varepsilon_1 k + \varepsilon\sqrt{k^2+\kappa^2}} \frac{\exp[-2\sqrt{k^2+\kappa^2}(H+a_2)]}{\sqrt{k^2+\kappa^2}}$$

$$\times T_n\left(\frac{\sqrt{k^2+\kappa^2}}{\kappa}\right) T_m\left(\frac{\sqrt{k^2+\kappa^2}}{\kappa}\right) dk \tag{14.99}$$

$$= \int_1^{\infty} \frac{\varepsilon t - \varepsilon_1\sqrt{t^2-1}}{\varepsilon t + \varepsilon_1\sqrt{t^2-1}} \frac{\exp[-2\kappa t(H+a_2)]}{\sqrt{t^2-1}} T_n(t) T_m(t) dt$$

When plate 1 has a constant surface potential and cylinder 2 has a constant surface charge density, the interaction energy $V^{\psi-\sigma}$ is given by Eq. (14.97) with $G_n(2)$ replaced by $H_n(2)$. When plate 1 has a constant surface charge density and cylinder 2 has constant surface potential, the interaction energy is given by Eq. (14.97) with β_{nm} replaced by $-\gamma_{nm}$.

Finally, we compare the image interactions between a hard cylinder and a hard plate with the usual image interaction between a line charge and a plate by taking the limit of $\kappa a_2 \to 0$ for the case where the surface charge density of plate 1 is always zero ($\psi_{o1} = 0$). In this limit, we have

$$V^{\sigma}(H) = \begin{cases} \dfrac{Q^2}{4\pi\varepsilon_r\varepsilon_o} K_0(2\kappa H), & (\varepsilon_{p1} = 0) \\ -\dfrac{Q^2}{4\pi\varepsilon_r\varepsilon_o} K_0(2\kappa H), & (\varepsilon_{p1} = 0) \end{cases} \tag{14.100}$$

where we have introduced the total charge $Q \equiv 2\pi a\sigma$ on cylinder 2 per unit length (i.e., the line charge density of cylinder 2), which is related to the unperturbed surface potential ψ_{o2} of cylinder 2 by

$$\psi_{o2} = \frac{Q}{2\pi\varepsilon_r\varepsilon_o\kappa a_2} \frac{K_0(\kappa a_2)}{K_1(\kappa a_2)} \tag{14.101}$$

Equation (14.101) is the screened image interaction between a line charge and an uncharged plate, both immersed in an electrolyte solution of Debye–Hückel parameter κ. We see that in the former case ($\varepsilon_{p1} = 0$), the interaction force is repulsion and the latter case ($\varepsilon_{p1} = \infty$) attraction. Further, in the absence of electrolytes ($\kappa \to 0$), we can show from Eq. (14.101) that the interaction force $-\partial V^{\sigma}/\partial H$ per unit length between plate 1 and cylinder 2 with $a_2 \to 0$ is given by

$$-\frac{\partial V^{\sigma}}{\partial H} = \frac{Q^2}{4\pi\varepsilon\varepsilon_o}\left(\frac{\varepsilon_r - \varepsilon_{p1}}{\varepsilon_r + \varepsilon_{p1}}\right)\frac{1}{H} \tag{14.102}$$

which exactly agrees with usual image force per unit length between a line charge and an uncharged plate [23].

REFERENCES

1. V. I. Smirnov, *Course of Higher Mathematics*, Vol. 4, Pergamon, Oxford, 1958, Chapter 4.
2. B. V. Derjaguin, *Kolloid Z.* 69 (1934) 155.
3. H. Ohshima, *Theory of Colloid and Interfacial Electric Phenomena*, Elsevier/Academic Press, Amsterdam, 2006.
4. H. Ohshima and T. Kondo, *J. Colloid Interface Sci.* 157 (1993) 504.
5. H. Ohshima and T. Kondo, *Colloid Polym. Sci.* 271 (1993) 1191.
6. H. Ohshima, *J. Colloid Interface Sci.* 162 (1994) 487.
7. H. Ohshima, *J. Colloid Interface Sci.* 168 (1994) 255.
8. H. Ohshima, *Adv. Colloid Interface Sci.* 53 (1994) 77.
9. H. Ohshima, *J. Colloid Interface Sci.* 170 (1995) 432.
10. H. Ohshima, *J. Colloid Interface Sci.* 176 (1995) 7.
11. H. Ohshima, E. Mishonova, and E. Alexov, *Biophys. Chem.* 57 (1996) 189.
12. H. Ohshima, *J. Colloid Interface Sci.* 198 (1998) 42.
13. H. Ohshima, *Colloid Polym. Sci.* 274 (1996) 1176.
14. H. Ohshima, *Colloid Polym. Sci* 277 (1999) 563.
15. H. Ohshima, T. Kondo, *J. Colloid Interface Sci.* 155 (1993) 499.
16. M. Abramowitz and I. A. Stegun, *Handbook of Mathematical Functions*, Dover, New York, 1965, p. 445.
17. D. Langbein, *Theory of van der Waals Attraction*, Springer, Berlin, 1974, p. 82.
18. R. Hogg, T. W. Healy, and D. W. Fuerstenau, *Trans. Faraday Soc.* 62 (1966) 1638.
19. G. R. Wiese and T. W. Healy, *Trans. Faraday Soc.* 66 (1970) 490.
20. S. S. Dukhin and J. Lyklema, *Colloid J. USSR Eng. Trans.* 51 (1989) 244.
21. J. Kijlstra, *J. Colloid Interface Sci.* 153 (1992) 30.
22. G. Kar, S. Chander, and T. S. Mika, *J. Colloid Interface Sci.* 44 (1973) 347.
23. L. D. Landau and E. M. Lifshitz, *Electrodynamics in Continuous Media*, Pergamon, New York, 1963.

15 Electrostatic Interaction Between Soft Particles

15.1 INTRODUCTION

In this chapter, we give approximate analytic expressions for the force and potential energy of the electrical double-layer interaction two soft particles. As shown in Fig. 15.1, a spherical soft particle becomes a hard sphere without surface structures, while a soft particle tends to a spherical polyelectrolyte when the particle core is absent. Expressions for the interaction force and energy between two soft particles thus cover various limiting cases that include hard particle/hard particle interaction, soft particle/hard particle interaction, soft particle/porous particle interaction, and porous particle/porous particle interaction.

15.2 INTERACTION BETWEEN TWO PARALLEL DISSIMILAR SOFT PLATES

Consider two parallel dissimilar soft plates 1 and 2 at a separation h between their surfaces immersed in an electrolyte solution containing N ionic species with valence z_i and bulk concentration (number density) n_i^∞ ($i = 1, 2, \ldots, N$) [1]. We assume that each soft plate consists of a core and an ion-penetrable surface charge layer of polyelectrolytes covering the plate core and that there is no electric field within the plate core. We denote by d_1 and d_2 the thicknesses of the surface charge layers of plates 1 and 2, respectively. The x-axis is taken to be perpendicular to the plates with the origin at the boundary between the surface charge layer of plate 1 and the solution, as shown in Fig. 15.2. We assume that each surface layer is uniformly charged. Let Z_1 and N_1, respectively, be the valence and the density of fixed-charge groups contained in the surface layer of plate 1, and let Z_2 and N_2 be the corresponding quantities for plate 2. Thus, the charge densities $\rho_{\text{fix}1}$ and $\rho_{\text{fix}2}$ of the surface charge layers of plates 1 and 2 are, respectively, given by

$$\rho_{\text{fix}1} = Z_1 e N_1 \tag{15.1}$$

$$\rho_{\text{fix}2} = Z_2 e N_2 \tag{15.2}$$

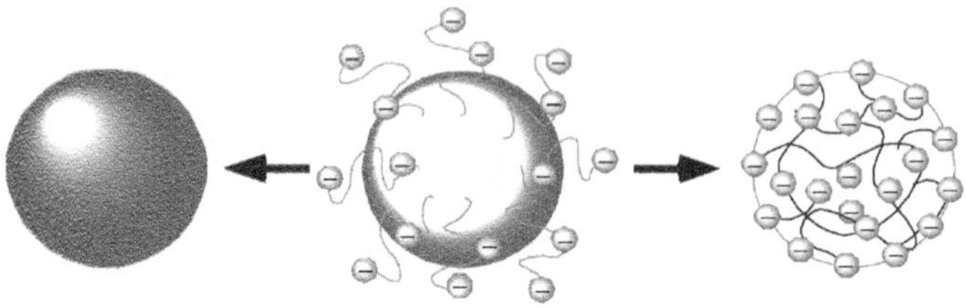

FIGURE 15.1 A soft sphere becomes a hard sphere in the absence of the surface layer of polyelectrolyte while it tends to a spherical polyelectrolyte (i.e., porous sphere) when the particle core is absent.

The Poisson–Boltzmann equations for the present system are then

$$\frac{d^2\psi}{dx^2} = -\frac{1}{\varepsilon_r\varepsilon_o}\sum_{i=1}^{M} z_i e n_i^{\infty} \exp\left(-\frac{z_i e\psi}{kT}\right) - \frac{\rho_{\text{fix1}}}{\varepsilon_r\varepsilon_o}, \qquad -d_1 < x < 0 \tag{15.3}$$

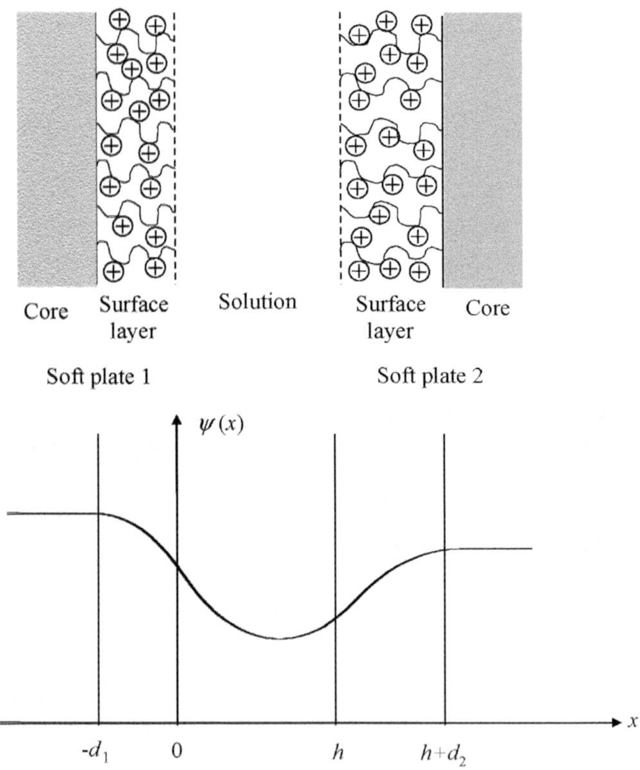

FIGURE 15.2 Interaction between two parallel soft plates 1 and 2 at separation h and the potential distribution $\psi(x)$ across plates 1 and 2, which are covered with surface charge layers of thicknesses d_1 and d_2, respectively.

$$\frac{d^2\psi}{dx^2} = -\frac{1}{\varepsilon_r\varepsilon_o}\sum_{i=1}^{M} z_i e n_i^{\infty} \exp\left(-\frac{z_i e\psi}{kT}\right), \quad 0 < x < h \tag{15.4}$$

$$\frac{d^2\psi}{dx^2} = -\frac{1}{\varepsilon_r\varepsilon_o}\sum_{i=1}^{M} z_i e n_i^{\infty} \exp\left(-\frac{z_i e\psi}{kT}\right) - \frac{\rho_{\rm fix2}}{\varepsilon_r\varepsilon_o}, \quad h < x < h + d_2 \tag{15.5}$$

where $\psi(x)$ is the electric potential at position x relative to the one at a point in the bulk solution far from the plates (the plates are actually surrounded by the electrolyte solution). We assume that the relative permittivity ε_r in the surface layers takes the same value as that in the electrolyte solution.

We consider the case where N_1 and N_2 are low. The Poisson–Boltzmann equations (15.3)–(15.5) can be linearized to give

$$\frac{d^2\psi}{dx^2} = \kappa^2\psi - \frac{\rho_{\rm fix1}}{\varepsilon_r\varepsilon_o}, \quad -d_1 < x < 0 \tag{15.6}$$

$$\frac{d^2\psi}{dx^2} = \kappa^2\psi, \quad 0 < x < h \tag{15.7}$$

$$\frac{d^2\psi}{dx^2} = \kappa^2\psi - \frac{\rho_{\rm fix2}}{\varepsilon_r\varepsilon_o}, \quad h < x < h + d_2 \tag{15.8}$$

where κ is the Debye–Hückel parameter (Eq. (1.10)).

The boundary conditions are

$$\psi(0^-) = \psi(0^+) \tag{15.9}$$

$$\psi(h^-) = \psi(h^+) \tag{15.10}$$

$$\left.\frac{d\psi}{dx}\right|_{x=0^-} = \left.\frac{d\psi}{dx}\right|_{x=0^+} \tag{15.11}$$

$$\left.\frac{d\psi}{dx}\right|_{x=h^-} = \left.\frac{d\psi}{dx}\right|_{x=h^+} \tag{15.12}$$

$$\left.\frac{d\psi}{dx}\right|_{x=-d_1^+} = 0 \tag{15.13}$$

$$\left.\frac{d\psi}{dx}\right|_{x=h+d_2^-} = 0 \tag{15.14}$$

Integration of Eqs. (15.6)–(15.8) with the above boundary conditions gives

$$\psi(x) = \frac{1}{\varepsilon_r\varepsilon_o\kappa^2}\left[\rho_{\rm fix1} + \frac{\{-\rho_{\rm fix1}\sinh[\kappa(h+d_2)] + \rho_{\rm fix2}\sinh(\kappa d_2)\}}{\sinh[\kappa(h+d_1+d_2)]}\cosh[\kappa(x+d_1)]\right],$$
$$-d_1 \le x \le 0 \tag{15.15}$$

$$\psi(x) = \frac{1}{\varepsilon_r\varepsilon_o\kappa^2}\left[\frac{\rho_{\text{fix1}}\sinh(\kappa d_1)\cosh[\kappa(h+d_2-x)] + \rho_{\text{fix2}}\sinh(\kappa d_2)\cosh[\kappa(x+d_1)]}{\sinh[\kappa(h+d_1+d_2)]}\right],$$

$$0 \le x \le h \tag{15.16}$$

$$\psi(x) = \frac{1}{\varepsilon_r\varepsilon_o\kappa^2}\left[\rho_{\text{fix2}} + \frac{\{-\rho_{\text{fix2}}\sinh[\kappa(h+d_1)] + \rho_{\text{fix1}}\sinh(\kappa d_1)\}}{\sinh[\kappa(h+d_1+d_2)]}\cosh[\kappa(h+d_2-x)]\right],$$

$$h \le x \le h+d_2 \tag{15.17}$$

When the potential $\psi(x)$ is low, the electrostatic force $P_{\text{pl}}(h)$ between the two parallel plates 1 and 2 at separation h per unit area can be calculated from Eq. (10.18), namely,

$$P_{\text{pl}}(h) = kT\sum_{i=1}^{N} n_i^{\infty}\left[\exp\left(-\frac{z_i e\psi(0)}{kT}\right) - 1\right] - \frac{1}{2}\varepsilon_r\varepsilon_o\left(\left.\frac{d\psi}{dx}\right|_{x=0}\right)^2 \tag{15.18}$$

which, for the low potential case, reduces to

$$P_{\text{pl}}(h) = \frac{1}{2}\varepsilon_r\varepsilon_o\left[\kappa^2\psi^2(0) - \left(\left.\frac{d\psi}{dx}\right|_{x=0}\right)^2\right] \tag{15.19}$$

The result is

$$P_{\text{pl}}(h) = \frac{1}{8\varepsilon_r\varepsilon_o\kappa^2}\left[\frac{\{\rho_{\text{fix1}}\sinh(\kappa d_1) + \rho_{\text{fix2}}\sinh(\kappa d_2)\}^2}{\sinh^2[\kappa(h+d_1+d_2)/2]} - \frac{\{\rho_{\text{fix1}}\sinh(\kappa d_1) - \rho_{\text{fix2}}\sinh(\kappa d_2)\}^2}{\cosh^2[\kappa(h+d_1+d_2)/2]}\right] \tag{15.20}$$

Integrating Eq. (15.18) with respect to h gives the potential energy $V_{\text{pl}}(h)$ of electrostatic interaction between the plates per unit area as a function of h:

$$V_{\text{pl}}(h) = \frac{1}{4\varepsilon_r\varepsilon_o\kappa^3}\left[\{\rho_{\text{fix1}}\sinh(\kappa d_1) + \rho_{\text{fix2}}\sinh(\kappa d_2)\}^2\left\{\coth\left(\frac{\kappa(h+d_1+d_2)}{2}\right) - 1\right\} - \left\{\rho_{\text{fix1}}\sinh(\kappa d_1) - \rho_{\text{fix2}}\sinh(\kappa d_2)\}^2\left\{1 - \tanh\left(\frac{\kappa(h+d_1+d_2)}{2}\right)\right\}\right] \tag{15.21}$$

For the special case of two similar soft plates carrying $Z_1 = Z_2 = Z$, $N_1 = N_2 = N$, and $d_1 = d_2 = d$ so that $\rho_{\text{fix1}} = \rho_{\text{fix2}} = \rho_{\text{fix}}$, Eqs. (15.20) and (15.21) reduce to

$$P_{\text{pl}}(h) = \frac{\rho_{\text{fix}}^2}{2\varepsilon_r\varepsilon_o\kappa^2}\frac{\sinh^2(\kappa d)}{\sinh^2[\kappa(h/2+d)]} \tag{15.22}$$

$$V_{\text{pl}}(h) = \frac{(ZeN)^2 \rightarrow \rho_{\text{fix}}^2\sinh^2(\kappa d)}{\varepsilon_r\varepsilon_o\kappa^3}\left\{\coth\left[\kappa\left(\frac{h}{2}+d\right)\right] - 1\right\} \tag{15.23}$$

When the magnitude of $\psi(x)$ is arbitrary, one must solve the original nonlinear Poisson–Boltzmann equations (15.3)–(15.5). Consider the case of two parallel similar plates in a symmetrical electrolyte with valence z and bulk concentration n. In this case, we need consider only the region $-d < x < h/2$ so that Eqs. (15.3)–(15.5) become

$$\frac{d^2y}{dx^2} = \kappa^2 \sinh y, \quad 0 < x < h/2 \tag{15.24}$$

$$\frac{d^2y}{dx^2} = \kappa^2(\sinh y - \sinh y_{DON}), \quad -d < x < 0 \tag{15.25}$$

with

$$y = \frac{ze\psi}{kT} \tag{15.26}$$

$$y_{DON} = \frac{ze\psi_{DON}}{kT} \tag{15.27}$$

where y, ψ_{DON}, and y_{DON} are, respectively, the scaled potential, the Donnan potential given by Eq. (20.18), and the scaled Donnan potential. The boundary conditions for $y(x)$ similar to Eqs. (15.9)–(15.14) are

$$\left.\frac{dy}{dx}\right|_{x=h/2} = 0 \tag{15.28}$$

$$y(0^-) = y(0^+) \tag{15.29}$$

$$\left.\frac{dy}{dx}\right|_{x=0^-} = \left.\frac{dy}{dx}\right|_{x=0^+} \tag{15.30}$$

$$\left.\frac{dy}{dx}\right|_{x=-d^+} = 0 \tag{15.31}$$

Equation (15.31) follows from the symmetry of the system. The solution to Eqs. (15.24) and (15.25) subject to the boundary conditions (15.28)–(15.29) takes a complicated form, involving numerical integration. For the case where $\kappa d \gtrsim 1$, soft plates can be approximated by porous membranes (see Chapter 13). In this case, the value of $y(h/2)$ can be calculated by solving the following coupled equations for two parallel porous membranes (see Eqs. (13.94) and (13.95)):

$$2\cosh y_{DON} + \frac{ZN_o}{zn}[y_o - y_{DON}] = 2\cosh y(h/2) \tag{15.32}$$

$$\cosh\left(\frac{y_o}{2}\right) = \cosh\left(\frac{y(h/2)}{2}\right) \cdot dc\left(\frac{\kappa h}{2} \cdot \cosh\left(\frac{y(h/2)}{2}\right) \cdot 1/\cosh\left(\frac{y(h/2)}{2}\right)\right) \tag{15.33}$$

where dc is a Jacobian elliptic function with modulus $1/\cosh(y(h/2))$ and $y_o \equiv y(0)$ is the scaled unperturbed potential at the front edge $x = 0$ of the surface layer and is given by Eq. (4.36).

The interaction force between two parallel similar plates per unit area $P_{pl}(h)$ is given by Eq. (13.96), namely,

$$P(h) = 4nkT \sinh^2\left(\frac{y_m}{2}\right) = 4nkT \sinh^2\left(\frac{y(h/2)}{2}\right) \tag{15.34}$$

A simple approximate analytic expression for $P_{pl}(h)$ can be obtained using the linear superposition approximation (LSA) (Chapter 11). In this approximation, $y(h/2)$ in Eq. (15.34) is approximated by the sum of the asymptotic values of the two scaled unperturbed potentials $y_s(x)$ that is produced by the respective plates in the absence of interaction. For two similar plates,

$$y(h/2) \approx 2y_s(h/2) \tag{15.35}$$

This approximation is correct in the limit of large κh. It follows from Eq. (1.37), the value of the unperturbed potential of a single plate at $x = h/2$ is given by

$$y_s(h/2) = 4 \operatorname{arctanh}\left[\tanh\left(\frac{y_o}{4}\right)e^{-\kappa h/2}\right] \tag{15.36}$$

where the scaled unperturbed surface potential y_o is given by the solution to Eqs. (15.32) and (15.33). Equation (15.36) becomes, for large κh,

$$y_s(h/2) \approx 4 \tanh\left(\frac{y_o}{4}\right)e^{-\kappa h/2} \tag{15.37}$$

Hence

$$y(h/2) = 8 \tanh\left(\frac{y_o}{4}\right)e^{-\kappa h/2} \tag{15.38}$$

For large κh, Eq. (15.36) asymptotes

$$P_{pl}(h) \approx 4nkT\left(\frac{y(h/2)}{2}\right)^2 \tag{15.39}$$

Substituting Eq. (15.38) into Eq. (15.39), we obtain

$$P_{el}(h) = 64 \tanh\left(\frac{y_o}{4}\right)^2 nkT \exp(-\kappa h) \tag{15.40}$$

The potential energy $V_{pl}(h)$ can be obtained by integrating Eq. (15.40) with the result

$$V_{el}(h) = \frac{64}{\kappa} \tanh\left(\frac{y_o}{4}\right)^2 nkT \exp(-\kappa h) \tag{15.41}$$

which is of the same form as that for hard plates, although the expressions for y_o are different for hard and soft plates.

15.3 INTERACTION BETWEEN TWO DISSIMILAR SOFT SPHERES

Consider the electrostatic interaction between two dissimilar spherical soft spheres 1 and 2 (Fig. 15.3). We denote by d_1 and d_2 the thicknesses of the surface charge layers of spheres 1 and 2, respectively. Let the radius of the core of soft sphere 1 be a_1 and that for sphere 2 be a_2. We imagine that each surface layer is uniformly charged. Let Z_1 and N_1, respectively, be the valence and the density of fixed-charge layer of sphere 1 and Z_2 and N_2 for sphere 2.

With the help of Derjaguin's approximation [2] (Eq. (12.2)), namely,

$$V_{\mathrm{sp}}(H) = \frac{2\pi a_1 a_2}{a_1 + a_2} \int_H^\infty V_{\mathrm{pl}}(h)dh \tag{15.42}$$

which is a good approximation if

$$\kappa a_1 \gg 1, \quad \kappa a_2 \gg 1, \quad H \ll a_1, \quad \text{and} \quad H \ll a_2 \tag{15.43}$$

one can calculate the interaction energy $V_{\mathrm{sp}}(H)$ between two dissimilar soft spheres 1 and 2 separated by a distance H between there surfaces via the corresponding interaction energy $V_{\mathrm{pl}}(h)$ between two parallel dissimilar plates. By substituting Eq. (15.20) into Eq. (15.42) we obtain [3]

$$\begin{aligned} V_{\mathrm{sp}}(H) = \frac{1}{\varepsilon_r \varepsilon_o \kappa^4}\left(\frac{\pi a_1 a_2}{a_1 + a_2}\right)\Bigg[&\{\rho_{\mathrm{fix1}} \sinh(\kappa d_1) + \rho_{\mathrm{fix2}} \sinh(\kappa d_2)\}^2 \\ &\times \ln\left(\frac{1}{1 - e^{-\kappa(H+d_1+d_2)}}\right) \\ &- \{\rho_{\mathrm{fix1}} \sinh(\kappa d_1) - \rho_{\mathrm{fix2}} \sinh(\kappa d_2)\}^2 \ln(1 + e^{-\kappa(H+d_1+d_2)})\Bigg] \end{aligned} \tag{15.44}$$

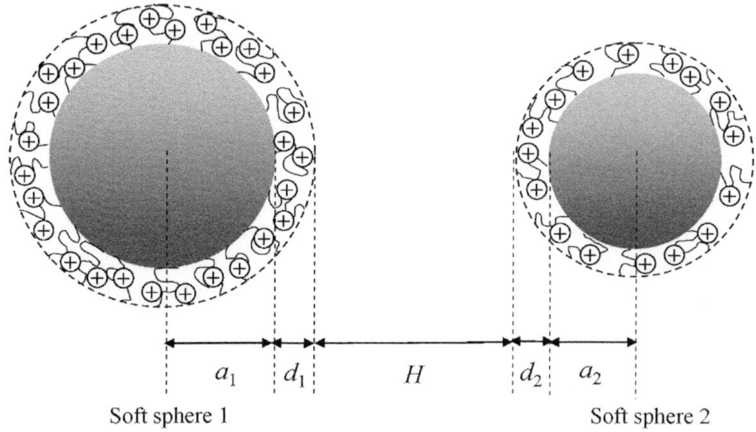

FIGURE 15.3 Interaction between two soft spheres 1 and 2 at separation H. Spheres 1 and 2 are covered with surface charge layers of thicknesses d_1 and d_2, respectively. The core radii of spheres 1 and 2 are a_1 and a_2, respectively.

For the special case of two similar soft spheres carrying $Z_1 = Z_2 = Z$, $N_1 = N_2 = N$, $d_1 = d_2 = d$ so that $\rho_{\text{fix}1} = \rho_{\text{fix}2} = \rho_{\text{fix}}$ and $a_1 = a_2 = a$, Eq. (15.44) reduces to

$$V_{\text{sp}}(H) = \frac{2\pi a \rho_{\text{fix}}^2 \sinh^2(\kappa d)}{\varepsilon_r \varepsilon_o \kappa^4} \ln\left(\frac{1}{1 - e^{-\kappa(H+2d)}}\right) \tag{15.45}$$

We now introduce the quantities

$$\sigma_1 = \rho_{\text{fix}1} d_1 = Z_1 e N_1 d_1, \tag{15.46}$$

$$\sigma_2 = \rho_{\text{fix}2} d_2 = Z_2 e N_2 d_2 \tag{15.47}$$

which are, respectively, the amounts of fixed charges contained in the surface layers per unit area on spheres 1 and 2. If we take the limit d_1, $d_2 \to 0$ and N_1, $N_2 \to \infty$, keeping the products $N_1 d_1$ and $N_2 d_2$ constant, then σ_1 and σ_2 reduce to the surface charge densities of two interacting hard plates without surface charge layers. In this limit, Eqs. (15.20),(15.21) and (15.44) reduce to

$$P_{\text{pl}}(h) = \frac{1}{8\varepsilon_r \varepsilon_o} \left[(\sigma_1 + \sigma_2)^2 \operatorname{cosech}^2\left(\frac{\kappa h}{2}\right) - (\sigma_1 - \sigma_2)^2 \operatorname{sech}^2\left(\frac{\kappa h}{2}\right) \right] \tag{15.48}$$

$$V_{\text{pl}}(h) = \frac{1}{4\varepsilon_r \varepsilon_o \kappa} \left[(\sigma_1 + \sigma_2)^2 \left\{ \coth\left(\frac{\kappa h}{2}\right) - 1 \right\} - (\sigma_1 - \sigma_2)^2 \left\{ 1 - \tanh\left(\frac{\kappa h}{2}\right) - 1 \right\} \right] \tag{15.49}$$

$$V_{\text{sp}}(H) = \frac{1}{\varepsilon_r \varepsilon_o \kappa^2} \left(\frac{\pi a_1 a_2}{a_1 + a_2}\right) \left[(\sigma_1 + \sigma_2)^2 \ln\left(\frac{1}{1 - e^{-\kappa H}}\right) - (\sigma_1 - \sigma_2)^2 \ln(1 + e^{-\kappa H}) \right] \tag{15.50}$$

Equations (15.49) and (15.50), respectively, agrees with the expression for the electrostatic interaction energy between two parallel hard plates at constant surface charge density and that for two hard spheres at constant surface charge density [4] (Eqs. (10.54) and (10.55)).

In order to see the effects of the thickness of the surface charge layer, we calculate the interaction energy $V_{\text{sp}}(H)$ via Eq. (15.44) for the case of two dissimilar soft spheres with fixed charges of like sign and that for spheres with fixed charges of unlike sign and illustrate the results calculated for several values of $\kappa d_1 = \kappa d_1 = \kappa d$ with σ_1 and σ_2 kept constant at $\sigma_2/\sigma_1 = 0.5$ (Fig. 15.4) and at $\sigma_2/\sigma_1 = -0.5$ (Fig. 15.5), showing a remarkable dependence of $V_{\text{sp}}(H)$ upon κd. This is because electrolyte ions can penetrate the surface charge layer, exerting the shielding effect on the fixed charges in the surface charge layers. Because of the ion penetration, the increase in potential inside the interacting plates due to their approach is much less than that for $\kappa d = 0$. In particular, when $\kappa d_1 \gg 1$ and $\kappa d_2 \gg 1$, being fulfilled for practical cases, the potential deep inside the plates remains constant, equal to the Donnan potential for the respective surface charge layers, which are given by Eq. (4.20), namely,

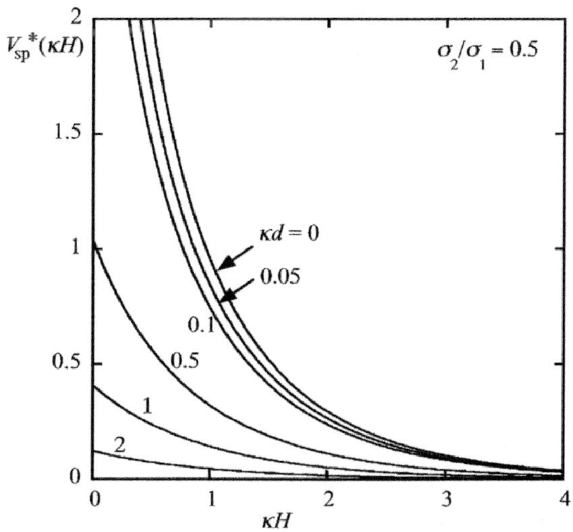

FIGURE 15.4 Scaled electrostatic interaction energy $V_{sp}^*(H) = \{\varepsilon_r\varepsilon_o\kappa^4(a_1+a_2)/\pi a_1 a_2 \sigma_1^2\}V_{sp}(H)$ between two dissimilar soft spheres with fixed charges of like sign as a function of scaled sphere separation κH calculated with Eq. (15.44) at various values of $\kappa d_1 = \kappa d_2 = \kappa d$, where σ_1 and σ_2 (defined by Eqs. (15.46) and (15.47)) are kept constant at $\sigma_2/\sigma_1 = 0.5$. From Ref. [3].

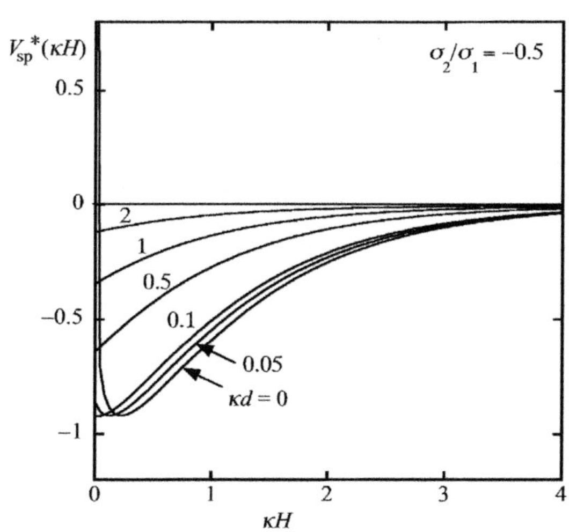

FIGURE 15.5 Scaled electrostatic interaction energy $V_{sp}^*(H) = \{\varepsilon_r\varepsilon_o\kappa^4(a_1+a_2)/\pi a_1 a_2 \sigma_1^2\}V_{sp}(H)$ between two dissimilar soft spheres with fixed charges of unlike sign as a function of scaled sphere separation κH calculated with Eq. (15.44) at various values of $\kappa d_1 = \kappa d_2 = \kappa d$, where σ_1 and σ_2 (defined by Eqs. (15.46) and (15.47)) are kept constant at $\sigma_2/\sigma_1 = -0.5$. From Ref. [3].

$$\psi_{\mathrm{DON1}} = \frac{\rho_{\mathrm{fix1}}}{\varepsilon_r \varepsilon_o \kappa^2} \tag{15.51}$$

$$\psi_{\mathrm{DON2}} = \frac{\rho_{\mathrm{fix2}}}{\varepsilon_r \varepsilon_o \kappa^2} \tag{15.52}$$

It is well known [4,5] that in the case of hard spheres ($d_1 = d_2 = 0$), the electrostatic force between two dissimilar spheres with charges of unlike sign is attractive for large κH but becomes repulsive at small κH, that is, there is a minimum in the interaction energy $V_{sp}(H)$ except when $\sigma_2/\sigma_1 = -1$. The case of nonzero κd_1 and κd_2, however, leads to quite different results. Figure 15.6 shows that except for very small κd_1 and κd_2, the interaction force is always attractive, that is, there is no minimum in $V_{sp}(H)$. On the other hand, it is repulsive for all κh when the fixed charges of spheres 1 and 2 are like sign as is seen in Fig. 15.4. Figure 15.6 shows results for $V_{sp}(H)$ calculated with Eq. (15.44) at various values of $\kappa d_1 = \kappa d_2 = \kappa d$ when ρ_{fix1} and ρ_{fix2} are kept constant at $\rho_{\mathrm{fix1}}/\rho_{\mathrm{fix2}} = 0.5$. It is seen that as κd_1 and κd_2 increase, the dependence of $V_{sp}(H)$ on κd_1 and κd_2 becomes smaller. The limiting form of $V_{sp}(H)$ is given later by Eq. (15.55).

Consider other limiting cases of Eqs. (15.20),(15.21) and (15.44).

(i) *Thick surface charge layers*

Consider the limiting case of $\kappa d_1 \gg 1$ and $\kappa d_2 \gg 1$. In this case, soft plates and soft spheres become planar polyelectrolytes and spherical

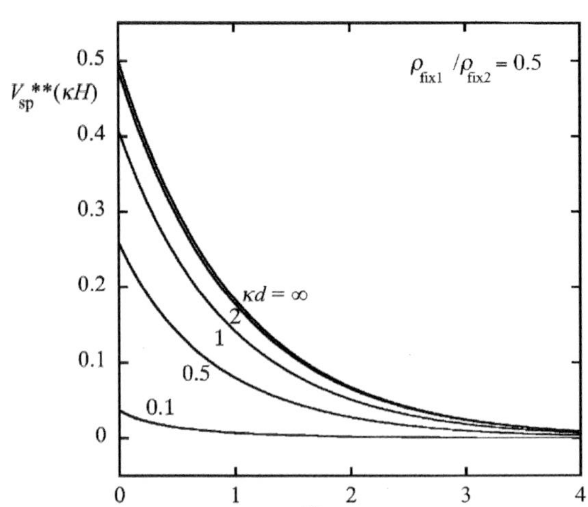

FIGURE 15.6 Scaled electrostatic interaction energy $V_{sp}^{**}(H) = \{\varepsilon_r \varepsilon_o \kappa^4 (a_1 + a_2)/\pi a_1 a_2 \rho_{\mathrm{fix1}}^2\} V_{sp}(H)$ between two dissimilar soft spheres with fixed charges of like sign as a function of scaled sphere separation κH calculated with Eq. (15.44) at various values of $\kappa d_1 = \kappa d_2 = \kappa d$, where ρ_{fix1} and ρ_{fix2} are kept constant at $\rho_{\mathrm{fix2}}/\rho_{\mathrm{fix1}} = 0.5$. From Ref. [3].

polyelectrolytes, respectively, and Eqs. (15.20),(15.21) and (15.44) reduce to

$$P_{\rm pl}(h) = \frac{\rho_{\rm fix1}\rho_{\rm fix2}}{2\varepsilon_{\rm r}\varepsilon_{\rm o}\kappa^2} e^{-\kappa h} \tag{15.53}$$

$$V_{\rm pl}(h) = \frac{\rho_{\rm fix1}\rho_{\rm fix2}}{2\varepsilon_{\rm r}\varepsilon_{\rm o}\kappa^3} e^{-\kappa h} \tag{15.54}$$

$$V_{\rm sp}(H) = \frac{1}{\varepsilon_{\rm r}\varepsilon_{\rm o}\kappa^4}\left(\frac{\pi a_1 a_2}{a_1 + a_2}\right)\rho_{\rm fix1}\rho_{\rm fix2}\, e^{-\kappa H} \tag{15.55}$$

It is seen that for thick surface charge layers, $P(h)$, $V_{\rm pl}(h)$, and $V_{\rm sp}(H)$ are always positive when Z_1 and Z_2 are of like sign while $V_{\rm pl}(h)$ and $V_{\rm sp}(H)$ are always negative when Z_1 and Z_2 are of unlike sign

Note that the following exact expression for the electrostatic interaction between two porous spheres (spherical polyelectrolytes) for the low charge density case has been derived [5,6] (Eq. (13.46)):

$$V_{\rm sp}(H) = \frac{\pi a_1 a_2 \rho_{\rm fix1}\rho_{\rm fix2}}{\varepsilon_{\rm r}\varepsilon_{\rm o}\kappa^4}\frac{e^{-\kappa H}}{H + a_1 + a_2} \times\left(1 + e^{-2\kappa a_1} - \frac{1 - e^{-2\kappa a_1}}{\kappa a_1}\right)\left(1 + e^{-2\kappa a_2} - \frac{1 - e^{-2\kappa a_2}}{\kappa a_2}\right) \tag{15.56}$$

or

$$V_{\rm sp}(H) = 4\pi\varepsilon_{\rm r}\varepsilon_{\rm o} a_1 a_2 \psi_{\rm o1}\psi_{\rm o2}\frac{e^{-\kappa H}}{H + a_1 + a_2} \tag{15.57}$$

with

$$\psi_{\rm o1} = \frac{\rho_{\rm fix1}}{2\varepsilon_{\rm r}\varepsilon_{\rm o}\kappa^2}\left(1 + e^{-2\kappa a_1} - \frac{1 - e^{-2\kappa a_1}}{\kappa a_1}\right) \tag{15.58}$$

$$\psi_{\rm o2} = \frac{\rho_{\rm fix2}}{2\varepsilon_{\rm r}\varepsilon_{\rm o}\kappa^2}\left(1 + e^{-2\kappa a_2} - \frac{1 - e^{-2\kappa a_2}}{\kappa a_2}\right) \tag{15.59}$$

where $\psi_{\rm o1}$ and $\psi_{\rm o2}$ are, respectively, the unperturbed surface potentials of soft spheres 1 and 2 at infinite separation. Equation (15.56), under the condition given by Eq. (15.43), tends to Eq. (15.55).

As in the case of two interacting soft plates, when the thicknesses of the surface charge layers on soft spheres 1 and 2 are very large compared with the Debye length $1/\kappa$, the potential deep inside the surface charge layer is practically equal to the Donnan potential (Eqs. (15.51) and (15.52)), independent of the particle separation H. In contrast to the usual electrostatic interaction models assuming constant surface potential or constant surface charge density of interacting particles, the electrostatic interaction between soft particles may be regarded as the Donnan potential-regulated interaction (Chapter 13).

(ii) *Interaction between soft sphere and porous sphere (spherical polyelectrolyte)*

Consider the case where sphere 1 is a soft sphere and sphere 2 is a porous sphere (spherical polyelectrolyte). By taking the limit $\kappa d_2 \gg 1$, we obtain from Eq. (15.44)

$$V_{sp}(H) = \frac{2}{\varepsilon_r \varepsilon_o \kappa^4}\left(\frac{\pi a_1 a_2}{a_1 + a_2}\right)\left[\rho_{fix1}\rho_{fix2}\sinh(\kappa d_1)e^{-\kappa(H+d_1)} + \frac{1}{4}\rho_{fix2}^2 e^{-2\kappa(H+d_1)}\right] \tag{15.60}$$

In the limit of $\kappa d_1 \gg 1$, Eq. (15.60) tends back to Eq. (15.55).

(iii) *Interaction between a soft sphere and a hard sphere*

Consider the case where sphere 1 is a soft sphere and sphere 2 is a hard sphere. By taking the limit $d_2 \to 0$ and $N_2 \to \infty$ with the product $\sigma_2 = Z_2 e N_2 d_2$ kept constant, we obtain from Eq. (15.44)

$$V_{sp}(H) = \frac{1}{\varepsilon_r \varepsilon_o \kappa^4}\left(\frac{\pi a_1 a_2}{a_1 + a_2}\right)\left[\{\rho_{fix1}\sinh(\kappa d_1) + \sigma_2\kappa\}^2 \ln\left(\frac{1}{1 - e^{-\kappa(H+d_1)}}\right) - \{\rho_{fix1}\sinh(\kappa d_1) - \sigma_2\kappa\}^2 \ln(1 + e^{-\kappa(H+d_1)})\right] \tag{15.61}$$

(iv) *Interaction between a spherical polyelectrolyte and a hard sphere*

If we further take the limit $\kappa d_1 \gg 1$ in Eq. (15.61), then we obtain the electrostatic interaction energy for the case where sphere 1 is a spherical polyelectrolyte and sphere 2 is a hard sphere, namely,

$$V_{sp}(H) = \frac{2}{\varepsilon_r \varepsilon_o \kappa^4}\left(\frac{\pi a_1 a_2}{a_1 + a_2}\right)\left(\rho_{fix1}\sigma_2\kappa\, e^{-\kappa H} + \frac{1}{8}\rho_{fix1}^2 e^{-2\kappa H}\right) \tag{15.62}$$

Note that the following exact expression for the electrostatic interaction between spherical polyelectrolyte 1 and hard sphere 2 has been derived [5,7] (see Chapter 14):

$$V_{sp}(H) = 4\pi\varepsilon_r\varepsilon_o \psi_{o1}\psi_{o2} a_1 a_2 \frac{e^{-\kappa H}}{H + a_1 + a_2} + 2\pi\varepsilon_r\varepsilon_o \psi_{o1}^2 a_1^2 e^{2\kappa a_1} \frac{1}{H + a_1 + a_2}\sum_{n=0}^{\infty}(2n+1) \times \frac{I_{n-1/2}(\kappa a_2) - (n+1)I_{n+1/2}(\kappa a_2)/\kappa a_2}{K_{n-1/2}(\kappa a_2) + (n+1)K_{n+1/2}(\kappa a_2)/\kappa a_2} K_{n+1/2}^2(\kappa(H + a_1 + a_2)) \tag{15.63}$$

where the unperturbed surface potentials ψ_{o1} and ψ_{o2} are given by Eqs. (15.58) and (1.76), respectively. Under the condition given by Eq. (15.43), Eq. (15.63) becomes

$$V_{sp}(H) = 4\pi\varepsilon_r\varepsilon_o \frac{a_1 a_2}{a_1 + a_2}\left(\psi_{o1}\psi_{o2}\, e^{-\kappa H} + \frac{1}{4}\psi_{o1}^2\, e^{-2\kappa H}\right) \tag{15.64}$$

15.4 INTERACTION BETWEEN TWO DISSIMILAR SOFT CYLINDERS

Consider the electrostatic interaction between two parallel dissimilar cylindrical soft particles 1 and 2. We denote by d_1 and d_2 the thicknesses of the surface charge layers of cylinders 1 and 2, respectively. Let the radius of the core of soft cylinder 1 be a_1 and that for soft cylinder 2 be a_2. We imagine that each surface layer is uniformly charged. Let Z_1 and N_1, respectively, be the valence and the density of fixed-charge layer of cylinder 1, and Z_2 and N_2 for cylinder 2.

Consider first the case of two parallel soft cylinders (Fig. 15.7). With the help of Derjaguin's approximation for two parallel cylinders [8,9] (Eq. (12.38)), namely,

$$V_{cy//}(H) = \sqrt{\frac{2a_1a_2}{a_1 + a_2}} \int_H^\infty V_{pl}(h)\frac{dh}{\sqrt{h - H}} \tag{15.65}$$

which is a good approximation for $\kappa a_1 \gg 1$, $\kappa a_2 \gg 1$, $H \ll a_1$, and $H \ll a_2$ (Eq. (15.43)), one can calculate the interaction energy $V_{cy//}(H)$ per unit length between two dissimilar soft cylinders 1 and 2 separated by a distance H between there

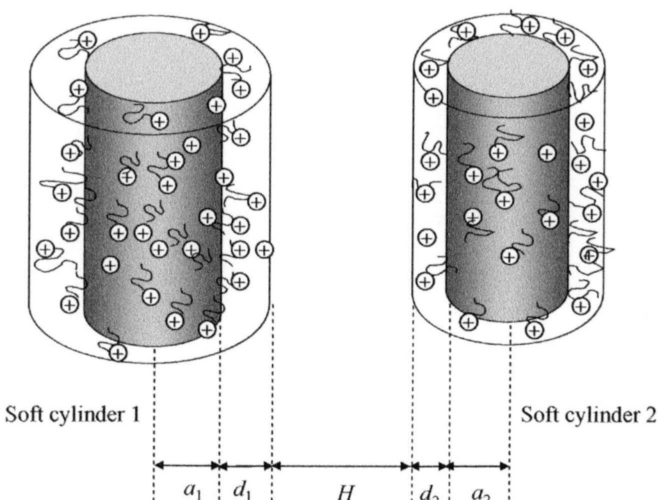

FIGURE 15.7 Interaction between two parallel soft cylinders 1 and 2 at separation H. Cylinders 1 and 2 are covered with surface charge layers of thicknesses d_1 and d_2, respectively. The core radii of cylinders 1 and 2 are a_1 and a_2, respectively.

surfaces via the corresponding interaction energy $V_{\rm pl}(h)$ per unit area between two parallel dissimilar soft plates at separation h. By substituting Eq. (15.20) into Eq. (15.65), we obtain [9]

$$V_{{\rm cy}/\!/}(H) = \frac{1}{2\varepsilon_{\rm r}\varepsilon_{\rm o}\kappa^{7/2}}\sqrt{\frac{2\pi a_1 a_2}{a_1+a_2}}[\{\rho_{\rm fix1}\sinh(\kappa d_1) + \rho_{\rm fix2}\sinh(\kappa d_2)\}^2 {\rm Li}_{1/2}(e^{-\kappa(H+d_1+d_2)}) + \{\rho_{\rm fix1}\sinh(\kappa d_1) - \rho_{\rm fix2}\sinh(\kappa d_2)\}^2 {\rm Li}_{1/2}(-e^{-\kappa(H+d_1+d_2)})] \tag{15.66}$$

where $\mathrm{Li}_s(z)$ is the polylogarithm function, defined by

$$\mathrm{Li}_s(z) = \sum_{k=1}^{\infty}\frac{z^k}{k^s} \tag{15.67}$$

For the special case of two similar soft cylinders carrying $Z_1 = Z_2 = Z$, $N_1 = N_2 = N$, $a_1 = a_2 = a$, $d_1 = d_2 = d$, and $\rho_{\rm fix1} = \rho_{\rm fix2} = \rho_{\rm fix}$, Eq. (15.66) reduces to

$$V_{{\rm cy}//}(H) = \frac{2\sqrt{\pi a}}{\varepsilon_{\rm r}\varepsilon_{\rm o}\kappa^{7/2}}\rho_{\rm fix}^2\sinh^2(\kappa d){\rm Li}_{1/2}(e^{-\kappa(H+2d)}) \tag{15.68}$$

The interaction force $P_{{\rm cy}//}(H)$ acting between two soft cylinders per unit length is given by $P_{{\rm cy}//}(H) = -dV_{{\rm cy}//}(H)/dH$, which gives [9]

$$P_{{\rm cy}/\!/}(H) = \frac{1}{2\varepsilon_{\rm r}\varepsilon_{\rm o}\kappa^{5/2}}\sqrt{\frac{2\pi a_1 a_2}{a_1+a_2}}\Big[\{\rho_{\rm fix1}\sinh(\kappa d_1) + \rho_{\rm fix2}\sinh(\kappa d_2)\}^2 \times {\rm Li}_{-1/2}(e^{-\kappa(H+d_1+d_2)}) + \{\rho_{\rm fix1}\sinh(\kappa d_1) - \rho_{\rm fix2}\sinh(\kappa d_2)\}^2 {\rm Li}_{-1/2}(-e^{-\kappa(H+d_1+d_2)})\Big] \tag{15.69}$$

Consider next the case of two crossed soft cylinders (Fig. 15.8). Derjaguin's approximation for two crossed cylinders under condition (15.43) is given by Eq. (12.48), namely,

$$V_{{\rm cy}\perp}(H) = 2\pi\sqrt{a_1 a_2}\int_H^{\infty} V_{\rm pl}(h)dh \tag{15.70}$$

By substituting Eq. (15.21) into Eq. (15.70), we obtain [9]

$$V_{{\rm cy}\perp}(H) = \frac{\pi\sqrt{a_1 a_2}}{\varepsilon_1\varepsilon_0\kappa^4}\left[\left\{\rho_{\rm fix1}\sinh(\kappa d_1) + \rho_{\rm fix2}\sinh(\kappa d_2)\right\}\ln\left(\frac{1}{1-e^{-\kappa(H+d_1+d_2)}}\right) - \{\rho_{\rm fix1}\sinh(\kappa d_1) - \rho_{\rm fix2}\sinh(\kappa d_2)\}^2\ln(1+e^{-\kappa(H+d_1+d_2)})\right] \tag{15.71}$$

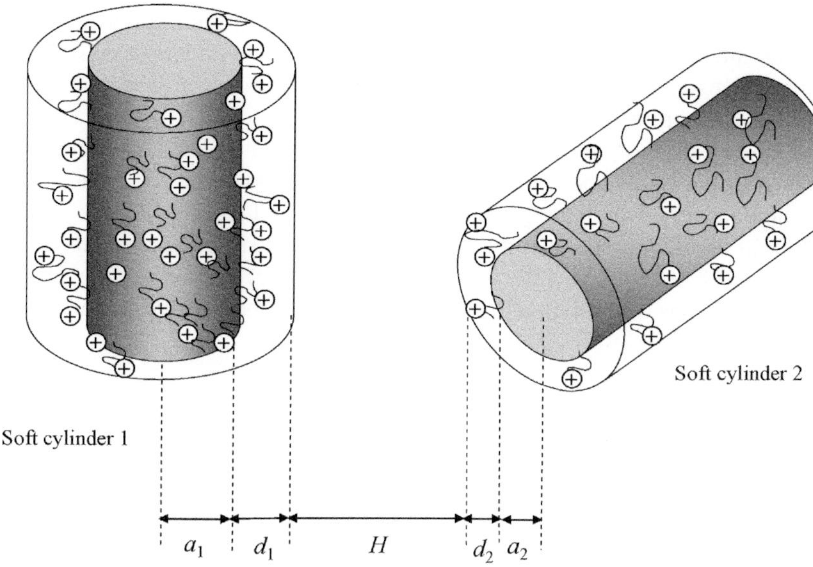

FIGURE 15.8 Interaction between two crossed soft cylinders 1 and 2 at separation H. Cylinders 1 and 2 are covered with surface charge layers of thicknesses d_1 and d_2, respectively. The core radii of cylinders 1 and 2 are a_1 and a_2, respectively.

For the special case of two similar soft cylinders carrying $Z_1 = Z_2 = Z$, $N_1 = N_2 = N$, $a_1 = a_2 = a$, $d_1 = d_2 = d$, and $\rho_{\text{fix}1} = \rho_{\text{fix}2} = \rho_{\text{fix}}$, Eq. (15.71) reduces to

$$V_{\text{cy}\perp}(H) = \frac{4\pi a}{\varepsilon_r \varepsilon_o \kappa^4} \rho_{\text{fix}}^2 \sinh^2(\kappa d) \ln\left(\frac{1}{1 - e^{-\kappa(H+2d)}}\right) \tag{15.72}$$

We introduce the quantities

$$\sigma_1 = \rho_{\text{fix}1} d_1 = Z_1 e N_1 d_1 \tag{15.73}$$

$$\sigma_2 = \rho_{\text{fix}2} d_2 = Z_2 e N_2 d_2 \tag{15.74}$$

which are, respectively, the amounts of fixed charges contained in the surface layers per unit area on cylinders 1 and 2. If we take the limit d_1, $d_2 \to 0$ and N_1, $N_2 \to \infty$, keeping the products $N_1 d_1$ and $N_2 d_2$ constant, then σ_1 and σ_2 reduce to the surface charge densities of two interacting hard cylinders without surface charge layers. In this limit, Eqs. (15.66) and (15.71) reduce to

$$V_{\text{cy}//}(H) = \frac{2}{\varepsilon_r \varepsilon_o \kappa^{3/2}} \sqrt{\frac{2\pi a_1 a_2}{a_1 + a_2}} \left[\left(\frac{\sigma_1 + \sigma_2}{2}\right)^2 \text{Li}_{1/2}(e^{-\kappa H}) - \left(\frac{\sigma_1 - \sigma_2}{2}\right)^2 \text{Li}_{1/2}(-e^{-\kappa H})\right] \tag{15.75}$$

$$V_{\text{cy}\perp}(H) = \frac{4\pi\sqrt{a_1 a_2}}{\varepsilon_r \varepsilon_o \kappa^2} \left[\left(\frac{\sigma_1 + \sigma_2}{2}\right)^2 \ln\left(\frac{1}{1 - e^{-\kappa H}}\right) - \left(\frac{\sigma_1 - \sigma_2}{2}\right)^2 \ln(1 + e^{-\kappa H})\right] \tag{15.76}$$

Equations (15.75) and (15.76), respectively, agree with the expression for the electrostatic interaction energy between two parallel hard cylinders at constant surface charge density and that for two crossed hard cylinders at constant surface charge density (Chapter 12).

When $\kappa d_1 \gg 1$ and $\kappa d_2 \gg 1$, being fulfilled for practical cases, the potential deep inside the plates remains constant, equal to the Donnan potential for the respective surface charge layers, which are given by Eqs. (15.51) and (15.52).

Where ψ_{o1} and ψ_{o2} are, respectively, the unperturbed surface potentials of hard cylinders 1 and 2 at infinite separation and $I_n(z)$ and $K_n(z)$ are, respectively, modified Bessel functions of the first and second kinds, ε_{pi} is the relative permittivity of cylinder i ($i = 1$ and 2).

Consider other limiting cases of Eqs. (2.101) and (2.105).

(i) *Thick surface charge layers*

In the limiting case of $\kappa d_1 \gg 1$ and $\kappa d_2 \gg 1$, soft cylinders become cylindrical polyelectrolytes and Eqs. (2.101) and (2.105) reduce to

$$V_{\text{cy}/\!/}(R) = \frac{1}{2\varepsilon_r\varepsilon_o\kappa^{7/2}}\sqrt{\frac{2\pi a_1 a_2}{a_1 + a_2}}\rho_{\text{fix1}}\rho_{\text{fix2}}\, e^{-\kappa H} \tag{15.77}$$

$$V_{\text{cy}\perp}(H) = \frac{\pi\sqrt{a_1 a_2}}{\varepsilon_r\varepsilon_o\kappa^4}\rho_{\text{fix1}}\rho_{\text{fix2}}\, e^{-\kappa H} \tag{15.78}$$

It is seen that for thick surface charge layers, $V_{\text{cy}/\!/}(H)$ and $V_{\text{cy}\perp}(H)$ are always positive when Z_1 and Z_2 are of like sign while they are always negative when Z_1 and Z_2 are of unlike sign.

Note that the following exact expression for the electrostatic interaction energy per unit area $V_{\text{cy}/\!/}(H)$ between two porous cylinders for the low charge density case has been derived [5,10] (Chapter 13):

$$V_{\text{cy}/\!/}(H) = 2\pi\varepsilon_r\varepsilon_o\psi_{o1}\psi_{o2}\frac{K_0(\kappa(H + a_1 + a_2))}{K_0(\kappa a_1)K_0(\kappa a_2)} \tag{15.79}$$

with

$$\psi_{oi} = \frac{\rho_{\text{fix}i}}{\varepsilon_r\varepsilon_o\kappa}a_i K_0(\kappa a_i)I_1(\kappa a_i), \quad (i = 1, 2) \tag{15.80}$$

where ψ_{o1} and ψ_{o2} are, respectively, the unperturbed surface potentials of cylindrical polyelectrolytes 1 and 2 at infinite separation. Equation (15.79), under the condition given by Eq. (15.43), tends to Eq. (15.77).

As in the case of soft spheres, when the thicknesses of the surface charge layers on soft cylinders 1 and 2 are very large compared with the Debye length $1/\kappa$, the potential deep inside the surface charge layer is practically equal to the Donnan potential (Eqs. (15.51) and (15.52)), independent of the particle separation H.

(ii) *Interaction between soft cylinder and cylindrical polyelectrolyte*

Consider the case where cylinder 1 is a soft cylinder and cylinder 2 is a cylindrical polyelectrolyte. By taking the limit $\kappa d_2 \gg 1$, we obtain from Eqs. (15.66) and (15.71)

$$V_{\mathrm{cy}//}(H) = \frac{1}{\varepsilon_r \varepsilon_o \kappa^{7/2}} \sqrt{\frac{2\pi a_1 a_2}{a_1 + a_2}} \left[\rho_{\mathrm{fix1}} \rho_{\mathrm{fix2}} \sinh(\kappa d_1) e^{-\kappa(H+d_1)} + \frac{1}{4\sqrt{2}} \rho_{\mathrm{fix2}}^2 \, e^{-2\kappa(H+d_1)} \right] \quad (15.81)$$

$$V_{\mathrm{cy}\perp}(H) = \frac{2\pi\sqrt{a_1 a_2}}{\varepsilon_r \varepsilon_o \kappa^4} \left[\rho_{\mathrm{fix1}} \rho_{\mathrm{fix2}} \sinh(\kappa d_1) e^{-\kappa(H+d_1)} + \frac{1}{8} \rho_{\mathrm{fix2}}^2 \, e^{-2\kappa(H+d_1)} \right] \quad (15.82)$$

In the limit of $\kappa d_1 \gg 1$, Eqs. (15.81) and (15.82) tend back to Eqs. (15.77) and (15.78), respectively.

(iii) *Interaction between a soft cylinder and a hard cylinder*

Consider the case where cylinder 1 is a soft cylinder and cylinder 2 is a hard cylinder. By taking the limit $d_2 \to 0$ and $N_2 \to \infty$ with the product $\sigma_2 = Z_2 e N_2 d_2$ kept constant, we obtain from Eqs. (15.66) and (15.71)

$$V_{\mathrm{cy}//}(H) = \frac{1}{2\varepsilon_r \varepsilon_o \kappa^{7/2}} \sqrt{\frac{2\pi a_1 a_2}{a_1 + a_2}} [\{\rho_{\mathrm{fix1}} \sinh(\kappa d_1) + \sigma_2 \kappa\}^2 \mathrm{Li}_{1/2}(-e^{-\kappa(H+d_1)}) + \{\rho_{\mathrm{fix1}} \sinh(\kappa d_1) - \sigma_2 \kappa\}^2 \mathrm{Li}_{1/2}(-e^{-\kappa(H+d_1)})] \quad (15.83)$$

$$V_{\mathrm{cy}\perp}(H) = \frac{\pi\sqrt{a_1 a_2}}{\varepsilon_r \varepsilon_o \kappa^4} \left[\left\{ \rho_{\mathrm{fix1}} \sinh(\kappa d_1) + \sigma_2 \kappa \right\}^2 \ln\left(\frac{1}{1 - e^{-\kappa(H+d_1)}} \right) - \{\rho_{\mathrm{fix1}} \sinh(\kappa d_1) - \sigma_2 \kappa\}^2 \ln(1 + e^{-\kappa(H+d_1)}) \right] \quad (15.84)$$

(iv) *Interaction between a porous cylinder and a hard cylinder*

If we further take the limit $\kappa d_1 \gg 1$ in Eqs. (15.82) and (15.84), then we obtain the electrostatic interaction energies for the case where cylinder 1 is a cylindrical polyelectrolyte and cylinder 2 is a hard cylinder, namely,

$$V_{\mathrm{cy}/\!/}(H) = \frac{1}{\varepsilon_r \varepsilon_o \kappa^{7/2}} \sqrt{\frac{2\pi a_1 a_2}{a_1 + a_2}} \left(\rho_{\mathrm{fix1}} \sigma_2 \kappa \, e^{-\kappa H} + \frac{1}{4\sqrt{2}} \rho_{\mathrm{fix1}}^2 \, e^{-2\kappa H} \right) \quad (15.85)$$

$$V_{\mathrm{cy}\perp}(H) = \frac{2\pi\sqrt{a_1 a_2}}{\varepsilon_r \varepsilon_o \kappa^4}\left(\rho_{\mathrm{fix1}}\sigma_2 \kappa\, e^{-\kappa H} + \frac{1}{8}\rho_{\mathrm{fix1}}^2\, e^{-2\kappa H}\right) \tag{15.86}$$

Note that the following exact expression for the electrostatic interaction between porous cylinder (cylindrical polyelectrolyte) 1 and hard cylinder 2 has been derived [5,10]:

$$\begin{aligned} V_{\mathrm{cy}/\!/}(H) &= 2\pi\varepsilon_r\varepsilon_o\psi_{o1}\psi_{o2}\frac{K_0(\kappa(H+a_1+a_2))}{K_0(\kappa a_1)K_0(\kappa a_2)} \\ &\quad -\pi\varepsilon_r\varepsilon_o\psi_{o1}^2 a_1\, e^{2\kappa a_1} \\ &\times \sum_{n=-\infty}^{\infty}\frac{I'_n(\kappa a_2)-(\varepsilon_{p2}|n|/\varepsilon\kappa a_2)I_n(\kappa a_2)}{K'_n(\kappa a_2)-(\varepsilon_{p2}|n|/\varepsilon\kappa a_2)K_n(\kappa a_2)}K_n^2(\kappa(H+a_1+a_2)) \end{aligned} \tag{15.87}$$

with

$$\psi_{o1} = \frac{\rho_{\mathrm{fix1}}}{\varepsilon_r\varepsilon_o\kappa}a_i K_0(\kappa a_1)I_1(\kappa a_1) \tag{15.88}$$

$$\psi_{o2} = \frac{\sigma_2}{\varepsilon_r\varepsilon_o\kappa}\frac{K_0(\kappa a_2)}{K_1(\kappa a_2)}, \quad (i = 1, 2) \tag{15.89}$$

where ε_{p2} is relative permittivity of hard cylinder 2. Under the condition given by Eq. (15.43), Eq. (15.87) becomes Eq. (15.85).

REFERENCES

1. H. Ohshima, K. Makino, and T. Kondo, *J. Colloid Interface Sci.* 116 (1987) 196.
2. B. V. Derjaguin, *Kolloid Z.* 69 (1934) 155.
3. H. Ohshima, *J. Colloid Interface Sci.* 328 (2008) 3.
4. G. R. Wiese and T. W. Healy, *Trans. Faraday Soc.* 66 (1970) 490.
5. H. Ohshima, *Theory of Colloid and Interfacial Electric Phenomena*, Elsevier/Academic Press, 1968.
6. H. Ohshima and T. Kondo, *J. Colloid Interface Sci.* 155 (1993) 499.
7. H. Ohshima, *J. Colloid Interface Sci.* 168 (1994) 255.
8. M. J. Sparnaay, *Recueil* 78 (1959) 680.
9. H. Ohshima and A. Hyono, *J. Colloid Interface Sci.* 332 (2009) 251.
10. H. Ohshima, *Colloid Polym. Sci.* 274 (1996) 1176.

16 Electrostatic Interaction Between Nonuniformly Charged Membranes

16.1 INTRODUCTION

As a model for the electrostatic interaction between biological cells or their model particles, we have developed a theory of the double-layer interaction between ion--penetrable charged membranes in Chapter. This theory, unlike the conventional models for interactions of colloidal particles at constant surface potential or constant surface charge density, assumes that the potential far inside the membranes remains constant at the Donnan potential during interaction. We thus term this type of interaction, the Donnan potential-regulated interaction. In Chapter 13, we confined ourselves mainly to uniformly charged membranes. Since, however, in the membranes of real cells the distribution of membrane-fixed charges may actually be nonuniform rather than uniform, it is necessary to extend our theory to cover the case of arbitrary distributions of membrane-fixed charges. In this chapter, we shall consider the electrostatic interaction of ion-penetrable multilayered membranes as a model for real cell membranes.

16.2 BASIC EQUATIONS

Consider two parallel planar ion-penetrable membranes 1 and 2, which may not be identical, at separation h in a symmetrical electrolyte solution of valence z and bulk concentration n (Fig. 16.1). We take an x-axis perpendicular to the membranes with its origin at the surface of membrane 1. The electric potential $\psi(x)$ at position x between the membranes (relative to the bulk solution phase, where $\psi(x)$ is set equal to zero) is assumed to be small so that the linearized Poisson–Boltzmann equation can be employed. Membranes 1 and 2, respectively, consist of N and M layers. All the layers are perpendicular to the x-axis. Let the thickness and the density of membrane-fixed charges of the ith layer of membrane j ($j = 1$, 2) be $d_i^{(j)}$ and $\rho_i^{(j)}$. The linearized Poisson–Boltzmann equation for the ith layer

---

Biophysical Chemistry of Biointerfaces By Hiroyuki Ohshima

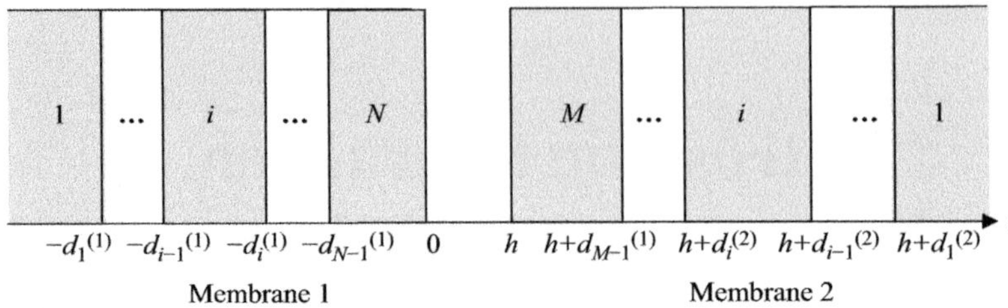

FIGURE 16.1 Interaction of two ion-penetrable membranes 1 and 2 consisting of N and M layers, respectively.

of membrane j is

$$\frac{d^2\psi}{dx^2} = \kappa^2\psi - \frac{\rho_i^{(1)}}{\varepsilon_r\varepsilon_o}, \quad -d_{i-1}^{(1)} < x < -d_i^{(1)},\ 1 \le i \le N \tag{16.1}$$

$$\frac{d^2\psi}{dx^2} = \kappa^2\psi - \frac{\rho_i^{(2)}}{\varepsilon_r\varepsilon_o}, \quad h + d_i^{(2)} < x < h + d_{i-1}^{(2)},\ 1 \le i \le M \tag{16.2}$$

and

$$\frac{d^2\psi}{dx^2} = \kappa^2\psi, \quad 0 < x < h \tag{16.3}$$

with κ being the Debye–Hückel parameter of the electrolyte solution, we have assumed that the relative permittivity ε_r takes the same value in the regions inside and outside the membranes, and we have defined $d_N^{(1)} = d_M^{(2)} = 0$ and $d_0^{(1)} = d_0^{(2)} = \infty$. It is convenient to introduce the Donnan potential $\psi_{\text{DON},i}^{(j)}$ of the respective layers. Since the Donnan potential satisfies $d^2\psi/dx^2 = 0$, we have from Eqs. (16.1) and (16.2)

$$\psi_{\text{DON},i}^{(j)} = \frac{\rho_i^{(j)}}{\varepsilon_r\varepsilon_o\kappa^2} \tag{16.4}$$

Now we assume that the distribution of mobile electrolyte ions is always at thermodynamic equilibrium within the membranes as well as in the surrounding solution, then the potential far inside the membranes remains constant during interaction. In the present system the potentials far inside membranes 1 and 2, respectively, remain constant at $\psi_{\text{DON},1}^{(1)}$ and at $\psi_{\text{DON},2}^{(2)}$.

16.3 INTERACTION FORCE

The interaction force $P(h)$ driving the two membranes apart per unit area in the low potential case is given by Eq. (8.25), namely,

$$P(h) = \frac{1}{2}\varepsilon_r\varepsilon_o\kappa^2\left[\psi^2(0) - \left(\left.\frac{d\psi}{dx}\right|_{x=0^+}\right)^2\right] \tag{16.5}$$

The solution to Eqs. (16.1)–(16.3) subject to the boundary conditions that ψ and its derivative are, respectively, continuous at the boundaries $x=-d_i^{(1)}$ and $x=h+d_i^{(2)}$ can easily be obtained. Since only the values of ψ and its derivative at $x=0$ are required to calculate $P(h)$ via Eq. (16.5), we give explicitly the solution to Eq. (16.3) below.

$$\psi(x) = \psi_{o1}\, e^{-\kappa x} + \psi_{o2}\, e^{-\kappa(h-x)} \tag{16.6}$$

where ψ_{o1} and ψ_{o2} are, respectively, given by

$$\begin{aligned}\psi(x) &= \frac{1}{2}\psi_{\mathrm{DON},1}^{(1)} \exp\left(-\kappa d_1^{(1)}\right) + \frac{1}{2}\sum_{i=2}^{N-1}\psi_{\mathrm{DON},i}^{(1)}\left\{\exp\left(-\kappa d_i^{(1)}\right) - \exp\left(-\kappa d_{i-1}^{(1)}\right)\right\}\\ &\quad + \frac{1}{2}\psi_{\mathrm{DON},N}^{(1)}\left\{1 - \exp\left(-\kappa d_{N-1}^{(1)}\right)\right\}\end{aligned} \tag{16.7}$$

and

$$\begin{aligned}\psi_{o1} &= \frac{1}{2}\psi_{\mathrm{DON},1}^{(2)} \exp\left(-\kappa d_1^{(2)}\right) + \frac{1}{2}\sum_{i=2}^{M-1}\psi_{\mathrm{DON},i}^{(2)}\left\{\exp\left(-\kappa d_i^{(2)}\right) - \exp\left(-\kappa d_{i-1}^{(2)}\right)\right\}\\ &\quad + \frac{1}{2}\psi_{\mathrm{DON},M}^{(2)}\left\{1 - \exp\left(-\kappa d_{M-1}^{(2)}\right)\right\}\end{aligned} \tag{16.8}$$

Evaluating ψ and its derivative at $x=0$ via Eq. (16.6) and substituting the result into Eq. (16.5), we obtain

$$P(h) = 2\varepsilon_r\varepsilon_o\kappa^2\psi_{o1}\psi_{o2}\, e^{-\kappa h} \tag{16.9}$$

It is interesting to note that ψ_{o1} and ψ_{o2}, are, respectively, the unperturbed surface potentials of membranes 1 and 2 (i.e., the surface potentials at $h=\infty$) and that Eq. (16.9) states that the interaction force is proportional to the product of the unperturbed surface potentials of the interacting membranes. This is generally true for the Donnan potential-regulated interaction between two ion-penetrable membranes in which the distribution of the membrane-fixed charges far inside the membranes is uniform but may be arbitrary in the region near the membrane surfaces (see Eq. (8.28)).

Our treatment is based upon the assumption that the Donnan potential of the innermost layer of each membrane $\left(\psi_{\mathrm{DON},1}^{(1)}\text{ and }\psi_{\mathrm{DON},1}^{(2)}\right)$ remains constant during interaction. Let us examine the relative contributions from each layer of membrane 1 to the surface potential ψ_{o1}. From Eq. (16.7) it follows that the contribution from the innermost layer, that is, a semi-infinite layer having the Donnan potential $y_{\mathrm{DON},1}^{(1)}$ occupying the region $-\infty<x<-d_1$ is seen to be

$$\frac{1}{2}\psi_{\mathrm{DON},1}^{(1)} \exp\left(-\kappa d_1^{(1)}\right) \tag{16.10}$$

The contribution of the ith layer, having a finite thickness $d_i^{(1)} - d_{i-1}^{(1)}$ $(2 \leq i \leq N-1)$ and the Donnan potential $\psi_{\mathrm{DON},i}^{(1)}$, can thus be expressed as the difference in contribution between two semi-infinite layers, both having the same Donnan potential $\psi_{\mathrm{DON},i}^{(1)}$, occupying the regions $-\infty < x < -d_i^{(1)}$ and $-\infty < x < -d_{i-1}^{(1)}$, that is,

$$\frac{1}{2}\psi_{\mathrm{DON},i}^{(1)}\left\{\exp\left(-\kappa d_i^{(1)}\right) - \exp\left(-\kappa d_{i-1}^{(1)}\right)\right\} \tag{16.11}$$

For the outermost layer (the Nth layer), we have

$$\frac{1}{2}\psi_{\mathrm{DON},N}^{(1)}\left\{1 - \exp\left(-\kappa d_{N-1}^{(1)}\right)\right\} \tag{16.12}$$

16.4 ISOELECTRIC POINTS WITH RESPECT TO ELECTROLYTE CONCENTRATION

Equations (16.10)–(16.12) show that owing to the factor $\exp\left(-\kappa d_i^{(1)}\right)$, the contribution of the ith layer $(1 \leq i \leq N-1)$ becomes negligible if $\kappa d_i^{(1)} \gg 1$. In other words, the unperturbed surface potential ψ_{o1} is determined mainly by the contribution from layers located within the depth of $1/\kappa$ from the membrane surface. This implies that variation of the electrolyte concentration (and thus of $1/\kappa$) changes the relative contributions to the unperturbed surface potentials from the respective layers of the interacting membranes so that the interaction force may also alter with electrolyte concentration. It is of particular interest to note that if layers forming the region within the depth of $1/\kappa$ are not of the same sign, the surface potential may alter even its sign with changing electrolyte concentration. This implies the existence of isoelectric points with respect to variation of the electrolyte concentration. Note, however, that the isoelectric point introduced here is the value of the electrolyte concentration at which the unperturbed surface potential becomes zero and does not have the usual meaning that the total (net) amount of membrane-fixed charges is zero.

It immediately follows from the existence of isoelectric points that if the two interacting membranes have different isoelectric points, then the interaction force alters its sign at these isoelectric points (note that the interaction force between membranes with the same isoelectric point does not change its sign).

An example is given in Fig. 16.2, which shows the interaction force between two membranes, each consisting of with two sublayers, in a monovalent symmetrical electrolyte solution as a function of the electrolyte concentration. Here membranes 1 and 2 have different isoelectric points. It is seen that the interaction force is positive (i.e., repulsive) at concentrations less than 0.07 M and higher than 0.18 M and negative (i.e., attractive) between 0.07 and 0.18 M, corresponding to the isoelectric points at 0.07 M for membrane 1 and 0.18 M for membrane 2. Membranes consisting of more than two layers may exhibit more than one isoelectric point. The

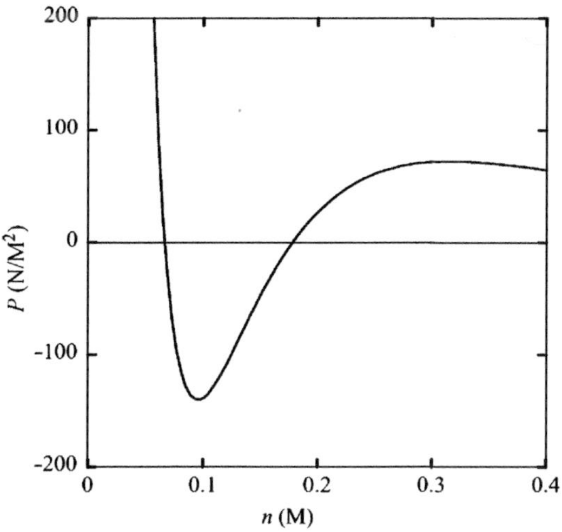

FIGURE 16.2 Electrostatic force P between two ion-penetrable membranes, each consisting of two layers, immersed in a monovalent symmetrical electrolyte solution as a function of the electrolyte concentration n (M). The Donnan potentials at $n=0.1$ M in the respective layers are as follows. $\psi^{(1)}_{\mathrm{DON},1} = -20$ mV, $\psi^{(1)}_{\mathrm{DON},2} = +15$ mV, $\psi^{(2)}_{\mathrm{DON},1} = -10$ mV, $\psi^{(2)}_{\mathrm{DON},2} = +10$ mV, $d^{(1)}_1 = 0.1$ nm, $d^{(2)}_1 = 0.5$ nm, and $h = 2.5$ nm, $\varepsilon_r = 78.5$, $T = 298$ K. $P > 0$ corresponds to repulsion and $P < 0$ to attraction. (From Ref. 1.)

interaction force between such membranes will show complicated behavior on variation of the electrolyte concentration. The present study has revealed that the sign of the interaction force $P(h)$ is determined not by the net or total amount of membrane-fixed charges but, rather, depends strongly on the isoelectric points of the interacting membranes. This behavior is not predicted by the conventional interaction models assuming constant surface potential or constant surface charge density. We would therefore suggest a possibility of selective aggregation in the system of ion-penetrable membranes with different isoelectric points by using the difference between their isoelectric points.

Equations (16.1) and (16.2) can be generalized to the case of arbitrary fixed--charge distributions in membranes 1 and 2. Let the membrane-charge distribution be $\rho^{(1)}(x)$ in membrane 1 ($-\infty < x < 0$) and that in membrane 2 ($h < x < +\infty$) be $\rho^{(2)}(x)$, respectively. Then Eqs. (16.1) and (16.2) become

$$\frac{d^2\psi}{dx^2} = \kappa^2\psi - \frac{\rho^{(1)}(x)}{\varepsilon_r\varepsilon_o}, \qquad -\infty < x < 0 \tag{16.13}$$

$$\frac{d^2\psi}{dx^2} = \kappa^2\psi - \frac{\rho^{(2)}(x)}{\varepsilon_r\varepsilon_o}, \qquad h < x < +\infty \tag{16.14}$$

Equations (16.13), (16.14) and (16.3) under appropriate boundary conditions can be solved to give

$$\psi(x) = \psi_1(x) + \psi_2(x) \tag{16.15}$$

where $\psi_1(x)$ and $\psi_2(x)$ are unperturbed potentials produced by membranes 1 and 2 in the absence of interaction and are given by

$$\psi_1(x) = \frac{1}{2\varepsilon_r\varepsilon_o\kappa}\left[\int_x^0 e^{-\kappa(t-x)}\rho^{(1)}(t)dt + \int_{-\infty}^x e^{-\kappa(x-t)}\rho^{(1)}(t)dt\right], \quad -\infty < x < 0 \tag{16.16}$$

$$\psi_1(x) = \psi_{o1}\, e^{-\kappa x}, \quad 0 < x < \infty \tag{16.17}$$

$$\psi_2(x) = \frac{1}{2\varepsilon_r\varepsilon_o\kappa}\left[\int_x^\infty e^{-\kappa(t-x)}\rho^{(2)}(t)dt + \int_h^x e^{-\kappa(x-t)}\rho^{(2)}(t)dt\right], \quad h < x < +\infty \tag{16.18}$$

$$\psi_1(x) = \psi_{o1}\, e^{-\kappa(h-x)}, \quad -\infty < x < \mathrm{h} \tag{16.19}$$

with

$$\psi_{o1} = \frac{1}{2\varepsilon_r\varepsilon_o\kappa}\int_{-\infty}^0 e^{\kappa x}\rho^{(1)}(x)dx \tag{16.20}$$

$$\psi_{o2} = \frac{1}{2\varepsilon_r\varepsilon_o\kappa}\int_h^\infty e^{-\kappa(x-h)}\rho^{(2)}(x)dx \tag{16.21}$$

where ψ_{o1} and ψ_{o2} are, respectively, the unperturbed surface potentials of membranes 1 and 2. The interaction force $P(h)$ per unit area between membranes 1 and 2 is given by the same equation as Eq. (16.9).

REFERENCE

1. R. Tokugawa, K. Makino, H. Ohshima, and T. Kondo, *Biophys. Chem.* 40 (1991) 77.

17 Electrostatic Repulsion Between Two Parallel Soft Plates After Their Contact

17.1 INTRODUCTION

We have so far considered the interaction between soft particles for particle separations before their surface charge layers come into contact. In this section, we consider the interaction after the surface charge layers come into contact. We consider the interaction forces or energies between colloidal particles covered with graft polymers (i.e., polymer brush layers). When these particles approach each other, steric repulsion is acting between the surface charge layers on the respective particles. A theory of de Gennes [1] on the interaction between particles coated by uncharged polymer brush layers assumes that when the two brushes come into contact, they are squeezed against each other but they do not interdigitate. If brushes are charged, then electrostatic repulsion between brushes must also be taken into account in addition to the steric repulsion. In this section, we extend the theory of de Gennes [1] to the case of charged polymer brushes and obtain an expression for the electrostatic repulsion between the surface charge layers of the plates after they come into contact. We restrict ourselves to the case where the brush layers are squeezed against each other but they do not interdigitate. We will thus not consider electrostatic interactions for the case where the polymer layers on interacting particles penetrate each other.

17.2 REPULSION BETWEEN INTACT BRUSHES

Consider two parallel identical plates coated with a charged polymer brush layer of intact thickness d_o at separation h immersed in a symmetrical electrolyte solution of valence z and bulk concentration n as shown in Fig. 17.1a [2]. We assume that dissociated groups of valence Z are uniformly distributed in the intact brush layer at a number density of N_o. We first obtain the potential distribution in the system when the two brushes are not in contact ($h > 2d_o$). We take an x-axis perpendicular to the brushes with its origin 0 at the core surface of the left plate so that the region

Biophysical Chemistry of Biointerfaces By Hiroyuki Ohshima

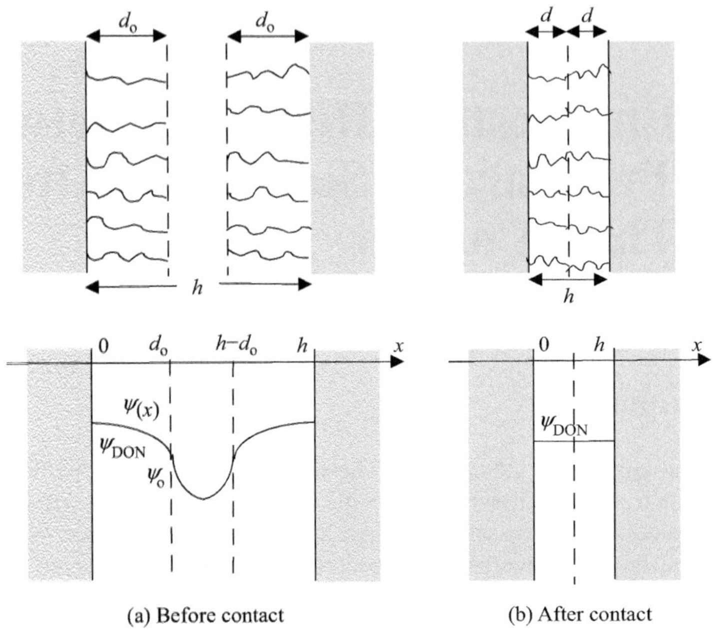

FIGURE 17.1 Interaction between two identical parallel plates covered with a charged polymer brush layer and potential distribution across the brush layers. (a) Before the two brushes come into contact ($h \geq 2d_o$). (b) After the two brushes come into contact ($h < 2d_o$). d_o is the thickness of the intact brush layer. From Ref. 2.

$d_o < x < h - d_o$ is the electrolyte solution and the regions $0 < x < d_o$ and $h - d_o < x < h$ are the brush layers. We also assume that the brush layers are penetrable to electrolyte ions as well as water molecules. As a result of the symmetry of the system we need consider only the regions $d_o < x < h/2$ and $0 < x < d_o$.

The electrostatic interaction force $P_e(h)$ per unit area between the two charged brush layers at separation h when they are separated ($h \geq 2d_o$) is given by the osmotic pressure at the midpoint between the plates $x = h/2$ minus that in the bulk solution phase, namely,

$$P_e(h) = 4nkT \sinh^2[y(h/2)/2], \quad h \geq 2d_o \tag{17.1}$$

where $y(h/2)$ can be obtained by solving Eqs. (32) and (33).

17.3 REPULSION BETWEEN COMPRESSED BRUSHES

Next we consider the situation after the two brushes come into contact. In this case the thickness of each compressed brush layer, which we denote by d ($\leq d_o$), is always half the separation h, namely,

$$h = 2d \leq 2d_o \tag{17.2}$$

Since the density of dissociated groups increases as h decreases, it is now a function of h. We denote by $N(h)$ the density of dissociated groups after contact. Following de Gennes [1], we assume that when the two brushes come into contact, they are squeezed against each other but they do not interdigitate (Fig. 17.1b). We also assume that the brushes are contracted in such a way that the density of dissociated groups is always uniform over the compressed brush layers. That is, the product of $N(h)$ and d is constant and always equal to $N_o d_o$, namely,

$$N(h)d = N(h)\frac{h}{2} = N_o d_o \tag{17.3}$$

The potential inside the brush layer is equal to the Donnan potential with the charge density $ZeN(h)$. We denote by $\psi_{DON}(h)$ the Donnan potential in the compressed brush layer, which is a function of h. Since the potential inside the compressed brush layers is constant independent x (but depends on the separation h) (Fig. 17.1b), the Poisson–Boltzmann equation in the brush layer becomes

$$\frac{d^2y}{dx^2} = \kappa^2\left[\sinh y - \frac{ZN(h)}{2zn}\right] = 0 \tag{17.4}$$

which is equivalent to the condition of electroneutrality in the brush layer, namely,

$$\rho_{el}(h) + \rho_{fix}(h) = 0 \tag{17.5}$$

where $\rho_{el}(h)$ is the charge density of fixed charges and $\rho_{fix}(h)$ is the charge density resulting from the electrolyte ions, both being functions of h, and they are given by

$$\rho_{fix}(h) = zeN(h) \tag{17.6}$$

$$\rho_{el}(h) = -2zen \sinh[y_{DON}(h)] \tag{17.7}$$

In Eq. (17.7), $y_{DON}(h) = ze\psi_{DON}(h)/kT$ is the scaled Donnan potential in the compressed brush layer. We thus obtain from Eq. (17.7)

$$\sinh(y_{DON}(h)) = \frac{ZN(h)}{2zn} = \frac{ZN_o d_o}{znh} \tag{17.8}$$

or equivalently

$$y_{DON}(h) = \ln\left[\frac{ZN(h)}{2zn} + \sqrt{\left(\frac{ZN(h)}{2zn}\right)^2 + 1}\right] = \ln\left[\frac{ZN_o d_o}{znh} + \sqrt{\left(\frac{ZN_o d_o}{znh}\right)^2 + 1}\right] \tag{17.9}$$

where Eq. (17.3) has been used in the last step in Eqs. (17.8) and (17.9). Note that the scaled Donnan potential $y_{\mathrm{DON}}(h)$ of the compressed brush layer is related to that of the intact brush layer as

$$\sinh(y_{\mathrm{DON}}(h)) = \sinh(y_{\mathrm{DON}}^{(0)})\left(\frac{2d_{\mathrm{o}}}{h}\right) \tag{17.10}$$

The interaction force $P_{\mathrm{e}}(h)$ per unit area between the two brush layers after contact may be calculated from Eq. (17.1) by replacing $y(h/2)$ with $y_{\mathrm{DON}}(h)$, namely,

$$P_{\mathrm{e}}(h) = 4nkT\sinh^2[y_{\mathrm{DON}}(h)/2], \quad h \le 2d_{\mathrm{o}} \tag{17.11}$$

which gives, by using Eq. (17.8),

$$P_{\mathrm{e}}(h) = 2nkT\left[\sqrt{1+\left(\frac{ZN_{\mathrm{o}}d_{\mathrm{o}}}{znh}\right)^2}-1\right], \quad h \le 2d_{\mathrm{o}} \tag{17.12}$$

Equation (2.136) may be rewritten in terms of the Donnan potential of the intact polymer brush layers $y_{\mathrm{DON}}^{(0)}$ with the help of Eq. (17.8) as

$$P_{\mathrm{e}}(h) = 2nkT\left[\sqrt{1+(\sinh y_{\mathrm{DON}}^{(0)})^2\left(\frac{2d_{\mathrm{o}}}{h}\right)^2}-1\right], \quad h \le 2d_{\mathrm{o}} \tag{17.13}$$

In Fig. 17.2, we give some results of the calculation of the interaction force $P_{\mathrm{e}}(h)$ acting between two compressed brush layers as a function of reduced separation κh for several values of ZN_{o}/zn at scaled brush thickness $\kappa d_{\mathrm{o}} = 10$. Calculation was made with the help of Eq. (17.12) for $h \le 2d_{\mathrm{o}}$ and Eq. (17.1) for $h \ge 2d_{\mathrm{o}}$. We see that the interaction force is always repulsive before and after the two brushes come into contact.

It is interesting to note that in contrast to the case of $h \ge 2d_{\mathrm{o}}$, where the interaction force is essentially an exponential function of κh, the interaction force after contact of the brush layers ($h \le 2d_{\mathrm{o}}$) is a function of the inverse power of h. In order to see this more clearly we consider the following limiting cases.

a. For highly charge brushes, that is, $|y_{\mathrm{DON}}| \gg 1$ or $|ZN_{\mathrm{o}}/zn| \gg 1$, Eq. (17.13) becomes

$$P_{\mathrm{e}}(h) = 2nkT|\sinh(y_{\mathrm{DON}}^{(0)})|\left(\frac{2d_{\mathrm{o}}}{h}\right) = \frac{2|Z|N_{\mathrm{o}}d_{\mathrm{o}}kT}{zh} \tag{17.14}$$

That is, the interaction force is proportional to $1/h$. It must be stressed that the interaction force in this case becomes independent of the electrolyte concentration n.

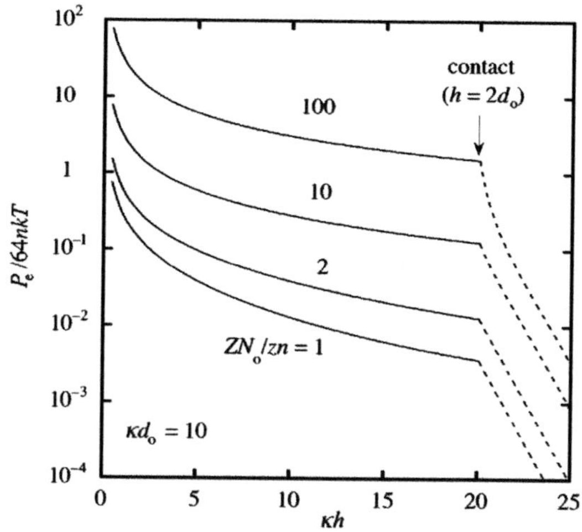

FIGURE 17.2 Scaled repulsion $P_e/64nkT$ per unit area between two parallel plates covered with a polymer brush layer as a function of scaled separation κh for $zN_o/\nu n = 1$, 2, 10, and 100 (i.e., $y_{DON} = 0.4812$, 0.8814, 2.312, and 4.605, respectively) at $\kappa d_o = 10$. Solid lines, after contact ($h < 2d_o$). Dotted lines, before contact ($h \geq 2d_o$). From Ref. 2.

b. For weakly charge brushes, that is, $|y_{DON}| \ll 1$ or $|ZN_o/zn| \ll 1$, on the other hand, Eq. (17.13) becomes

$$P_e(h) = nkT(y_{DON}^{(0)})^2 \left(\frac{2d_o}{h}\right)^2 = \frac{(ZN_o d_o)^2 kT}{z^2 n h^2} \tag{17.15}$$

which is proportional to $1/h^2$ and to $1/n$.

c. In the limit of small separations $h \to 0$, we again have Eq. (17.14), namely,

$$P_e(h) \to \frac{2|Z|N_o d_o kT}{zh}, \quad h \to 0 \tag{17.16}$$

If, further, the total charge amount σ per unit area, namely

$$\sigma = ZN_o d_o = ZNd \tag{17.17}$$

(which is independent of h) is introduced, then we have that

$$P_e(h) \to \frac{2|\sigma|kT}{zeh}, \quad h \to 0 \tag{17.18}$$

This result agrees with the limiting behavior of the interaction force per unit area between two parallel identical plates with constant surface charge density σ [3, 4].

According to de Gennes' theory [1], the repulsive force $P_s(h)$ per unit area between two uncharged brushes of intact thickness d_o at separation h is given by

$$P_s(h) = \frac{kT}{s^3}\left[\left(\frac{2d_o}{h}\right)^{9/4} - \left(\frac{h}{2d_o}\right)^{3/4}\right], \quad h \leq 2d_o \tag{17.19}$$

where each grafted polymer is considered as a chain of blobs of size s. Note that Eq. (17.19) is correct to within a numerical prefactor of order unity. The first term on the right-hand side of Eq. (17.19) comes from the osmotic repulsion between the brushes and the second term from the elastic energy of the chains. When brushes are charged, it is expected that the steric force $P_s(h)$ and electrostatic force $P_e(h)$ are both acting between the charged brush layers under compression. That is, the total repulsive force $P(h)$ per unit area between the charged compressed brush layers is given by the sum of $P_e(h)$ and $P_s(h)$, namely,

$$P(h) = P_e(h) + P_s(h) \tag{17.20}$$

Now we compare the magnitudes of $P_e(h)$ and $P_s(h)$ with the help of Eqs. (17.13) and (17.19). We can expect that N_o and $1/s^3$ are of the same order of magnitude. Indeed, if each brob in the polymer chain has one elementary electric charge, then we exactly have $N_o = 1/s^3$ (with $|Z| = 1$). It can be shown that for highly charged brushes with $|ZN_o/zn| \gg 1$, $P_e(h)$ can be comparable in magnitude to $P_s(h)$, while otherwise $P_e(h)$ gives relatively small contribution. The condition $|ZN_o/zn| \gg 1$ is

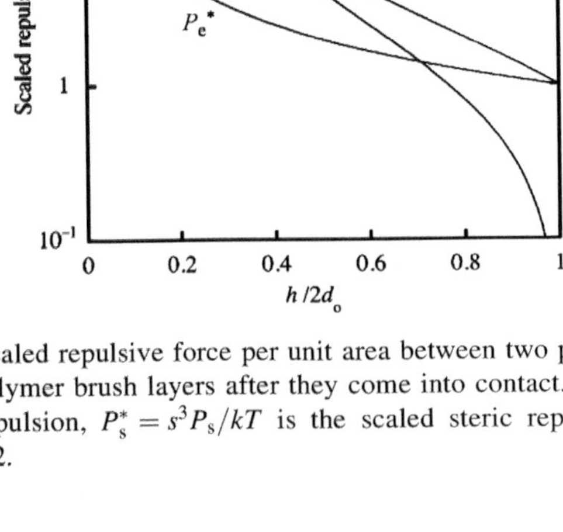

FIGURE 17.3 Scaled repulsive force per unit area between two parallel plates covered with compressed polymer brush layers after they come into contact. $P_e^* = P_e/N_o kT$ is the scaled electrical repulsion, $P_s^* = s^3 P_s/kT$ is the scaled steric repulsion and their sum $P_e^* + P_s^*$. From Ref. 2.

fulfilled for most practical cases. As a typical example, we may put $s \approx 1$ nm and thus $N_o \approx 1/s^3 \approx 10^{27}\,\mathrm{m}^{-3}$. Thus, if, for instance, $n = 10^{-2}\,\mathrm{M} = 6 \times 10^{24}\,\mathrm{m}^{-3}$, then we have $N/n \approx 1.7 \times 10^2$ so that $P_e\,(h)$ may be approximated by Eq. (17.14), which is independent of the electrolyte concentration n. In Fig. 17.3, we compare scaled repulsive forces $P_s^*(h)$ and $P_e^*(h)$ and their sum $P_s^*(h) + P_e^*(h)$ as a function of $h/2d_o$ for $|Z| = 1$ and $z = 1$, where $P_e^*(h) = P_e/N_okT$ and $P_s^*(h) = s^3 P_s/kT$. We see that P_e can be comparable in magnitude to P_s and can even exceed P_s especially when the brushes are weakly compressed.

REFERENCES

1. P. G. de Gennes, *Adv. Colloid Interface Sci.* 27 (1987) 189.
2. H. Ohshima, *Colloid Polym. Sci.* 277 (1999) 535.
3. D. McCormik, S.L. Carnie, and D. Y. C. Chan, *J. Colloid Interface Sci.* 169 (1995) 177.
4. J. N. Israelachvili, *Intermolecular and Surface Forces, 2nd edition*, Academic Press, New York, 1992.

18 Electrostatic Interaction Between Ion-Penetrable Membranes in a Salt-Free Medium

18.1 INTRODUCTION

Electric behaviors of colloidal particles in a salt-free medium containing counterions only are quite different from those in electrolyte solutions, as shown in Chapter 6. In this chapter, we consider the electrostatic interaction between two ion-penetrable membranes (i.e., porous plates) in a salt-free medium [1].

18.2 TWO PARALLEL HARD PLATES

Before considering the interaction between two ion-penetrable membranes, we here treat the interaction between two similar ion-impenetrable hard plates 1 and 2 carrying surface charge density σ at separation h in a salt-free medium containing counterions only (Fig. 18.1) [2]. We take an x-axis perpendicular to the plates with its origin on the surface of plate 1. As a result of the symmetry of the system, we need consider only the region $0 \leq x \leq h/2$. Let the average number density and the valence of counterions be n_o and $-z$, respectively. Then we have from electroneutrality condition that

$$\sigma = zen_o \cdot \frac{h}{2} \tag{18.1}$$

or

$$n_o = \frac{2\sigma}{zeh} \tag{18.2}$$

Note that n_o, which is a function of h, is proportional to $1/h$. We set the equilibrium electric potential $\psi(x)$ to zero at points where the volume charge density $\rho_{el}(x)$ resulting from counterions equals its average value ($-zen$).

Biophysical Chemistry of Biointerfaces By Hiroyuki Ohshima

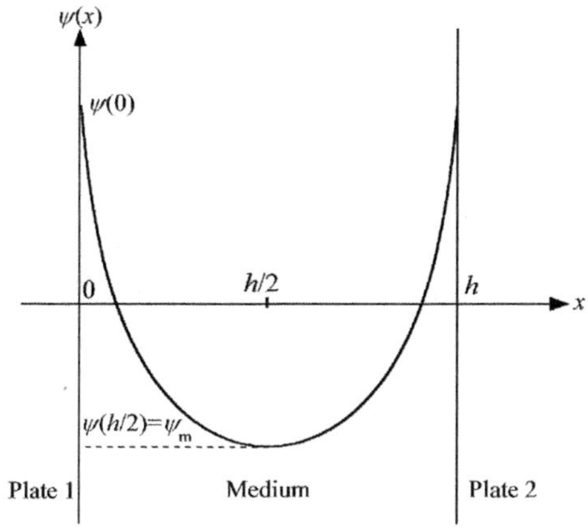

FIGURE 18.1 Schematic representation of the electrostatic interaction between two parallel identical hard plates separated by a distance h between their surfaces.

The Poisson equation is thus given by

$$\frac{d^2\psi}{dx^2} = -\frac{\rho_{el}}{\varepsilon_r\varepsilon_o}, \quad 0 < x < h/2 \tag{18.3}$$

Here we have assumed that the relative permittivity ε_r is assumed to take the same value inside and outside the membrane. We also assume that the distribution of counterions $n(x)$ obeys a Boltzmann distribution, namely,

$$n(x) = n_o \exp\left[-\frac{-ze\psi(x)}{kT}\right] = n_o \exp\left[\frac{ze\psi(x)}{kT}\right] \tag{18.4}$$

Thus, the charge density $\rho_{el}(x)$ is given by

$$\rho_{el}(x) = zen_o \exp\left[\frac{ze\psi(x)}{kT}\right] \tag{18.5}$$

Thus, we obtain the following Poisson–Boltzmann equation:

$$\frac{d^2y}{dx^2} = \kappa^2 e^y, \quad 0 < x < h/2 \tag{18.6}$$

with

$$y(x) = \frac{ze\psi(x)}{kT} \tag{18.7}$$

and

$$\kappa = \left(\frac{z^2 e^2 n_o}{\varepsilon_r \varepsilon_o kT}\right)^{1/2} = \left(\frac{2ze\sigma}{\varepsilon_r \varepsilon_o kTh}\right)^{1/2} \tag{18.8}$$

where $y(r)$ is the scaled equilibrium potential, κ is the Debye–Hückel parameter in the present system. The boundary conditions for Eq. (18.6) are

$$\left.\frac{dy}{dx}\right|_{x=0} = -\left(\frac{ze}{kT}\right)\frac{\sigma}{\varepsilon_r \varepsilon_o} = -\frac{\kappa^2 h}{2} \tag{18.9}$$

$$\left.\frac{dy}{dx}\right|_{x=h/2} = 0 \tag{18.10}$$

Equation (18.9), which implies that the influence of the electric field within the plates can be neglected, is consistent with Eq. (18.1). Equation (18.10) comes from the result of symmetry of the system.

Integration of Eq. (18.6) subject to Eq. (18.10) yields

$$\frac{dy}{dx} = -\sqrt{2}\kappa \exp\left(\frac{y_m}{2}\right)\tan\left[\exp\left(\frac{y_m}{2}\right)\frac{\kappa}{\sqrt{2}}\left(\frac{h}{2} - x\right)\right] \tag{18.11}$$

with

$$y_m = y(h/2) \tag{18.12}$$

where y_m is the scaled potential at the midpoint between the plates. Equation (18.11) is further integrated to give

$$y(x) = -\ln\left[\cos^2\left\{\exp\left(\frac{y_m}{2}\right)\frac{\kappa}{\sqrt{2}}\left(\frac{h}{2} - x\right)\right\}\right] + y_m \tag{18.13}$$

By combining Eqs. (18.9) and (18.12), we obtain the following transcendental equation for y_m:

$$\tan\left[\frac{\kappa h}{2\sqrt{2}}\exp\left(\frac{y_m}{2}\right)\right] = \frac{\kappa h}{2\sqrt{2}}\exp\left(-\frac{y_m}{2}\right), \tag{18.14}$$

The electrostatic interaction force acting between plates 1 and 2 per unit area can be calculated from an excess osmotic pressure at the midpoint between the plates, namely,

$$P(h) = n(h/2)kT = n_o kT\, e^{y_m} = \frac{2\sigma}{h}\left(\frac{kT}{ze}\right)e^{y_m} \tag{18.15}$$

In the limit of small κh, it follows from Eq. (18.14) that

$$y_m \to 0 \quad \text{as } h \to 0 \tag{18.16}$$

so that Eq. (18.15) gives

$$P(h) \to \frac{2\sigma}{h}\left(\frac{kT}{ze}\right) \tag{18.17}$$

In the opposite limit of large κh, it follows from Eq. (18.14) that

$$\frac{\kappa h}{2\sqrt{2}}\exp\left(\frac{y_m}{2}\right) \to \frac{\pi}{2} \quad \text{as } h \to \infty \tag{18.18}$$

Thus by substituting Eq. (18.8) into Eq. (18.15), we have

$$P(h) \to \frac{2\pi^2 \varepsilon_r \varepsilon_o}{h^2}\left(\frac{kT}{ze}\right)^2 \quad \text{as } h \to \infty \tag{18.19}$$

Equations (18.17) and (18.19) show that $P(h)$ is proportional to $1/h$ for small κh but to $1/h^2$ at large κh.

18.3 TWO PARALLEL ION-PENETRABLE MEMBRANES

Now consider two parallel identical ion-penetrable membranes 1 and 2 at separation h immersed in a salt-free medium containing only counterions. Each membrane is fixed on a planar uncharged substrates (Fig. 18.2). We obtain the electric potential distribution $\psi(x)$. We assume that fixed charges of valence Z are distributed in the membrane of thickness d with a number density of N (m^{-3}) so that the fixed-charge density ρ_{fix} within the membrane is given by

$$\rho_{\text{fix}} = ZeN \tag{18.20}$$

We take an x-axis perpendicular to the membranes with its origin on the surface of membrane 1. Let the average number density and the valence of counterions be n_o and $-v$, respectively. Then we have from electroneutrality condition that

$$ZeNd = zen_o\left(\frac{h}{2} + d\right) \tag{18.21}$$

which is rewritten as

$$\frac{zn_o}{ZN} = \frac{2d}{h + 2d} \tag{18.22}$$

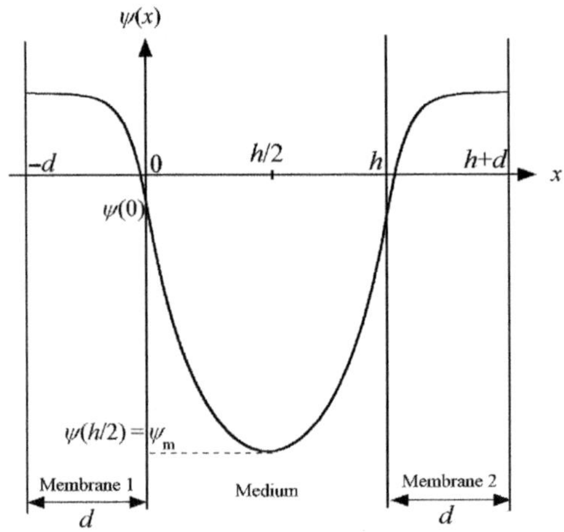

FIGURE 18.2 Schematic representation of the electrostatic interaction between two parallel identical ion-penetrable membranes (porous plates) of thickness d separated by a distance h between their surfaces. Each membrane is fixed on an uncharged planar substrate.

Note that n_o becomes proportional to $1/h$ for $h \gg d$. We set the equilibrium electric potential $\psi(x)$ to zero at points where the volume charge density $\rho_{el}(x)$ resulting from counterions equals its average value $(-zen)$.

The Poisson equations are thus given by

$$\frac{d^2\psi}{dx^2} = -\frac{\rho_{el} + \rho_{fix}}{\varepsilon_r \varepsilon_o}, \quad -d < x < 0 \tag{18.23}$$

$$\frac{d^2\psi}{dx^2} = -\frac{\rho_{el}}{\varepsilon_r \varepsilon_o}, \quad 0 < x \leq h/2 \tag{18.24}$$

Here we have assumed that the relative permittivity ε_r is assumed to take the same value inside and outside the membrane. We also assume that the distribution of counterions $n(x)$ obeys Eq. (18.4) and thus the charge density $\rho_{el}(x)$ is given by Eq. (18.5).Thus, we obtain the following Poisson–Boltzmann equations for the scaled potential $y(x) = ze\psi(x)/kT$:

$$\frac{d^2y}{dx^2} = \kappa^2 \left(e^y - \frac{ZN}{zn_o} \right), \quad -d < x < 0 \tag{18.25}$$

$$\frac{d^2y}{dx^2} = \kappa^2 e^y, \quad 0 < x < h/2 \tag{18.26}$$

with

$$\kappa = \left(\frac{z^2 e^2 n_o}{\varepsilon_r \varepsilon_o kT} \right)^{1/2} = \left(\frac{2ze^2 ZN}{\varepsilon_r \varepsilon_o kT} \cdot \frac{d}{h + 2d} \right)^{1/2} \tag{18.27}$$

where κ is the Debye–Hückel parameter in the present system. With the help of Eq. (18.21), Eq. (18.25) is rewritten as

$$\frac{d^2y}{dx^2} = \kappa^2\left(e^y - \frac{h+2d}{2d}\right), \quad -d < x < 0 \tag{18.28}$$

Equation (18.27) shows that the Debye–Hückel parameter κ depends on the fixed charge density ZeN in the membrane and the membrane separation h (κ is proportional to $1/h^{1/2}$ for $h \gg d$ and the h dependence of κ becomes small for $h \ll d$), unlike the case of media containing electrolyte solution, in which case κ is essentially constant independent of ZeN and h. The boundary conditions for Poisson–Boltzmann equations (18.26) and (18.28) are

$$\left.\frac{dy}{dx}\right|_{x=-d} = 0 \tag{18.29}$$

$$\left.\frac{dy}{dx}\right|_{x=h/2} = 0 \tag{18.30}$$

$$y(0^-) = y(0^+) \tag{18.31}$$

$$\left.\frac{dy}{dx}\right|_{x=0^-} = \left.\frac{dy}{dx}\right|_{x=0^+} \tag{18.32}$$

As a result of symmetry of the system, we need only to consider the region $-d \le x \le h/2$. For the region inside the membrane $-d \le x \le 0$, Eq. (18.28) subject to Eq. (18.29) is solved to give

$$\frac{dy}{dx} = -\sqrt{2}\kappa\sqrt{\exp(y) - \exp(y(-d)) - \left(\frac{h+2d}{2d}\right)\{y - y(-d)\}} \tag{18.33}$$

Equation (18.33) can further be integrated to

$$-\sqrt{2}\kappa(x+d) = \int_{y(-d)}^{y(x)} \frac{dy}{\sqrt{\exp(y) - \exp(y(-d)) - (h+2d/2d)\{y - y(-d)\}}} \tag{18.34}$$

which gives $y(x)$ as an implicit function of x in the region $-d \le x \le 0$.

For the region $0 \le x \le h/2$, integration of Eq. (18.26) subject to Eq. (18.30) yields (see Eq. (18.11))

$$\frac{dy}{dx} = -\sqrt{2}\kappa\exp\left(\frac{y_m}{2}\right)\tan\left[\exp\left(\frac{y_m}{2}\right)\frac{\kappa}{\sqrt{2}}\left(\frac{h}{2} - x\right)\right] \tag{18.35}$$

where $y_m = y(h/2)$is the potential at the midpoint between the membranes. Equation (18.35) is further integrated to give

$$y(x) = -\ln\left[\cos^2\left\{\exp\left(\frac{y_m}{2}\right)\frac{\kappa}{\sqrt{2}}\left(\frac{h}{2} - x\right)\right\}\right] + y_m \tag{18.36}$$

Equations (18.34) and (18.36) contain unknown potential values $y_m = y(h/2)$, $y(-d)$, and $y(0)$ that can be determined by continuity conditions (18.31) and (18.32) at $x = 0$, namely,

$$y(0) = -\ln\left[\cos^2\left\{\exp\left(\frac{y_m}{2}\right)\frac{\kappa h}{2\sqrt{2}}\right\}\right] + y_m \tag{18.37}$$

$$-\sqrt{2}\kappa d = \int_{y(-d)}^{y(0)} \frac{dy}{\sqrt{\exp(y) - \exp(y(-d)) - (h + 2d/2d)\{y - y(-d)\}}} \tag{18.38}$$

$$e^{y_m} - e^{y(-d)} - \left(\frac{h + 2d}{2d}\right)\{y(0) - y(-d)\} = 0 \tag{18.39}$$

where the following condition must be satisfied:

$$0 \leq \frac{\kappa h}{2\sqrt{2}}\exp\left(\frac{y_m}{2}\right) < \frac{\pi}{2} \tag{18.40}$$

The solution to coupled Eqs. (18.37)–(18.39) determines the values of $y_m = y(h/2)$, $y(-d)$, and $y(0)$ as functions of the membrane separation h.

The potential distribution can be calculated with Eqs. (18.34) and (18.36) with the help of the values of $y(-d)$ and y_m. The electrostatic interaction force acting between membranes 1 and 2 per unit area can be calculated from

$$\begin{aligned} P(h) = n(h/2)kT &= n_o kT\, e^{y_m} \\ &= \frac{ZN}{z}\left(\frac{2d}{h + 2d}\right)kT\, e^{y_m} \end{aligned} \tag{18.41}$$

where the value of y_m can be obtained by solving Eqs. (18.37)–(18.39). Note that both n_o and y_m depend on h.

In Figs 18.3–18.5, the results of some calculations are given for the counter-ion concentration $n(x)$ (Fig. 18.3), potential distribution $\psi(x)$ (Fig. 18.4), and the interaction force per unit area $P(h)$ (Fig. 18.5). Calculations are made for water at 25°C ($\varepsilon_r = 78.55$), $Z = z = 1$ and the membrane thicknesses $d = 0$, 1, and 5 nm with the amount of fixed charges σ in the membrane per unit area kept constant at 0.2 C/m^2, with σ defined by

$$\sigma = ZeNd \tag{18.42}$$

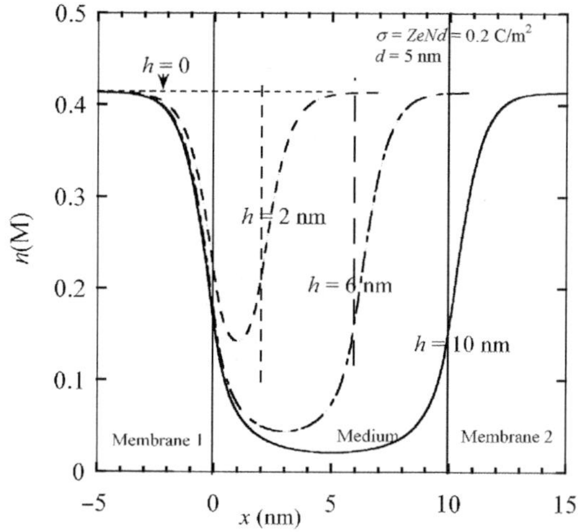

FIGURE 18.3 Distributions of counter-ion concentration $n(x)$ across two parallel identical ion-penetrable membranes of thickness d separated by a distance h between their surfaces in a salt-free medium. Distributions were calculated at $h=0$, 2, 6, and 10 nm for $Z=z=1$, $d=5$ nm, and the charge amount per unit area $\sigma=ZeNd=0.2\,\mathrm{C/m^2}$ in water at 25°C ($\varepsilon_r=78.55$). From Ref. [1].

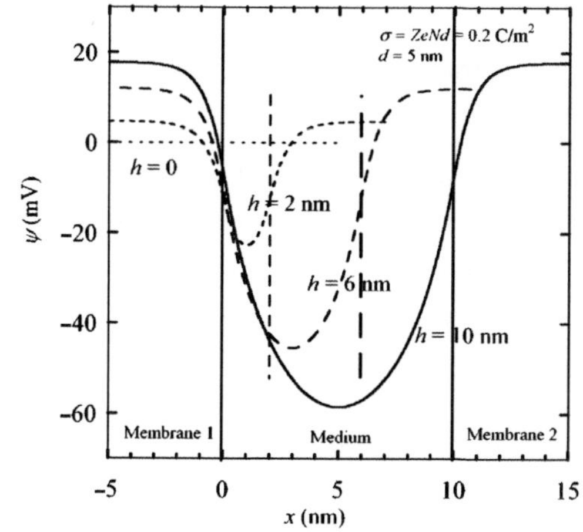

FIGURE 18.4 Distribution of electric potential $\psi(x)$ across two parallel identical ion-penetrable membranes of thickness d separated by a distance h between their surfaces in a salt-free medium. Calculated at $h=0$, 2, 6, and 10 nm for $Z=z=1$, $d=0$, 1, and 5 nm and the charge amount per unit area $\sigma=ZeNd$ kept constant at $\sigma=0.2\,\mathrm{C/m^2}$ in water at 25°C ($\varepsilon_r=78.55$). From Ref. [1].

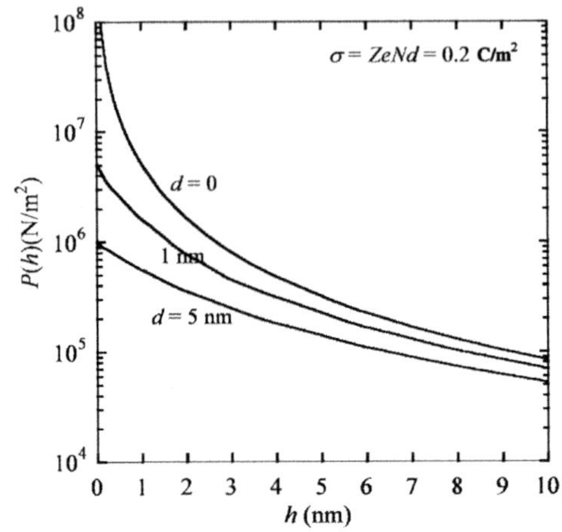

FIGURE 18.5 Repulsive force $P(h)$ per unit area between two parallel identical ion-penetrable membranes of thickness d separated by a distance h between their surfaces in a salt-free medium. Calculated for $Z=z=1$, $d=0$, 1, and 5 nm with the charge amount per unit area $\sigma=ZeNd$ kept constant at $\sigma=0.2\,\mathrm{C/m^2}$ in water at 25°C ($\varepsilon_r=78.55$). From Ref. [1].

Figures 18.3–18.5 demonstrate that $n(x)$, $\psi(x)$, and $P(h)$ depend significantly on h and/or d. In the limiting case in which $d\to 0$ and $N\to\infty$ with $\sigma=ZeNd$ kept constant so that the membrane becomes a planar surface carrying a charge density σ. In this limit, Eq. (18.27) for the Debye–Hückel parameter tends to Eq. (18.8) and the electroneutrality condition (Eq. (18.21)) tends to Eq. (18.1). Thus, Eqs. (18.37)–(18.39) reduce to the single transcendental Eq. (18.14) for two parallel planar surfaces. Figure 18.4 demonstrate how the potential distribution $\psi(x)$ changes with the membrane separation h. Figure 18.4 shows that the potential in the region deep inside the membrane is almost equal to the Donnan potential ψ_{DON}, which is given by setting the right-hand side of Eq. (18.25) or Eq. (18.28) equal zero:

$$\psi_{\mathrm{DON}}=\frac{kT}{ze}\ln\left(\frac{ZN}{zn_o}\right)=\frac{kT}{ze}\ln\left(\frac{h+2d}{2d}\right) \tag{18.43}$$

Equation (18.43) shows that ψ_{DON} depends on h and decreases in magnitude with decreasing h, tending to zero in the limit $h\to 0$. Note that in the limiting case of $h\to 0$, the potential $\psi(x)$ becomes zero everywhere outside and inside the membranes, as is seen in Fig. 18.4. In the region where the potential is equal to the Donnan potential, the electroneutrality condition holds so that the membrane-fixed charges (ZeN) are completely neutralized by the charges of counterions penetrating the membrane interior ($-zen(x)$). Figure 18.3 indeed shows that in the region deep inside the membrane the concentration $n(x)$ of counterions is practically equal to the concentration N of membrane-fixed charges (which is independent of h). It is seen from Fig. 18.4, on the other hand, that the potential deep inside the membrane (which is almost equal to the Donnan potential) changes significantly with h. This is because the Donnan potential ψ_{DON} is a function of h (Eq. (18.43)).

Figure 18.5 shows the dependence of the interaction force $P(h)$ on the membrane separation h. The interaction force is shown to be always repulsive and decreases in magnitude with increasing h, tending to zero as $h \to \infty$. It can be shown that in the case where $d \to 0$ and $N \to \infty$ with $\sigma = ZeNd$ kept constant, $P(h)$ becomes proportional to $1/h$ as $h \to 0$, namely,

$$P(h) \to \frac{2\sigma}{h}\left(\frac{kT}{ze}\right) \quad \text{as } h \to 0 \tag{18.44}$$

which agrees with the case of two charged planar surfaces in a salt-free medium (Eq. (16.17)). It is of interest to note that Eq. (18.44) agrees also with the limiting force expression for the case of electrolyte solutions (Eq. (9.203)). This is because for very small h, the interaction force is determined only by counterions irrespective of whether coions are present (salt-containing media) or absent (salt-fee media). For the case of finite d, the interaction force $P(h)$ remains finite in the limit of $h \to 0$. The value of $P(0)$ can easily be obtained from Eq. (18.41) by noting that in this limit $zn_o = ZN$ and $y_m = 0$, namely,

$$P(h) \to n_o kT \to \frac{ZN}{z}kT \quad \text{as } h \to 0 \tag{18.45}$$

For large values of h, on the other hand, it follows from Eq. (18.40) that $(\kappa h/2\sqrt{2})\exp[y_m/2]$ must tend to $\pi/2$. Thus in this limit we have from Eq. (18.41) that

$$P(h) \to \frac{2\pi^2 \varepsilon_r \varepsilon_o}{h^2}\left(\frac{kT}{ze}\right)^2 \quad \text{as} \quad h \to \infty \tag{18.46}$$

It is of interest to note that this limiting expression for $P(h)$ is independent of the membrane-fixed charges ZeN and the membrane thickness d. This limiting form thus agrees with that for the interaction between two charged planar surfaces in a salt-free medium (Eq. 18.19).

Equation (18.38) involves numerical integration. Here we give an approximate expression without numerical integration for Eq. (18.38). Since the largest contribution to the integrand of Eq. (18.38) comes from the region near $y = y(-d)$, we expand the integrand around $y = y(-d)$ so that the right-hand side of Eq. (18.38) can be approximated by

$$\int_{y(-d)}^{y(0)} \frac{dy}{\sqrt{e^y - e^{y(-d)} - (h + 2d/2d)\{y - y(-d)\}}} \approx \frac{-2\sqrt{y(-d) - y(0)}}{\sqrt{(h + 2d/2d) - e^{y(-d)}}} \tag{18.47}$$

By using this result, Eqs. (18.37)–(18.41) reduce to the single transcendental equation for y_m,

$$\left(\frac{Q}{1+Q}\right)\left\{1-\left(\frac{2d}{h+2d}\right)e^{y_m}\right\}$$
$$=\ln\left[\left\{\frac{1+(h/2d+1)Q\,e^{-y_m}}{1+Q}\right\}\cos^2\left\{\left(\frac{2Qd}{h+2d}\right)^{1/2}\left(\frac{h}{2d}\right)\exp\left(\frac{y_m}{2}\right)\right\}\right] \quad (18.48)$$

with

$$0\leq\left(\frac{2Qd}{h+2d}\right)^{1/2}\left(\frac{h}{2d}\right)\exp\left(\frac{y_m}{2}\right)<\frac{\pi}{2} \quad (18.49)$$

where Q is defined by

$$Q=\frac{zZe^2Nd^2}{2\varepsilon_r\varepsilon_o kT} \quad (18.50)$$

and is related to κ by

$$\left(\frac{2Qd}{h+2d}\right)^{1/2}\left(\frac{h}{2d}\right)=\frac{\kappa h}{2\sqrt{2}}$$

Note that in the limit of $d \to 0$ (and $N \to \infty$) with σ kept constant, Eq. (18.48) reduces correctly to Eq. (18.14).

REFERENCES

1. H. Ohshima, *J. Colloid Interface Sci.* 260 (2003) 339.
2. J.N. Israelachivil, *Intermolecular Forces*, Academic Press, New York, 1991, Sec. 12.7.

19 van der Waals Interaction Between Two Particles

19.1 INTRODUCTION

A neutral molecule can have a fluctuating instantaneous electric dipole. This fluctuation originates from quantum mechanics. When two molecules approach toward each other, attractive intermolecular forces, known as the van der Waals forces, act due to interactions between the fluctuating dipoles. A most remarkable characteristic of the van der Waals interaction is that the additivity of interactions approximately holds. The interaction energy between two particles in a vacuum may thus be calculated approximately by a summation (or by an integration) of the interaction energies for all molecular pairs formed by two molecules belonging to different particles [1–4] (Fig. 19.1). That is, the interaction energy V between two particles 1 and 2, containing N_1 and N_2 molecules per unit volume, respectively, can be obtained by integration of the interaction energy u between two volume elements dV_1 and dV_2 at separation r, containing $N_1\,dV_1$ and $N_2\,dV_2$ molecules, respectively, over the volumes V_1 and V_2 of particles 1 and 2 (Fig. 19.2), namely,

$$V = \int_{V_1}\int_{V_2} u(r) N_1 N_2 \, dV_1 \, dV_2 \tag{19.1}$$

As a result, although the van der Waals interaction energy between two single molecules is small, the interaction energy between two particles consisting many molecules can be large, as shown below.

19.2 TWO MOLECULES

The potential energy $u(r)$ of the van der Waals forces between two similar molecules at separation r between their centers in a vacuum is given by [1–4]

$$u(r) = -\frac{C}{r^6} \tag{19.2}$$

Biophysical Chemistry of Biointerfaces By Hiroyuki Ohshima

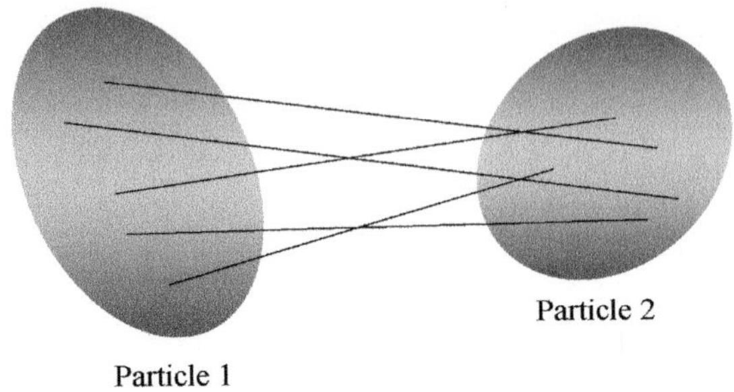

FIGURE 19.1 Interaction between two particles 1 and 2.

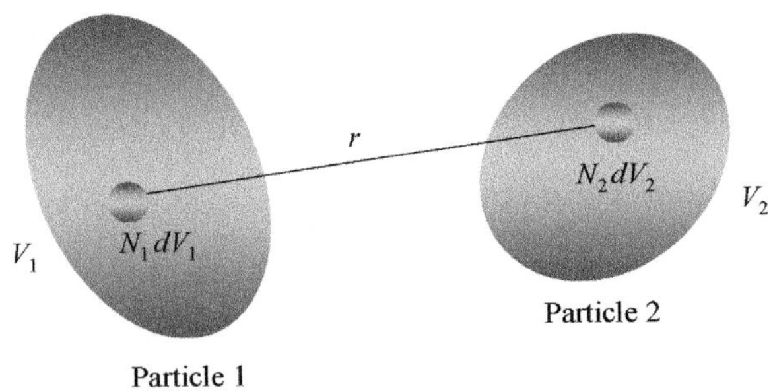

FIGURE 19.2 Integration of the interaction energy between two volume elements dV_1 and dV_2, containing $N_1\,dV_1$ and $N_2\,dV_2$ molecules, respectively, over the volumes V_1 and V_2 of particles 1 and 2.

with

$$C = \frac{3\alpha^2 h\nu}{4(4\pi\varepsilon_o)^2} \tag{19.3}$$

where C is called the London–van der Waals constant, α and ν are, respectively, the electronic polarizability and the characteristic frequency of the molecule, and h is the Planck's constant. Thus, $h\nu$ corresponds to the ionization energy of the molecule. Equation (19.2) holds when the intermolecular distance is shorter than the characteristic wavelength $2\pi/\nu$. At large distances, $u(r)$ becomes proportional to $1/r^7$ instead of $1/r^6$ (retardation effect).

For two interacting dissimilar molecules 1 and 2, $u(r)$ is given by

$$u(r) = -\frac{C_{12}}{r^6} \tag{19.4}$$

with

$$C_{12} = \frac{3\alpha_1\alpha_2 h}{2(4\pi\varepsilon_o)^2}\left(\frac{\nu_1\nu_2}{\nu_1 + \nu_2}\right) \tag{19.5}$$

where α_i and ν_i are, respectively, the electronic polarizability and the characteristic frequency of molecule i ($i = 1, 2$). To a good approximation, Eq. (19.5) may be replaced by

$$C_{12} \approx \frac{3\alpha_1\alpha_2 h\sqrt{\nu_1}\sqrt{\nu_2}}{4(4\pi\varepsilon_o)^2} = \sqrt{C_1}\sqrt{C_2} \tag{19.6}$$

with

$$C_1 = \frac{3\alpha_1^2 h\nu_1}{4(4\pi\varepsilon_o)^2} \tag{19.7}$$

and

$$C_2 = \frac{3\alpha_2^2 h\nu_2}{4(4\pi\varepsilon_o)^2} \tag{19.8}$$

19.3 A MOLECULE AND A PLATE

The interaction energy $V(h)$ between a point-like molecule and an infinitely large plate of thickness d and molecular density N separated by a distance h in a vacuum (Fig. 19.3) can be calculated by integrating the molecular interaction energy $u(r)$ (Eq. (19.2)) between the molecule and a ring of radius r and volume $2\pi r dr dx$ (in which the number of molecules is equal to $N{\cdot}2\pi r dr\, dx$) over the entire volume of the plate, namely,

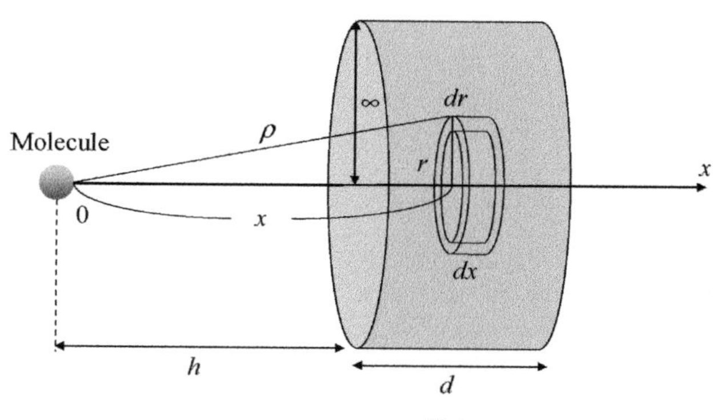

FIGURE 19.3 A point-like molecule and a plate of thickness d separated by a distance h.

$$V(h) = \int_{x=h}^{h+d} \int_{r=0}^{\infty} u(\rho) N \cdot 2\pi r dr \, dx \tag{19.9}$$

with

$$u(\rho) = -\frac{C_{12}}{\rho^6} = -\frac{C_{12}}{(r^2 + x^2)^3} \tag{19.10}$$

where C_{12} is the London–van der Waals constant for the interaction between the single molecule and a molecule within the sphere. After integrating Eq. (19.9), we obtain

$$V(h) = -\frac{\pi C_{12} N}{6} \left[\frac{1}{h^3} - \frac{1}{(h+d)^3} \right] \tag{19.11}$$

In particular, as $d \to \infty$, Eq. (19.11) becomes

$$V(h) = -\frac{\pi C_{12} N}{6h^3} \tag{19.12}$$

which is the interaction energy between a molecule and an infinitely thick plate.

19.4 TWO PARALLEL PLATES

The interaction energy $V(h)$ per unit area between two parallel similar plates of thickness d separated by distance h between their surfaces in a vacuum (Fig. 19.4) can be obtained by integrating Eq. (19.11)

$$V(h) = \int_{h}^{h+d} \left\{ -\frac{\pi C N}{6} \left[\frac{1}{h^3} - \frac{1}{(h+d)^3} \right] \right\} N dh \tag{19.13}$$

which gives

$$V(h) = -\frac{A}{12\pi} \left[\frac{1}{h^2} - \frac{2}{(h+d)^2} + \frac{1}{(h+2d)^2} \right] \tag{19.14}$$

where

$$A = \pi^2 C N^2 \tag{19.15}$$

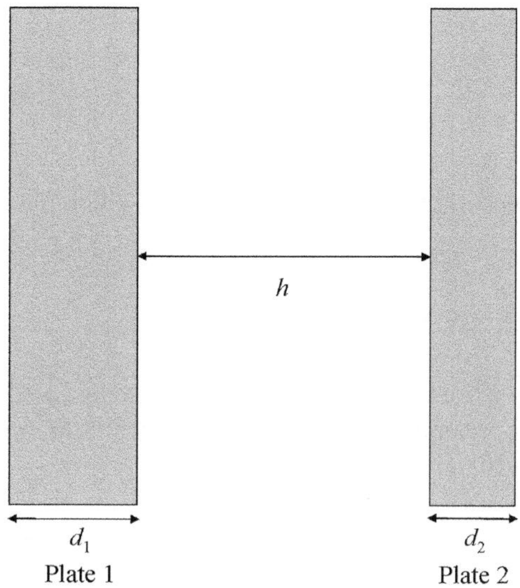

FIGURE 19.4 Two parallel plates of thicknesses d_1 and d_2 separated by a distance h.

is called the Hamaker constant. Note that the Hamaker constant given by Eq. (19.15) can be applied to particles of arbitrary shape (plates, spheres, or cylinders).

Similarly, the interaction energy $V(h)$ between two parallel dissimilar plates 1 and 2 per unit area in a vacuum, which have molecular densities N_1 and N_2, London–van der Waals constant C_1 and C_2, and thicknesses d_1 and d_2, respectively, can be obtained from Eq. (4.11)

$$V(h) = -\frac{A_{12}}{12\pi}\left[\frac{1}{h^2} - \frac{1}{(h+d_1)^2} - \frac{1}{(h+d_2)^2} + \frac{1}{(h+d_1+d_2)^2}\right] \quad (19.16)$$

with

$$A_{12} = \sqrt{A_1}\sqrt{A_2} = \pi\sqrt{C_1}\sqrt{C_2}N_1N_2 \quad (19.17)$$

$$A_1 = \pi^2 C_1 N_1^2 \quad (19.18)$$

$$A_2 = \pi^2 C_2 N_2^2 \quad (19.19)$$

where A_i is the Hamaker constant for the interaction between similar particles of substance i ($i = 1, 2$) and A_{12} is that for dissimilar particles of substances 1 and 2. Equation (19.17) follows from approximate Eq. (19.6).

For small h ($h \ll d_1, d_2$), Eq. (19.16) tends to

$$V(h) = -\frac{A_{12}}{12\pi h^2} \tag{19.20}$$

which is the interaction energy between two infinitely thick plates 1 and 2.

19.5 A MOLECULE AND A SPHERE

The interaction energy $V(R)$ between a single molecule and a sphere of radius a and molecular density N separated by a distance R between their centers (Fig. 19.5) can be calculated by using Eq. (19.4):

$$V(R) = \int_{r=0}^{a} \int_{\theta=0}^{\pi} \int_{\phi=0}^{2\pi} Nu(\rho) r^2 \sin\theta \, d\theta \, d\phi \, dr \tag{19.21}$$

with

$$u(\rho) = -\frac{C_{12}}{\rho^6} = -\frac{C_{12}}{(r^2 + R^2 - 2rR\cos\theta)^3} \tag{19.22}$$

where C_{12} is the London–van der Waals constant for the interaction between the single molecule and a molecule within the sphere. With the help of the following formula:

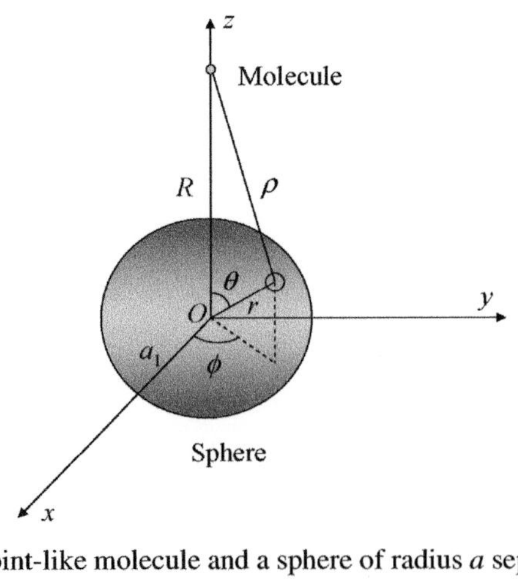

FIGURE 19.5 A point-like molecule and a sphere of radius a separated by a distance R.

$$\int_0^{\pi} \frac{\sin\theta\, d\theta}{(r^2 + R^2 - 2rR\cos\theta)^3} = \frac{2(R^2 + r^2)}{(R^2 - r^2)^4} \tag{19.23}$$

Eq. (19.21) gives

$$V(R) = -\frac{4\pi C_{12}N}{3} \frac{a^3}{(R^2 - a^2)^3} \tag{19.24}$$

19.6 TWO SPHERES

The interaction energy $V(R)$ between two dissimilar spheres 1 and 2, having radii a_1 and a_2, respectively, separated by a distance R between their centers (Fig. 19.6) can be calculated by using Eq. (19.23)

$$\begin{aligned} V(R) &= \int_{r=0}^{a_1}\int_{\theta=0}^{\pi}\int_{\phi=0}^{2\pi} N_1 \left[-\frac{4\pi C_{12}N_2}{3} \frac{a_2^3}{(\rho^2 - a_2^2)^3} \right] r^2 \sin\theta d\theta d\phi dr \\ &= \int_{r=0}^{a_1}\int_{\theta=0}^{\pi}\int_{\phi=0}^{2\pi} N_1 \left[-\frac{4\pi C_{12}N_2}{3} \frac{a_2^3}{(r^2 + R^2 - 2rR\cos\theta - a_2^2)^3} \right] r^2 \sin\theta d\theta d\phi dr \end{aligned} \tag{19.25}$$

where N_1 and N_2 are, respectively, the numbers of molecules within spheres 1 and 2.

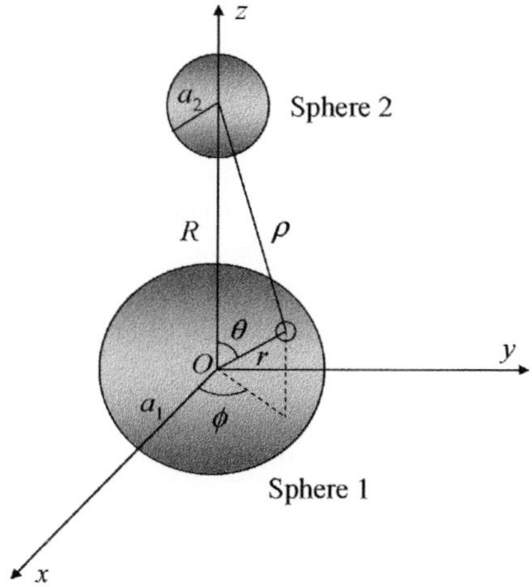

FIGURE 19.6 Two spheres of radii a_1 and a_2 separated by a distance R between their centers.

With the help of the following formula:

$$\int_0^{\pi} \frac{\sin\theta\, d\theta}{(r^2+R^2-2rR\cos\theta-a_2^2)^3} = \frac{1}{4rR}\left[\frac{1}{\{a_2^2-(r-R)^2\}^2} - \frac{1}{\{a_2^2-(r+R)^2\}^2}\right] \tag{19.26}$$

Eq. (19.25) becomes

$$V(R) = -\frac{A_{12}}{6}\left\{\frac{2a_1a_2}{R^2-(a_1+a_2)^2} + \frac{2a_1a_2}{R^2-(a_1-a_2)^2} + \ln\left[\frac{R^2-(a_1+a_2)^2}{R^2-(a_1-a_2)^2}\right]\right\} \tag{19.27}$$

where A_{12} is given by Eq. (19.17). For the special case of two identical spheres ($a_1 = a_2 = a$, $C_1 = C_2 = C$, and $N_1 = N_2 = N$), Eq. (19.27) becomes

$$V(R) = -\frac{A}{6}\left\{\frac{2a^2}{R^2-4a^2} + \frac{2a^2}{R^2} + \ln\left(1-\frac{4a^2}{R^2}\right)\right\} \tag{19.28}$$

where A is given by Eq. (19.15).

For small H ($H \ll a_1, a_2$), Eq. (19.27) tends to

$$V(H) = -\frac{A_{12}a_1a_2}{6(a_1+a_2)H} \tag{19.29}$$

with

$$H = R-(a_1+a_2) \tag{19.30}$$

where H is the closest distance between the surfaces of two spheres 1 and 2. For the special case of two identical spheres, Eq. (19.29) gives

$$V(H) = -\frac{Aa}{12H} \tag{19.31}$$

Equation (19.29) can also be obtained with the help of Derjaguin's approximation [5] (see Chapter 12), that is, by substituting Eq. (19.20) into Eq. (12.2).

In the opposite limit of large ($R \gg a_1 + a_2$), Eq. (19.27) tends to

$$\begin{aligned} V(R) &= -\frac{16A_{12}a_1^3a_2^3N_1N_2}{9R^6} \\ &= -\frac{(\frac{4}{3}\pi a_1^3N_1)(\frac{4}{3}\pi a_2^3N_2)C_{12}}{R^6} \end{aligned} \tag{19.32}$$

which is consistent with Eq. (19.4).

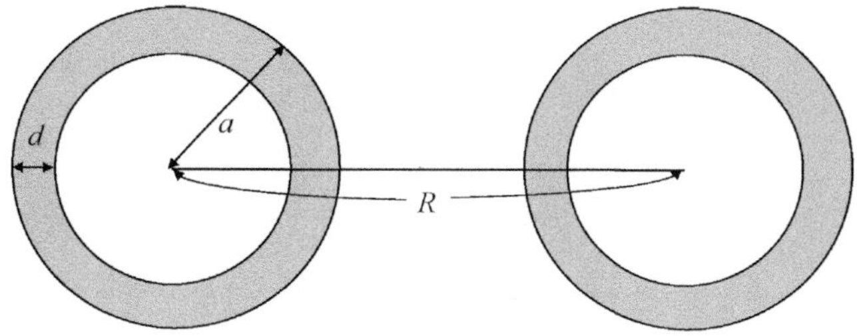

FIGURE 19.7 Interaction between two shells of outer radius a and thickness d separated by a distance R between their centers.

The interaction energy between two dissimilar shells separated by R, having outer radius a and thicknesses d (Fig. 19.7), can be calculated by

$$V(R) = V(a, a, R) - 2V(a, a-d, R) + V(a-d, a-d, R) \tag{19.33}$$

where $V(a_1, a_2, R)$ is given by Eq. (19.27).

19.7 A MOLECULE AND A ROD

The interaction energy between a point-like molecule and an infinitely long thin rod separated by a distance R (Fig. 19.8) can be calculated by

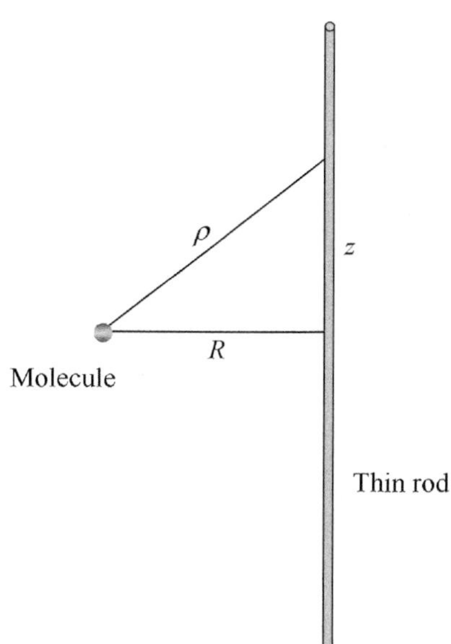

FIGURE 19.8 A point-like molecule and a thin rod separated by a distance R.

$$V(R) = \int_{z=-\infty}^{\infty} Nu(\rho)dz \tag{19.34}$$

where

$$u(\rho) = -\frac{C_{12}}{\rho^6} = -\frac{C_{12}}{(z^2 + R^2)^3} \tag{19.35}$$

is the interaction energy between the single molecule and a molecule within the rod, N is the line density of molecules along the rod, and C_{12} is the corresponding London–van der Waals constant. By carrying out the integration in Eq. (19.34), we obtain

$$V(R) = -\frac{3\pi N C_{12}}{8R^5} \tag{19.36}$$

19.8 TWO PARALLEL RODS

The interaction energy $V(R)$ per unit length between two dissimilar thin rods 1 and 2, having molecular densities N_1 and N_2, respectively, separated by a distance R between their axes (Fig. 19.9) can be given by

$$V(R) = -\frac{3\pi N_1 N_2 C_{12}}{8R^5} \tag{19.37}$$

19.9 A MOLECULE AND A CYLINDER

The interaction energy $U(R)$ between a single molecule and an infinitely long cylinder of radius a and molecular density N separated by a distance R (Fig. 19.10) can be calculated by

$$V(R) = \int_{r=0}^{a} \int_{\theta=0}^{2\pi} \int_{z=-\infty}^{\infty} Nu(\rho) r \, d\theta \, dz \, dr \tag{19.38}$$

where

$$u(\rho) = -\frac{C_{12}}{\rho^6} = -\frac{C_{12}}{(r^2 + R^2 - 2rR\cos\theta + z^2)^3} \tag{19.39}$$

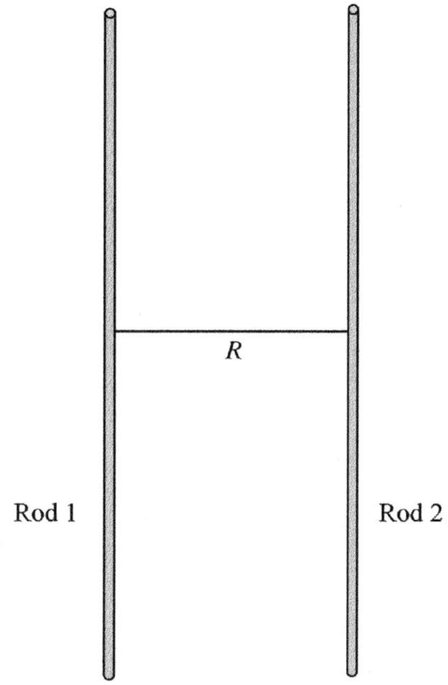

FIGURE 19.9 Two parallel thin rods separated by a distance R.

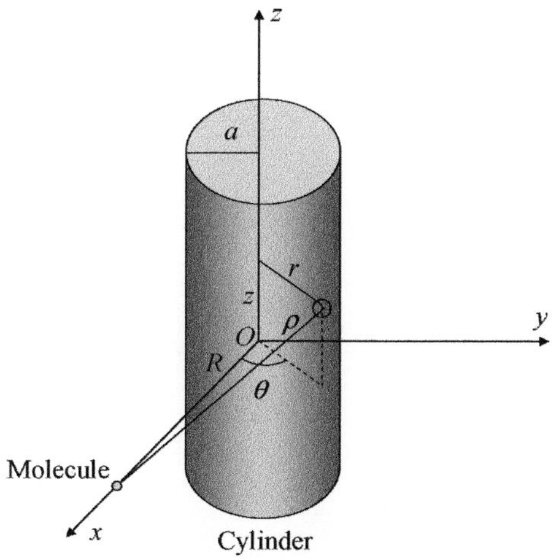

FIGURE 19.10 A point-like molecule and a cylinder of radius a separated by a distance R.

is the interaction energy between M and a molecule within the cylinder and C_{12} is the corresponding London–van der Waals constant. We use the following formulas:

$$\int_{-\infty}^{\infty} \frac{dz}{(r^2 + R^2 - 2rR\cos\theta + z^2)^3} = \frac{3\pi}{8(r^2 + R^2 - 2rR\cos\theta)^{5/2}} \tag{19.40}$$

$$\int_0^{2\pi} \frac{d\theta}{(r^2 + R^2 - 2rR\cos\theta)^{5/2}} = \frac{2\pi}{R^5} F\left(\frac{5}{2}, \frac{5}{2}, 1; \left(\frac{r}{R}\right)^2\right) \tag{19.41}$$

where $F(\alpha, \beta, \gamma; x)$ is the hypergeometric function defined by

$$F(\alpha, \beta, \gamma, z) = \frac{\Gamma(\gamma)}{\Gamma(\alpha)\Gamma(\beta)} \sum_{n=0}^{\infty} \frac{\Gamma(\alpha + n)\Gamma(\beta + n)z^n}{\Gamma(\gamma + n)n!} \tag{19.42}$$

Then, we obtain

$$V(R) = -\frac{3\pi^2 C_{12} N a^2}{8R^5} F\left(\frac{5}{2}, \frac{5}{2}, 2; \left(\frac{a}{R}\right)^2\right) \tag{19.43}$$

which has been derived by Kirsch [6].

19.10 TWO PARALLEL CYLINDERS

The interaction energy $V(R)$ per unit length between two parallel dissimilar cylinders 1 and 2, having radii a_1 and a_2, molecular densities N_1 and N_2, respectively, separated by a distance R between their axes (Fig. 19.11) can be calculated by

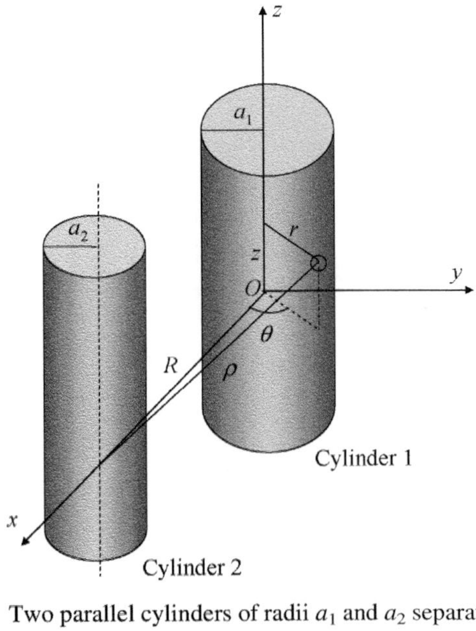

FIGURE 19.11 Two parallel cylinders of radii a_1 and a_2 separated by a distance R.

$$V(R)=\int_{r=0}^{a_1}\int_{\theta=0}^{2\pi}N_1\left[-\frac{3\pi^2C_{12}N_2a_2^2}{8R^5}F\left(\frac{5}{2},\frac{5}{2},2;\left(\frac{a_2}{R}\right)^2\right)\right]r^2\,d\theta\,dr \tag{19.44}$$

With the help of the following formula:

$$\int_0^{2\pi}\frac{d\theta}{(r^2+R^2-2rR\cos\theta)^{5/2+n}}=\frac{2\pi}{R^{2n+5}}F\left(\frac{5}{2}+n,\frac{5}{2}+n,1;\left(\frac{r}{R}\right)^2\right) \tag{19.45}$$

Eq. (19.44) becomes

$$V(R)=-\frac{2A_{12}}{3R}\sum_{m=0}^{\infty}\sum_{n=0}^{\infty}\frac{\Gamma^2(m+n+1/2)}{m!n!(m-1)!(n-1)!}\left(\frac{a_1}{R}\right)^{2m}\left(\frac{a_2}{R}\right)^{2n} \tag{19.46}$$

where A_{12} is given by Eq. (19.17). For the special case of two identical cylinders ($a_1=a_2=a$, $C_1=C_2=C$, $N_1=N_2=N$, and $A_{12}=A$), Eq. (19.46) becomes

$$V(R)=-\frac{2A}{3R}\sum_{m=0}^{\infty}\sum_{n=0}^{\infty}\frac{\Gamma^2(m+n+1/2)}{m!n!(m-1)!(n-1)!}\left(\frac{a}{R}\right)^{2(m+n)} \tag{19.47}$$

where A is given by Eq. (19.15).

For small $H=R-(a_1+a_2)$ ($H\ll a_1$ and a_2), Eq. (19.47) tends to

$$V(H)=-\frac{A_{12}}{12\sqrt{2}H^{3/2}}\left(\frac{a_1a_2}{a_1+a_2}\right)^{1/2} \tag{19.48}$$

For the special case of two identical cylinders per unit length, Eq. (19.48) gives

$$V(H)=-\frac{A\sqrt{a}}{24H^{3/2}} \tag{19.49}$$

Equation (19.48) can also be obtained with the help of Derjaguin's approximation [7,8] (see Chapter 12), that is, by substituting Eq. (19.20) into Eq. (12.38).

In the opposite limit of large R ($R\gg a_1+a_2$), Eq. (19.47) tends to

$$\begin{aligned}V(R)&=-\frac{3\pi a_1^2a_2^2A_{12}}{8R^5}\\&=-\frac{3\pi(\pi a_1^2N_1)(\pi a_2^2N_2)C_{12}}{8R^5}\end{aligned} \tag{19.50}$$

which is consistent with Eq. (19.37).

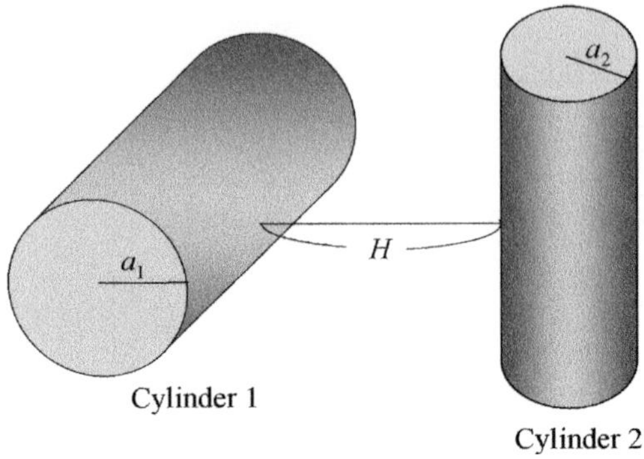

FIGURE 19.12 Two crossed cylinders of radii a_1 and a_2 separated by a distance H.

19.11 TWO CROSSED CYLINDERS

With the help of Derjaguin's approximation (see Chapter 12), one can derive the van der Waals interaction energy between two crossed cylinders of radii a_1 and a_2 at separation H between their surfaces (Fig. 19.12). By substituting Eq. (19.20) into Eq. (12.48), we obtain

$$V(H) = -\frac{A\sqrt{a_1 a_2}}{6H} \tag{19.51}$$

which is applicable for small H ($H \ll a_1, a_2$).

19.12 TWO PARALLEL RINGS

Consider two parallel rings 1 and 2 of radii r_1 and r_2, respectively, separated by a distance h (Fig. 19.13). The two rings are of the same substance. Let $\tilde{N}$ be the number of molecules on each ring, that is, $\tilde{N}$ is the line density of molecules on each ring. The distance ρ between a molecule in ring 1 and that in ring 2 is given by

$$\begin{aligned}\rho &= \sqrt{(r_1 \sin\theta_1 - r_2 \sin\theta_2)^2 + (r_1 \cos\theta_1 - r_2 \cos\theta_2)^2 + h^2} \\ &= \sqrt{r_1^2 + r_2^2 - 2r_1 r_2 \cos(\theta_1 - \theta_2) + h^2}\end{aligned} \tag{19.52}$$

For the van der Waals interaction energy between these two rings, Eq. (19.1) becomes

$$\begin{aligned}V(r_1, r_2, h) &= \tilde{N}^2 \int_0^{2\pi}\int_0^{2\pi} u(\rho) r_1\, d\theta_1 r_2\, d\theta_2 \\ &= \tilde{N}^2 \int_0^{2\pi}\int_0^{2\pi} \left(-\frac{C}{\rho^6}\right) r_1\, d\theta_1 r_2\, d\theta_2\end{aligned} \tag{19.53}$$

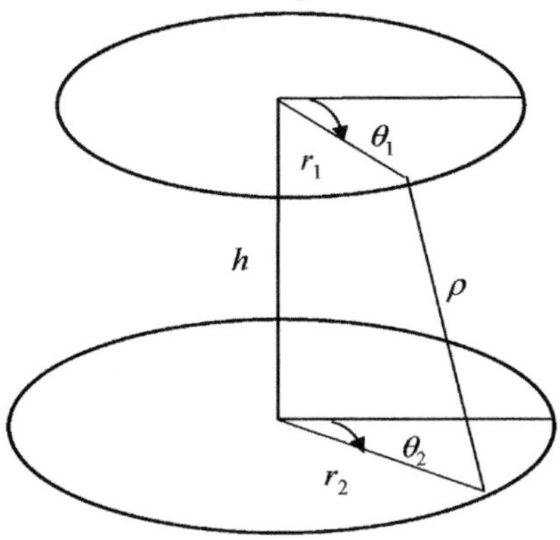

FIGURE 19.13 Two parallel rings of radii r_1 and r_2, respectively, separated by a distance h.

By substituting Eq. (19.52) into Eq. (19.4), we obtain

$$V(r_1, r_2, h) = -\frac{4\pi^2 C\tilde{N}^2 r_1 r_2[\{h^2 + (r_1 + r_2)^2\}\{h^2 + (r_1 - r_2)^2\} + 6r_1^2 r_2^2]}{[h^2 + (r_1 + r_2)^2]^{5/2}[h^2 + (r_1 - r_2)^2]^{5/2}} \tag{19.54}$$

Note that in the limit of large h, Eq. (19.54) tends to

$$V(r_1, r_2, h) = -\frac{(2\pi r_1 \tilde{N})(2\pi r_2 \tilde{N})C}{h^6} \tag{19.55}$$

which is the interaction energy between a point-like particle having $2\pi r_1 \tilde{N}$ molecules and that having $2\pi r_2 \tilde{N}$ molecules. Note also that for two identical parallel rings ($r_1 = r_2 = r$) in the limit of very small h, Eq. (19.53) becomes

$$V(r_1, r_2, h) = -\frac{3\pi C\tilde{N}^2}{8h^5} \cdot 2\pi r \tag{19.56}$$

It is interesting to note that Eq. (19.56) agrees with the energy between two parallel rods of length $2\pi r$ carrying molecules of line density $\tilde{N}$ [9].

19.13 TWO PARALLEL TORUS-SHAPED PARTICLES

A torus-shaped particle can serve as a model for biocolloids such as biconcave human red blood cells. The interaction between two parallel biconcave-shaped red

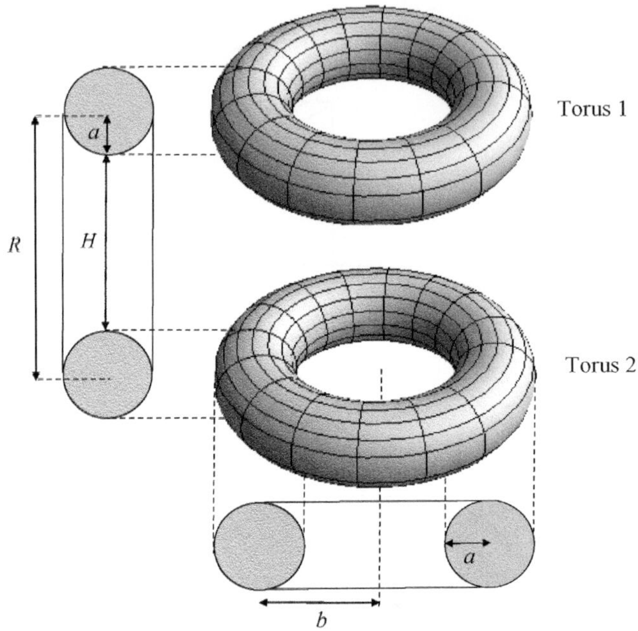

FIGURE 19.14 Two parallel torus-shaped particles at separation R between their centers. a is the radius of the tube and b is the distance from the center of the tube to the center of the torus. From Ref. [9].

blood cells can be approximated by the interaction between two parallel torus--shaped particles.

Consider two parallel similar torus-shaped particles at separation R between their centers (Fig. 19.14) [9]. A torus can be constructed by joining the ends of a cylindrical tube. We denote the radius of the tube by a and the distance from the center of the tube to the center of the torus by b. The length of the cylindrical tube is thus $2\pi b$ and the closest distance H between the surfaces of these tori is $H = R - 2a$. Let N be the number of molecules contained within the particles. The interaction energy $V(R)$ between two parallel tori can be calculated with the help of Eqs. (19.1) and (19.54) (Fig. 19.15), namely,

$$V(R) = \int_{\rho_1=0}^{a} \int_{\rho_2=0}^{a} \int_{\phi_1=0}^{2\pi} \int_{\phi_2=0}^{2\pi} V(r_1, r_2, h)\rho_1 \rho_2 \, d\rho_1 \, d\rho_2 \, d\phi_1 \, d\phi_2 \tag{19.57}$$

with

$$r_1 = b - \rho_1 \cos \phi_1 \tag{19.58}$$

$$r_2 = b - \rho_2 \cos \phi_2 \tag{19.59}$$

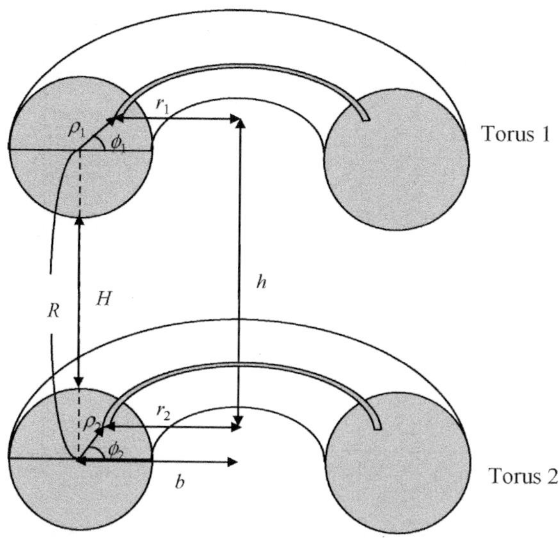

FIGURE 19.15 Cross sections of two parallel tori. From Ref. [9].

$$h = R + \rho_1 \, sin \, \phi_1 - \rho_2 \, sin \, \phi_2 \tag{19.60}$$

where the line density $\tilde{N}$ in $V(r_1, r_2, h)$ must be replaced by the volume density N.

By substituting Eq. (19.54) into Eq. (19.57), we obtain

$$V(R) = -4A \int_{\rho_1=0}^{a} \int_{\rho_2=0}^{a} \int_{\phi_1=0}^{2\pi} \int_{\phi_2=0}^{2\pi} \left(\frac{r_1 r_2 [\{h^2 + (r_1 + r_2)^2\}\{h^2 + (r_1 - r_2)^2\} + 6 r_1^2 r_2^2]}{[h^2 + (r_1 + r_2)^2]^{5/2} [h^2 + (r_1 - r_2)^2]^{5/2}} \right)$$
$$\times \rho_1 \rho_2 d\rho_1 d\rho_2 d\phi_1 d\phi_2 \tag{19.61}$$

Some of the results of the calculation of $V(R)$ based on Eq. (19.61) are shown in Fig. 19.16.

As is well known, a torus and the corresponding cylindrical tube of radius a and length $2\pi b$ have the same surface area ($2\pi a \cdot 2\pi b$) and interior volume ($\pi a^2 \cdot 2\pi b$) (the Pappus–Guldinus theorem). Therefore, the interaction energy between two parallel tori is expected to be close to that between the corresponding two parallel cylinders. It follows from Eq. (19.47) that the interaction energy $V(R)$ between two parallel similar cylinders of radius a and length $2\pi b$ is given by

$$V(R) = -\frac{2A}{3R} \sum_{m=0}^{\infty} \sum_{n=0}^{\infty} \frac{\Gamma^2(n + m + 1/2)}{m!n!(m-1)!(n-1)!} \left(\frac{a}{R}\right)^{2(m+n)} \times 2\pi b \tag{19.62}$$

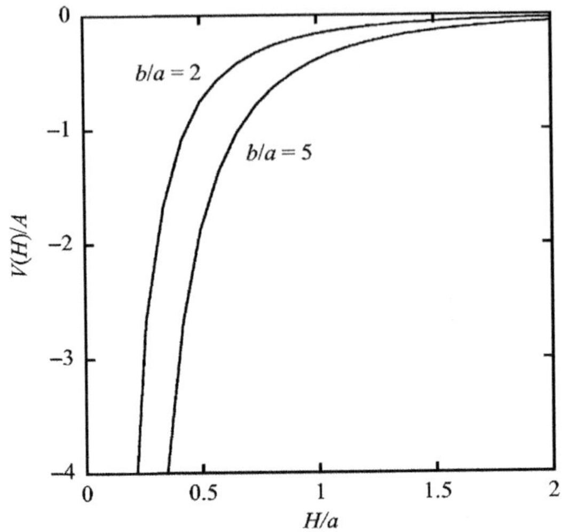

FIGURE 19.16 Scaled van der Waals interaction energy $V(H)/A$ between two parallel tori as a function of $H = R - 2a$ calculated with Eq. (19.61) at $b/a = 2$ and 5. From Ref. [9].

For small $H/a < 2$, Eqs. (19.61) and (19.62) give results very close to each other. The relative difference between Eqs. (19.61) and (19.62) is less than 1% at $H/a \leq 0.3$ for $b/a = 2$ and less than 1% at $H/a \leq 1.2$ for $b/a = 5$. Indeed, in the limit of small H/a, both Eqs. (19.61) and (19.62) tend to the same limiting value

$$V(R) = -\frac{A\sqrt{a}}{24H^{3/2}} \times 2\pi b \tag{19.63}$$

which also agrees with the results obtained by applying Derjaguin's approximation to the interaction between two parallel cylinders (Eq. (19.49)) [7,8].

In the opposite limit of large H ($H/a \gg 1$), however, Eqs. (19.61) and (19.62) show different limiting values. Equation (19.61) tends to

$$\begin{aligned} V(R) &= -4A \int_{\rho_1=0}^{a} \int_{\rho_2=0}^{a} \int_{\phi_1=0}^{2\pi} \int_{\phi_2=0}^{2\pi} \left(\frac{r_1 r_2}{R^6}\right) \rho_1 \rho_2 \, d\rho_1 \, d\rho_2 \, d\phi_1 \, d\phi_2 \\ &= -\frac{4A\pi^2 a^4 b^2}{R^6} \\ &= = -\frac{C\{N \cdot (\pi a^2 \cdot 2\pi b)\}}{R^6} \end{aligned} \tag{19.64}$$

which agrees with the interaction energy between two point-like particles consisting of $N \cdot (\pi a^2 \cdot 2\pi b)$ molecules contained in the torus volume $\pi a^2 \cdot 2\pi b$. Equation (19.62), on the other hand, for large H ($H/a \gg 1$), tends to

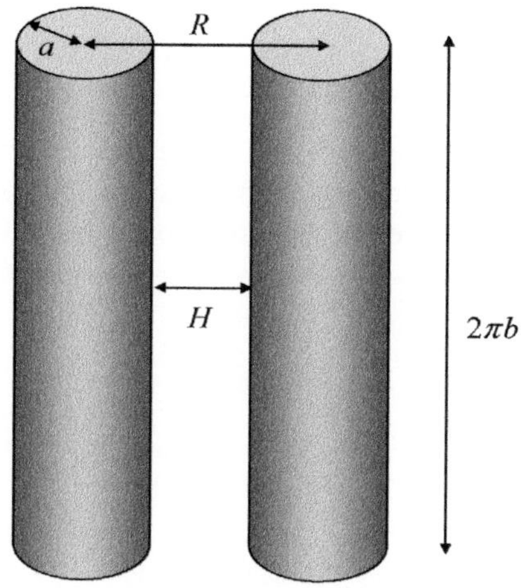

FIGURE 19.17 Interaction between two parallel cylinder of radius a and length $2\pi b$ separated by a distance R.

$$\begin{aligned} V(R) &= -\frac{3\pi A a^4}{8R^5} \cdot 2\pi b \\ &= -\frac{3\pi C(\pi a^2 N)^2}{8R^5} \cdot 2\pi b \end{aligned} \tag{19.65}$$

which agrees with the interaction energy between two thin rods of length $2\pi b$ consisting of $\pi a^2 N$ molecules per unit length (see Eq. (19.37)) (Fig. 19.17). We thus see that for large R, $V(R)$ becomes proportional to $1/R^6$ for two tori but to $1/R^5$ for two parallel cylinders.

19.14 TWO PARTICLES IMMERSED IN A MEDIUM

The equations for the interactions between particles so far derived in this chapter are based on the assumption that the interacting particles are interacting in a vacuum. If the medium between the particles is no longer a vacuum but a second substance (e.g., water), we have to account for the fact that each particle replaces an equal volume of water. It follows from the assumption of the additivity of the van der Waals interaction energies that Hamaker constant A_{132} for the interaction between two different particles of substances 1 and 2 immersed in a medium of substance 3 is

$$A_{132} = \left(\sqrt{A_1} - \sqrt{A_3}\right)\left(\sqrt{A_2} - \sqrt{A_3}\right) \tag{19.66}$$

$$|A_{132}| = \sqrt{A_{131}}\sqrt{A_{232}} \tag{19.67}$$

where A_1 and A_2 are, respectively, referred to the interaction between two similar particles of substances 1 and 2 immersed in a vacuum. In particular, the Hamaker constant for the interaction between two similar particles of substance 1 immersed in a medium of substance 3 is

$$A_{131} = \left(\sqrt{A_1} - \sqrt{A_3}\right)^2 \tag{19.68}$$

Note that while A_{131} is always positive, A_{132} may be positive or negative depending on the relative magnitudes of A_1, A_2, and A_3. Thus, if two particles of substances 1 and 2 are interacting in medium 3, then A_{12} appearing in Eqs. (19.16),(19.27), (19.32),(19.46) and (19.48) must be replaced by A_{132}.

19.15 TWO PARALLEL PLATES COVERED WITH SURFACE LAYERS

It is possible to derive the interaction energy between particles with multilayer structures [4,10,11]. Consider the interaction between two parallel similar plates separated by a distance h (Fig. 19.18). Let the thicknesses of the plate core and the

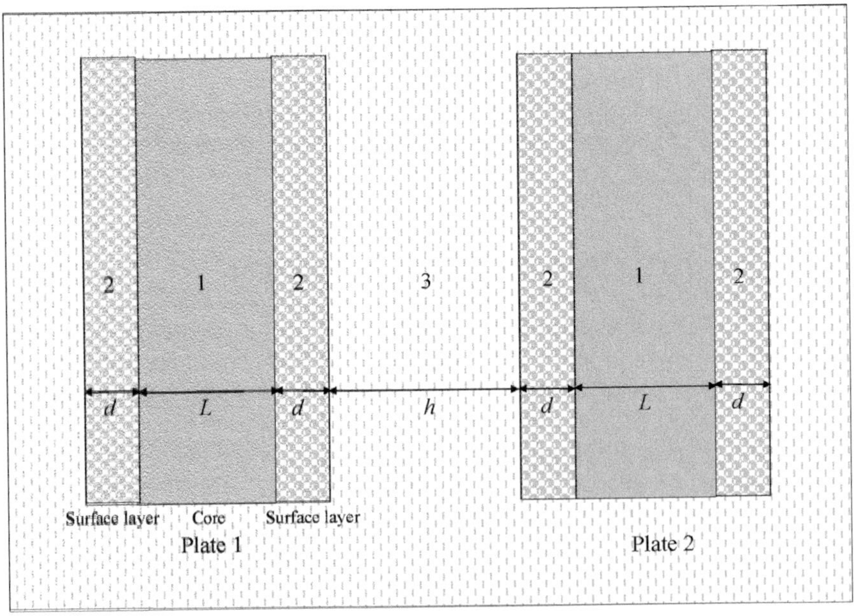

FIGURE 19.18 Two parallel similar plates of substance 1 and thickness L covered by a surface layer of substance 2 and thickness d separated at a separation h in a medium 3.

surface layer be L and d, respectively. The interaction energy per unit area between the plates is thus given by

$$V(h) = -\frac{1}{12\pi}\left[\frac{A_{232}}{h^2} - \frac{2A_{123}}{(h+d)^2} + \frac{A_{121}}{(h+2d)^2} + \frac{2A_{123}}{(h+d+L)^2} - \frac{2(A_{232}+A_{121})}{(h+d+L)^2} + \frac{A_{121}}{(h+2d+2L)^2} + \frac{2A_{123}}{(h+3d+L)^2} - \frac{2A_{123}}{(h+3d+2L)^2} + \frac{A_{232}}{(h+4d+2L)^2}\right] \quad (19.69)$$

For very small h ($h \ll d$ and $h \gg L$), Eq. (19.59) becomes the interaction energy between the two surface layers of substance 2 in a medium 3, namely,

$$V(h) = -\frac{A_{232}}{12\pi h^2} \quad (19.70)$$

For large h such that $L \gg h \gg d$, Eq. (19.59) approaches the interaction between the two plate cores of substance 1 in a medium 3, namely,

$$V(h) = -\frac{(A_{232} - 2A_{123} + A_{121})}{12\pi h^2} = -\frac{A_{131}}{12\pi h^2} \quad (19.71)$$

REFERENCES

1. H. C. Hamaker, *Physica* 4 (1937) 1058.
2. E. J. W. Verwey and J. Th. G. Overbeek, *Theory of the Stability of Lyophobic Colloids*, Elsevier/Academic Press, Amsterdam, 1948.
3. D. Langbein, *Theory of van der Waals Attraction*, Springer, Berlin, 1974.
4. J. N. Israelachvili, *Intermolecular and Surface Forces, 2nd edition*, Elsevier/Academic Press, London, 1992.
5. B. V. Derjaguin, *Kolloid Z.* 69 (1934) 155.
6. V. A. Kirsch, *Adv. Colloid Interface Sci.* 104 (2003) 311.
7. M. J. Sparnaay, *Recueil* 78 (1959) 680.
8. H. Ohshima and A. Hyono, *J. Colloid Interface Sci.*, 333 (2009) 202.
9. H. Ohshima and A. Hyono, *J. Colloid Interface Sci.*, 332 (2009) 251.
10. M. J. Vold, *J. Colloid Sci.* 16 (1961) 1.
11. B. Vincent, *J. Colloid Interface Sci.* 42 (1973) 270.

20 DLVO Theory of Colloid Stability

20.1 INTRODUCTION

The stability of colloidal systems consisting of charged particles can be essentially explained by the Derjaguin–Landau–Verwey–Overbeek (DLVO) theory [1–7]. According to this theory, the stability of a suspension of colloidal particles is determined by the balance between the electrostatic interaction and the van der Waals interaction between particles. A number of studies on colloid stability are based on the DLVO theory. In this chapter, as an example, we consider the interaction between lipid bilayers, which serves as a model for cell–cell interactions [8, 9]. Then, we consider some aspects of the interaction between two soft spheres, by taking into account both the electrostatic and van der Waals interactions acting between them.

20.2 INTERACTION BETWEEN LIPID BILAYERS

Lipid bilayer membranes have been employed extensively as an experimental model of biological membranes. We take the dipalmitoylphosphatidylcholine (DPPC)/water system as an example to calculate the intermembrane interactions. On the basis of the DLVO theory [1–7], we consider two interactions, that is, the electrostatic repulsive double-layer interaction and the van der Waals attractive interaction. In addition to these two interactions, we take into account the steric interaction between the hydration layers formed on the membrane surface. DPPC forms bilayers when it is dissolved in water. Inoko et al. [10] have made X-ray diffraction studies on the effects of various cations upon the DPPC/water system. They have found remarkable effects of $CaCl_2$ and $MgCl_2$: depending upon the concentration of $CaCl_2$, four different states appear, which they term I, II, III, and IV, and three states I, III, and IV appear in the case of $MgCl_2$. Figure 20.1 illustrates the four states in the case of $CaCl_2$. In pure water, the system consists of a lamellar phase with a period of 6.45 nm and excess water (state I). Addition of 1.3 mM $CaCl_2$ destroys the lamellar structure and makes it swell into the excess water (state II). The

Biophysical Chemistry of Biointerfaces By Hiroyuki Ohshima

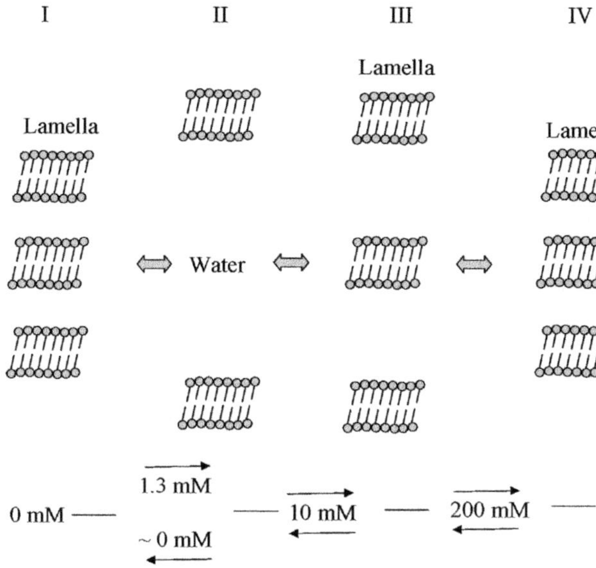

FIGURE 20.1 Schematic representation of the four states I, II, III, and IV of the DPPC/water system observed by Inoko et al. [10]. Values in units of mM give the $CaCl_2$ concentration at which the transition between two states occurs.

lamellar phase, however, reappears when the concentration of $CaCl_2$ increases (state III): a partially disordered lamellar phase with a repeat distance of about 15 nm comes out at a concentration of about 10 mM, and then the degree of order of the lamellar phase is improved and the repeat distance decreases with increasing $CaCl_2$ concentration. Above 200 mM, the repeat distance becomes constant at 6.45 nm (state IV), the same value as in state I. When the concentration is reduced, the transitions IV → III and III → II occur at the same concentration as mentioned above, but the transition II → I takes place at a concentration close to 0 mM. In other words, there is a hysteresis in the transition I → II. As shown below, the appearance of the above four or three states in the DPPC/water system by addition of $CaCl_2$ or $MgCl_2$ can be explained well by the DLVO theory [8, 9].

Consider the potential energy V between two adjacent bilayer membranes per unit area, which may be decomposed into three parts:

$$V = V_e + V_v + V_h \tag{20.1}$$

where V_e is the potential of the electrostatic interaction caused by the adsorption of Ca^{2+} or Mg^{2+} ions, V_v is the potential energy of the van der Waals interaction, and V_h is the potential energy of the repulsive interaction of hydration layers. The DPPC membrane is expected to be electrically neutral in pure water. It will be positively charged by adsorption of divalent cations (Ca^{2+} or Mg^{2+}) on its surfaces. We assume that there are N_{max} sites available for adsorption of cation per unit area and that the cation adsorption is of the Langmuir type. Then, the number N of adsorbed cations per unit area is given by

$$N = \frac{N_{\text{max}} Kn \exp(-2e\psi_o/kT)}{1 + Kn \exp(-2e\psi_o/kT)} \tag{20.2}$$

where n and K are, respectively, the bulk concentration of Ca^{2+} or Mg^{2+} and the binding constant of Ca^{2+} or Mg^{2+} ions, and ψ_o is the surface potential of the bilayer membrane. The surface charge density $\sigma = 2eN$ on the bilayer membrane is thus given by

$$\sigma = \frac{2eN_{\text{max}} Kn \exp(-2e\psi_o/kT)}{1 + Kn \exp(-2e\psi_o/kT)} \tag{20.3}$$

Adsorption of cations causes the electrostatic repulsion between the bilayer membranes. The electrostatic repulsive force $P_e(h)$, which is given by $P_e(h) = -dV_e(h)/dl$, acting between two adjacent membranes is directly expressed as (see Eq. (9.26))

$$P_e(h) = nkT\left[\exp(-2e\psi_m/kT) + 2\exp(e\psi_m/kT) - 3\right] \tag{20.4}$$

where $\psi(h/2)$ is the potential at the midpoint between the two adjacent bilayers and is obtained by solving the following Poisson–Boltzmann equation for the electric potential $\psi(x)$:

$$\frac{d^2\psi}{dx^2} = -\frac{2e(n_{\text{Ca}} + n_{\text{Mg}})}{\varepsilon_r\varepsilon_o}\left[\exp\left(-\frac{2e\psi}{kT}\right) - \exp\left(\frac{e\psi}{kT}\right)\right] \tag{20.5}$$

where the x-axis is taken perpendicular to the bilayers with its origin at the surface of one bilayer so that the region $0 < x < h$ corresponds to the electrolyte solution between the two adjacent bilayers. Because of the periodic structure of the lamellar phase, we have

$$\left.\frac{d\psi}{dx}\right|_{x=0^+} = 0 \tag{20.6}$$

$$\left.\frac{d\psi}{dx}\right|_{x=0^+} = -\frac{\sigma}{\varepsilon_r\varepsilon_o} \tag{20.7}$$

$$\left.\frac{d\psi}{dx}\right|_{x=h/2} = 0 \tag{20.8}$$

We assume that the van der Waals interactions of one membrane with the others are additive. Then, the expressions for V_v are given below.

$$V_v(h) = \sum_{m=0}^{\infty} v_v(h + m(h + d)) \tag{20.9}$$

with

$$v_v(h) = -\frac{A}{12\pi}\left[\frac{1}{h^2} + \frac{1}{(h + 2d)^2} - \frac{2}{(h + d)^2}\right] \tag{20.10}$$

where $v(h)$ is the potential energy of the van der Waals interaction per unit area between two parallel membranes of thickness d at a separation h and A is the Hamaker constant. The van der Waals force $P_v(h) = -dV_v/dh$ is given by

$$P_v(h) = \sum_{m=0}^{\infty} (m + 1)p_v(h + m(h + d)) \tag{20.11}$$

with

$$p_v(h) = -\frac{A}{6\pi}\left[\frac{1}{h^3} + \frac{1}{(h + 2d)^3} - \frac{2}{(h + d)^3}\right] \tag{20.12}$$

We assume that each membrane has adsorbed water layers on its surface. Contact of these hydration layers will cause a repulsive force, whose potential V_h is approximately given by

$$V_v(h) = \begin{cases} \infty, & h < h_o \\ 0, & h \geq h_o \end{cases} \tag{20.13}$$

The critical concentration for the I → II·III transition is determined by

$$\left.\frac{d(V_e + V_v)}{dh}\right|_{h=h_o} = 0 \tag{20.14}$$

or

$$P_e(h_o) + P_v(h_o) = 0 \tag{20.15}$$

Here, we put $h_o = 2.25$ nm [8, 9]. In Ref. [9], the condition (20.15) was applied to a mixed aqueous solution of $CaCl_2$ and $MgCl_2$ for various concentrations and the three unknown parameters, that is, the Hamaker constant A for the DPPC/water system, and K_{Ca} and K_{Mg}, which are the adsorption constants for Ca^{2+} and Mg^{2+} ions, respectively, were determined. The results are

$$A = (3.6 \pm 0.8) \times 10^{-21}\,\mathrm{J}, \quad K_{Ca} = 21 \pm 9\,\mathrm{M}^{-1}, \quad K_{Mg} = 2.5 \pm 0.7\,\mathrm{M}^{-1} \tag{20.16}$$

Using these values, we have calculated the potential energy V of interaction per μm^2 per pair of adjacent membranes as a function of membrane separation h and $CaCl_2$ concentration n. Results are shown in Fig. 20.2. The curve for $n = 0$ mM in Fig. 20.2 shows that V becomes minimum at $h = h_o$ in accordance with the observation. The minimum of V at $h = h_o$ disappears in the curve for $n = 1.3$ mM (i.e., $V_e + V_v$ has a maximum at $h = h_o$). This curve has a secondary minimum at larger h. Its depth is too shallow to maintain the observed arrangement of membranes in the lamellar phase. Therefore, the curve for $n = 1.3$ mM corresponds to the disordered dispersion of membranes as was observed experimentally at $n = 1.3$ mM, that is, the state II. The minimum of V is deepened and correspondingly the position of potential minimum shifts to smaller h with increasing n as seen in the curves for $n = 50$ mM and $n = 150$ mM, which therefore correspond to state III. With a further increase of n, the curve again has a sharp minimum at $h = h_o$ as shown by the curve for $n = 500$ mM, which should correspond to state IV. The curve for $n = 0.05$ mM has a special character: it has a sharp minimum at $h = h_o$ and a very shallow minimum at very large h and a maximum between them. Here, we assume that the

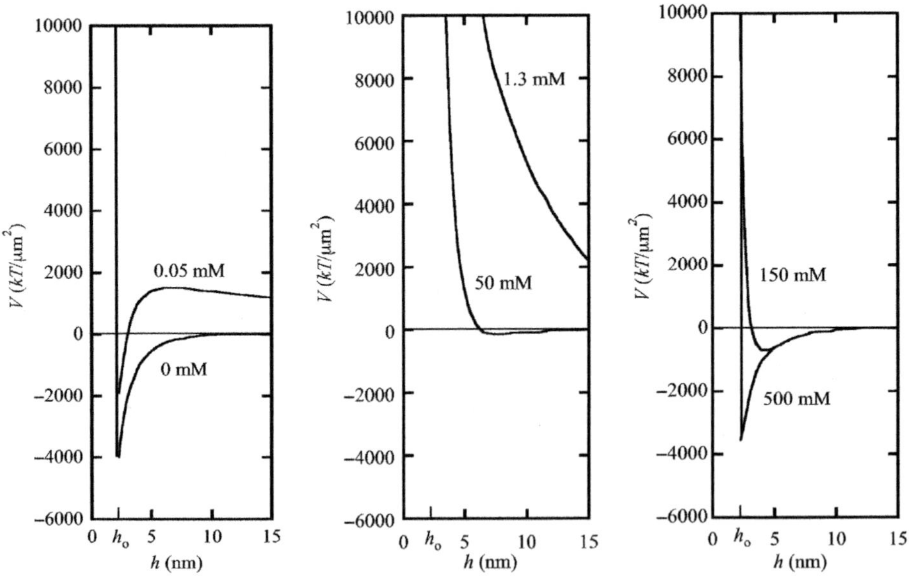

FIGURE 20.2 The interaction energy V per pair of two adjacent DPPC membranes as a function of h at various $CaCl_2$ concentrations, calculated with $A = 3.6 \times 10^{-21}$ J, $K_{Ca} = 21$ M^{-1}, and $T = 278$K.

transition between two states cannot take place beyond the maximum or the potential barrier. The potential barrier disappears at 1 mM as shown by curve for 1.3 mM (as seen from the side of $h = h_o$) and thus the transition I → II occurs at this concentration when n increases. On the other hand, it disappears at about 0 mM (as seen from the side of large h) and the transition II → I occurs at this concentration when n decreases. This behavior of V explains the hysteresis observed in the I → II transition.

In the case of $MgCl_2$, there appear only three states I, III, and IV and the transition I → III occurs at $n = 5$ mM. This can be explained as follows. Since K_{Mg} is much smaller than K_{Ca}, $V(H)$ for $n = 5$ mM, where a maximum at $h = h_o$, also has a minimum at large h. This minimum is much deeper than that for $CaCl_2$ and is deep enough to preserve the order of the lamellar structure. That is, the disordered state II does not appear in the case of $MgCl_2$.

20.3 INTERACTION BETWEEN SOFT SPHERES

In the preceding section, we have treated the interaction between lipid bilayers, where we have regarded a lipid bilayer as a hard plate without surface structures. In this section, we consider the total interaction energy between two identical soft spheres of radius a at separation H (Fig. 20.3). Each particle is covered with a layer of polyelectrolytes of thickness d. Let the density and valence of ionized groups are, respectively, Z and N, so that the density of the fixed charges within the surface charge layer is $\rho_{\text{fix}} = ZeN$. With the help of Eq. (15.45), the total interaction energy $V(H)$ is given by

$$V_e(H) = \frac{2\pi a \rho_{\text{fix}}^2 \sinh^2(\kappa d)}{\varepsilon_r \varepsilon_o \kappa^4} \ln\left(\frac{1}{1 - e^{-\kappa(H+2d)}}\right) \tag{20.17}$$

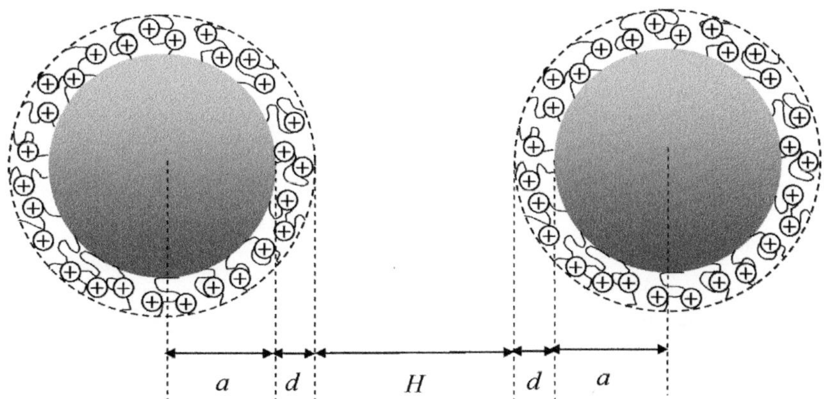

FIGURE 20.3 Interaction between two similar spherical soft particles.

We treat the case in which the number density of molecules within the surface layer is negligibly small compared to that in the particle core. The van der Walls interaction energy is dominated by the interaction between the particle cores, while the contribution from the surface charge layer can be neglected. Thus, we have (from Eq. (19.31))

$$V_v(H) = -\frac{Aa}{12(H+2d)} \tag{20.18}$$

where A is the Hamaker constant for the interaction between the particle cores in the electrolyte solution. The total interaction energy $V(H) = V_e(H) + V_v(H)$ is thus

$$V(H) = \frac{2\pi a \rho_{\text{fix}}^2 \sinh^2(\kappa d)}{\varepsilon_r \varepsilon_o \kappa^4} \ln\left(\frac{1}{1 - e^{-\kappa(H+2d)}}\right) - \frac{A}{12(H+2d)} \tag{20.19}$$

which is a good approximation provided that

$$H \ll a \quad \text{and} \quad \kappa a \gg 1 \tag{20.20}$$

We rewrite Eq. (20.19) in terms of the unperturbed surface potential ψ_o (see Eq. (1.27)), which is given by, under condition given by Eq. (20.20),

$$\psi_o = \frac{\sigma}{\varepsilon_r \varepsilon_o \kappa}\left(\frac{1 - e^{-\kappa d}}{2\kappa d}\right) \tag{20.21}$$

Then, Eq. (20.19) becomes

$$V_e(H) = 2\pi \varepsilon_r \varepsilon_o a \psi_o^2 e^{2\kappa d} \ln\left(\frac{1}{1 - e^{-\kappa(H+2d)}}\right) \tag{20.22}$$

The total potential energy V is thus given by

$$V(H) = 2\pi \varepsilon_r \varepsilon_o a \psi_o^2 e^{2\kappa d} \ln\left(\frac{1}{1 - e^{-\kappa(H+2d)}}\right) - \frac{A}{12(H+2d)} \tag{20.23}$$

At a maximum or a minimum of a plot of $V(H)$ versus H, we have $dV(H)/dH = 0$, namely,

$$-\kappa \cdot 2\pi \varepsilon_r \varepsilon_o a \psi_o^2 e^{2\kappa d} \left(\frac{e^{-\kappa(H+2d)}}{1 - e^{-\kappa(H+2d)}}\right) + \frac{A}{12(H+2d)^2} = 0 \tag{20.24}$$

which can be rewritten as

$$f(\kappa H) = Z \tag{20.25}$$

where $f(\kappa H)$ and Z are defined by

$$f(\kappa H) = \frac{[\kappa(H + 2d)]^2}{e^{\kappa(H+2d)} - 1} \tag{20.26}$$

$$Z = \frac{A\kappa}{24\pi\varepsilon_r\varepsilon_o a\psi_o^2 e^{2\kappa d}} \tag{20.27}$$

We plot $F(\kappa H)$ as a function of κH for several values of κd in Fig. 20.4, showing that $F(\kappa H)$ takes the maximum value of 0.6476 at $\kappa H = 1.5936 - 2\kappa d$. (Discussions based on a plot of $F(\kappa H)$ versus κH were found in Refs. [11] and [12].)

There are five possible situations (Fig. 20.5a, b, and c and Fig. 20.6a and b). Consider first the situations in Fig. 20.5, that is,

Case (i) $0 \leq 2\kappa d < 1.5936$

In this situation, $F(\kappa H)$ shows a maximum ($F(\kappa H) = 0.6476$) in the region $\kappa H \geq 0$. In case (i), we have further three possible cases.

Case (a) $Z \geq 0.6476$

In this case, Eq. (20.25) has no root, so $V(\kappa H)$ decreases monotonically as κH decreases.

Case (b) $F(0) \leq Z < 0.6476$

Here, $F(0)$ is given by

$$F(0) = \frac{(2\kappa d)^2}{e^{2\kappa d} - 1} \tag{20.28}$$

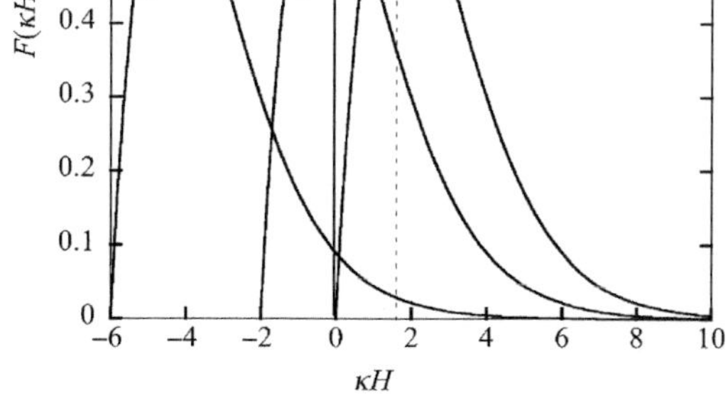

FIGURE 20.4 Plot of $F(\kappa H)$ as a function of κH for $\kappa d = 0$, 1, and 3.

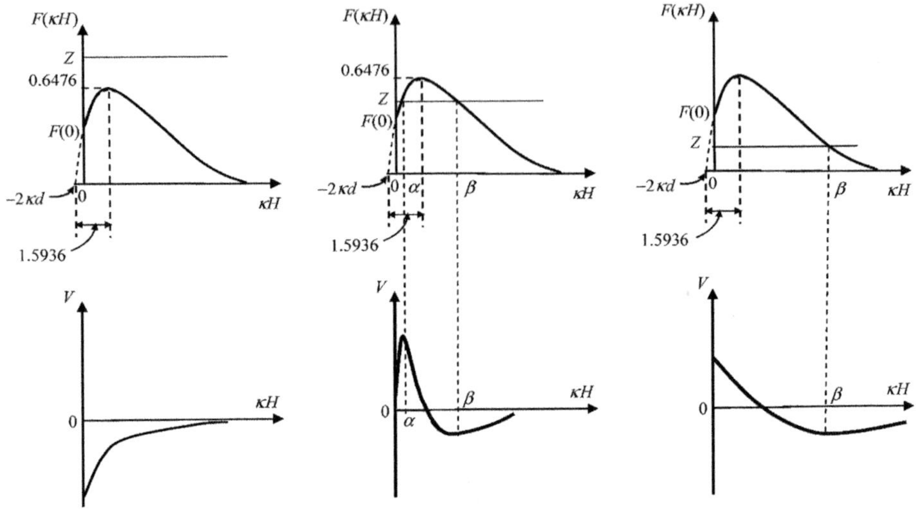

(a) $Z \geq 0.6476$ **(b)** $F(0) \leq Z < 0.6476$ **(c)** $Z < F(0)$

FIGURE 20.5 $F(\kappa H)$ and $V(\kappa H)$ at $0 \leq 2\kappa d < 1.5936$ for three cases (a), (b), and (c).

In this case, Eq. (20.25) has two roots $\kappa H = \alpha$ and $\kappa H = \beta$ ($\alpha < \beta$), and $V(\kappa H)$ shows a maximum at $\kappa H = \alpha$ and a minimum at $\kappa H = \beta$.

Case (c) $Z < F(0)$

In this case, Eq. (20.25) has one root $\kappa H = \beta$ and $V(\kappa H)$ shows a minimum at $\kappa H = \beta$.

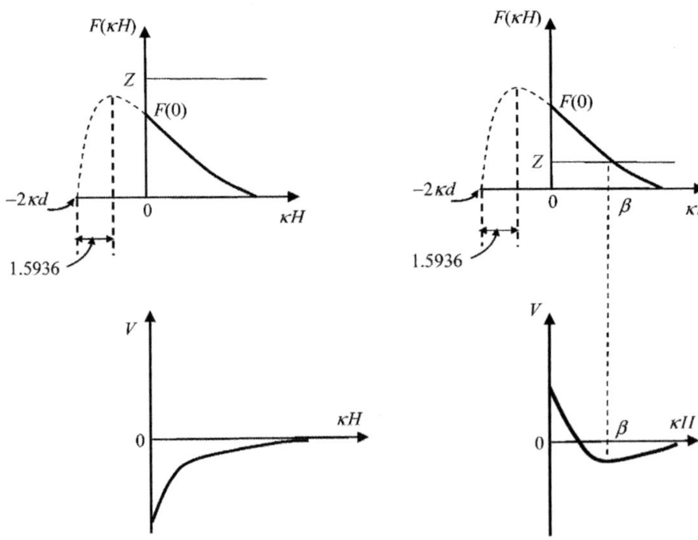

(a) $Z \geq F(0)$ **(b)** $Z < F(0)$

FIGURE 20.6 $F(\kappa H)$ and $V(\kappa H)$ at $2\kappa d \geq 1.5936$ for two cases (a) and (b).

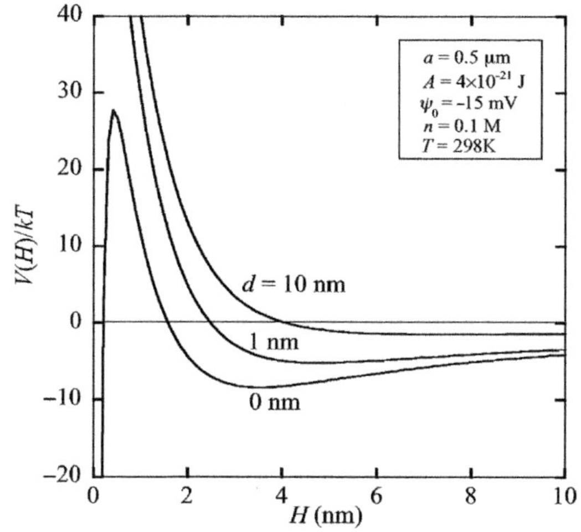

FIGURE 20.7 Potential energy $V(H)$ between two similar soft spheres at separation H, calculated with $A = 4 \times 10^{-21}$ J, $\psi_o = -15$ mV, $n = 0.1$ M, and $T = 298$K.

Next, consider the situations in Fig. 20.6, that is,

Case (ii) $2\kappa d \geq 1.5936$

In this situation, $F(\kappa H)$ shows no maximum, decreasing monotonically as κH increases in the region $\kappa H \geq 0$. In this situation, we have further two possible cases.

Case (a) $Z \geq F(0)$

In this case, Eq. (20.25) has no root, so $V(\kappa H)$ decreases monotonically as κH decreases.

Case (b) $Z < F(0)$

In this case, Eq. (20.25) has one root $\kappa H = \beta$ and $V(\kappa H)$ shows a minimum at $\kappa H = \beta$.

Figure 20.7 shows some examples of the calculation of the interaction energy $V(H)$ between two similar soft spheres at separation H, calculated with $a = 0.5$ μm, $A = 4 \times 10^{-21}$ J, $\psi_o = -15$ mV, $n = 0.1$ M, and $T = 298$K. The curve for $d = 0$ nm corresponds to Fig. 20.5b and curves for $d = 1$ and 10 nm to Fig. 20.6b.

Note that we have considered only the region $H \geq 0$, in which the surface charge layers of the interacting soft spheres are not in contact. We have to consider other interactions after contact of the surface layers, as discussed in the previous chapter.

REFERENCES

1. B. V. Derjaguin and L. D. Landau, *Acta Physicochim.* 14 (1941) 633.
2. E. J. W. Verwey and J. Th. G. Overbeek, *Theory of the Stability of Lyophobic Colloids*, Elsevier/Academic Press, Amsterdam, 1948.

3. B. V. Derjaguin, *Theory of Stability of Colloids and Thin Films*, Consultants Bureau, New York, 1989.
4. J. N. Israelachvili, *Intermolecular and Surface Forces, 2nd edition*, Academic Press, New York, 1992.
5. J. Lyklema, *Fundamentals of Interface and Colloid Science: Solid–Liquid Interfaces*, Vol. 2, Academic Press, New York, 1995.
6. H. Ohshima, in: H. Ohshima and K. Furusawa (Eds.), *Electrical Phenomena at Interfaces: Fundamentals, Measurements, and Applications, 2nd edition, revised and expanded*, Dekker, New York, 1998, Chapter 3.
7. T. F. Tadros (Ed.), *Colloid Stability: The Role of Surface Forces, Part 1*, Wiley-VCH, Weinheim, 2007.
8. H. Ohshima and T. Mitsui, *J. Colloid Interface Sci.* 63 (1978) 525.
9. H. Ohshima, Y. Inoko, and T. Mitsui, *J. Colloid Interface Sci.* 86 (1982) 57.
10. Y. Inoko, T. Yamaguchi, K. Furuya, and T. Mitsui, Biochim. Biophys. Acta 413 (1975) 24.
11. B. Chu, *Molecular Forces*, Wiley, New York, 1967.
12. M. J. Sparnaay, in: P. Sherman (Ed.), *Rheology of Emulsions*, Macmillan, New York, 1963, p. 30.

PART III
Electrokinetic Phenomena at Interfaces

21 Electrophoretic Mobility of Soft Particles

21.1 INTRODUCTION

When an external electric field is applied to a suspension of charged colloidal particles in a liquid, the electric field accelerates the particles and at the same time a viscous force exerted by the liquid on the particles tends to retard the particles. In the stationary state, these two forces are balanced with each other so that the particles move with a constant velocity in the liquid. This phenomenon, which is called electrophoresis, depends strongly on the thickness (or the Debye length) $1/\kappa$ of the electrical diffuse double layer formed around the charged particle relative to the particle size and the zeta potential ζ [1–29]. The zeta potential is defined as the potential at the plane where the liquid velocity relative to the particle is zero. This plane, which is called the slipping plane or the shear plane, does not necessarily coincide with the particle surface. Only if the slipping plane is located at the particle surface, the zeta potential ζ becomes equal to the surface potential ψ_o. In the following we treat the case where $\zeta = \psi_o$ and where particles are symmetrical (planar, spherical, or cylindrical). Usually, we treat the case where the magnitude of the applied electric field $\boldsymbol{E}$ is not very large. In such cases, the velocity $\boldsymbol{U}$ of the particles, which is called electrophoretic velocity, is proportional to $\boldsymbol{E}$ in magnitude. The magnitude ratio of $\boldsymbol{U}$ to $\boldsymbol{E}$ is called electrophoretic mobility μ, which is defined by $\mu = U/E$ (where $U = |\boldsymbol{U}|$ and $E = |\boldsymbol{E}|$). Before dealing with the electrophoretic mobility of soft particles, we briefly discuss the electrophoretic mobility of hard particles in the next section.

21.2 BRIEF SUMMARY OF ELECTROPHORESIS OF HARD PARTICLES

The most widely employed formula relating the electrophoretic mobility μ of a colloidal particle in an electrolyte solution to its zeta potential ζ is the following Smoluchowski's formula [1]:

$$\mu = \frac{\varepsilon_r \varepsilon_o}{\eta} \zeta \tag{21.1}$$

Biophysical Chemistry of Biointerfaces By Hiroyuki Ohshima

Here ε_r and η are, respectively, the relative permittivity and the viscosity of the electrolyte solution. This formula, however, is the correct limiting mobility equation for very large particles and is valid irrespective of the shape of the particle provided that the dimension of the particle is much larger than the Debye length $1/\kappa$ (where κ is the Debye–Hückel parameter, defined by Eq. (1.8)) and thus the particle surface can be considered to be locally planar. For a sphere with radius a, this condition is expressed by $\kappa a \gg 1$. In the opposite limiting case of very small spheres ($\kappa a \gg 1$), the mobility–zeta potential relationship is given by Hückel's equation [2],

$$\mu = \frac{2\varepsilon_r\varepsilon_o}{3\eta}\zeta \tag{21.2}$$

Henry [3] derived the mobility equations for spheres of radius a and an infinitely long cylinder of radius a, which are applicable for low ζ and any value of κa. Henry's equation for the electrophoretic mobility μ of a spherical colloidal particle of radius a with a zeta potential ζ is expressed as:

$$\mu = \frac{\varepsilon_r\varepsilon_o}{\eta}\zeta f(\kappa a) \tag{21.3}$$

with

$$f(\kappa a) = 1 - e^{\kappa a}\{5E_7(\kappa a) - 2E_5(\kappa a)\} \tag{21.4}$$

where $f(\kappa a)$ is called Henry's function and $E_n(\kappa a)$ is the exponential integral of order n. As $\kappa a \to \infty$, $f(\kappa a) \to 1$ and Eq. (21.3) tends to Smoluchowski's equation (21.1), while if $\kappa a \to 0$, then $f(\kappa a) \to 2/3$ and Eq. (21.3) becomes Hückel's equation (21.2). Ohshima [17] has derived the following simple approximate formula for Henry's function $f(\kappa a)$ with relative errors less than 1%:

$$f(\kappa a) = \frac{2}{3}\left[1 + \frac{1}{2(1 + (2.5/\kappa a\{1 + 2\exp(-\kappa a)\}))^3}\right] \tag{21.5}$$

For the case of a cylindrical particle, the electrophoretic mobility depends on the orientation of the particle with respect to the applied electric field. When the cylinder is oriented parallel to the applied electric field, its electrophoretic mobility μ is given by Smoluchowski's equation (1), namely,

$$\mu_{//} = \frac{\varepsilon_r\varepsilon_o}{\eta}\zeta \tag{21.6}$$

If, on the other hand, the cylinder is oriented perpendicularly to the applied field, then the mobility depends not only on ζ but also on the value of κa. Henry [3] showed that $f(\kappa a)$ for a cylindrical particle of radius a oriented perpendicularly to the applied field is given by

$$\mu_{\perp} = \frac{\varepsilon_r\varepsilon_o}{\eta}\zeta f(\kappa a) \tag{21.7}$$

where $f(\kappa a)$ is Henry's function for a cylinder. Ohshima [19] obtained an approximate formula for Henry's function for a cylinder,

$$f(\kappa a) = \frac{1}{2}\left[1 + \frac{1}{(1 + (2.55/\kappa a\{1 + \exp(-\kappa a)\}))^2}\right] \tag{21.8}$$

For a cylindrical particle oriented at an arbitrary angle between its axis and the applied electric field, its electrophoretic mobility averaged over a random distribution of orientation is given by $\mu_{\text{av}} = \mu_{//}/3 + 2\mu_{\perp}/3$ [9]. The above expressions for the electrophoretic mobility are correct to the first order of ζ so that these equations are applicable only when ζ is low. The readers should be referred to Ref. [4–8, 10–13, 22, 27, 29] for the case of particles with arbitrary zeta potential, in which case the relaxation effects (i.e., the effects of the deformation of the electrical double layer around particles) become appreciable.

21.3 GENERAL THEORY OF ELECTROPHORETIC MOBILITY OF SOFT PARTICLES

This section deals with a general theory of electrophoresis of soft particles and approximate analytic expressions for the mobility of soft particles [30–51]. This theory unites the electrophoresis theories of hard particles [1–29] and of polyelectrolytes [52], since a soft particle tends to a hard particle in the absence of the polymer layer and to a polyelectrolyte in the absence of the particle.

Consider a spherical soft particle, that is, a charged spherical particle covered with an ion-penetrable layer of polyelectrolytes moving with a velocity $\boldsymbol{U}$ in a liquid containing a general electrolyte in an applied electric field $\boldsymbol{E}$. We assume that the uncharged particle core of radius a is coated with an ion-penetrable layer of polyelectrolytes of thickness d and that ionized groups of valence Z are distributed within the polyelectrolyte layer at a uniform density N so that the polyelectrolyte later is uniformly charged at a constant density $\rho_{\text{fix}} = ZeN$. The polymer-coated particle has thus an inner radius a and an outer radius $b \equiv a + d$ (Fig. 21.1). The origin of the spherical polar coordinate system (r, θ, φ) is held fixed at the center of the particle core and the polar axis $(\theta = 0)$ is set parallel to $\boldsymbol{E}$. Let the electrolyte be composed of N ionic mobile species of valence z_i, bulk concentration (number density) n_i^∞, and drag coefficient λ_i $(i = 1, 2, \ldots, M)$. The drag coefficient λ_i of the ith ionic species is further related to the limiting conductance Λ_i^o of that ionic species by

$$\lambda_i = \frac{N_\text{A} e^2 |z_i|}{\Lambda_i^\text{o}} \tag{21.9}$$

where N_A is the Avogadro's number. We adopt the model of Debye–Bueche [53] that the polymer segments are regarded as resistance centers distributed uniformly in the polyelectrolyte layer, exerting frictional forces on the liquid flowing in the polyelectrolyte layer. We give below the general theory of electrophoresis of soft particles [29, 40, 42, 45].

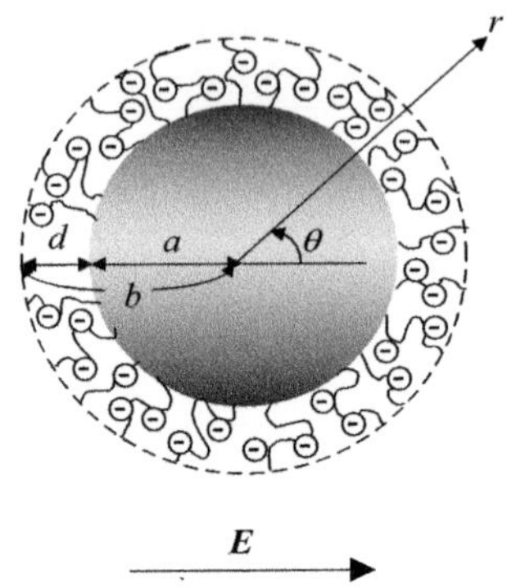

FIGURE 21.1 A soft particle in an external applied electric field $\boldsymbol{E}$. a = radius of the particle core and d = thickness of the polyelectrolyte layer coating the particle core. $b = a + d$.

The fundamental electrokinetic equations for the liquid velocity $\boldsymbol{u}(\boldsymbol{r})$ at position $\boldsymbol{r}$ relative to the particle ($\boldsymbol{u}(\boldsymbol{r}) \to -\boldsymbol{U}$ as $r \equiv |\boldsymbol{r}| \to \infty$) and the velocity of the ith ionic species $\boldsymbol{v}_i$ are the same as those for rigid spheres except that the Navier–Stokes equations for $\boldsymbol{u}(\boldsymbol{r})$ become different for the regions outside and inside the surface layer, namely,

$$\eta \nabla \times \nabla \times \boldsymbol{u} + \nabla p + \rho_{\mathrm{el}} \nabla \psi = \boldsymbol{0}, \quad r > b \tag{21.10}$$

$$\eta \nabla \times \nabla \times \boldsymbol{u} + \gamma \boldsymbol{u} + \nabla p + \rho_{\mathrm{el}} \nabla \psi = \boldsymbol{0}, \quad a < r < b \tag{21.11}$$

Here Eq. (21.10) is the usual Navier–Stokes equation. The term $\gamma \boldsymbol{u}$ on the left--hand side of Eq. (21.11) represents the frictional forces exerted on the liquid flow by the polymer segments in the polyelectrolyte layer, and γ is the frictional coefficient. If it is assumed that each resistance center corresponds to a polymer segment, which in turn is regarded as a sphere of radius a_{p} and the polymer segments are distributed at a uniform volume density of N_{p} in the polyelectrolyte layer, then each polymer segment exerts the Stokes resistance $6\pi\eta a_{\mathrm{p}} \boldsymbol{u}$ on the liquid flow in the polyelectrolyte layer so that

$$\gamma = 6\pi\eta a_{\mathrm{p}} N_{\mathrm{p}} \tag{21.12}$$

Similarly, the Poisson equation relating the charge density $\rho_{\mathrm{el}}(\boldsymbol{r})$ resulting from the mobile charged ionic species and the electric potential $\psi(\boldsymbol{r})$ are also different for the different regions,

$$\Delta \psi(\boldsymbol{r}) = -\frac{\rho_{\mathrm{el}}(\boldsymbol{r})}{\varepsilon_{\mathrm{r}} \varepsilon_o}, \quad r > b \tag{21.13}$$

$$\Delta \psi(\boldsymbol{r}) = -\frac{\rho_{\mathrm{el}}(\boldsymbol{r}) + ZeN}{\varepsilon_{\mathrm{r}} \varepsilon_{\mathrm{o}}}, \quad a < r < b \tag{21.14}$$

where the relative permittivity ε_r is assumed to take the same value both inside and outside the polyelectrolyte layer. Note that the right-hand side of Eq. (21.14) contains the contribution of the fixed charges of density $\rho_{\text{fix}} = ZeN$ in the polyelectrolyte layer.

For a weak applied filed $\boldsymbol{E}$, the electrokinetic equations are linearized to give

$$\eta\nabla\times\nabla\times\nabla\times\boldsymbol{u} = \sum_{i=1}^{N}\nabla\delta n_i\times\nabla n_i^{(0)}, \quad r > b \tag{21.15}$$

$$\eta\nabla\times\nabla\times\nabla\times u + \gamma\nabla\times\boldsymbol{u} = \sum_{i=1}^{N}\nabla\delta n_i\times\nabla n_i^{(0)}, \quad a < r < b \tag{21.16}$$

and

$$\nabla\cdot\left(n_i^{(0)}u - \frac{1}{\lambda_i}n_i^{(0)}\nabla\delta\mu_i\right) = 0 \tag{21.17}$$

where $n_i^{(0)}(r)$ is the equilibrium concentration (number density) of the ith ionic species ($n_i^{(0)} \rightarrow n_i^{\infty}$ as $r \rightarrow \infty$), and $\delta n_i^{(0)}(\boldsymbol{r})$ and $\delta\mu_i(\boldsymbol{r})$ are, respectively, the deviation of $n_i^{(0)}(r)$ and that of the electrochemical potential $\mu_i(\boldsymbol{r})$ of the ith ionic species due to the applied field $\boldsymbol{E}$.

Symmetry considerations permit us to write

$$\boldsymbol{u}(\boldsymbol{r}) = \left(-\frac{2}{r}h(r)E\cos\theta, \frac{1}{r}\frac{d}{dr}(rh(r))E\sin\theta, 0\right) \tag{21.18}$$

$$\delta\mu_i(\boldsymbol{r}) = -z_ie\phi_i(r)E\cos\theta \tag{21.19}$$

where $E = |\boldsymbol{E}|$. The fundamental electrokinetic equations can be transformed into equations for $h(r)$ and $\phi_i(r)$ as

$$L(Lh) = G(r), \quad r > b \tag{21.20}$$

$$L(Lh - \lambda^2 h) = G(r), \quad a < r < b \tag{21.21}$$

$$L\phi_i = \frac{dy}{dr}\left(z_i\frac{d\phi_i}{dr} - \frac{2\lambda_i}{e}\frac{h}{r}\right) \tag{21.22}$$

with

$$\lambda = (\gamma/\eta)^{1/2} \tag{21.23}$$

$$L \equiv \frac{d}{dr}\frac{1}{r^2}\frac{d}{dr}r^2 = \frac{d^2}{dr^2} + \frac{2}{r}\frac{d}{dr} - \frac{2}{r^2} \tag{21.24}$$

$$G(r) = -\frac{e}{\eta r}\frac{dy}{dr}\sum_{i=1}^{M} z_i^2 n_i^\infty \, e^{-z_i y}\phi_i \tag{21.25}$$

where $y = e\psi^{(0)}/kT$ is the scaled equilibrium potential outside the particle core and the reciprocal of λ, that is, $1/\lambda$, is called the electrophoretic softness.

The boundary conditions for $\psi(\boldsymbol{r})$, $\boldsymbol{u}(\boldsymbol{r})$, and $\boldsymbol{v}_i(\boldsymbol{r})$ are as follows.

(i) $\psi(\boldsymbol{r})$ and $\Delta\psi(\boldsymbol{r})$ are continuous at $r = b$,

$$\psi(b^-, \theta) = \psi(b^+, \theta) \tag{21.26}$$

$$\left.\frac{\partial\psi(r,\theta)}{\partial r}\right|_{r=b^-} = \left.\frac{\partial\psi(r,\theta)}{\partial r}\right|_{r=b^+}$$

$$\left.\frac{\partial\psi(r,\theta)}{\partial \theta}\right|_{r=b^-} = \left.\frac{\partial\psi(r,\theta)}{\partial \theta}\right|_{r=b^+} \tag{21.27}$$

where the continuity of $\Delta\psi(\boldsymbol{r})$ results from the assumption that the relative permittivity ε_r takes the same value both inside and outside the polyelectrolyte layer.

(ii) $\boldsymbol{u} = (u_r, u_\theta, 0) = \mathbf{0}$ at $r = a$ (21.28)

(iii) $\boldsymbol{u} \to -\boldsymbol{U} = (-U\cos\theta, U\sin\theta, 0)$ as $r \to \infty$ (21.29)

(iv) The normal and tangential components of $\boldsymbol{u}$ are continuous at $r = b$,

$$u_r(b^-) = u_r(b^+) \tag{21.30}$$

$$u_\theta(b^-) = u_\theta(b^+) \tag{21.31}$$

(v) The normal and tangential components of the stress tensor $\boldsymbol{\sigma}^{\mathrm{T}}$, which is the sum of the hydrodynamic stress $\boldsymbol{\sigma}^{\mathrm{H}}$ and the Maxwell stress $\boldsymbol{\sigma}^{\mathrm{E}}$, are continuous at $r = b$. From the continuity of $\Delta\psi(\boldsymbol{r})$, the normal and tangential components of $\boldsymbol{\sigma}^{\mathrm{E}}$ are continuous at $r = b$. Therefore, the normal and tangential components of $\boldsymbol{\sigma}^{\mathrm{H}}$ must be continuous at $r = b$ so that the pressure $p(\boldsymbol{r})$ is continuous at $r = b$.

(vi) The electrolyte ions cannot penetrate the particle core, namely,

$$\boldsymbol{v}_i \cdot \boldsymbol{n}\big|_{r=a} = 0 \tag{21.32}$$

where $\boldsymbol{n}$ is the unit normal outward from the surface of the particle core.

(vii) In the stationary state the net force acting on the particle (the particle core plus the polyelectrolyte layer) or an arbitrary volume enclosing the particle must be zero. Consider a large sphere S of radius r containing the particle. The radius r is taken to be sufficiently large so that the net electric charge within S is zero. There is then no electric force acting on S and we need consider only hydrodynamic force $\boldsymbol{F}_{\mathrm{H}}$, which must be zero, namely,

$$\boldsymbol{F}_{\mathrm{H}} = \int_S \left(\sigma_{rr}^{\mathrm{H}} \cos\theta - \sigma_{r\theta}^{\mathrm{H}} \sin\theta \right) dS(\boldsymbol{E}/E) = \mathbf{0}, \quad r \to \infty \tag{21.33}$$

where the integration is carried out over the surface of the sphere S.

The above boundary conditions are expressed in terms of $h(r)$ and $\phi_i(r)$ as

$$h = \frac{dh}{dr} = 0 \quad \text{at } r = a \tag{21.34}$$

$$h(b^+) = h(b^-) \tag{21.35}$$

$$\left.\frac{dh}{dr}\right|_{r=b^+} = \left.\frac{dh}{dr}\right|_{r=b^-} \tag{21.36}$$

$$\left.\frac{d^2h}{dr^2}\right|_{r=b^+} = \left.\frac{d^2h}{dr^2}\right|_{r=b^-} \tag{21.37}$$

$$\frac{d}{dr}\left[r(L - \lambda^2)h \right] \Big|_{r=b^-} = \left.\frac{d}{dr}(rLh)\right|_{r=b^+} \tag{21.38}$$

$$\left.\frac{d\phi_i}{dr}\right|_{r=a} = 0 \tag{21.39}$$

The equilibrium potential $\psi^{(0)}(r)$ satisfies Eqs. (4.50) and (4.51).
The electrophoretic mobility $\mu = U/E$ (where $U = |\boldsymbol{U}|$) can be calculated from

$$\mu = 2 \lim_{r \to \infty} \frac{h(r)}{r} \tag{21.40}$$

with the result [45]

$$\mu = \frac{b^2}{9} \int_b^\infty \left[3\left(1 - \frac{r^2}{b^2}\right) - \frac{2L_2}{L_1}\left(1 - \frac{r^3}{b^3}\right) \right] G(r)dr + \frac{2L_3}{3\lambda^2 L_1} \int_a^\infty \left(1 + \frac{r^3}{2b^3}\right) G(r)dr$$
$$- \frac{2}{3\lambda^2} \int_a^b \left[1 - \frac{3a}{2\lambda^2 b^3 L_1} \{ (L_3 + L_4 \lambda r) \cosh[\lambda(r-a)] \right.$$
$$\left. - (L_4 + L_3 \lambda r) \sinh[\lambda(r-a)] \} \right] G(r)dr \tag{21.41}$$

with

$$L_1 = \left(1 + \frac{a^3}{2b^3} + \frac{3a}{2\lambda^2 b^3} - \frac{3a^2}{2\lambda^2 b^4}\right)\cosh[\lambda(b-a)] - \left(1 - \frac{3a^2}{2b^2} + \frac{a^3}{2b^3} + \frac{3a}{2\lambda^2 b^3}\right)\frac{\sinh[\lambda(b-a)]}{\lambda b} \tag{21.42}$$

$$L_2 = \left(1 + \frac{a^3}{2b^3} + \frac{3a}{2\lambda^2 b^3}\right)\cosh[\lambda(b-a)] + \frac{3a^2}{2b^2}\frac{\sinh[\lambda(b-a)]}{\lambda b} - \frac{3a}{2\lambda^2 b^3} \tag{21.43}$$

$$L_3 = \cosh[\lambda(b-a)] - \frac{\sinh[\lambda(b-a)]}{\lambda b} - \frac{a}{b} \tag{21.44}$$

$$L_4 = \sinh[\lambda(b-a)] - \frac{\cosh[\lambda(b-a)]}{\lambda b} + \frac{\lambda a^2}{3b} + \frac{2\lambda b^2}{3a} + \frac{1}{\lambda b} \tag{21.45}$$

21.4 ANALYTIC APPROXIMATIONS FOR THE ELECTROPHORETIC MOBILITY OF SPHERICAL SOFT PARTICLES

21.4.1 Large Spherical Soft Particles

We derive approximate mobility formulas for the simple but important case where the potential is arbitrary but the double-layer potential still remains spherically symmetrical in the presence of the applied electric field (the relaxation effect is neglected). Further we treat the case where the following conditions hold

$$\kappa a \gg 1, \lambda a \gg 1 (\text{and thus } \kappa b \gg 1, \lambda b \gg 1) \text{ and } \kappa d \gg 1 \text{ and } \lambda d \gg 1 \tag{21.46}$$

which are satisfied for most practical cases. For the case in which the electrolyte is symmetrical with a valence z and bulk concentration n, if $\kappa d \gg 1$, then the potential inside the polyelectrolyte layer can be approximated by (see Eq. (4.39))

$$\psi^{(0)}(r) = \psi_{\mathrm{DON}} + (\psi_o - \psi_{\mathrm{DON}})e^{-\kappa_m|r-b|}, \quad (a < r < b) \tag{21.47}$$

with

$$\psi_{\mathrm{DON}} = \left(\frac{kT}{ze}\right)\ln\left[\frac{ZN}{2zn} + \left\{\left(\frac{ZN}{2zn}\right)^2 + 1\right\}^{1/2}\right] \tag{21.48}$$

$$\psi_o = \frac{kT}{ze}\left(\ln\left[\frac{ZN}{2zn} + \left\{\left(\frac{ZN}{2zn}\right)^2 + 1\right\}^{1/2}\right] + \frac{2zn}{ZN}\left[1 - \left\{\left(\frac{ZN}{2zn}\right)^2 + 1\right\}^{1/2}\right]\right) \tag{21.49}$$

$$\kappa_{\mathrm{m}} = \kappa\left[1 + \left(\frac{ZN}{2zn}\right)^2\right]^{1/4} \tag{21.50}$$

where κ_{m} is the Debye–Hückel parameter in the surface layer and ψ_{DON} and $\psi_{\mathrm{o}} = \psi(b)$ are, respectively, the Donnan potential and the surface potential of the surface charge layer. By using the potential distribution given by Eq. (21.47), from Eq. (21.41) we obtain [45]

$$\mu = \frac{\varepsilon_{\mathrm{r}}\varepsilon_{\mathrm{o}}}{\eta}\frac{\psi_{\mathrm{o}}/\kappa_{\mathrm{m}} + \psi_{\mathrm{DON}}/\lambda}{1/\kappa_{\mathrm{m}} + 1/\lambda} f\left(\frac{d}{a}\right) + \frac{ZeN}{\eta\lambda^2} \tag{21.51}$$

with

$$f\left(\frac{d}{a}\right) = \frac{2}{3}\left(1 + \frac{a^3}{2b^3}\right) = \frac{2}{3}\left\{1 + \frac{1}{2(1 + d/a)^3}\right\} \tag{21.52}$$

We plot $f(d/a)$ in Fig. 21.2. In the limit of $d \gg a$ (Fig. 21.2), in which case $f(d/a) \rightarrow 2/3$, Eq. (21.51) reduces to

$$\mu = \frac{2\varepsilon_{\mathrm{r}}\varepsilon_{\mathrm{o}}}{3\eta}\frac{\psi_{\mathrm{o}}/\kappa_{\mathrm{m}} + \psi_{\mathrm{DON}}/\lambda}{1/\kappa_{\mathrm{m}} + 1/\lambda} + \frac{ZeN}{\eta\lambda^2} \tag{21.53}$$

For the low potential case, Eq. (21.53) further reduces to

$$\mu = \frac{ZeN}{\eta\lambda^2}\left[1 + \frac{2}{3}\left(\frac{\lambda}{\kappa}\right)^2\left(\frac{1 + \lambda/2\kappa}{1 + \lambda/\kappa}\right)\right] \tag{21.54}$$

which agrees with Hermans–Fujita's equation for the electrophoretic mobility of a spherical polyelectrolyte (a soft particle with no particle core) [52], as will be shown later.

In the opposite limit of $d \ll a$ (Fig. 21.2), in which case $f(d/a) \rightarrow 1$, Eq. (21.51) reduces to

$$\mu = \frac{\varepsilon_{\mathrm{r}}\varepsilon_{\mathrm{o}}}{\eta}\frac{\psi_{\mathrm{o}}/\kappa_{\mathrm{m}} + \psi_{\mathrm{DON}}/\lambda}{1/\kappa_{\mathrm{m}} + 1/\lambda} + \frac{ZeN}{\eta\lambda^2} \tag{21.55}$$

which, for low potentials, reduces to

$$\mu = \frac{ZeN}{\eta\lambda^2}\left[1 + \left(\frac{\lambda}{\kappa}\right)^2\left(\frac{1 + \lambda/2\kappa}{1 + \lambda/\kappa}\right)\right] \tag{21.56}$$

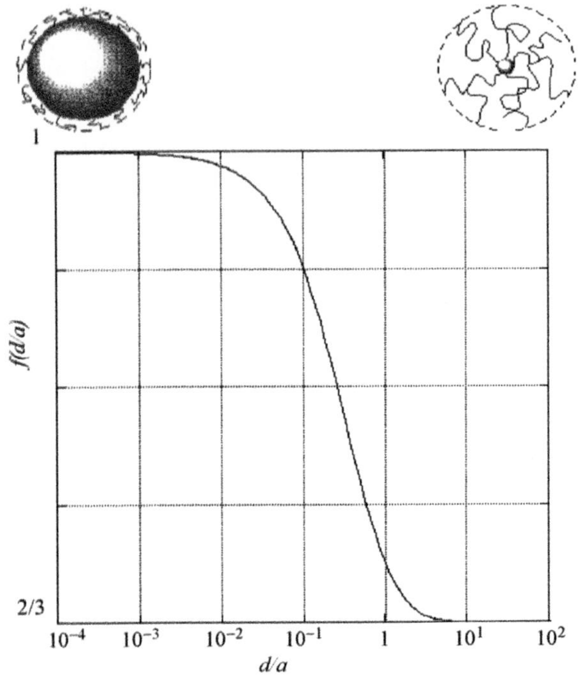

FIGURE 21.2 $f(d/a)$, defined by Eq. (21.52), as a function of d/a. From Ref. [40].

Equation (21.63) covers a plate-like soft particle. Indeed, in the limiting case of $a \to \infty$, the general mobility expression (Eq. (21.41)) reduces to

$$\mu = \frac{\varepsilon_r \varepsilon_o}{\eta} \frac{1}{\cosh(\lambda d)} \left[\psi^{(0)}(-d) + \lambda \int_{-d}^{0} \psi^{(0)}(x) \sinh \lambda(x+d) dx \right] + \frac{ZeN}{\eta \lambda^2} \left\{ 1 - \frac{1}{\cosh(\lambda d)} \right\} \tag{21.57}$$

which is the general mobility expression of a plate-like soft particle. For the case of $\kappa d \gg 1$ and $\lambda d \gg 1$, Eq. (21.57) gives to Eq. (21.55) after Eq. (21.47) is substituted. Note that Eq. (21.57) can also be derived from the following one-dimensional Navier–Stokes equations

$$\eta \frac{d^2 u}{dx^2} + \rho_{el}(x) E = 0, \quad 0 < x < +\infty \tag{21.58}$$

$$\eta \frac{d^2 u}{dx^2} - \gamma u + \rho_{el}(x) E = 0, \quad -d < x < 0 \tag{21.59}$$

together with the Poisson–Boltzmann equations

$$\frac{d^2\psi}{dx^2} = -\frac{\rho_{el}(x)}{\varepsilon_r\varepsilon_o}, \quad 0 < x < +\infty \tag{21.60}$$

$$\frac{d^2\psi}{dx^2} = -\frac{\rho_{el}(x) + ZeN}{\varepsilon_r\varepsilon_o}, \quad -d < x < 0 \tag{21.61}$$

where $x = r - a$ denotes the distance from the particle core.

In Fig. 21.2 we plot the function $f(d/a)$, which varies from 2/3 to 1, as d/a increases. For $d \ll a$ ($f(d/a) \approx 1$), the polyelectrolyte layer can be regarded as planar, while for $d \gg a$ ($f(d/a) \approx 2/3$), the soft particle behaves like a spherical polyelectrolyte with no particle core.

Equation (21.51) consists of two terms: the first term is a weighted average of the Donnan potential ψ_{DON} and the surface potential ψ_o. It should be stressed that only the first term is subject to the shielding effects of electrolytes, tending zero as the electrolyte concentration n increases, while the second term does not depend on the electrolyte concentration. In the limit of high electrolyte concentrations, all the potentials vanish and only the second term of the mobility expression remains, namely,

$$\mu \to \mu^{\infty} = \frac{ZeN}{\eta\lambda^2} \tag{21.62}$$

Equation (21.62) shows that as $\kappa \to \infty$, μ tends to a nonzero limiting value μ^{∞}. This is a characteristic of the electrokinetic behavior of soft particles, in contrast to the case of the electrophoretic mobility of hard particles, which should reduces to zero due to the shielding effects, since the mobility expressions for rigid particles (Chapter 3) do not have μ^{∞}. The term μ^{∞} can be interpreted as resulting from the balance between the electric force acting on the fixed charges $(ZeN)\boldsymbol{E}$ and the frictional force $\gamma\boldsymbol{u}$, namely,

$$(ZeN)\boldsymbol{E} + \gamma\boldsymbol{u} = \boldsymbol{0} \tag{21.63}$$

from which Eq. (21.62) follows.

In order to see this more clearly, in Fig. 21.3b and a we give a schematic representation of the liquid velocity $u(x)$ $(= |\boldsymbol{u}(x)|)$ as well as potential distributions $\psi(x)$ around a soft particle, x being given by $x = r - a$, across the surface charge layer, in comparison with those for a hard particle. Figure 21.3c shows the electrophoretic mobility μ as a function of electrolyte concentration n for a soft particle and a hard particle. It is seen that the liquid velocity $u(x)$ in the middle region of the surface charge layer (the plateau region) (Fig. 21.3b), where the potential is almost equal to the Donnan potential ψ_{DON} (Fig. 21.3a), gives μ^{∞}. It is this term that yields the term independent of the electrolyte concentration. We also note that $1/\lambda$ can be

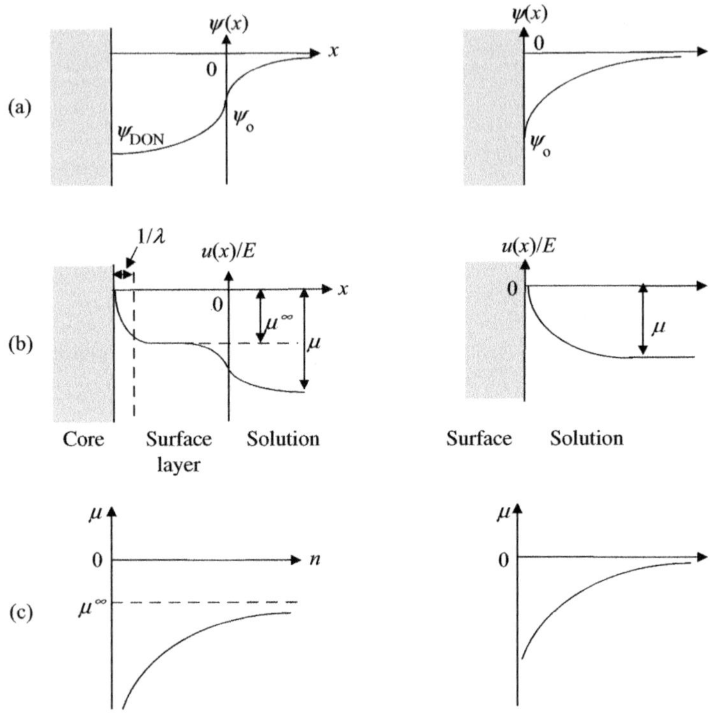

FIGURE 21.3 Schematic representation of liquid velocity distribution $u(x)$ (b) as well as potential distribution $\psi(x)$ (a) around a soft particle, and the electrophoretic mobility μ as a function of electrolyte concentration n (c).

interpreted as the distance between the slipping plane at the core surface at $r=a$ ($x=0$) and the bulk phase of the surface layer (the plateau region in Fig. 21.3b), where $u(x)/E$ is given by $-\mu^{\infty}$. Figure 21.3a shows that a small sift of the slipping plane causes no effects on the mobility value. Therefore, the mobility of soft particles is insensitive to the position of the slipping plane and the zeta potential loses its meaning.

21.4.2 Weakly Charged Spherical Soft Particles

When the fixed-charge density ZeN is low, it can be shown [17] that the mobility is given by

$$\mu = \frac{ZeNb^2 f_0}{2\eta}\left[\frac{a^3}{6b}\int_b^\infty \frac{e^{-\kappa(r-b)}}{r^3}dr - \frac{a^3 b}{2}\left(1 - \frac{2L_2}{3L_1} + \frac{2L_3}{\lambda^2 b^2 L_1}\right)\int_b^\infty \frac{e^{-\kappa(r-b)}}{r^5}dr\right.$$
$$\left. - \frac{2}{3\kappa b} + \frac{L_3}{\lambda^2 b^2 L_1}\left(1 + \frac{a^3}{2b^3} + \frac{1}{\kappa b} + \frac{1}{\kappa^2 b^2}\right) + \frac{2L_2}{3\kappa b L_1}\left(1 + \frac{1}{\kappa b}\right)\right]$$

$$
\begin{aligned}
& -\frac{ZeNa^2(1+\kappa b)}{4\eta\lambda^3 b^3(1+\kappa a)} e^{-\kappa(b-a)} \left[\left(\frac{L_3+L_4}{L_1}\right) \left\{ \left(1-\frac{1}{\kappa a}\right) f_1(\kappa,\lambda) \right. \right. \\
& \left. + \left(1+\frac{1}{\kappa a}\right) f_1(-\kappa,\lambda) \right\} - \left(\frac{L_3-L_4}{L_1}\right) \left\{ \left(1-\frac{1}{\kappa a}\right) f_1(\kappa,-\lambda) \right. \\
& \left. \left. + \left(1+\frac{1}{\kappa a}\right) f_1(-\kappa,-\lambda) \right\} \right] - \frac{ZeN(1+\kappa b)}{3\eta\lambda^2(1+\kappa a)} e^{-\kappa(b-a)} \left[\left(1-\frac{L_3}{L_1}\right) \right. \\
& \times \left\{ \left(1-\frac{1}{\kappa a}\right) f_2(\kappa) + \left(1+\frac{1}{\kappa a}\right) f_2(-\kappa) \right\} - \frac{L_3}{L_1} \left\{ \left(1-\frac{1}{\kappa a}\right) f_3(\kappa) \right. \\
& \left. \left. + \left(1+\frac{1}{\kappa a}\right) f_3(-\kappa) \right\} \right]
\end{aligned} \tag{21.64}
$$

where

$$
f_0 = 1 - \frac{1}{\kappa b} + \frac{(1-\kappa a)(1+\kappa b)}{(1+\kappa a)\kappa b} e^{-2\kappa(b-a)} \tag{21.65}
$$

$$
\begin{aligned}
f_1(\kappa,\lambda) = & \frac{\kappa a^3}{2} \int_a^b \frac{e^{-(\kappa+\lambda)(r-a)}}{r^3} dr - \frac{3a^3}{2\lambda} \int_a^b \frac{e^{-(\kappa+\lambda)(r-a)}}{r^5} dr + \frac{3}{2\lambda a} + \frac{\kappa}{\kappa+\lambda} \\
& - \left\{ \frac{1}{\lambda b}\left(1+\frac{a^3}{2b^3}\right) + \frac{\kappa}{\kappa+\lambda} \right\} e^{-(\kappa+\lambda)(b-a)}
\end{aligned} \tag{21.66}
$$

$$
f_2(\kappa) = \frac{3a^4}{2} \int_a^b \frac{e^{-\kappa(r-a)}}{r^5} dr + \frac{a}{b} e^{-\kappa(b-a)} \left(1+\frac{a^3}{2b^3}\right) - \frac{3}{2} \tag{21.67}
$$

$$
f_3(\kappa) = \frac{a}{2b} e^{-\kappa(b-a)} \left(1+\frac{a^3}{2b^3}+\frac{3}{\kappa b}+\frac{3}{\kappa^2 b^2}\right) - \frac{3a^3}{4b^3}\left(1+\frac{2}{\kappa a}+\frac{2}{\kappa^2 a^2}\right) \tag{21.68}
$$

Consider various limiting behaviors of the electrophoretic mobility μ of soft particles.

(i) Low electrolyte concentration limit. In the limit $\kappa \to 0$, Eq. (21.64) becomes

$$
\mu = \frac{2(b^3-a^3)\rho_{\text{fix}}}{9\eta b} \left(\frac{L_2}{L_1} + \frac{3L_3}{2\lambda^2 b^2 L_1}\right) \tag{21.69}
$$

Note that Eq. (21.69) can be directly derived from the condition of balance between the electric force and the drag force acting on the particle, by using an expression for the drag force acting on a particle covered with an uncharged polymer layer derived by Masliyah et al. [54]. That is, Eq. (21.69) can be rewritten as

$$\mu = \frac{Q}{D_H} \tag{21.70}$$

with

$$Q = \frac{4}{3}\pi(b^3 - a^3)\rho_{\text{fix}} \tag{21.71}$$

$$D_{\text{H}} = 6\pi\eta a\left[\frac{a}{b}\left(\frac{L_2}{L_1} + \frac{3L_3}{2\lambda^2 b^2 L_1}\right)\right]^{-1} \tag{21.72}$$

where Q is the total particle charge and D_{H} is the drag coefficient of a soft particle [21]. If, further, $\lambda a \gg 1$, then D_{H} approaches the Stokes drag $6\pi\eta b$ so that Eq. (21.70) becomes

$$\mu = \frac{2(b^3 - a^3)ZeN}{9\mu b} \tag{21.73}$$

(ii) High electrolyte concentration limit. In the limit $\kappa \to \infty$, Eq. (21.64) becomes

$$\mu = \frac{ZeN}{\eta\lambda^2}\left(1 + \frac{a^3}{2b^3}\right)\left\{\frac{L_3}{L_1} - \frac{1}{3} + \frac{a(1-\lambda b)}{4\lambda^2 b^3}\left(\frac{L_3 - L_4}{L_1}\right)e^{\lambda(b-a)}\right\} \tag{21.74}$$

In the further limit of $\lambda a \to \infty$, Eq. (21.74) tends to Eq. (21.62).

(iii) In the limit $\lambda \to 0$ and/or $a \to b$, the polyelectrolyte-coated particle becomes a hard particle of radius a with no polyelectrolyte layer and the electrophoretic mobility becomes Henry's equation (21.4) for a hard particle.

(iv) Spherical polyelectrolyte. In the limit $a \to 0$, Eq. (21.64) tends to

$$\mu = \frac{ZeN}{\eta\lambda^2}\left[1 + \frac{1}{3}\left(\frac{\lambda}{\kappa}\right)^2\left(1 + e^{-2\kappa b} - \frac{1 - e^{-2\kappa b}}{\kappa b}\right)\right.$$
$$+\frac{1}{3}\left(\frac{\lambda}{\kappa}\right)^2 \frac{1 + 1/\kappa b}{(\lambda/\kappa)^2 - 1}\left\{\left(\frac{\lambda}{\kappa}\right)\frac{1 + e^{-2\kappa b} - (1 - e^{-2\kappa b})/\kappa b}{(1 + e^{-2\lambda b})/(1 - e^{-2\lambda b}) - 1/\lambda b}\right.$$
$$\left.\left. -(1 - e^{-2\kappa b})\right\}\right] \tag{21.75}$$

For $\lambda b \gg 1$ and $\kappa b \gg 1$, Eq. (21.75) reduces to Eq. (21.54).

(v) Plate-like soft particle. In the limit $a \to \infty$, Eq. (21.64) tends to

$$\mu = \frac{ZeN}{2\eta\kappa^2}\left[(1-e^{-2\kappa d})\left\{1+\frac{\kappa}{\lambda}\tanh(\lambda d)\right\}\right.$$
$$\left. + \left\{\frac{1}{1-(\lambda/\kappa)^2}\right\}\left\{1+e^{-2\kappa d}-\frac{\kappa}{\lambda}(1-e^{-2\kappa d})\tanh(\lambda d)-\frac{2e^{-2\kappa d}}{\cosh(\lambda d)}\right\}\right]$$
$$+\frac{ZeN}{\eta\lambda^2}(1-e^{-2\kappa d})\left\{1-\frac{1}{\cosh(\lambda d)}\right\} \quad (21.76)$$

(vi) Large soft particles. For the case where $\kappa a \gg 1$, $\lambda a \gg 1$, $\kappa(b-a) \equiv \kappa d \gg 1$, and $\lambda(b-a) \equiv \lambda d \gg 1$, Eq. (21.64) tends to

$$\mu = \frac{ZeN}{\eta\lambda^2}\left[1+\frac{2}{3}\left(\frac{\lambda}{\kappa}\right)^2\left(\frac{1+\lambda/2\kappa}{1+\lambda/\kappa}\right)\left(1+\frac{a^3}{2b^3}\right)\right] \quad (21.77)$$

Note that Eq. (21.51) reduces to Eq. (21.77) for the low potential case. Further, Eq.(21.77) reduces to Eq. (21.54) for a thick polyelectrolyte layer ($a \ll d$), and to Eq. (21.56) for a thin polyelectrolyte layer ($a \gg d$).

(vii) Semisoft particle. In the limiting case of $\lambda \to \infty$, where the liquid flow inside the polyelectrolyte layer vanishes, the soft particle becomes a hard particle of radius b carrying zeta potential $\psi^{(0)}(b)$ and Eq. (21.64) becomes

$$\begin{aligned}\mu &= \frac{ZeNb^2 f_0}{2\eta}\left[\frac{a^3}{6b^3}e^{\kappa b}\{E_3(\kappa b)-E_5(\kappa b)\}+\frac{2}{3\kappa^2 b^2}\right] \\ &= \frac{ZeNf_0}{2\eta\kappa^2}\left[\frac{2}{3}\left(1+\frac{a^3}{2b^3}\right)-\left(\frac{a^3}{b^3}\right)e^{\kappa b}\{5E_7(\kappa b)-2E_5(\kappa b)\}\right]\end{aligned} \quad (21.78)$$

Note that Eq. (21.78) corresponds to the limit of $\lambda \to \infty$, but still assumes that electrolyte ions can penetrate the polyelectrolyte layer. Therefore, the particle in this limiting case is neither a "perfectly" hard particle nor a "perfectly" soft particle. We term a particle of this type a "semisoft" particle. The polyelectrolyte layer coating a "semisoft" particle is ion permeable but there is no liquid flow inside the polyelectrolyte layer.

21.4.3 Cylindrical Soft Particles

As in the case of electrophoresis of hard cylindrical particles, the electrophoretic mobility of a soft cylinder depends on the orientation of the particle [44, 47].

Consider an infinitely long cylindrical colloidal particle moving with a velocity $\boldsymbol{U}$ in a liquid containing a general electrolyte in an applied electric field $\boldsymbol{E}$. By solving the fundamental electrokinetic equations, we finally obtain the following general expression for the electrophoretic mobility $\mu_\perp$ of a soft cylinder in a transverse field [47]:

$$\mu_\perp = \frac{b^2}{8}\int_b^\infty \left[\left\{1 + \frac{2L_3}{\lambda b L_2(\lambda b)}\right\}\left(1 - \frac{r^2}{b^2}\right) + \frac{2r^2}{b^2}\ln\left(\frac{r}{b}\right)\right] G(r)dr$$

$$+ \frac{L_1(\lambda b)}{2\lambda^2 L_2(\lambda b)}\int_a^\infty \left(1 + \frac{r^3}{2b^3} \rightarrow \frac{r^2}{b^2}\right) G(r)dr$$

$$+ \frac{1}{2\lambda^2}\int_a^b \lambda r\left[L_1(\lambda r) - \frac{L_1(\lambda b)}{L_2(\lambda b)} L_2(\lambda r)\right] G(r)dr \tag{21.79}$$

where

$$L_1(x) = I_0(\lambda a)K_1(x) + K_0(\lambda a)I_1(x) - 1/x \tag{21.80}$$

$$L_2(x) = \left\{I_0(\lambda a) + \frac{a^2}{b^2} I_2(\lambda a)\right\}K_1(x) + \left\{K_0(\lambda a) + \frac{a^2}{b^2}K_2(\lambda a)\right\}I_1(x) \tag{21.81}$$

$$L_3 = \left\{I_0(\lambda a) + \frac{a^2}{b^2} I_2(\lambda a)\right\}K_0(\lambda b) - \left\{K_0(\lambda a) + \frac{a^2}{b^2}K_2(\lambda a)\right\}I_0(\lambda b) + \frac{2}{(\lambda b)^2} \tag{21.82}$$

where $I_n(z)$ and $K_n(z)$ are, respectively, the nth-order modified Bessel functions of the first and second kinds.

In the limit $a \rightarrow 0$, the particle core vanishes and the particle becomes a cylindrical polyelectrolyte (a porous charged cylinder) of radius b. For the low potential case, Eq.(21.79) gives

$$\mu_\perp = \frac{ZeN}{\eta\lambda^2} + \frac{ZeNb}{2\eta\kappa}\left[I_1(\kappa b)K_0(\kappa b) + \frac{\kappa}{\kappa^2 - \lambda^2}\left\{\kappa I_0(\kappa b) - \lambda\frac{I_0(\lambda b)}{I_1(\lambda b)}I_1(\kappa b)\right\}K_1(\kappa b)\right] \tag{21.83}$$

Consider next the case where $\lambda a \gg 1$, $\kappa a \gg 1$ (and thus $\lambda b \gg 1$, $\kappa b \gg 1$) and $\lambda d = \lambda(b-a) \gg 1$, $\kappa d = \kappa(b-a) \gg 1$ and the relaxation effect is negligible, and where the electrolyte is symmetrical with a valence z and bulk concentration n We obtain from Eq. (21.79)

$$\mu_\perp = \frac{\varepsilon_r\varepsilon_o}{\eta}\frac{\psi_o/\kappa_m + \psi_{DON}/\lambda}{1/\kappa_m + 1/\lambda} f\left(\frac{d}{a}\right) + \frac{ZeN}{\eta\lambda^2} \tag{21.84}$$

with

$$f\left(\frac{d}{a}\right) = \frac{1}{2}\left[1 + \left(\frac{a}{b}\right)^2\right] = \frac{1}{2}\left[1 + \frac{1}{(1+d/a)^2}\right] \tag{21.85}$$

where ψ_{DON} is the Donnan potential in the polyelectrolyte layer, ψ_o is the potential at the boundary $r=b$, which is the surface potential of the soft particle, and κ_m is the Debye–Hückel parameter of the polyelectrolyte layer. The limiting forms of Eq. (21.84) for the two cases $d/a \ll 1$ and $d/a \gg 1$ are given by

$$\mu_\perp = \frac{\varepsilon_r\varepsilon_o}{\eta}\frac{\psi_o/\kappa_m + \psi_{DON}/\lambda}{1/\kappa_m + 1/\lambda} + \frac{ZeN}{\eta\lambda^2}, \quad (d \ll a) \tag{21.86}$$

$$\mu_\perp = \frac{\varepsilon_r\varepsilon_o}{2\eta}\frac{\psi_o/\kappa_m + \psi_{DON}/\lambda}{1/\kappa_m + 1/\lambda} + \frac{ZeN}{\eta\lambda^2}, \quad (d \gg a) \tag{21.87}$$

The electrophoretic mobility $\mu_{//}$ of an infinitely long cylindrical soft particle in a tangential field for the case where $\lambda a \gg 1$, $\kappa a \gg 1$, $\lambda d \gg 1$, $\kappa d \gg 1$, is given by Eq. (21.55), namely,

$$\mu_{//} = \frac{\varepsilon_r\varepsilon_o}{\eta}\frac{\psi_o/\kappa_m + \psi_{DON}/\lambda}{1/\kappa_m + 1/\lambda} + \frac{ZeN}{\eta\lambda^2} \tag{21.88}$$

21.5 ELECTROKINETIC FLOW BETWEEN TWO PARALLEL SOFT PLATES

Consider the steady flow of an incompressible aqueous solution of a symmetrical electrolyte through a pore between two parallel fixed similar plates, 1 and 2, at separation h, under the influence of a uniform electric field E and a uniform pressure gradient P, both applied parallel to the plates [83]. The plates are covered by ion--penetrable surface charge layers of thickness d. The plate areas are assumed large enough for edge effects to be neglected. We take the x-axis to be perpendicular to the plates with its origin at the surface of plate 1 and the y-axis to be parallel to the plates so that E and P are both in the y-direction (Fig. 21.4) and are given by $E = -\partial\Psi/\partial y$ and $P = -\partial p/\partial y$, where Ψ is the electric potential and p is the pressure. In this case the Navier–Stokes equations to be solved are

$$\eta\frac{d^2u}{dx^2} + \rho_{el}(x)E + P = 0, \quad 0 < x \leq \frac{h}{2} \tag{21.89}$$

$$\eta\frac{d^2u}{dx^2} - \gamma u(x) + \rho_{el}(x)E + P = 0, \quad -d < x < 0 \tag{21.90}$$

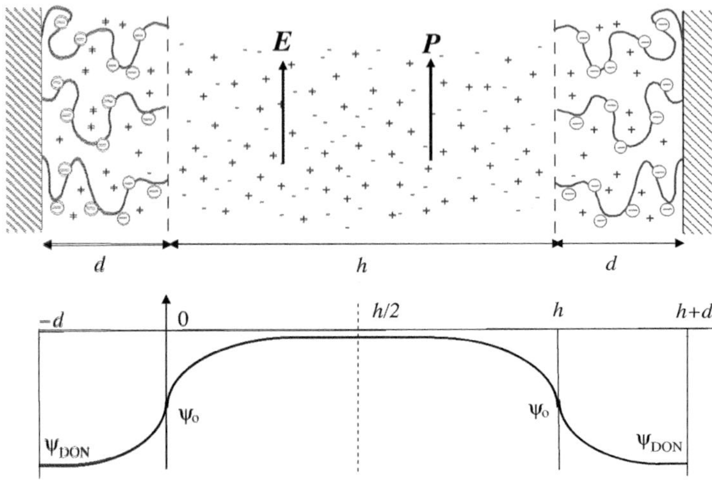

FIGURE 21.4 Schematic representation of a channel of two parallel similar plates 1 and 2 at separation h covered by ion-penetrable surface charge layers of thickness d under the influence of an electric field E and a pressure gradient P and potential distribution.

We can write the electric potential $\Psi(x,y)$ as the sum of the equilibrium electric potential $\psi(x)$ and the potential of the applied field $-Ey$, namely,

$$\Psi(x, y) = \psi(x) - Ey \tag{21.91}$$

We assume that charged groups of valence Z are distributed in the surface charge layers at a uniform density N. Since end effects are neglected, the fluid velocity is in the y-direction, depending only on x, and the present system can be considered to be at thermodynamic equilibrium with respect to the x-direction. We define $u(x)$ as the fluid velocity in the y-direction and $\rho_{el}(x)$ as the volume charge density of electrolyte ions, both independent of y.

The Navier–Stokes equations (21.89) and (21.90) can be solved to give

$$u(x) = u(0) + \frac{\varepsilon_r\varepsilon_o E}{\eta}\{\psi(x) - \psi(0)\} - \frac{P}{2\eta}x^2 + \frac{Ph}{2\eta}x, \quad 0 \le x \le \frac{h}{2} \tag{21.92}$$

$$\begin{aligned} u(x) = {} & \frac{\varepsilon_r\varepsilon_o E}{\eta}\left[\psi(x) - \frac{\cosh(\lambda x)}{\cosh(\lambda d)}\psi(-d) + \lambda\int_{-d}^{x}\psi(x')\sinh(\lambda(x - x'))dx' \right. \\ & \left. - \frac{\lambda}{\cosh(\lambda d)}\int_{-d}^{0}\psi(x')\cosh(\lambda x)dx \cdot \sinh(\lambda(x + d))\right] \\ & - \frac{ZeNE - P}{\eta\lambda^2}\left[1 - \frac{\cosh(\lambda x)}{\cosh(\lambda d)}\right] + \frac{Ph}{2\eta\lambda}\frac{\sinh(\lambda(x + d))}{\cosh(\lambda d)}, \quad -d \le x \le 0 \end{aligned} \tag{21.93}$$

Let us consider the situation where $P = 0$ and h is infinitely large, and obtain the liquid velocity at a large distance from plate 1. This velocity is called the electroosmotic velocity, which we denote by U_{eo}. By setting $P = 0$ and $h \to \infty$ in Eqs. (21.92) and (21.93) and noting that $\psi(x) \to 0$ far from the plate, we find that

$$\begin{aligned} U_{eo} &= u(0) - \frac{\varepsilon_r \varepsilon_o}{\eta}\psi(0)E \\ &= -\frac{\varepsilon_r \varepsilon_o E}{\eta}\frac{1}{\cosh(\lambda d)}\left[\psi(-d) + \lambda\int_{-d}^{0}\psi(x)\sinh(\lambda(x+d))dx\right] \\ &\quad -\frac{ZeNE}{\eta\lambda^2}\left[1 - \frac{1}{\cosh(\lambda d)}\right] \end{aligned} \quad (21.94)$$

Note that the ratio $-U_{eo}/E$ agrees exactly with the expression for the electrophoretic mobility μ given by Eq. (21.94), namely,

$$\mu = -\frac{U_{eo}}{E} \quad (21.95)$$

The volume flow per unit time, V, transported by electroosmosis is obtained by integrating $u(x)$, namely,

$$V = 2\int_{-d}^{h/2} u(x)|_{P=0}\,dx \quad (21.96)$$

The liquid flow accompanies the flow of electrolyte ions, that is, an electric current. The density of the electric current $i(x)$, which is in the y-direction, is given by the flow of the positive charges minus that of the negative charges, namely,

$$i(x) = ze[n_+(x)v_+(x) - n_-(x)v_-(x)] \quad (21.97)$$

where $v_+(x)$ and $v_-(x)$ are, respectively, the velocity of cations and that of anions, both independent of y. The flow of electrolyte ions is caused by the liquid flow (convection) and the gradient of the electrochemical potential of the ions (conduction). Let μ_+ and μ_- be, respectively, the electrochemical potentials of cations and anions, which are given by

$$\mu_\pm(x, y) = \mu_\pm^o \pm ze\Psi(x) + kT\ln n_\pm(x) \quad (21.98)$$

where $\mu_\pm^o$ are constant. By substituting Eq. (21.91) into Eq. (21.98), we obtain

$$\mu_\pm(x, y) = \mu_\pm^o \pm ze[\psi(x) - Ey] + kT\ln n_\pm(x) \quad (21.99)$$

Then the ionic velocities $v_+(x)$ and $v_-(x)$ can be expressed as

$$v_\pm(x) = u(x) - \frac{1}{\lambda_\pm}\frac{\partial \mu_\pm}{\partial y} \tag{21.100}$$

which gives, after Eq. (21.99) is substituted

$$v_\pm(x) = u(x) \pm \frac{zeE}{\lambda_\pm} \tag{21.101}$$

Substituting Eq. (21.101) into Eq. (21.97), we obtain

$$i(x) = \rho_{\mathrm{el}}(x)u(x) + z^2e^2\left[\frac{n_+(x)}{\lambda_+} + \frac{n_-(x)}{\lambda_-}\right]E \tag{21.102}$$

The first term corresponds to the convection current and the second to the conduction current. The total electric current I flowing in the y-direction is then given by

$$I = 2\left[\int_{-d}^{0} i(x)dx + \int_{0}^{h/2} i(x)dx\right] \tag{21.103}$$

The streaming potential E_{st} is equal to the value of E such that $I = 0$.

We use the approximate Eq. (21.93) to calculate the electroosmotic velocity U_{eo}, the volume flow V, the electric current I, and the streaming potential E_{st}. The results are

$$U_{\mathrm{eo}} = -\mu E \tag{21.104}$$

$$V = U_{\mathrm{eo}}h = -\mu Eh \tag{21.105}$$

$$I = (KE - \mu P)h \tag{21.106}$$

$$E_{\mathrm{st}} = \frac{\mu}{K}P \tag{21.107}$$

Equations (21.102)–(21.111) are all related to the approximate expression (21.55) for the electrophoretic mobility μ. It can easily be shown that the following Onsager's relation is satisfied between electroosmosis and streaming potential.

$$\left.\frac{E}{P}\right|_{I=0} = -\left.\frac{V}{I}\right|_{P=0} = \frac{\mu}{K} \tag{21.108}$$

As mentioned earlier, in the limit of $1/\lambda \to 0$ ($\lambda \to \infty$), a soft particle becomes a hard particle, we call $1/\lambda$ the softness parameter. Debye and Bueche [53], on the other hand, showed that the parameter $1/\lambda$ is a shielding length, because beyond this distance the liquid flow $u(x)$ vanishes. As is also seen in Fig. 21.3b, however, $u(x)$ exhibits that under the applied fields E and P the liquid flow $u(x)$ remains a nonzero value beyond the distance of order $1/\lambda$. Indeed, it follows from Eq. (21.93) that this nonzero value is given by

$$-\frac{ZeNE - P}{\eta\lambda^2} \tag{21.109}$$

The reason for the existence of the nonzero value of the electrophoretic mobility μ is the same as that for the electroosmotic velocity. Note that only when $E=0$, $P=0$, and the third and forth terms on the left-hand side of Eq. (21.93) cancel with each other, Eq. (21.93) becomes

$$\eta\frac{d^2u}{dx^2} - \gamma u(x) = \mathbf{0}, \qquad -d < x < 0 \tag{21.110}$$

whose solution is

$$u(x) \propto \exp(\lambda x) \tag{21.111}$$

This implies the liquid velocity $u(x)$ vanishes beyond the distance $1/\lambda$ and the parameter $1/\lambda$ has the meaning of the shielding length. Such a situation is possible when, for example, a constant velocity gradient field is applied in the y-direction outside the surface layer (Fig. 21.5).

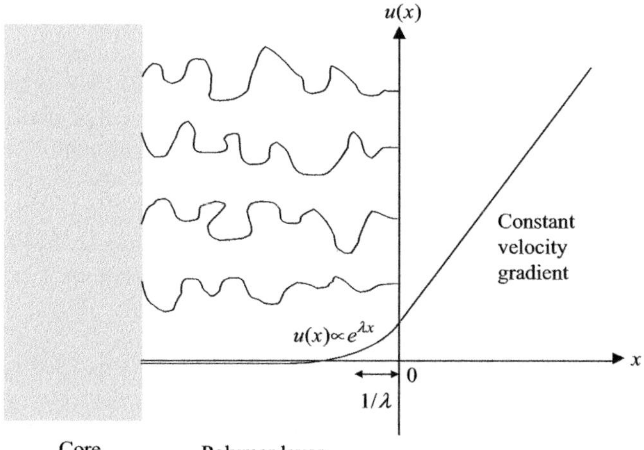

FIGURE 21.5 Liquid velocity distribution across a polymer layer under constant velocity gradient field.

21.6 SOFT PARTICLE ANALYSIS OF THE ELECTROPHORETIC MOBILITY OF BIOLOGICAL CELLS AND THEIR MODEL PARTICLES

We give below some examples of the analysis of the electrophoretic mobility of biological cells and their model particles on the basis of the electrophoresis theory of soft particles [55–58]. Equation (21.55) (or Eq. (21.51)) directly relates the measured values of electrophoretic mobility to the density of fixed charges ZN and the parameter $1/\lambda$. Since Eq. (21.55) contains two unknown parameters, ZN and $1/\lambda$, we calculated the electrophoretic mobility as a function of the electrolyte concentration n in the suspending medium for various ZN and $1/\lambda$ in order to determine the values of ZN and $1/\lambda$ by a curve fitting procedure. Equation (21.55) involves two parameters (ZeN and $1/\lambda$), the later of which can be considered to characterize the "softness" of the polyelectrolyte layer, as mentioned before, because in the limit of $1/\lambda \to 0$, the particle becomes a rigid particle. Experimentally, these parameters may be determined from a plot of measured mobility values of a soft particle as a function of electrolyte concentration by a curve fitting procedure. A number of experimental studies have been carried out in order to analyze experimental mobility date via. Eq. (21.55) (or Eq. (21.51)). Note that N and n are given in units of m^{-3}. If one uses the units of M, then N and n must be replaced by $1000N_AN$ and N_An, where N_A is Avogadro's number.

21.6.1 RAW117 Lymphosarcoma Cells and Their Variant Cells

Makino et al. [56] measured the electrophoretic mobilities of the cells of malignant lymphosarcoma cell line RAW117-P and its variant H10 with a highly metastatic property to the liver at various ionic strengths. The cells of parental cell line (RAW117-P) show higher mobility values in magnitude than those of its variant line (RAW117-H10) in the whole range of electrolyte concentration measured. The mobility data obtained have been analyzed via Eq. (21.55). By a curve fitting procedure, we have found that the best-fit curves (shown as solid lines in Fig. 21.6) are obtained with $ZN = -0.04$ M and $1/\lambda = 1.5$ nm for RAW117-P cells and $ZN = -0.03$ M and $1/\lambda = 1.7$ nm for RAW117-H10 cells. It is observed that the theoretical curves and the experimental data points are in good agreement with each other over a wide range of electrolyte concentration. At very low electrolyte concentrations, however, theoretical curves deviate from the experimental data, that is, Eq. (21.55) overestimates the magnitude of the mobility. This can be explained as follows. The particle fixed charges located over the depth of order $1/\kappa_m$ ($1/\kappa$) from the surface layer/solution boundary contribute to the mobility. Thus, as the ionic strength decreases ($1/\kappa$ increases), one can obtain information on the fixed charges in the deeper interior of the surface layer. This suggests that positive fixed charges possibly arising from dissociation of amino groups, for example, are present in the deep interior of the surface charge layer. The above analysis demonstrates that the mobility difference between RAW117-P and RAW117-H10 cells is attributed to the differences both in $1/\lambda$ and in charge density ZeN in their surface layers. Among

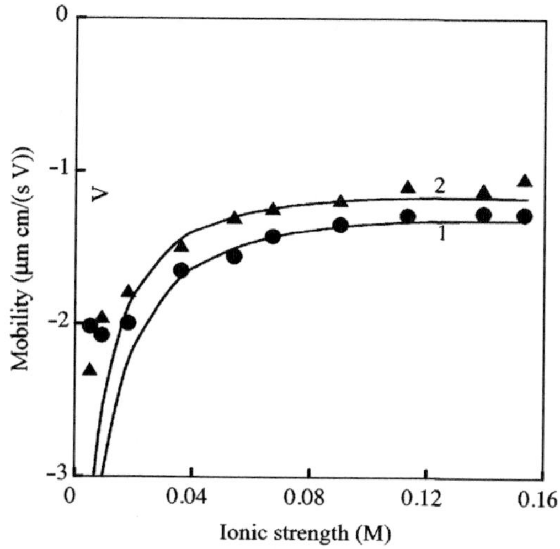

FIGURE 21.6 Electrophoretic mobilities of RAW117-P cells and RAW117-H10 cells. Symbols are experimental data measured as a function of the ionic strength in the suspending medium at pH 7.4 and 37°C: ●, RAW117-P cells; △, RAW117-H10 cells. Solid curves are theoretical results calculated via Eq. (21.55) with $ZN = -0.04$ M and $1/\lambda = 1.5$ nm (curve 1), and $ZN = -0.03$ M and $1/\lambda = 1.7$ nm (curve 2). (From Ref. 56.)

these differences, the difference in N is corresponds to the difference in sialic acid contents in both cells. The difference in $1/\lambda$ may possibly reflect a change of packing state of polymer chains in the cell surface layer. The result (obtained by the curve fitting procedure) that $1/\lambda$ is greater for RAW117-H10 cells than for RAW117-P cells implies that the friction exerted by the surface layer on the liquid flow around the cells is less for RAW117-H10 than for RAW117-P. In other words, the surface layer of RAW117-H10 cells is "softer" than that of RAW117-P cells.

21.6.2 Poly(*N*-Isopropylacrylamide) Hydrogel-Coated Latex

Poly(*N*-isopropylacrylamide) hydrogel is known to be temperature sensitive, exhibiting a phase transition at about 32°C (1); that is, it swells below 32°C and shrinks above 32°C. In the present paper, we report results of the measurements and the analysis of the electrophoretic mobility of a latex particles composed of a poly(styrene-*co*-ethylhexyl methacrylate) core and a poly(*N*-isopropylacrylamide) hydrogel layer in an electrolyte solution as a function of the ionic strength and temperature of the suspending media. Although poly(*N*-isopropylacrylamide) is itself electrically neutral, the hydrogel layer on the particle carries fixed negative charges, arising from dissociation of sulfate groups in the hydrogel layer, since potassium persulfate is used as an initiator of soap-free emulsion copolymerization.

Figure 21.7 shows that the electrophoretic mobility of the latex particles was measured in a phosphate buffer solution at pH 7.4 at four different temperatures,

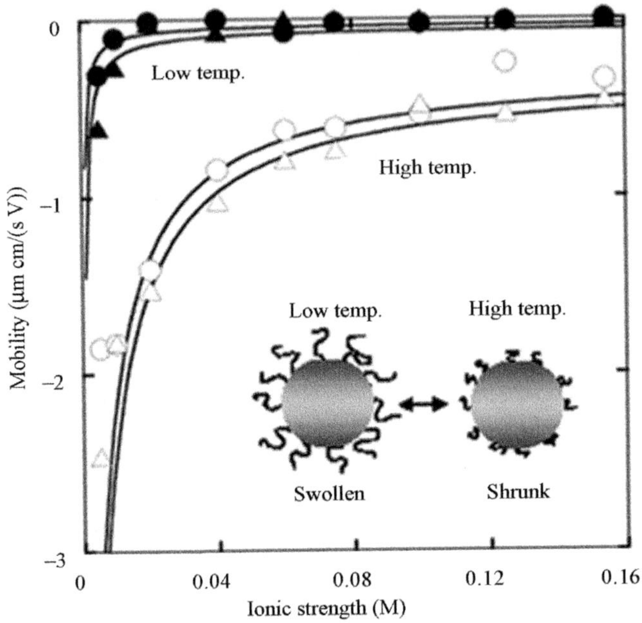

FIGURE 21.7 Electrophoretic mobility of latex particles with temperature-sensitive poly (*N*-isopropylacrylamide) hydrogel layers as a function of ionic strength at pH = 7.4 at several values of the temperature of the suspending media. Symbols represent experimental data: ●, 25°C; ▲, 30°C (below the phase transition temperature); ○, 35°C; △, 40°C (above the phase transition temperature). Solid curves are theoretical results calculated via Eq. (21.55) with $ZN = -0.0015$ M and $1/\lambda = 1.2$ nm at 25°C; $ZN = -0.0025$ M and $1/\lambda = 1.2$ nm at 30°C, $ZN = -0.03$ M and $1/\lambda = 0.9$ nm both at 35 and 40°C. (From Ref. 55.)

25, 30, 35, and 40°C (above and below the phase transition temperature of the poly (*N*-isopropylacrylamide) hydrogel [55]. A mobility expression employed here (Eq. (21.55)) to analyze the experimental data contains two unknown parameters (N and $1/\lambda$). We have thus measured the mobility as a function of the ionic strength of the suspending media to estimate the values of these parameters by a curve fitting procedure. Figure 21.7 shows results of the calculation of the mobility as a function of the electrolyte concentration on the basis of Eq. (21.55). Theoretical curves calculated via Eq. (21.55) with a single value of each of N and $1/\lambda$ are found to be in good agreement with the experimental data over a wide range of the ionic strength unless it is too low. The deviation at very low ionic strengths at 35 and 40°C, where Eq. (21.55) yields overestimation in magnitude, may be due to the fact that the hydrogel layer suffers an additional minor expansion due to reduction of the ionic shielding effects resulting in a decrease in N. The overall agreement between theory and experiment implies that hydrogel layer-coated latex particles show the electrophoretic behavior of "soft particles" described by Eq. (21.55) and also enables us to estimate the values of the unknown parameters N and $1/\lambda$ via a curve fitting procedure.

The best-fit curves (shown as solid lines in Fig. 21.3) at different temperatures are obtained with $ZN = -0.0015$ M and $1/\lambda = 1.2$ nm at 25°C; $ZN = -0.0025$ M and $1/\lambda = 1.2$ nm at 30°C, $ZN = -0.03$ M and $1/\lambda = 0.9$ nm both at 35 and 40°C. The best-fit values of ZN and $1/\lambda$ are thus the same for 35 and 40°C. Note first that the observed small difference in the mobility values between 35 and 40°C can be explained by the temperature dependence of the relative permittivity εr and viscosity η of the suspending media, that is ($\varepsilon_r = 75.1$ and $\eta = 0.719 \times 10^{-3}$ mPa s at 35°C, and $\varepsilon_r = 73.4$ and $\eta = 0.653 \times 10^{-3}$ mPa s at 40°C). This implies that both the values of N and $1/\lambda$ do not change appreciably in the temperature range 35–40°C. On the other hand, the large mobility increase from 30 to 35°C across the phase transition temperature (which correlates with the large change in size of the latex particles) is due mainly to an increase in the charge density zeN, that is, from $ZN = -0.0025$ M at 30°C to $ZN = -0.03$ M at 35°C with a slight decrease of $1/\lambda$ (1.2 nm at 30°C to 0.9 nm at 35°C). That is, below the phase transition temperature, where the hydrogel layer swells, the charge density decreases, while above the phase transition temperature, the hydrogel layer shrinks and the charge density increases, causing a large change in the mobility. Note that the mobility given by Eq. (21.55) is determined not by the total amount of fixed charges carried by a colloidal particle but by their volume density. Thus, below the phase transition temperature, when the hydrogel layer swells, the decreased charge density leads to a large reduction of the mobility, even if the total amount of the fixed charges remains constant. The minor difference in the estimated values of N between 25 and 30°C ($ZN = -0.0015$ M at 25°C to $ZN = -0.0025$ M at 30°C) may also be due to a small change in the thickness of the hydrogel layer in this temperature range.

The value of $1/\lambda$, which is the reciprocal of the friction parameter λ, decreases as the drag exerted by the hydrogel layer on the liquid flow increases. In the limit of $1/\lambda \to 0$, Eq. (21.55) tends to the well-known Smoluchowski's mobility formula for "hard particles." In other words, as $1/\lambda$ increases, the hydrogel layer on the particle becomes "softer." That is, the parameter $1/\lambda$ can be considered to characterize the "softness" of the hydrogel layer on the particle. The observed reduction of the "softness parameter" $1/\lambda$ (1.2 nm at 30°C to 0.9 nm at 35°C) implies that the hydrogel layer becomes "harder," which is in accordance with the observed shrinkage of the hydrogel.

Note that Makino et al. [57] found that Eq. (21.108) holds between the electroosmotic velocity U_∞ on a poly(N-isopropylacrylamide) hydrogel-coated solid surface and the electrophoretic mobility μ of a poly(N-isopropylacrylamide) hydrogel-coated latex particle.

21.7 ELECTROPHORESIS OF NONUNIFORMLY CHARGED SOFT PARTICLES

The surface potential of a uniformly charged soft particle in an electrolyte solution increase in magnitude with decreasing electrolyte concentration. This is not the case for a nonuniformly charged soft particle. The surface charge layer consists of a

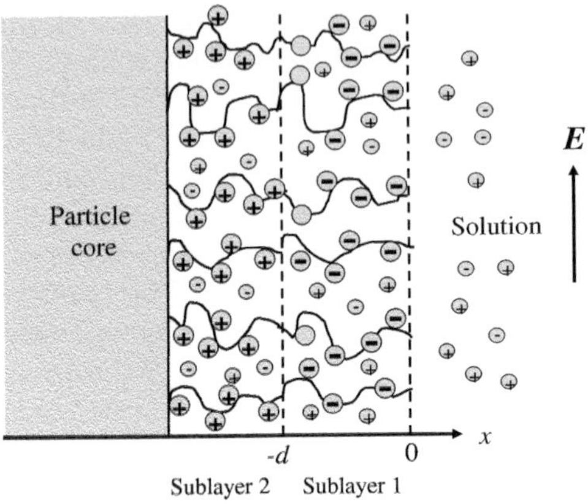

FIGURE 21.8 A surface charge layer consisting of two oppositely charged sublayers 1 and 2.

positively charge inner sublayer and a negatively charged outer sublayer (Fig. 21.8), then the surface potential is negative at high electrolyte concentrations but may become positive at low concentrations. The reason for the sign reversal of the surface potential can be explained as follows. It is the fixed charges located through the depth (form the membrane surface) of the order of the Debye length $1/\kappa$ that mainly contribute to the membrane surface potential. At high electrolyte concentrations, $1/\kappa$ is small so that the sign of the surface potential is determined mainly by the sign of the fixed charges present in the outer layer. On the other hand, at low electrolyte concentrations, $1/\kappa$ is large and the contribution from the fixed charges in the inner layer becomes appreciable. Thus, if the inner layer is sufficiently highly charged, then at low electrolyte concentrations the sign of the surface potential may coincide with the sign of the fixed charges in the inner layer.

The sign reversal takes place also in the electrophoretic mobility of a non-uniformly charged soft particles, as shown in this section. We treat a large soft particle. The x-axis is taken to be perpendicular to the soft surface with its origin at the front edge of the surface layer (Fig. 21.8). The soft surface consists of the outer layer ($-d < x < 0$) and the inner layer ($x < -d$), where the inner layer is sufficiently thick so that the inner layer can be considered practically to be infinitely thick. The liquid flow $u(x)$ and equilibrium electric potential $\psi(x)$ satisfy the following planar Navier–Stokes equations and the Poisson–Boltzmann equations [39]:

$$\eta \frac{d^2 u}{dx^2} + \rho_{\mathrm{el}}(x) E = 0, \quad x > 0 \tag{21.112}$$

$$\eta \frac{d^2 u}{dx^2} - \gamma u + \rho_{\mathrm{el}}(x) E = 0, \quad x < 0 \tag{21.113}$$

$$\frac{d^2\psi}{dx^2} = -\frac{\rho_{el}(x)}{\varepsilon_r\varepsilon_o}, \quad x > 0 \tag{21.114}$$

$$\frac{d^2\psi}{dx^2} = -\frac{\rho_{el}(x) + \rho_{fix}(x)}{\varepsilon_r\varepsilon_o}, \quad x < 0 \tag{21.115}$$

where $\rho_{fix}(x)$ is the density of the fixed charges within the surface layer and λ takes the same value in both inner sublayer and outer sublayer. Equation (21.115) reduces to Eq. (21.61) for the case of a uniformly charged soft surface ($\rho_{fix}(x) = ZeN$). It can be shown that integration of Eqs. (21.112)–(21.115) gives

$$\begin{aligned}\mu = {} & \frac{\varepsilon_r\varepsilon_o}{\eta}\frac{1}{\cosh(\lambda d)}\left[\psi(-d) + \lambda\int_{-d}^{0}\psi(x)\sinh\lambda(x+d)dx\right] \\ & + \frac{1}{\eta\lambda\cosh(\lambda d)}\int_{-d}^{0}\rho_{fix}(x)\sinh\lambda(x+d)dx\end{aligned} \tag{21.116}$$

with $\lambda = (\gamma/\eta)^{1/2}$. If $\rho_{fix}(x) = ZeN$, Eq. (21.116) reduces to Eq. (21.57) for a uniformly charged surface layer. When $d \gg 1/\kappa$, $1/\lambda$, Eq. (21.116) tends to

$$\mu = \frac{\varepsilon_r\varepsilon_o\overline{\psi(x)}}{\eta} + \frac{\overline{\rho_{fix}(x)}}{\eta\lambda^2} \tag{21.117}$$

where

$$\overline{f(x)} = \lambda\int_{-\infty}^{0} f(x)e^{\lambda x}\,dx \tag{21.118}$$

is a weighted average of $f(x)$ over the surface charge layer (the lower limit of the integration $-d$ has been replaced by $-\infty$ with negligible error.

Further, we assume that $\psi(x)$ is small enough to obey the following linearized form of Poisson–Boltzmann equations:

$$\frac{d^2\psi}{dx^2} = \kappa^2\psi, \quad x > 0 \tag{21.119}$$

$$\frac{d^2\psi}{dx^2} = \kappa^2\psi - \frac{\rho_{fix}(x)}{\varepsilon_r\varepsilon_o}, \quad x < 0 \tag{21.120}$$

Consider the case where the surface charge layer consists of two sublayers (Fig. 21.8). The outer sublayer (sublayer 1) carries ionized groups of valence Z_1

and number density N_1, while the inner sublayer (sublayer 2) carries ionized groups of valence Z_2 and number density N_2. Let the thickness of layer 1 be d. Then, we have

$$\rho_{fix}(x) = \begin{cases} Z_1 e N_1, & -d \le x \le 0 \\ Z_2 e N_2, & x \le -d \end{cases} \tag{21.121}$$

where Z_1eN_1 and Z_2eN_2 are, respectively, the charge densities of fixed charges in sublayers 1 and 2. The solution to Eqs. (21.119) and (21.120) subject to appropriate boundary conditions is given by

$$\psi(x) = \frac{1}{2}\psi_{\mathrm{DON1}}\left\{e^{-\kappa x} - e^{-\kappa(x+d)}\right\} + \frac{1}{2}\psi_{\mathrm{DON2}}\, e^{-\kappa(x+d)}, \quad x \ge 0 \tag{21.122}$$

$$\psi(x) = \psi_{\mathrm{DON1}}\left[1 - \frac{1}{2}\left\{e^{\kappa x} + e^{-\kappa(x+d)}\right\}\right] + \frac{1}{2}\psi_{\mathrm{DON2}}\, e^{-\kappa(x+d)}, \quad -d \le x \le 0 \tag{21.123}$$

$$\psi(x) = \frac{1}{2}\psi_{\mathrm{DON1}}\left\{e^{\kappa(x+d)} - e^{\kappa}\right\} + \psi_{\mathrm{DON2}}\left\{1 - \frac{1}{2}e^{\kappa(x+d)}\right\}, \quad x \le -d \tag{21.124}$$

with

$$\psi_{\mathrm{DON1}} = \frac{Z_1 e N_1}{\varepsilon_r \varepsilon_o \kappa^2} \tag{21.125}$$

$$\psi_{\mathrm{DON2}} = \frac{Z_2 e N_2}{\varepsilon_r \varepsilon_o \kappa^2} \tag{21.126}$$

where ψ_{DON1} and ψ_{DON2} are, respectively, the Donnan potentials of sublayers 1 and 2. The surface potential, which we define as the potential at $x = 0$, that is, at the front surface of the surface charge layer, is calculated by setting $x = 0$ in Eq. (21.122) or (21.123), namely,

$$\begin{aligned} \psi(0) &= \frac{1}{2}\psi_{\mathrm{DON1}}(1 - e^{-\kappa d}) + \frac{1}{2}\psi_{\mathrm{DON2}}\, e^{-\kappa d} \\ &= \frac{1}{2\varepsilon_r\varepsilon_o\kappa^2}\left[Z_1 e N_1 (1 - e^{-\kappa d}) + Z_1 e N_2\, e^{-\kappa d}\right] \end{aligned} \tag{21.127}$$

By substituting Eqs. (21.121) and (21.124) into Eq. (21.117), we obtain

$$\mu = \frac{Z_1 e N_1}{\eta \lambda^2}\left[1+\left(\frac{\lambda}{\kappa}\right)^2\left(\frac{1+\lambda/2\kappa}{1+\lambda/\kappa}\right)\right] + \frac{(Z_1 e N_1 - Z_2 e N_2)}{\eta \lambda^2}\left[\left(\frac{\lambda}{\kappa}\right)^2\left\{\frac{\lambda}{2(\kappa-\lambda)}e^{-\kappa d}\right\} - \left\{\frac{\kappa^2}{(\kappa^2-\lambda^2)}e^{-\lambda d}\right\}\right] \tag{21.128}$$

If the surface charge layer is uniformly charged, that is, $Z_1 N_1 = Z_2 N_2$, then Eq. (21.128) reduces to Eq. (21.56).

Consider the case where these two layers are oppositely charged, that is, the outer sublayer (sublayer 1) is negatively charged ($Z_1 < 0$) whereas the inner sublayer (sublayer 2) is positively charged ($Z_2 > 0$). Some results of the calculation based on Eq. (21.25) are given in Fig. 21.9. The values of parameters employed are $Z_1 N_1 = -0.2$ M and $Z_2 N_2 = +0.1$ M. Figure 21.9 depicts the electrophoretic mobility μ in a 1:1 symmetrical electrolyte ($v = 1$) as a function of the electrolyte concentration n for several values of d with $1/\lambda$ kept constant at 1 nm.

In order to analyze the dependence of the mobility on the electrolyte concentration, it is convenient to rewrite the mobility expression (21.128) as the sum of the

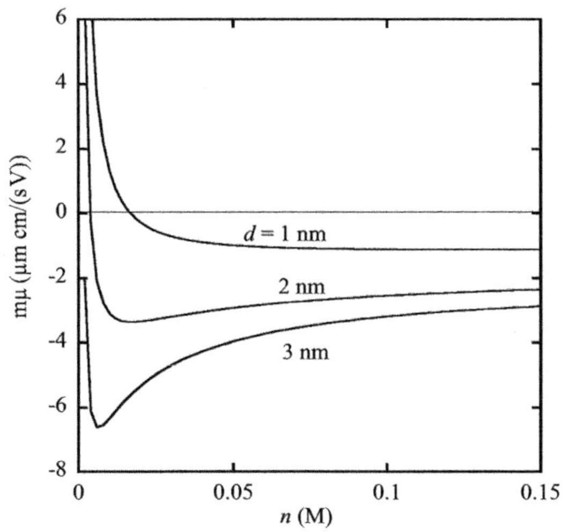

FIGURE 21.9 Electrophoretic mobility μ as a function of electrolyte concentration n for various values of d at $1/\lambda = 1$ nm, calculated via Eq. (21.130) with $z = 1$, $Z_1 N_1 = -0.2$ M, $Z_2 N_2 = +0.1$ M, $\eta = 0.89$ mPa s, $\varepsilon_r = 78.5$, and $T = 298$ K. Curves: (1) $d = 1$ nm; (2) $d = 2$ nm; (3) $d = 3$ nm.

contributions from sublayer 1 and sublayer 2, which we denote by μ_1 and μ_2, respectively, namely,

$$\mu = \mu_1 + \mu_2 \tag{21.129}$$

with

$$\mu_1 = \frac{Z_1 e N_1}{\eta \lambda^2}\left[1 + \left(\frac{\lambda}{\kappa}\right)^2 \left\{\frac{1 + \lambda/2\kappa}{1 + \lambda/\kappa} + \frac{\lambda}{2(\kappa - \lambda)} e^{-\kappa d}\right\} - \frac{\kappa^2}{(\kappa^2 - \lambda^2)} e^{-\lambda d}\right] \tag{21.130}$$

$$\mu_2 = -\frac{Z_2 e N_2}{\eta \lambda^2}\left[\left(\frac{\lambda}{\kappa}\right)^2 \frac{\lambda}{2(\kappa - \lambda)} e^{-\kappa d} - \frac{\kappa^2}{(\kappa^2 - \lambda^2)} e^{-\lambda d}\right] \tag{21.131}$$

Variation of the mobility with electrolyte concentration n is caused not only by the shielding effects of electrolyte ions but also by the change in the relative contributions from the fixed charges in layer 1 and from those in layer 2. The latter effects correspond to the fact that fixed charges located through the depth of around $1/\kappa$ mainly contributes to the mobility.

At low electrolyte concentrations n (i.e., large $1/\kappa$), because of the increase of the contribution from the positively charged inner sublayer (sublayer 2), the mobility becomes positive. Indeed, in the limit of small n (or large $1/\kappa$),

$$\mu_1 \to 0 \quad \text{and} \quad \mu_2 \to \frac{Z_2 e N_2}{2\eta \kappa^2} \tag{21.132}$$

so that the sign of the mobility always coincides with the sign of sublayer 2 (i.e., the sign of Z_2).

Next consider the limiting case of large n (small $1/\kappa$). In this limit,

$$\mu_1 \to \frac{Z_1 e N_1}{\eta \lambda^2}(1 - e^{-\lambda d}) \quad \text{and} \quad \mu_2 \to \frac{Z_2 e N_2}{\eta \lambda^2} e^{-\lambda d} \tag{21.133}$$

so that Eq. (21.129) tends to

$$\mu \to \frac{Z_1 e N_1}{\eta \lambda^2}(1 - e^{-\lambda d}) + \frac{Z_2 e N_2}{\eta \lambda^2} e^{-\lambda d} \tag{21.134}$$

Note that except for the case of $\lambda \to \infty$, which corresponds to Smoluchowski's equation (21.1), the mobility tends to a nonzero limiting value. Only for the case of

$\lambda \to \infty$ $(1/\lambda \to 0)$, the mobility tends to zero as n increases. Also note that the limiting mobility value for $1/\lambda \neq 0$ may become positive or negative. If

$$\frac{|Z_2|N_2}{|Z_1|N_1} < e^{\lambda d} - 1 \qquad (21.135)$$

then μ becomes negative, whereas if

$$\frac{|Z_2|N_2}{|Z_1|N_1} > e^{\lambda d} - 1 \qquad (21.136)$$

then μ becomes positive.

Makino et al. [63] measured the electrophoretic mobility of four types M_1–M_4 of hydrophilic gel microcapsules containing water prepared by an interfacial polymerization method. Each type of microcapsules has membranes of different compositions. The results of the analysis of the measured mobilitiy values on the basis of Eqs. (21.128) are given in Figs 21.10 and 21.11.

21.8 OTHER TOPICS OF ELECTROPHORESIS OF SOFT PARTICLES

Varoqui [70] presented a theory of the electrophoretic mobility of an uncharged polymer-coated particle by taking into account the effect of polymer segment

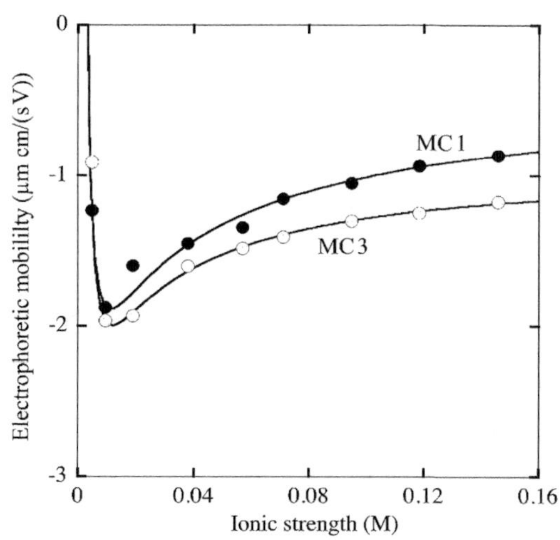

FIGURE 21.10 Electrophoretic mobility of MC 1 and MC 3. Symbols are experimental data with MC 1 (●) and MC 3 (○) measured as a function of the ionic strength in the suspending medium at pH 7.4 and at 37°C. Solid curves are theoretical ones calculated with $Z_1N_1 = -0.111$ M, $Z_2N_2 = 0.041$ M, $1/\lambda = 0.505$ nm, $d = 1.706$ nm (MC 1), and with $Z_1N_1 = -0.048$ M, $Z_2N_2 = 0.048$ M, $1/\lambda = 1.230$ nm, Z_1N_1, $d = 3.758$ nm (MC 3). (From. Ref. 63.)

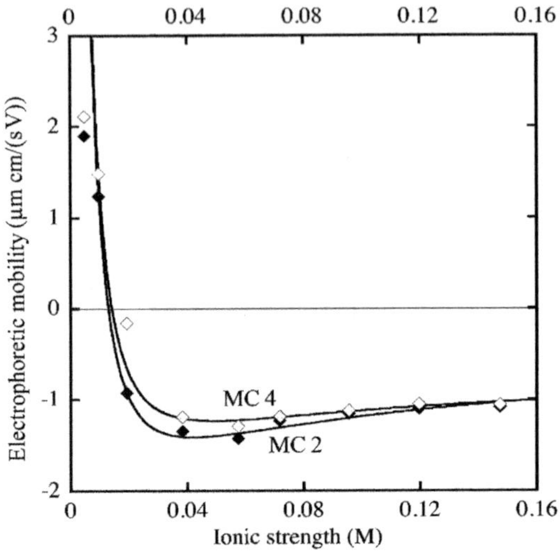

FIGURE 21.11 Electrophoretic mobility of MC 2 and MC 4. Symbols are experimental data with MC 2 (◆) and MC 4 (◇) measured as a function of the ionic strength in the suspending medium at pH 7.4 and at 37°C. Solid curves are theoretical ones calculated with $Z_1N_1 = -0.140$ M, $Z_2N_2 = 0.166$ M, $1/\lambda = 0.471$ nm, $d = 2.090$ nm (MC 2), and with $Z_1N_1 = -0.084$ M, $Z_2N_2 = 0.159$ M, $1/\lambda = 0.778$ nm, $d = 2.751$ nm (MC 4). (From Ref. 63.)

distribution. Ohshima [71] extended Varoqui's theory to the case of polyelectrolyte-coated particles. The case where the polyelectrolyte layer is not fully ion-penetrable is considered in Ref. 72. Saville [73] considered the relaxation effects in electrophoresis of soft particles. Dukhin et al. [49] pointed out the importance of the degree of dissociation of charged groups in the polyelectrolyte layer. Quite recently, Duval and coworkers [74–77] proposed a new model of diffuse soft particle. In this mode, a diffuse soft particle consists of an uncharged impenetrable core and a charged diffuse polyelectrolyte layer. The diffuse character of the polyelectrolyte layer is defined by a gradual distribution of the density of polymer segments in the interspatial region separating the core from the bulk electrolyte solution. For numerical calculations of the electrophoretic mobility based on more rigorous theories, the readers are referred to studies of Hill et al. [78–81] and Lopez-Garcia et al. [82–83] as well as Duval and Ohshima [77].

REFERENCES

1. M. von Smoluchowski,in: L. Greatz (Ed.), *Handbuch der Elektrizität und des Magnetismus*, Vol. 2, Barth, Leipzig, 1921, p. 366.
2. E. Hückel, *Phys. Z.* 25 (1924) 204.
3. D. C. Henry, *Proc. R. Soc. London, Ser. A* 133 (1931) 106.
4. J. Th. G. Overbeek, *Kolloid-Beihefte* 54 (1943) 287.

5. F. Booth, *Proc. R. Soc. London, Ser. A* 203 (1950) 514.
6. P. H. Wiersema, A. L. Loeb, and J. Th. G. Overbeek, *J. Colloid Interface Sci.* 22 (1966) 78.
7. S. S. Dukhin and N. M. Semenikhin, *Kolloid Zh.* 32 (1970) 360.
8. S. S. Dukhin and B.V. Derjaguin, in: E. Matievic (Ed.), *Surface and Colloid Science*, Vol. 7, Wiley, New York, 1974.
9. A. de Keizer, W. P. J. T. van der Drift, and J. Th. G. Overbeek, *Biophys. Chem.* 3 (1975) 107.
10. R. W. O'Brien and L. R. White, *J. Chem. Soc., Faraday Trans. 2* 74 (1978) 1607.
11. R. W. O'Brien and R. J. Hunter, *Can. J. Chem.* 59 (1981) 1878.
12. R. J. Hunter, *Zeta Potential in Colloid Science*, Academic Press, New York, 1981.
13. H. Ohshima, T. W. Healy, and L. R. White, *J. Chem. Soc., Faraday Trans. 2* 79 (1983) 1613.
14. T. G. M. van de Ven, *Colloid Hydrodynamics*, Academic Press, New York, 1989.
15. R. J. Hunter, *Foundations of Colloid Science*, Vol. 2, Clarendon Press/Oxford University Press, Oxford, 1989, Chapter 13.
16. S. S. Dukhin, *Adv. Colloid Interface Sci.* 44 (1993) 1.
17. H. Ohshima, *J. Colloid Interface Sci.* 168 (1994) 269.
18. J. Lyklema, *Fundamentals of Interface and Colloid Science, Solid–Liquid Interfaces*, Vol. 2, Academic Press, New York, 1995.
19. H. Ohshima, *J. Colloid Interface Sci.* 180 (1996) 299.
20. H. Ohshima, in: H. Ohshima and K. Furusawa (Eds.), *Electrical Phenomena at Interfaces, Fundamentals, Measurements, and Applications, 2nd edition, revised and expanded*, Dekker, New York, 1998.
21. A. V. Delgado (Ed.), *Electrokinetics and Electrophoresis*, Dekker, New York, 2000.
22. H. Ohshima, *J. Colloid Interface Sci.* 239 (2001) 587.
23. H. Ohshima, in: A. Hubbard (Ed.), *Encyclopedia in Surface and Colloid Science*, Marcel Dekker, New York, 2002, p. 1834.
24. H. Ohshima, in: A. M. Spasic and J.-P. Hsu (Eds.), *Finely Dispersed Particles: Micro-, Nano-, and Atto-Engineering*, CRC press, 2005 p. 27.
25. H. Ohshima, *J. Colloid Interface Sci.* 263 (2003) 337.
26. H. Ohshima, *J. Colloid Interface Sci.* 275 (2004) 665.
27. H. Ohshima, *Colloids Surf. A: Physicochem. Eng. Aspects* 267 (2005) 50.
28. H. Ohshima, in: M. Akay (Ed.), *Wiley Encyclopedia of Biomedical Engineering*, John Wiley & Sons, Inc., 2006.
29. H. Ohshima, *Theory of Colloid and Interfacial Electric Phenomena*, Elsevier/Academic Press, 2006.
30. E. Donath and V. Pastushenko, *Bioelectrochem. Bioenerg.* 6 (1979) 543.
31. I. S. Jones, *J. Colloid Interface Sci.* 68 (1979) 451.
32. R. W. Wunderlich, *J. Colloid Interface Sci.* 88 (1982) 385.
33. S. Levine, M. Levine, K. A. Sharp, and D. E. Brooks, *Biophys. J.* 42 (1983) 127.
34. K. A. Sharp and D. E. Brooks, *Biophys. J.* 47 (1985) 563.
35. H. Ohshima and T. Kondo, *Colloid Polym. Sci.* 264 (1986) 1080.

36. H. Ohshima and T. Kondo, *J. Colloid Interface Sci.* 116 (1987) 305.
37. H. Ohshima and T. Kondo, *J. Colloid Interface Sci.* 130 (1989) 281.
38. H. Ohshima and T. Kondo, *Biophys. Chem.* 39 (1991) 191.
39. H. Ohshima and T. Kondo, *Biophys. Chem.* 46 (1993) 145.
40. H. Ohshima, *J. Colloid Interface Sci.* 163 (1994) 474.
41. H. Ohshima, *Colloids Surf. A: Physicochem. Eng. Aspects* 103 (1995) 249.
42. H. Ohshima, *Adv. Colloid Interface Sci.* 62 (1995) 189.
43. H. Ohshima, *Electrophoresis* 16 (1995) 1360.
44. H. Ohshima, *Colloid Polym Sci.* 275 (1997) 480.
45. H. Ohshima, *J. Colloid Interface Sci.* 228 (2000) 190.
46. H. Ohshima, *J. Colloid Interface Sci.* 229 (2000) 140.
47. H. Ohshima, *Colloid Polym. Sci.* 279 (2001) 88.
48. H. Ohshima, *J. Colloid Interface Sci.* 233 (2001) 142.
49. S. S. Dukhin, R. Zimmermann, and C. Werner, *J. Colloid Interface Sci.* 286 (2005) 761.
50. H. Ohshima, *Electrophoresis* 27 (2006) 526.
51. H. Ohshima, *Colloid Polym. Sci.* 285 (2007) 1411.
52. J. J. Hermans and H. Fujita, *Koninkl. Ned. Akad. Wetenschap. Proc.* B58 (1955) 182.
53. P. Debye and A. Bueche, *J. Chem. Phys.* 16 (1948) 573.
54. J. H. Masliyah, G. Neale, K. Malysa, and T. G. M. van de Ven, *Chem. Sci.* 42 (1987) 245.
55. H. Ohshima, K. Makino, T. Kato, K. Fujimoto, T. Kondo, and H. Kawaguchi, *J. Colloid Interface Sci.* 159 (1993) 512.
56. K. Makino, T. Taki, M. Ogura, S. Handa, M. Nakajima, T. Kondo, and H. Ohshima, *Biophys. Chem.* 47 (1993) 261.
57. K. Makino, K. Suzuki, Y. Sakurai, T. Okano, and H. Ohshima, *J. Colloid Interface Sci.* 174 (1995) 400.
58. T. Mazda, K. Makino, and H. Ohshima, *Colloids Surf. B* 5 (1995) 75; Erratum, Colloids Surf. B 10 (1998) 303.
59. S. Takashima and H. Morisaki, *Colloids Surf. B* 9 (1997) 205.
60. R. Bos, H. C. van der Mei, and H. J. Busscher, *Biophys. Chem.* 74 (1998) 251.
61. A. Larsson, M. Rasmusson, and H. Ohshima, *Carbohydrate Res.* 317 (1999) 223.
62. H. Morisaki, S. Nagai, H. Ohshima, E. Ikemoto, and K. Kogure, *Microbiology* 145 (1999) 2797.
63. K. Makino, M. Umetsu, Y. Goto, A. Kikuchi, Y. Sakurai, T. Okano, and H. Ohshima, *J. Colloid Interface Sci.* 218 (1999) 275.
64. M. Rasmusson, B. Vincent, and N. Marston, *Colloid Polym. Sci.* 278 (2000) 253.
65. H. Hayashi, S. Tsuneda, A. Hirata, and H. Sasaki, *Colloids Surf. B* 22 (2001) 149.
66. M. J. Garcia-Salinas, M. S. Romero-Cano, and F. J. de las Nieves, *Prog. Colloid Polym. Sci.* 118 (2001) 180.
67. P. J. M. Kiers, R. Bos, H. C. van der Mei, and H. J. Busscher, *Microbiology* 147 (2001) 757.
68. J. A. Molina-Bolivar, F. Galisteo-Gonzalez, and R. Hidalgo-Alvalez, *Colloids Surf. B: Biointerfaces* 21 (2001) 125.
69. H. Ohshima and T. Kondo, *J. Colloid Interface Sci.* 125 (1990) 443.

70. R. Varoqui, *Nouv. J. Chim.* 6 (1982) 187.
71. H. Ohshima, *J. Colloid Interface Sci.* 185 (1997) 269.
72. H. Ohshima and K. Makino, *Colloids Surf. A* 109 (1996) 71.
73. D. A. Saville, *J. Colloid Interface Sci.* 222 (2000) 137.
74. J. F. L. Duval and H. P. van Leeuwen, *Langmuir* 20 (2004) 10324.
75. J. F. L. Duval, *Langmuir* 21 (2005) 3247.
76. L. P. Yezek, J. F. L. Duval, and H. P. van Leeuwen, *Langmuir* 21 (2005) 6220.
77. J. F. L. Duval and H. Ohshima, *Langmuir* 22 (2006) 3533.
78. R. J. Hill, D. A. Saville, and W. B. Russel, *J. Colloid Interface Sci.* 258 (2003) 56.
79. R. J. Hill, D. A. Saville, and W. B. Russel, *J. Colloid Interface Sci.* 263 (2003) 478.
80. R. J. Hill and D. A. Saville, *Colloids Surf. A* 267 (2005) 31.
81. R. J. Hill, *Phys. Rev. E* 70 (2004) 051406.
82. J. J. Lopez-Garcia, C. Grosse, and J. J. Horno, *J. Colloid Interface Sci.* 265 (2003) 327.
83. J. J. Lopez-Garcia, C. Grosse, and J. J. Horno, *J. Colloid Interface Sci.* 265 (2003) 341.

22 Electrophoretic Mobility of Concentrated Soft Particles

22.1 INTRODUCTION

In this chapter, we extend the electrokinetic theory of soft particles (Chapter 21), which is applicable for dilute suspensions, to cover the case of concentrated suspensions [1–3] on the basis of Kuwabara's cell model [4], which has been applied to theoretical studies of various electrokinetic phenomena in concentrated suspensions of hard colloidal particles [5–23].

22.2 ELECTROPHORETIC MOBILITY OF CONCENTRATED SOFT PARTICLES

Consider a concentrated suspension of charged spherical soft particles moving with a velocity $\boldsymbol{U}$ in a liquid containing a general electrolyte in an applied electric field $\boldsymbol{E}$. We assume that the particle core of radius a is coated with an ion-penetrable layer of polyelectrolytes with a thickness d. The polyelectrolyte-coated particle has thus an inner radius a and an outer radius $b \equiv a + d$. We employ a cell model [4] in which each particle is surrounded by a concentric spherical shell of an electrolyte solution, having an outer radius c such that the particle/cell volume ratio in the unit cell is equal to the particle volume fraction ϕ throughout the entire dispersion (Fig. 22.1), namely,

$$\phi = (b/c)^3 \tag{22.1}$$

The origin of the spherical polar coordinate system (r, θ, φ) is held fixed at the center of one particle and the polar axis ($\theta = 0$) is set parallel to $\boldsymbol{E}$. Let the electrolyte be composed of M ionic mobile species of valence z_i and drag coefficient λ_i ($i = 1, 2, \ldots, M$), and let n_i^∞ be the concentration (number density) of the ith ionic species in the electroneutral solution. We also assume that fixed charges are distributed with a density of ρ_{fix}. We adopt the model of Debye–Bueche where the polymer segments are regarded as resistance centers distributed in the polyelectrolyte

Biophysical Chemistry of Biointerfaces By Hiroyuki Ohshima

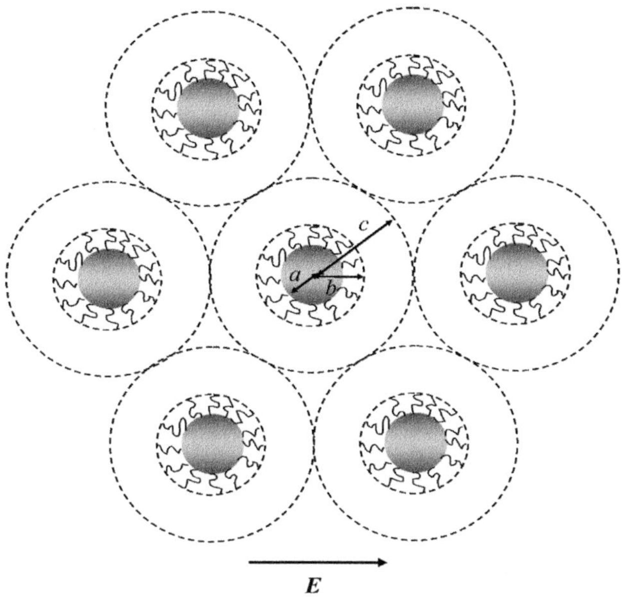

FIGURE 22.1 Spherical particles in concentrated suspensions in a cell model. Each sphere is surrounded by a virtual shell of outer radius c. The particle volume fraction ϕ is given by $\phi = (b/c)^3$. The volume fraction of the particle core of radius a is given by $\phi_c = (a/c)^3$.

layer, exerting frictional forces on the liquid flowing in the polyelectrolyte layer (Chapter 21).

The fundamental equations for the flow velocity of the liquid $\boldsymbol{u}(\boldsymbol{r})$ at position $\boldsymbol{r}$ and that of the ith ionic species $\boldsymbol{v}_i(\boldsymbol{r})$ are the same as those for the dilute case (Chapter 5) except that Eq. (5.10) applies to the region $b < r < c$ (not to the region $r > b$). The boundary conditions for $\boldsymbol{u}(\boldsymbol{r})$ and $\boldsymbol{v}_i(\boldsymbol{r})$ are also the same as those for the dilute case, but we need additional boundary conditions to be satisfied at $r = c$. According to Kuwabara's cell model [4], we assume that at the outer surface of the unit cell ($r = c$) the liquid velocity is parallel to the electrophoretic velocity $\boldsymbol{U}$ of the particle,

$$\boldsymbol{u} \cdot \boldsymbol{n}\big|_{r=c^-} = -U\cos\theta \tag{22.2}$$

where $\boldsymbol{n}$ is the unit normal outward from the unit cell, $U = |\boldsymbol{U}|$, and that the vorticity ω is 0 at $r = c$:

$$\omega = \nabla \times \boldsymbol{u} = \boldsymbol{0} \quad \text{at} \quad r = c \tag{22.3}$$

We also assume that at the outer surface of the unit cell ($r = c$), the gradient of the electric potential ψ is parallel to the applied electric field [6], namely,

$$\nabla\psi \cdot \boldsymbol{n}\big|_{r=c^-} = -E\cos\theta \quad \text{at} \quad r = c \tag{22.4}$$

The boundary conditions for $\boldsymbol{u}(\boldsymbol{r})$ and $\boldsymbol{v}_i(\boldsymbol{r})$ can be written in terms of $h(r)$ and $\phi_i(r)$ (defined by Eqs. (21.18) and (21.19)) as

$$h(c) = \frac{cU}{2E} \tag{22.5}$$

$$Lh(r)|_{r=c} = 0 \tag{22.6}$$

$$\eta \frac{d}{dr}(rLh)\bigg|_{r=c} - \rho_{\mathrm{el}}^{(0)}(c)Y(c) = 0, \tag{22.7}$$

$$\frac{dY}{dr}\bigg|_{r=c} = 1 \tag{22.8}$$

where $\rho_{\mathrm{el}}(c)$ is the equilibrium charge density at $r = c$.

The electrophoretic mobility μ of spherical soft particles in a concentrated suspension is defined by $\mu = U/E$. It must be mentioned here that the electrophoretic mobility μ in this chapter is defined with respect to the externally applied electric field $\boldsymbol{E}$ so that the boundary condition (22.8) has been employed following Levine and Neale [5]. There is another way of defining the electrophoretic mobility in the concentrated case, where the mobility μ^* is defined as $\mu^* = U/\langle E \rangle$, $\langle E \rangle$ being the magnitude of the average electric field $\langle \boldsymbol{E} \rangle$ within the suspension [8, 19–21]. It follows from the continuity condition of electric current that $K^*\langle \boldsymbol{E} \rangle = K^\infty \boldsymbol{E}$, where K^* and K^∞ are, respectively, the electric conductivity of the suspension and that of the electrolyte solution in the absence of the particles. Thus, μ and μ^* are related to each other by $\mu^*/\mu = K^*/K^\infty$. For the dilute case, there is no difference between μ and μ^*.

It follows from Eq. (22.5) that

$$\mu = \frac{2h(c)}{c} \tag{22.9}$$

By evaluating $h(r)$ at $r = c$, we obtain the following general expression for the electrophoretic mobility μ of soft particles in a concentrated suspension [1, 3]:

$$\mu = \frac{b^2}{9}\int_b^c \left[3\left(1 - \frac{r^2}{b^2}\right) - \frac{2L_2}{L_1}\left(1 - \frac{r^3}{b^3}\right)\right.$$
$$\left. -\phi\left\{\left(3 - \frac{2L_2}{L_1} - \frac{9a}{L_1\lambda^2 b^3}\right)\left(1 - \frac{r^3}{b^3}\right) - \frac{3}{5}\left(1 - \frac{r^5}{b^5}\right)\right\}\right] G(r)dr$$

$$+\frac{2}{3\lambda^2}\int_a^c \left[(1-\phi)\frac{L_5}{L_1}\left(1+\frac{r^3}{2b^3}\right)-3\phi\right]G(r)dr$$

$$-\frac{2}{3\lambda^2}(1-\phi)\int_a^b \left[1-\frac{3a}{2\lambda^2 b^3 L_1}\left\{(L_5+L_6\,\lambda r)\cosh[\lambda(r-a)]\right.\right.$$

$$\left.\left.-(L_6+L_5\lambda r)\sinh[\lambda(r-a)]\right\}\right]G(r)dr$$

$$-\frac{2b^2\rho_{\mathrm{el}}^{(0)}(c)Y(c)}{9\eta c}\left(1+\frac{1}{\phi}-\frac{9}{5\phi^{2/3}}-\frac{\phi}{5}+\frac{L_7}{\phi}\right) \qquad (22.10)$$

where L_1–L_4 are given by Eqs. (21.42)–(21.45) and L_5–L_7 are defined by

$$L_5 = L_3 + \frac{3\phi}{1-\phi}\left\{\cosh[\lambda(b-a)]-\frac{\sinh[\lambda(b-a)]}{\lambda b}\right\} \qquad (22.11)$$

$$L_6 = L_4 + \frac{3\phi}{1-\phi}\left\{\sinh[\lambda(b-a)]-\frac{\cosh[\lambda(b-a)]}{\lambda b}\right\} \qquad (22.12)$$

$$L_7 = (1-\phi)^2\left(\frac{L_2}{L_1}-1\right)-\frac{9\phi^2}{\lambda^2 b^2}+\frac{27\phi^2 a}{2\lambda^2 b^3 L_1}+\frac{6(\phi+1/2)^2 L_3}{\lambda^2 b^2 L_1} \qquad (22.13)$$

Consider several limiting cases.

1. In the limit $\phi \to 0$ (i.e., $c \to \infty$ so that $\rho_{\mathrm{el}}^{(0)}(c)$ becomes the bulk value and thus $\rho_{\mathrm{el}}^{(0)}(c) = 0$), Eq. (6.11) tends to the mobility expression of a spherical soft particle for the dilute case (Eq. (21.41)).
2. In the limit $a \to b$, the polyelectrolyte layer vanishes and the soft particle becomes a rigid particle. Indeed, in this limit Eq. (22.10) tends to

$$\mu = \frac{b^2}{9}\int_b^c \left[1-3\left(\frac{r}{b}\right)^2+2\left(\frac{r}{b}\right)^3-\phi\left\{\frac{2}{5}-\left(\frac{r}{b}\right)^3+\frac{3}{5}\left(\frac{r}{b}\right)^5\right\}\right]G(r)dr$$

$$-\frac{2b^2\rho_{\mathrm{el}}^{(0)}(c)Y(c)}{9\eta c}\left[1+\frac{1}{\phi}-\frac{9}{5\phi^{2/3}}-\frac{\phi}{5}\right] \qquad (22.14)$$

This agrees with the mobility expression for a rigid particle with a radius b in concentrated suspensions obtained in previous papers [5, 12].

3. In the limit $\lambda \to \infty$, the polyelectrolyte-coated particle behaves like a rigid particle with a radius b and the slipping plane is shifted outward from $r = a$ to $r = b$. In this limit, Eq. (22.10) tends to Eq. (22.14).
4. In the limit $\lambda \to 0$, the polyelectrolyte layer vanishes and the particle becomes a rigid particle with a radius a. In this limit, Eq. (22.10) tends to Eq. (22.14) with b replaced by a, as expected.

We derive approximate mobility formulas for the simple but important case where the double-layer potential remains spherically symmetrical in the presence of the applied electric field (the relaxation effect is neglected). In this case, it can be shown that Eq. (22.10) tends to, in the limit $\kappa \to \infty$,

$$\mu \to \mu^{\infty} = \frac{\rho_{\mathrm{fix}}}{\eta\lambda^2}\left(\frac{1 + a^3/2b^3}{1-\phi_{\mathrm{c}}}\right)\left\{(1-\phi)\frac{L_5}{L_1} - 2\phi\right\}, \quad \text{as } \kappa \to \infty \tag{22.15}$$

where

$$\phi_{\mathrm{c}} = (a/c)^3 \tag{22.16}$$

is the volume fraction of the particle core. The existence of the nonvanishing term of the mobility at very high electrolyte concentrations is a characteristic of soft particles (Chapter 21).

We derive an approximate expression for the electrophoretic mobility of spherical polyelectrolytes for the case of low potentials. In this case, the equilibrium potential $\psi^{(0)}(r)$ is given by

$$\psi^{(0)}(r) = \frac{\rho_{\mathrm{fix}}}{\varepsilon_r\varepsilon_o\kappa^2}\left[1 - \left\{\frac{\Omega + 1/\kappa b}{\coth(\kappa b) + \Omega}\right\}\frac{b\sinh(\kappa r)}{r\sinh(\kappa b)}\right], \quad 0 \le r \le b \tag{22.17}$$

$$\psi^{(0)}(r) = \frac{\rho_{\mathrm{fix}}}{\varepsilon_r\varepsilon_o\kappa^2}\left\{\frac{\coth(\kappa b) - 1/\kappa b}{\coth(\kappa b) + \Omega}\right\}\frac{b\{\coth[\kappa(c-r)] - \sinh[\kappa(c-r)]/\kappa c\}}{r\{\coth[\kappa(c-b)] - \sinh[\kappa(c-b)]/\kappa c\}},$$

$$b \le r \le c \tag{22.18}$$

where

$$\Omega = \frac{\sinh[\kappa(c-b)] - \cosh[\kappa(c-b)]/\kappa c}{\cosh[\kappa(c-b)] - \sinh[\kappa(c-b)]/\kappa c} = \frac{1 - \kappa c \cdot \tanh[\kappa(c-b)]}{\tanh[\kappa(c-b)] - \kappa c} \tag{22.19}$$

With the help of Eqs. (22.17) and (22.18), we obtain

$$\mu = \frac{\rho_{\text{fix}}}{\eta\lambda^2}\left[1+\frac{2}{3}\left(\frac{\lambda}{\kappa}\right)^2\left(\frac{\coth(\kappa b)-1/\kappa b}{\coth(\kappa b)+\Omega}\right)\left\{1-\phi-\frac{\phi}{\kappa b}\left(\Omega+\frac{1}{\kappa b}\right)\right\}\right.$$
$$\left.+\frac{2}{3}(1-\phi)\left(\frac{\lambda}{\kappa}\right)^2\frac{1}{(\lambda/\kappa)^2-1}\left\{\frac{\Omega+1/\kappa b}{\coth(\kappa b)+\Omega}\right\}\left\{\left(\frac{\lambda}{\kappa}\right)\frac{\coth(\kappa b)-1/\kappa b}{\coth(\kappa b)-1/\lambda b}-1\right\}\right] \tag{22.20}$$

This is the required expression for the mobility of spherical polyelectrolytes in concentrated suspension for low potentials.

When $\phi \to 0$, Eq. (22.20) reduces to Eq. (21.75). When $\kappa b \gg 1$ and $\lambda b \gg 1$, Eq. (22.20) becomes

$$\mu = \frac{\rho_{\text{fix}}}{\eta\lambda^2}\left[1+\frac{2(1-\phi)}{3}\left(\frac{\lambda}{\kappa}\right)^2\frac{1+2\kappa/\lambda}{1+\kappa/\lambda}\right] \tag{22.21}$$

which, for $\phi \to 0$, reduces to Eq. (21.56).

For the case where

$$\lambda a \gg 1,\ \kappa a \gg 1\ (\text{and thus } \lambda b \gg 1,\ \kappa b \gg 1)\ \text{ and}$$
$$\lambda d = \lambda(b-a) \gg 1,\ \kappa d = \kappa(b-a) \gg 1 \tag{22.22}$$

we obtain from Eq. (22.10)

$$\mu = \frac{\varepsilon_r\varepsilon_o}{\eta}\frac{\psi_o/\kappa_m+\psi_{\text{DON}}/\lambda}{1/\kappa_m+1/\lambda}f\left(\frac{d}{a},\phi\right)+\frac{\rho_{\text{fix}}}{\eta\lambda^2}, \tag{22.23}$$

where

$$f\left(\frac{d}{a},\phi\right) = \frac{2}{3}\left[1+\frac{1}{2}\left(\frac{a}{b}\right)^3\right]\left(\frac{1-\phi}{1-\phi_c}\right) = \frac{2}{3}\left[1+\frac{1}{2(1+d/a)^3}\right]\frac{(1-\phi)}{1-\phi/(1+d/a)^3} \tag{22.24}$$

Equation (22.23) is the required approximate expression for the electrophoretic mobility of soft particles in concentrated suspensions when the condition [22.22] (which holds for most cases) is satisfied. In the limit $\phi \to 0$, Eq. (22.23) tends to Eq. (21.51) for the dilute case. For low potentials, Eq. (22.23) reduces to

$$\mu = \frac{\rho_{\text{fix}}}{\eta\lambda^2}\left[1+\left(\frac{\lambda}{\kappa}\right)^2\frac{1+2\kappa/\lambda}{1+\kappa/\lambda}f\left(\frac{d}{a},\phi\right)\right] \tag{22.25}$$

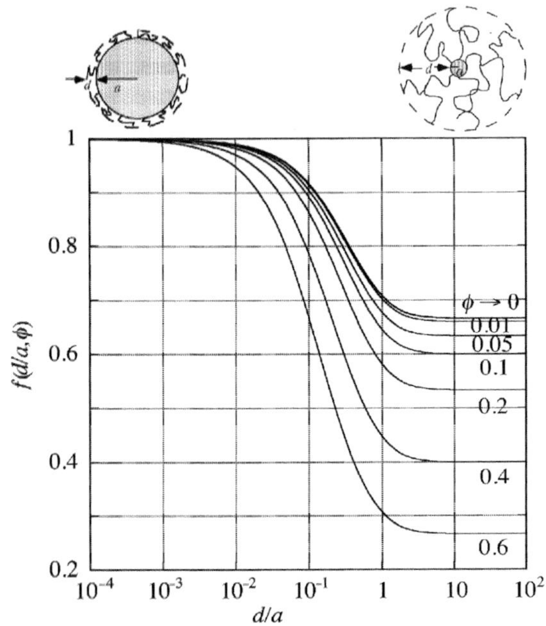

FIGURE 22.2 $f(d/a, \phi)$, defined by Eq. (4.27), as a function of d/a. From Ref. [22].

In the limit of $d \gg a$, Eq. (22.25) reduces to Eq. (22.21) for spherical polyelectrolytes, as expected. In the limit of very high electrolyte concentrations ($\kappa \to \infty$), Eqs. (22.20) and (22.25) give the following limiting value:

$$\mu \to \mu^{\infty} = \frac{\rho_{\text{fix}}}{\eta\lambda^2}, \quad \text{as } \kappa \to \infty \tag{22.26}$$

In Fig. 22.2, we plot the function $f(d/a, \phi)$, given by Eq. (22.24), as a function of d/a for several values of ϕ. The function $f(d/a, \phi)$ tends to 1, as d/a decreases, whereas it becomes $2(1-\phi)/3$ as d/a increases. We see that soft particles with $d/a > 5$ behave like spherical polyelectrolytes ($a = 0$). The limiting forms of the mobility for the two cases $d/a \ll 1$ and $d/a \gg 1$ are given by

$$\mu = \frac{\varepsilon_r\varepsilon_o}{\eta}\frac{\psi_o/\kappa_m + \psi_{\text{DON}}/\lambda}{1/\kappa_m + 1/\lambda} + \frac{\rho_{\text{fix}}}{\eta\lambda^2}, \quad d \ll a \tag{22.27}$$

$$\mu = \frac{2\varepsilon_r\varepsilon_o(1-\phi)}{3\eta}\frac{\psi_o/\kappa_m + \psi_{\text{DON}}/\lambda}{1 + \kappa_m + 1/\lambda} + \frac{\rho_{\text{fix}}}{\eta\lambda^2}, \quad d \gg a \tag{22.28}$$

22.3 ELECTROOSMOTIC VELOCITY IN AN ARRAY OF SOFT CYLINDERS

Consider the electroosmotic liquid velocity $\boldsymbol{U}$ in an array of parallel soft cylinders in a liquid containing a general electrolyte in an applied electric field $\boldsymbol{E}$[2] (Fig. 22.3). The velocity of the liquid flow, $\boldsymbol{U}$, is parallel to the applied field $\boldsymbol{E}$. We assume that the cylinder core of radius a is coated with an ion-penetrable layer of polyelectrolytes with a thickness d. The polyelectrolyte-coated cylinder has thus an inner radius a and an outer radius $b \equiv a + d$. We employ a cell model [4] in which each cylinder is surrounded by a fluid envelope of an outer radius c such that the cylinder/cell volume ratio in the unit cell is equal to the cylinder volume fraction ϕ, namely,

$$\phi = (b/c)^2 \tag{22.29}$$

The porosity ε is thus given by

$$\varepsilon = 1 - \phi = 1 - (b/c)^2 \tag{22.30}$$

The origin of the cylindrical coordinate system (r, θ, z) is held fixed on the axis of one cylinder. The polar axis ($\theta = 0$) is set parallel to the applied electric field $\boldsymbol{E}$ so that $\boldsymbol{E}$ is perpendicular to the cylinder axis. Let the electrolyte be composed of M ionic mobile species of valence z_i and drag coefficient λ_i ($i = 1, 2, \ldots, M$), and n_i^∞ be the concentration (number density) of the ith ionic species in the

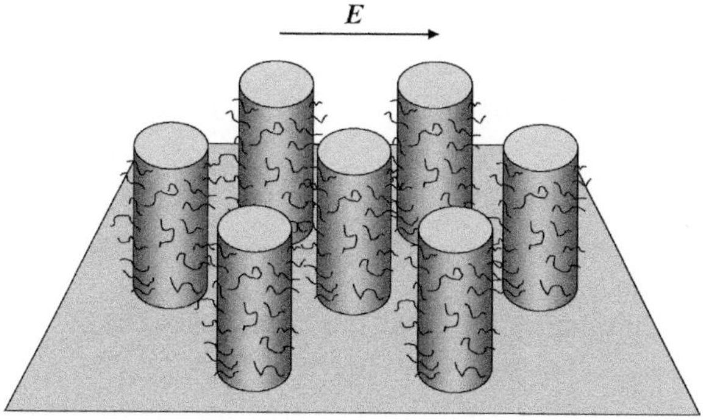

FIGURE 22.3 Electroosmosis in an array of parallel soft cylinders (polyelectrolyte-coated cylinders), which consist of the core of radius a covered with a layer of polyelectrolytes of thickness d. Each cylinder is surrounded by a fluid envelope of outer radius c. The volume fraction of the cylinders is given by $(b/c)^2$, where $b = a + d$, and the porosity ε is given by $\varepsilon = 1 - \phi = 1 - (b/c)^2$. The volume fraction of the particle core of radius a is given by $\phi_c = (a/c)^3$. The liquid flow $\boldsymbol{U}$, which is parallel to the applied electric field $\boldsymbol{E}$, is normal to the axes of the cylinders.

electroneutral solution. We assume that relative permittivity ε_r takes the same value both inside and outside the polyelectrolyte layer. We also assume that fixed charges are distributed within the polyelectrolyte layer with a density of $\rho_{fix}(r)$. If dissociated groups of valence Z are distributed with a constant density (number density) N within the polyelectrolyte layer, then we have $\rho_{fix} = ZeN$, where e is the elementary electric charge.

The general expression for electroosmotic velocity is [2]

$$\frac{U}{E} = \frac{1}{2\lambda^2}\int_a^c \left[\left\{\frac{L_1(\lambda b)}{L_2(\lambda b)} + \phi\left(\frac{L_1(\lambda b)}{L_2(\lambda b)} + \frac{2}{\lambda b L_2(\lambda b)}\right)\right\}(1+\phi)\left(1+\frac{r^2}{b^2}\right) - 2\phi\right] G(r)dr$$

$$+\frac{1}{2\lambda^2}\int_a^b \left[(1+\phi)\lambda r\left\{L_1(\lambda r) - \frac{L_1(\lambda b)}{L_2(\lambda b)}L_2(\lambda r)\right\} + 2\phi\left\{1 - \frac{rL_1(\lambda r)}{bL_2(\lambda b)}\right\}\right] G(r)dr$$

$$+\frac{b^2}{8}\int_b^c \left[\left\{1 + \frac{2}{\lambda b}\left(\frac{L_3}{L_2(\lambda b)} - \phi L_4\right)\right\}\left(1-\frac{r^2}{b^2}\right) + \frac{2r^2}{b^2}\ln\left(\frac{r}{b}\right)\right.$$

$$\left. -\frac{\phi}{2}\left(1-\frac{r^2}{b^2}\right)^2\right] G(r)dr - \frac{b^2\rho_{el}^{(0)}(c)Y(c)}{4\eta c}\left\{1 - \frac{3}{4\phi} - \frac{\phi}{4} - \frac{\ln(\phi)}{2\phi} + \frac{L_5}{\lambda b}\right\} \tag{22.31}$$

where L_1–L_3 are given by Eqs. (21.80)–(21.82) and L_4 and L_5 are defined as

$$L_4 = -\frac{L_3}{L_2(\lambda b)} - \frac{2L_3}{\lambda b L_1(\lambda b) L_2(\lambda b)} + \frac{2}{L_1(\lambda b)}\{I_0(\lambda a)K_0(\lambda b) - K_0(\lambda a)I_0(\lambda b)\} \tag{22.32}$$

$$L_5 = -\frac{(1-\phi)L_3}{\phi L_2(\lambda b)} + \frac{2L_1(\lambda b)}{\lambda b L_2(\lambda b)}\left(\frac{1}{\phi} + \phi + 2\right) + \frac{4(1+\phi)}{\lambda^2 b^2 L_2(\lambda b)} - \frac{4\phi}{\lambda b} + (1-\phi)L_4 \tag{22.33}$$

where $I_n(z)$ and $K_n(z)$ are, respectively, the nth-order modified Bessel functions of the first and second kinds.

Consider several limiting cases. In the limit $\phi \to 0$, Eq. (22.31) becomes

$$\frac{U}{E} = \frac{b^2}{8}\int_b^\infty \left[\left\{1 + \frac{2L_3}{\lambda b L_2(\lambda b)}\right\}\left(1-\frac{r^2}{b^2}\right) + \frac{2r^2}{b^2}\ln\left(\frac{r}{b}\right)\right] G(r)dr$$

$$+\frac{L_1(\lambda b)}{2\lambda^2 L_2(\lambda b)}\int_a^\infty \left(1+\frac{r^2}{b^2}\right)G(r)dr + \frac{1}{2\lambda^2}\int_a^b \lambda r\left[L_1(\lambda r) - \frac{L_1(\lambda b)}{L_2(\lambda b)}L_2(\lambda r)\right] G(r)dr \tag{22.34}$$

which agrees with the general expression for the electrophoretic mobility $\mu_\perp$ of a cylindrical soft cylinder in a transverse field (Eq. (21.79)).

In the limit $a \to b$, the polyelectrolyte layer vanishes and the soft cylinder becomes a rigid cylinder. Indeed, in this limit Eq. (22.31) tends to

$$\frac{U}{E} = \frac{b^2}{8}\int_b^c \left[1 - \frac{r^2}{b^2}\left\{1 - 2\ln\left(\frac{r}{b}\right)\right\} - \frac{b^2}{2c^2}\left(1 - \frac{r^2}{b^2}\right)^2\right] G(r)dr$$
$$- \frac{b^2 \rho_{\text{el}}^{(0)}(c) Y(c)}{4\eta c}\left[1 - \frac{3c^2}{4b^2} - \frac{b^2}{4c^2} + \frac{c^2}{b^2}\ln\left(\frac{c}{b}\right)\right]. \tag{22.35}$$

This agrees with the electrophoretic mobility expression $\mu_\perp$ for a rigid cylinder with a radius b in concentrated suspensions in a transverse field obtained in a previous paper [9, 17].

In the limit $\lambda \to \infty$, the polyelectrolyte-coated cylinder behaves like a rigid cylinder with a radius b and the slipping plane is shifted outward from $r = a$ to $r = b$. In this limit, Eq. (22.31) tends also to Eq. (22.35). In the opposite limit of $\lambda \to 0$, the polyelectrolyte layer vanishes and the cylinder becomes a rigid cylinder with a radius a. In this limit, Eq. (22.31) tends to Eq. (22.35) with b replaced by a, as expected.

Finally, in the limit $a \to 0$, the cylinder core vanishes and the cylinder becomes a cylindrical polyelectrolyte. In this limit, Eq. (22.35) becomes

$$\frac{U}{E} = \frac{1}{2\lambda^2}\int_0^c \left\{(1+\phi)\left(1 + \frac{r^2}{b^2}\right) - 2\phi\right\} G(r)dr + \frac{(1-\phi)}{2\lambda^2}\int_0^b \left\{\frac{rI_1(\lambda r)}{bI_1(\lambda b)} - 1\right\} G(r)dr$$
$$+ \frac{b^2}{8}\int_b^c \left[\left\{1 - \frac{2(1-\phi)I_0(\lambda b)}{\lambda b I_1(\lambda b)}\right\}\left(1 - \frac{r^2}{b^2}\right) + \frac{2r^2}{b^2}\ln\left(\frac{r}{b}\right) - \frac{\phi}{2}\left(1 - \frac{r^2}{b^2}\right)^2\right] G(r)dr$$
$$- \frac{b^2 \rho_{\text{el}}^{(0)}(c) Y(c)}{4\eta c}\left[1 - \frac{3}{4\phi} - \frac{\phi}{4} - \frac{\ln(\phi)}{2\phi} + \frac{2}{\lambda^2 b^2}\left(\frac{1}{\phi} - \phi + 2\right) + \frac{I_0(\lambda b)}{\lambda b I_1(\lambda b)}\left(\frac{1}{\phi} + \phi - 2\right)\right] \tag{22.36}$$

In this case, the equilibrium potential $\psi^{(0)}(r)$ can be obtained from the linearized Poisson–Boltzmann equation $\Delta\psi^{(0)} = \kappa^2 \psi^{(0)}$, namely,

$$\psi^{(0)}(r) = \frac{\rho_{\text{fix}}}{\varepsilon_r \varepsilon_o \kappa^2}\left[1 - \kappa b\left\{K_1(\kappa b) - \frac{K_1(\kappa c)}{I_1(\kappa c)} I_1(\kappa b)\right\} I_0(\kappa r)\right], \quad 0 \le r \le b \tag{22.37}$$

$$\psi^{(0)}(r) = \kappa b \frac{\rho_{\text{fix}}}{\varepsilon_r \varepsilon_o \kappa^2} I_1(\kappa b)\left\{K_0(\kappa r) + \frac{K_1(\kappa c)}{I_1(\kappa c)} I_0(\kappa r)\right\}, \quad b \le r \le c \tag{22.38}$$

Substituting the above equations into the general formula (22.36) yields

$$\frac{U}{E} = \frac{\rho_{\rm fix}}{\eta\lambda^2} + \frac{\rho_{\rm fix}(1-\phi)b}{2\eta\kappa} I_1(\kappa b)\left\{K_0(\kappa b) + \frac{K_1(\kappa c)}{I_1(\kappa c)} I_0(\kappa b)\right\}$$
$$+ \frac{\rho_{\rm fix}(1-\phi)\kappa b}{2\eta(\kappa^2-\lambda^2)}\left\{K_1(\kappa b) - \frac{K_1(\kappa c)}{I_1(\kappa c)} I_1(\kappa b)\right\}\left\{I_0(\kappa b) - \frac{\lambda I_0(\lambda b)}{\kappa I_1(\lambda b)} I_1(\kappa b)\right\}$$
$$- \frac{\rho_{\rm fix}}{\eta}\frac{\phi\lambda^2}{\kappa^2} I_1(\kappa b)\left\{K_1(\kappa b) - \frac{K_1(\kappa c)}{I_1(\kappa c)} I_1(\kappa b)\right\}. \tag{22.39}$$

When $\phi \to 0$, Eq. (22.39) reduces to

$$\frac{U}{E} = \frac{\rho_{\rm fix}}{\eta\lambda^2} + \frac{\rho_{\rm fix} b}{2\eta\kappa}\left[I_1(\kappa b)K_0(\kappa b) + \frac{\kappa^2}{\kappa^2-\lambda^2}\left\{I_0(\kappa b) - \frac{\lambda I_0(\lambda b)}{\kappa I_1(\lambda b)} I_1(\kappa b)\right\}K_1(\kappa b)\right] \tag{22.40}$$

which agrees with the electrophoretic mobility expression $\mu_\perp$ for a cylindrical polyelectrolyte in a transverse field for the dilute case (Eq. (21.83)). If, further, $\kappa b \gg 1$ and $\lambda b \gg 1$, then Eq. (22.40) becomes

$$\frac{U}{E} = \frac{\rho_{\rm fix}}{\eta\lambda^2}\left[1 + \frac{(1-\phi)}{2}\left(\frac{\lambda}{\kappa}\right)^2 \frac{1+\lambda/2\kappa}{1+\lambda/\kappa}\right] \tag{22.41}$$

Finally, we consider the important case where the double-layer potential remains cylindrically symmetrical in the presence of the applied electric field (the relaxation effect is neglected) and where

$$\lambda a \gg 1,\ \kappa a \gg 1 \text{ (and thus } \lambda b \gg 1,\ \kappa b \gg 1) \text{ and}$$
$$\lambda d = \lambda(b-a) \gg 1,\ \kappa d = \kappa(b-a) \gg 1 \tag{22.42}$$

In this case, we obtain [23]

$$\frac{U}{E} = \frac{\varepsilon_r\varepsilon_o}{\eta}\frac{\psi_o/\kappa_m + \psi_{\rm DON}/\lambda}{1/\kappa_m + 1/\lambda} f\left(\frac{d}{a}, \phi\right) + \frac{\rho_{\rm fix}}{\eta\lambda^2} \tag{22.43}$$

where

$$f\left(\frac{d}{a}, \phi\right) = \frac{1}{2}\left[1 + \left(\frac{a}{b}\right)^2\right]\left(\frac{1-\phi}{1-\phi_c}\right) = \frac{1}{2}\left[1 + \frac{1}{(1+d/a)^2}\right]\frac{(1-\phi)}{1-\phi/(1+d/a)^2}. \tag{22.44}$$

In the limit $\phi \to 0$, Eq. (22.43) tends to Eq. (21.84) for $\mu_\perp$ of soft cylinders for the dilute case. For low potentials, Eq. (22.43) further reduces to

$$\frac{U}{E} = \frac{\rho_{\rm fix}}{\eta\lambda^2}\left[1 + \left(\frac{\lambda}{\kappa}\right)^2 \frac{1+2\kappa/\lambda}{1+\kappa/\lambda} f\left(\frac{d}{a}, \phi\right)\right] \tag{22.45}$$

REFERENCES

1. H. Ohshima, *J. Colloid Interface Sci.* 225 (2000) 233.
2. H. Ohshima, *Colloids Surf. A: Physicochem. Eng. Sci.* 192 (2001) 227.
3. H. Ohshima, *Colloids Surf. A: Physicochem. Eng. Sci.* 195 (2001) 1295.
4. S. Kuwabara, *J. Phys. Soc. Jpn.* 14 (1959) 527.
5. S. Levine and G. H. Neale, *J. Colloid Interface Sci.* 47 (1974) 520.
6. S. Levine and G. H. Neale, *J. Colloid Interface Sci.* 49 (1974) 330.
7. S. Levine, G. H. Neale, and N. Epstein, *J. Colloid Interface Sci.* 57 (1976) 424.
8. V. N. Shilov, N. I. Zharkih, and Yu. B. Borkovskaya, *Colloid J.* 43 (1981) 434.
9. M. W. Kozak and E. J. Davis, *J. Colloid Interface Sci.* 112 (1986) 403.
10. M. W. Kozak and E. J. Davis, *J. Colloid Interface Sci.* 127 (1989) 497.
11. M. W. Kozak and E. J. Davis, *J. Colloid Interface Sci.* 129 (1989) 166.
12. H. Ohshima, *J. Colloid Interface Sci.* 188 (1997) 481.
13. H. Ohshima, *J. Colloid Interface Sci.* 195 (1997) 137.
14. H. Ohshima, *J. Colloid Interface Sci.* 208 (1998) 295.
15. E. Lee, J. W. Chu, and J. P. Hsu, *Colloid Interface Sci.* 209 (1999) 240.
16. H. Ohshima, *J. Colloid Interface Sci.* 210 (1999) 397.
17. H. Ohshima, *J. Colloid Interface Sci.* 212 (1999) 443.
18. H. Ohshima and A. S. Dukhin, *J. Colloid Interface Sci.* 212 (1999) 449.
19. A. S. Dukhin, H. Ohshima, V. N. Shilov, and P. J. Goetz, *Langmuir* 15 (1999) 3445.
20. A. S. Dukhin, V. N. Shilov, and Yu. B. Borkovskaya, *Langmuir* 15 (1999) 3452.
21. A. S. Dukhin, V. N. Shilov, H. Ohshima, and P. J. Goetz, *Langmuir* 15 (1999) 6692.
22. H. Ohshima, *J. Colloid Interface Sci.* 218 (1999) 535.
23. H. Ohshima, *Adv. Colloid Interface Sci.* 88 (2000) 1.

23 Electrical Conductivity of a Suspension of Soft Particles

23.1 INTRODUCTION

Electrokinetic equations describing the electrical conductivity of a suspension of colloidal particles are the same as those for the electrophoretic mobility of colloidal particles and thus conductivity measurements can provide us with essentially the same information as that from electrophoretic mobility measurements. Several theoretical studies have been made on dilute suspensions of hard particles [1–3], mercury drops [4], and spherical polyelectrolytes (charged porous spheres) [5], and on concentrated suspensions of hard spherical particles [6] and mercury drops [7] on the basis of Kuwabara's cell model [8], which was originally applied to electrophoresis problem [9,10]. In this chapter, we develop a theory of conductivity of a concentrated suspension of soft particles [11]. The results cover those for the dilute case in the limit of very low particle volume fractions. We confine ourselves to the case where the overlapping of the electrical double layers of adjacent particles is negligible.

23.2 BASIC EQUATIONS

Consider spherical soft particles moving with a velocity U (electrophoretic velocity) in a liquid containing a general electrolyte in an applied electric field E. Each soft particle consists of the particle core of radius a covered with a polyelectrolyte layer of thickness d (Fig. 22.1). The radius of the soft particle as a whole is thus $b = a + d$. We employ a cell model [8] in which each sphere is surrounded by a concentric spherical shell of an electrolyte solution, having an outer radius of c such that the particle/cell volume ratio in the unit cell is equal to the particle volume fraction ϕ throughout the entire suspension (Fig. 22.1), namely,

$$\phi = (b/c)^3 \tag{23.1}$$

Let the electrolyte be composed of M ionic mobile species of valence z_i and drag coefficient λ_i ($i = 1, 2, \ldots, M$), and n_i^∞ be the concentration (number density) of

Biophysical Chemistry of Biointerfaces By Hiroyuki Ohshima

the ith ionic species in the electroneutral solution. We also assume that fixed charges are distributed with a constant density of ρ_{fix} in the polyelectrolyte layer. The fundamental electrokinetic equations are the same as those for the electrophoresis problem (Chapter 22).

23.3 ELECTRICAL CONDUCTIVITY

Consider a suspension of N_p identical spherical soft particles in a general electrolyte solution of volume V. We define the macroscopic electric field in the suspension $\langle \boldsymbol{E} \rangle$, which differs from the applied electric field E. The field $\langle \boldsymbol{E} \rangle$ may be regarded as the average of the gradient of the electric potential $\psi(\boldsymbol{r})$ $(=\psi^{(0)}(r)+\delta\psi(\boldsymbol{r}))$ in the suspension over the volume V, namely,

$$\langle \boldsymbol{E} \rangle = -\frac{1}{V}\int_V \nabla\psi(\boldsymbol{r})dV = -\frac{1}{V}\int_V \nabla\delta\psi(\boldsymbol{r})dV \tag{23.2}$$

where we have used the fact that the volume average of $\psi^{(0)}$ is zero. The electrical conductivity K^* of the suspension is defined by

$$\langle \boldsymbol{i} \rangle = K^* \langle \boldsymbol{E} \rangle \tag{23.3}$$

where $\langle \boldsymbol{i} \rangle$ is the net electric current in the suspension given by

$$\langle \boldsymbol{i} \rangle = \frac{1}{V}\int_V \boldsymbol{i}(\boldsymbol{r})dV \tag{23.4}$$

The electric field $\langle \boldsymbol{E} \rangle$ is different from the applied field E and these two fields are related to each other by continuity of electric current, namely,

$$K^* \langle \boldsymbol{E} \rangle = K^\infty E \tag{23.5}$$

where

$$K^\infty = \sum_{i=1}^{M} z_i^2 e^2 n_i^\infty / \lambda_i \tag{23.6}$$

is the conductivity of the electrolyte solution (in the absence of the soft particles). If further we put

$$C_i = -\frac{c^2}{3}\left(r\frac{d\phi_i}{dr} - \phi_i\right)_{r=c} \tag{23.7}$$

then, the ratio of K^* to K^∞ is given by

$$\frac{K^*}{K^\infty} = \left(1 + \frac{3\phi_c}{a^3} \frac{\sum_{i=1}^{M} z_i^2 n_i^\infty C_i / \lambda_i}{\sum_{i=1}^{M} z_i^2 n_i^\infty / \lambda_i}\right)^{-1} \tag{23.8}$$

We obtain C_i for the low potential case and substitute the result into Eq. (23.8), giving

$$\begin{aligned}\frac{K^*}{K^\infty} &= \frac{1-\phi_c}{1+\phi_c/2}\left(1 - \frac{\phi_c}{2(1-\phi_c)(1+\phi_c/2)} \frac{\sum_{i=1}^{M} z_i^3 n_i^\infty / \lambda_i}{\sum_{i=1}^{M} z_i^2 n_i^\infty / \lambda_i}\right. \\ &\quad \left. \times \int_a^c \left(1 + \frac{2r^3}{a^3}\right)\left(1 - \frac{a^3}{r^3}\right) \frac{dy}{dr} dr\right)^{-1} \\ &\approx \frac{1-\phi_c}{1+\phi_c/2}\left(1 + \frac{\phi_c}{2(1-\phi_c)(1+\phi_c/2)} \frac{\sum_{i=1}^{M} z_i^3 n_i^\infty / \lambda_i}{\sum_{i=1}^{M} z_i^2 n_i^\infty / \lambda_i}\right. \\ &\quad \left. \times \int_a^c \left(1 + \frac{2r^3}{a^3}\right)\left(1 - \frac{a^3}{r^3}\right) \frac{dy}{dr} dr\right)\end{aligned} \tag{23.9}$$

where

$$\phi_c = (a/c)^3 \tag{23.10}$$

is the volume fraction of the particle core, which differs from the particle volume fraction ϕ, defined by Eq. (23.1). Equation (23.9) is the required expression for the electrical conductivity of a concentrated suspension of spherical soft particles for low potentials. Note that in the low potential approximation, K^* does not depend on the frictional coefficient γ (or λ).

We consider several limiting cases of Eq. (23.9).

(i) In the limit of very low potentials, Eq. (23.9) tends to

$$\frac{K^*}{K^\infty} = \frac{1-\phi_c}{1+\phi_c/2} \tag{23.11}$$

which is Maxwell's relation [12, 13] respect to the volume fraction ϕ_c of the particle core. Note that in this case the conductivity is determined by the volume fraction ϕ_c of the particle core (not by the particle volume fraction ϕ) and the contribution from the polyelectrolyte layer vanishes.

(ii) For $a = b$, soft particles become hard particles with no polyelectrolyte layer. In this case $\phi = \phi_c$ and we find that Eq. (23.9) becomes a conductivity expression for a concentrated suspension of spherical hard particles.

(iii) For $a = 0$, soft particles becomes porous spheres spherical polyelectrolytes with no particle core. In this case $\phi_c \to 0$ and Eq. (23.9) reduces to

$$\frac{K^*}{K^\infty} = 1 \tag{23.12}$$

That is, in this limit the conductivity equals that in the absence of the particles so that spherical polyelectrolytes do not contribute to the conductivity.

Finally, we derive an approximate conductivity formula for the important case where

$$\lambda a \gg 1,\ \kappa a \gg 1(\text{and thus } \lambda b \gg 1,\ \kappa b \gg 1),\ \lambda(b-a) \gg 1,\ \kappa(b-a) \gg 1. \tag{23.13}$$

In this case the potential inside the polyelectrolyte layer is essentially equal to the Donnan potential ψ_{DON} except in the region very near the boundary $r = b$ between the polyelectrolyte layer and the surrounding solution, where ψ_{DON} is related ρ_{fix} by

$$\psi_{\mathrm{DON}} = \frac{\rho_{\mathrm{fix}}}{\varepsilon_r \varepsilon_o \kappa^2} \tag{23.14}$$

Then Eq. (23.9) becomes

$$\begin{aligned}\frac{K^*}{K^\infty} &= \frac{1-\phi_c}{1+\phi_c/2}\left(1 - \frac{\phi_c(1-a^3/b^3)(1+2b^3/a^3)}{2(1-\phi_c)(1+\phi_c/2)}\left(\frac{e\psi_{\mathrm{DON}}}{kT}\right)\frac{\sum_{i=1}^{N} z_i^3 n_i^\infty/\lambda_i}{\sum_{i=1}^{N} z_i^2 n_i^\infty/\lambda_i}\right)\\ &= \frac{1-\phi_c}{1+\phi_c/2}\left(1 - \frac{\phi(1-a^3/b^3)(1+a^3/2b^3)}{(1-\phi_c)(1+\phi_c/2)}\left(\frac{e\psi_{\mathrm{DON}}}{kT}\right)\frac{\sum_{i=1}^{N} z_i^3 n_i^\infty/\lambda_i}{\sum_{i=1}^{N} z_i^2 n_i^\infty/\lambda_i}\right)\end{aligned} \tag{23.15}$$

which is an approximate expression for K^*/K^∞ for low potentials under conditions (23.13).

REFERENCES

1. D. A. Saville, *J. Colloid Interface Sci.* 71 (1979) 477.
2. R. W. O'Brien, *J. Colloid Interface Sci.* 81 (1981) 234.
3. H. Ohshima, T. W. Healy, and L. R. White, *J. Chem. Soc., Faraday Trans. 2* 79 (1983) 1613.
4. H. Ohshima, T. W. Healy, and L. R. White, *J. Chem. Soc., Faraday Trans. 2* 80 (1984) 1643.
5. Y. C. Liu and H. J. Keh, *J. Colloid Interface Sci.* 212 (1999) 443.
6. H. Ohshima, *J. Colloid Interface Sci.* 212 (1999) 443.
7. H. Ohshima, *J. Colloid Interface Sci.* 218 (1999) 535.
8. S. Kuwabara, *J. Phys. Soc. Jpn.* 14 (1959) 527.
9. S. Levine and G. H. Neale, *J. Colloid Interface Sci.* 47 (1974) 520.
10. H. Ohshima, *J. Colloid Interface Sci.* 188 (1997) 481.
11. H. Ohshima, *J. Colloid Interface Sci.* 229 (2000) 307.
12. J. C. Maxwell, *A Treatise on Electricity and Magnetism*, Vol. 1, Art. 313, Dover, New York, 1954.
13. G. H. Neale and W. K. Nader, *AIChE J.* 19 (1973) 112.

24 Sedimentation Potential and Velocity in a Suspension of Soft Particles

24.1 INTRODUCTION

When charged spherical particles fall steadily under gravity, the electrical double layer around each particle loses its spherical symmetry because of the fluid motion (Fig. 24.1). This is known as the relaxation effect. A microscopic electric field arising from the distortion of the double layer reduces the falling velocity (called the sedimentation velocity) of the particle and the fields from the individual particles are then superimposed to give rise to a macroscopic field (called the sedimentation field), which is uniform for a homogeneous dispersed suspension [1–5]. Experimentally, the sedimentation field is observed as a sedimentation potential. Fundamental electrokinetic equations describing sedimentation phenomena are closely related to those of electrophoresis (Chapter 21). Keh and Liu [6] developed a theory of sedimentation in a dilute suspension of soft particles. In this chapter, on the basis of Kuwabara's cell model [7] (which has been applied to sedimentation phenomena of concentrated hard particles [8, 9]), we derive general expressions for the sedimentation velocity and sedimentation potential for concentrated suspensions of soft particles [10, 11]. This is an extension of the theory of Keh and Liu [6] to concentrated suspensions.

24.2 BASIC EQUATIONS

Consider spherical soft particles (i.e., polyelectrolyte-coated particles) falling steadily with a velocity $\boldsymbol{U}_{\mathrm{SED}}$ (sedimentation velocity) in a liquid containing a general electrolyte in a gravitational field $\boldsymbol{g}$. Each particle consists of the particle core of radius a covered with a polyelectrolyte layer of thickness d. The radius of the soft particle as a whole is thus $b = a + d$. We employ a cell model [7] in which each spherical soft particle is surrounded by a concentric spherical shell of an electrolyte solution, having an outer radius of c such that the particle/solution volume ratio in the unit cell is equal to the volume fraction ϕ of the soft particle (particle core + polyelectrolyte layer) throughout the entire suspension (Fig. 24.2), namely,

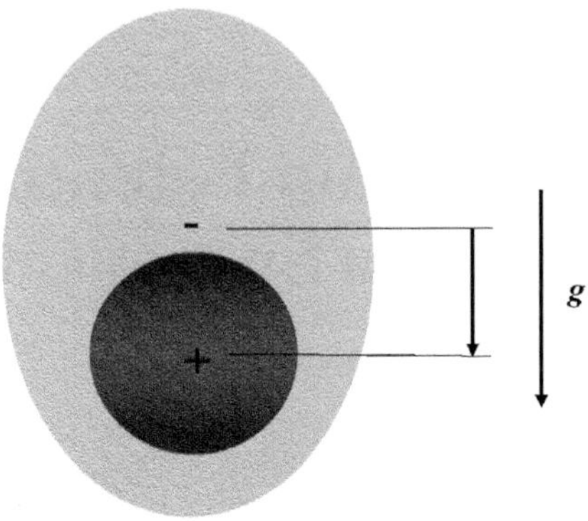

FIGURE 24.1 A spherical particle falling steadily in a gravitational field g. A microscopic electric field arises from the distortion of the double layer around the particle.

$$\phi = (b/c)^3 \tag{24.1}$$

We assume that the overlapping of electrical double layers of adjacent particles is negligible on the outer surface of the unit cell (at $r = c$). The origin of the spheri-

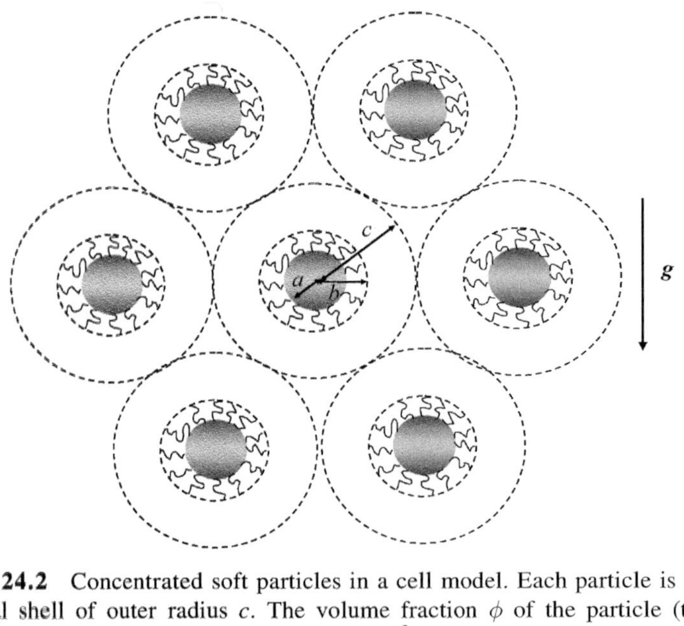

FIGURE 24.2 Concentrated soft particles in a cell model. Each particle is surrounded by a virtual shell of outer radius c. The volume fraction ϕ of the particle (the particle core + polyelectrolyte layer) is given by $\phi = (b/c)^3$. The volume fraction of the particle core is $\phi_c = (a/c)^3$.

cal coordinate system (r, θ, φ) is held fixed at the center of one particle. The polar axis ($\theta = 0$) is set parallel to $\boldsymbol{g}$. Let the electrolyte be composed of M ionic species of valence z_i and drag coefficient λ_i ($i = 1, 2, \ldots, M$), and n_i^{∞} be the concentration (number density) of the ith ionic species in the electroneutral solution. We also assume that fixed charges are distributed with a density of $\rho_{\text{fix}}(r)$ in the polyelectrolyte layer, which is a spherically symmetrical function of $r = |\boldsymbol{r}|$ only. If dissociated groups of valence Z is distributed with a constant density (number density) N within the polyelectrolyte layer, then we have$\rho_{\text{fix}} = ZeN$, where e is the elementary electric charge. We adopt the model of Debye–Bueche [14] where the polymer segments are regarded as resistance centers distributed in the polyelectrolyte layer, exerting frictional forces on the liquid flowing in the polyelectrolyte layer.

The fundamental electrokinetic equations for the flow velocity $\boldsymbol{u}(\boldsymbol{r}) = (u_r, u_\theta, u_\varphi)$ of the liquid at position $\boldsymbol{r}$ and that of the ith mobile ionic species $\boldsymbol{v}_i(\boldsymbol{r})$ are the same as those for the electrophoresis problem (Chapter 21) except that the Navier–Stokes equations involve the term $\rho_o \boldsymbol{g}$, where ρ_o is the mass density and the viscosity of the liquid, namely,

$$\eta \nabla \times \nabla \times \boldsymbol{u} + \nabla p + \rho_{\text{el}} \nabla \psi - \rho_o \boldsymbol{g} + \gamma \boldsymbol{u} = \boldsymbol{0}, \quad a < r < b \tag{24.2}$$

$$\eta \nabla \times \nabla \times \boldsymbol{u} + \nabla p + \rho_{\text{el}} \nabla \psi - \rho_o \boldsymbol{g} = \boldsymbol{0}, \quad b < r < c \tag{24.3}$$

We assume that the electrical double layer around the particle is only slightly distorted due to the gravitational field $\boldsymbol{g}$ (slow sedimentation). Then, we may write

$$n_i(\boldsymbol{r}) = n_i^{(0)}(r) + \delta n_i(\boldsymbol{r}) \tag{24.4}$$

$$\psi(\boldsymbol{r}) = \psi^{(0)}(r) + \delta \psi(\boldsymbol{r}) \tag{24.5}$$

$$\mu_i(\boldsymbol{r}) = \mu_i^{(0)} + \delta \mu_i(\boldsymbol{r}) \tag{24.6}$$

where the quantities with superscript (0) refer to those at equilibrium, that is, in the absence of $\boldsymbol{g}$, and $\mu_i^{(0)}$ is a constant independent of $\boldsymbol{r}$. The distribution of electrolyte ions at equilibrium $n_i^{(0)}(r)$ obeys the Boltzmann distribution and the equilibrium potential $\psi^{(0)}(r)$ in the electrolyte solution satisfies the Poisson–Boltzmann equation (Chapter 1).

Further, symmetry considerations permit us to write

$$\boldsymbol{u}(\boldsymbol{r}) = \left(-\frac{2}{r} h(r) g \cos\theta, \ \frac{1}{r}\frac{d}{dr}(rh(r)) g \sin\theta, \ 0 \right) \tag{24.7}$$

$$\delta \mu_i(\boldsymbol{r}) = -z_i e \phi_i(r) g \cos\theta \tag{24.8}$$

where $g = |\boldsymbol{g}|$. Here, $h(r)$ and $\phi_i(r)$ are functions of r only. It can be shown that the fundamental electrokinetic equations can be expressed in terms of h and ϕ_i as

$$L(Lh - \lambda^2 h) = G(r), \quad a < r < b \tag{24.9}$$

$$L(Lh) = G(r), \quad b < r < c \tag{24.10}$$

$$L\phi_i = \frac{dy}{dr}\left(z_i \frac{d\phi_i}{dr} - \frac{2\lambda_i}{e}\frac{h}{r}\right) \tag{24.11}$$

with

$$\lambda = (\gamma/\eta)^{1/2} \tag{24.12}$$

$$L \equiv \frac{d}{dr}\frac{1}{r^2}\frac{d}{dr}r^2 = \frac{d^2}{dr^2} + \frac{2}{r}\frac{d}{dr} - \frac{2}{r^2} \tag{24.13}$$

$$G(r) = -\frac{e}{\eta r}\frac{dy}{dr}\sum_{i=1}^{M} z_i^2 n_i^\infty \exp(-z_i y)\phi_i \tag{24.14}$$

where

$$y = \frac{e\psi^{(0)}}{kT} \tag{24.15}$$

is the scaled equilibrium potential.

The boundary conditions are the same as those for electrophoresis except that the force balance condition as described below. In the stationary state, the net force acting on the particle or the unit cell must be zero. Since the net electric charge within the unit cell is zero, there is no net electric force acting on the cell and one needs to consider only the gravitational force $\boldsymbol{F}_\mathrm{g}$ and the hydrodynamic force $\boldsymbol{F}_\mathrm{h}$ so that

$$\boldsymbol{F}_g + \boldsymbol{F}_h = 0 \tag{24.16}$$

The gravitational force $\boldsymbol{F}_\mathrm{g}$ acting on the particle and the liquid within the unit cell is given by

$$\boldsymbol{F}_g = \frac{4}{3}\pi a^3 \rho_\mathrm{c}\boldsymbol{g} + V_\mathrm{s}\rho_\mathrm{s}\boldsymbol{g} + \left[\frac{4}{3}\pi b^3 - \frac{4}{3}\pi a^3 - V_\mathrm{s}\right]\rho_\mathrm{o}\boldsymbol{g} \tag{24.17}$$

where ρ_c is the mass density of the particle, V_s is the total volume of the polyelectrolyte segments coating one particle (which differs from the volume of the surface layer $(4\pi b^3/3 - 4\pi a^3/3)$), and ρ_s is the mass density of the polyelectrolyte segment. The terms on the right-hand side of Eq. (24.17), respectively, correspond to the gravitational forces acting on the particle core, the polyelectrolyte, and the liquid contained in one unit cell. Equation (24.17) can be expressed in terms of $h(r)$ as

$$\frac{4}{3}\pi a^3 \Delta\rho_c + V_s \Delta\rho_s + \frac{4}{3}\pi c^2 \eta \frac{d}{dr}(rLh)\bigg|_{r=c} = 0 \tag{24.18}$$

where

$$\Delta\rho_c = \rho_c - \rho_o \tag{24.19}$$

$$\Delta\rho_s = \rho_s - \rho_o \tag{24.20}$$

According to Kuwabara's cell model [7], we assume that at the outer surface of the unit cell ($r = c$), the liquid velocity $\boldsymbol{u}$ is parallel to the electrophoretic velocity $\boldsymbol{U}$ of the particle,

$$\boldsymbol{U}_{\text{SED}} = \frac{2h(c)}{c}\boldsymbol{g} \tag{24.21}$$

where $U_{\text{SED}} = |\boldsymbol{U}_{\text{SED}}|$, and that the vorticity $\omega = \nabla \times \boldsymbol{u}$ is $\boldsymbol{0}$ at $r = c$, which is expressed as

$$Lh|_{r=c} = 0 \tag{24.22}$$

We assume that electrolyte ions cannot penetrate the particle core ($\boldsymbol{v}_i \cdot \hat{\boldsymbol{n}}|_{r=a} = 0$), which can be rewritten in terms of $\phi_i(r)$ as

$$\frac{d\phi_i}{dr}\bigg|_{r=a} = 0 \tag{24.23}$$

Also, we may write $\delta\mu_i = 0$ at $r = b$, since we have assumed that the overlapping of double layers is negligible on the outer surface of the unit cell so that $\delta\psi \approx 0$ and $\delta n_i \approx 0$ at $r = c$. This can be expressed as

$$\phi_i(c) = 0 \tag{24.24}$$

24.3 SEDIMENTATION VELOCITY OF A SOFT PARTICLE

The sedimentation velocity $\boldsymbol{U}_{\mathrm{SED}}$ is obtained from Eq. (24.21)

$$\begin{aligned}\boldsymbol{U}_{\mathrm{SED}} = \frac{b^2}{9}\int_b^c \Bigg[3\left(1-\frac{r^2}{b^2}\right) - \frac{2L_2}{L_1}\left(1-\frac{r^3}{b^3}\right) \\ -\phi\left\{\left(3-\frac{2L_2}{L_1}-\frac{9a}{L_1\lambda^2 b^3}\right)\left(1-\frac{r^3}{b^3}\right)-\frac{3}{5}\left(1-\frac{r^5}{b^5}\right)\right\}\Bigg]G(r)dr\cdot\boldsymbol{g} \\ +\frac{2}{3\lambda^2}\int_a^c\left[(1-\phi)\frac{L_5}{L_1}\left(1+\frac{r^3}{2b^3}\right)-3\phi\right]G(r)dr\cdot\boldsymbol{g} \\ -\frac{2}{3\lambda^2}(1-\phi)\int_a^b\Bigg[1-\frac{3a}{2\lambda^2 b^3 L_1}\{(L_5+L_6\lambda r)\cosh[\lambda(r-a)] \\ -(L_6+L_5\lambda r)\sinh[\lambda(r-a)]\}\Bigg]G(r)dr\cdot\boldsymbol{g} \\ +\frac{2c^3}{9\eta b}(\phi_c\Delta\rho_c+\phi_s\Delta\rho_s)\left(1+\phi-\frac{9}{5}\phi^{1/3}-\frac{\phi^2}{5}+L_7\right)\boldsymbol{g}\end{aligned} \tag{24.25}$$

where L_1–L_4 are given by Eqs. (21.42)–(21.45) and L_5–L_7 by Eqs. (22.11)–(22.13).

For an uncharged soft particle, that is, an uncharged polymer-coated particle ($\rho_{\mathrm{fix}}(r)=0$), we have

$$\boldsymbol{U}^0_{\mathrm{SED}} = \frac{2c^3}{9\eta b}(\phi_c\Delta\rho_c+\phi_s\Delta\rho_s)\left(1+\phi-\frac{9}{5}\phi^{1/3}-\frac{\phi^2}{5}+L_7\right)\boldsymbol{g} \tag{24.26}$$

If, further $\phi\to 0$, then Eq. (24.26) reduces to

$$\boldsymbol{U}^0_{\mathrm{SED}} = \frac{2c^3}{9\eta b}(\phi_c\Delta\rho_c+\phi_s\Delta\rho_s)\left(\frac{L_2}{L_1}+\frac{3L_3}{2\lambda^2 b^2 L_1}\right)\boldsymbol{g} \tag{24.27}$$

which agrees with the result of Masliyah et al. [12].

24.4 AVERAGE ELECTRIC CURRENT AND POTENTIAL

Consider a suspension of spherical soft particles in a general electrolyte solution of volume V. We define the macroscopic average electric field $\langle \boldsymbol{E}\rangle$ and current $\langle \boldsymbol{i}\rangle$ in the suspension of N_p soft particles in an electrolyte solution of volume V as

$$\langle \boldsymbol{E} \rangle = -\frac{1}{V}\int_V \nabla \psi(\boldsymbol{r})dV = -\frac{1}{V}\int_V \nabla \delta\psi(\boldsymbol{r})dV \tag{24.28}$$

$$\langle \boldsymbol{i} \rangle = \frac{1}{V}\int_V \boldsymbol{i}(\boldsymbol{r})dV \tag{24.29}$$

where $\boldsymbol{i}$ is the current density at position $\boldsymbol{r}$. The electrical conductivity K^* of the suspension? is defined by

$$\langle \boldsymbol{i} \rangle = K^* \langle \boldsymbol{E} \rangle \tag{24.30}$$

The electric field $\langle \boldsymbol{E} \rangle$ is different from the applied field $\boldsymbol{E}$ and these two fields are related to each other by continuity of electric current, namely,

$$\langle \boldsymbol{i} \rangle = K^* \langle \boldsymbol{E} \rangle = K^\infty \boldsymbol{E}, \tag{24.31}$$

where

$$K^\infty = \sum_{i=1}^{M} z_i^2 e^2 n_i^\infty / \lambda_i \tag{24.32}$$

is the conductivity of the electrolyte solution (in the absence of the particles). If we further put

$$C_i = -\frac{c^2}{3}\left(r\frac{d\phi_i}{dr} - \phi_i\right)_{r=c} = -\frac{c^3}{3}\frac{d\phi_i}{dr}\bigg|_{r=c} \tag{24.33}$$

then $\langle \boldsymbol{i} \rangle$ may be written as

$$\langle \boldsymbol{i} \rangle = K^\infty \left(\langle \boldsymbol{E} \rangle - \frac{4\pi N_p}{V} \frac{\sum_{i=1}^{M} z_i^2 n_i^\infty C_i^{(2)} / \lambda_i}{\sum_{i=1}^{M} z_i^2 n_i^\infty / \lambda_i} \boldsymbol{g} \right) \tag{24.34}$$

24.5 SEDIMENTATION POTENTIAL

The sedimentation field $\boldsymbol{E}_{\text{SED}}$ may be obtained by setting $\langle \boldsymbol{i} \rangle$ equal to be zero in Eq. (24.34), namely,

$$\boldsymbol{E}_{\text{SED}} = \frac{3\phi}{b^3} \frac{\sum_{i=1}^{M} z_i^2 n_i^\infty C_i / \lambda_i}{\sum_{i=1}^{M} z_i^2 n_i^\infty / \lambda_i} \boldsymbol{g} = \frac{3\phi}{b^3} \sum_{i=1}^{M} \frac{z_i^2 e^2 n_i^\infty C_i}{\lambda_i} \boldsymbol{g} \tag{24.35}$$

with

$$K^{\infty} = \sum_{i=1}^{M} z_i^2 e^2 n_i^{\infty} / \lambda_i \tag{24.36}$$

For low potentials, it can be shown that $\boldsymbol{E}_{\mathrm{SED}}$ is related to the electrophoretic mobility μ of concentrated soft particles (22.10) by

$$\boldsymbol{E}_{\mathrm{SED}} = -\frac{(1-\phi_{\mathrm{c}})}{(1+\phi_{\mathrm{c}}/2)} \frac{(\phi_{\mathrm{c}}\Delta\rho_{\mathrm{c}} + \phi_{\mathrm{s}}\Delta\rho_{\mathrm{s}})}{K^{\infty}} \mu \boldsymbol{g} \tag{24.37}$$

with

$$\phi_{\mathrm{s}} = \frac{V_{\mathrm{s}}}{\frac{4}{3}\pi c^3} \tag{24.38}$$

$$\phi_{\mathrm{c}} = (a/c)^3 \tag{24.39}$$

where ϕ_{s} is the volume fraction of the polyelectrolyte segments and ϕ_{c} is the volume fraction of the particle core, which differs from ϕ (Eq. (24.1)). Equation (24.37) is the required Onsager relation between sedimentation potential $\boldsymbol{E}_{\mathrm{SED}}$ and electrophoretic mobility μ, both depending on ϕ, in concentrated suspensions of spherical soft particles.

In the limit $\phi \to 0$ (dilute suspensions), Eq. (24.9) tends to

$$\boldsymbol{E}_{\mathrm{SED}} = -\frac{(\phi_{\mathrm{c}}\Delta\rho_{\mathrm{c}} + \phi_{\mathrm{s}}\Delta\rho_{\mathrm{s}})}{K^{\infty}} \mu \boldsymbol{g} \tag{24.40}$$

which agrees with the result of Keh and Liu [6] for an Onsager relation for a dilute suspension of soft particles. We thus see that the prefactor

$$\frac{1-\phi_{\mathrm{c}}}{1+\phi_{\mathrm{c}}/2} \tag{24.41}$$

which is a correction factor for concentrated suspensions, agrees with Maxwell's relation with respect to ϕ_{c}. The ratio (Eq. (24.41)) equals the conductivity ratio K^*/K^{∞} for soft particles with low potentials, where K^* is the electrical conductivity of the suspension (Eq. (7.11)). It is to be noted that ϕ_{c} (instead of ϕ) appears in the prefactor. This corresponds to the fact that the polyelectrolyte layer does not contribute to K^* for the case of low potentials.

In the limit $a \to 0$, we have that $\phi_c \to 0$ and soft particles become spherical polyelectrolytes. In this limit, Eq. (24.37) tends to

$$\boldsymbol{E}_{\text{SED}} = -\frac{\phi_s \Delta\rho_s}{K^\infty}\mu \boldsymbol{g} \tag{24.42}$$

Interestingly, the prefactor (24.41) disappears in Eq. (24.42), implying that Eq. (24.42) can be applied for both dilute and concentrated cases. This is again because the polyelectrolyte layer does not contribute to the electrical conductivity K^* for low potentials, that is, in this case $K^* = K^\infty$.

In the limit $a = b$, the polyelectrolyte layer vanishes so that $\phi_c = \phi$ and $\phi_s = 0$. In this case, Eq. (24.40) gives the following result for sedimentation potential for concentrated suspension of hard particles of radius a for low zeta potentials, as expected:

$$\boldsymbol{E}_{\text{SED}} = -\frac{\phi(1-\phi)}{(1+\phi/2)}\frac{\mu}{K^\infty}\boldsymbol{g} \tag{24.43}$$

In the limit $\lambda \to 0$, we have $V_s = 0$ (and thus $\phi_s = 0$). This is the same situation as case $a = b$ and Eq. (24.40) again tends to Eq. (24.43).

For the case where $\lambda \to \infty$ and $\rho_c = \rho_s$, soft particles may become hard particles of radius b. Note, however, that in the polyelectrolyte layer $\boldsymbol{u} = \boldsymbol{0}$ but ionic flow $\boldsymbol{v}_i$ is not $\boldsymbol{0}$. That is, in this case, there exist ionic flows in the polyelectrolyte layer. Thus, this case differs from the case of a concentrated suspension of hard particles of radius b, in which case liquid and ionic flows are both absent within the particle.

We give below an approximate expression for $\boldsymbol{E}_{\text{SED}}$ for the case where the electrolyte is symmetrical with valence z and bulk concentration (number density) n^∞, and the fixed charge density is constant, $\rho_{\text{fix}}(r) = \rho_{\text{fix}} = \text{constant}$. We consider the case where

$$\lambda a \gg 1,\ \kappa a \gg 1 \text{ (and thus } \lambda b \gg 1,\ \kappa b \gg 1\text{) and}$$
$$\lambda d = \lambda(b-a) \gg 1,\ \kappa d = \kappa(b-a) \gg 1 \tag{24.44}$$

In this case, it can be shown that with the help of Eq. (22.23), $\boldsymbol{E}_{\text{SED}}$ is given by

$$\boldsymbol{E}_{\text{SED}} = -\frac{(1-\phi_c)}{(1+\phi_c/2)}\frac{(\phi_c\Delta\rho_c + \phi_s\Delta\rho_s)}{K^\infty}\left[\frac{\varepsilon_r\varepsilon_o}{\eta}\frac{\psi_o/\kappa_m + \psi_{\text{DON}}/\lambda}{1/\kappa_m + 1/\lambda}f\left(\frac{d}{a},\phi\right) + \frac{\rho_{\text{fix}}}{\eta\lambda^2}\right]\boldsymbol{g} \tag{24.45}$$

For the case where $a\,?\,d$, in particular, Eq. (24.45) reduces to

$$\boldsymbol{E}_{\text{SED}} = -\frac{(1-\phi_c)}{(1+\phi_c/2)}\frac{(\phi_c\Delta\rho_c + \phi_s\Delta\rho_s)}{K^\infty}\left[\frac{\varepsilon_r\varepsilon_o}{\eta}\frac{\psi_o/\kappa_m + \psi_{\text{DON}}/\lambda}{1/\kappa_m + 1/\lambda} + \frac{\rho_{\text{fix}}}{\eta\lambda^2}\right]\boldsymbol{g} \tag{24.46}$$

24.6 ONSAGER'S RECIPROCAL RELATION

We show below that the Onsager relation (Eq. (24.37)) is the special case of the more general Onsager relation between sedimentation potential and electrophoretic mobility derived on the basis of the thermodynamics of irreversible processes [14–16]. According to the thermodynamics of irreversible processes, we may generally write

$$\begin{aligned}\langle \boldsymbol{i} \rangle &= L_{11}\langle \boldsymbol{E} \rangle + L_{12}\boldsymbol{g} \\ \langle \boldsymbol{j} \rangle &= L_{21}\langle \boldsymbol{E} \rangle + L_{22}\boldsymbol{g}\end{aligned} \tag{24.47}$$

for the average electric current density $\langle \boldsymbol{i} \rangle$ and the average mass flux $\langle \boldsymbol{j} \rangle$, where L_{ij} are constants. It follows from the Onsager relation that $L_{12} = L_{21}$. Further, we may write (see Eq. (26) of Ref. 16)

$$\langle \boldsymbol{j} \rangle = \boldsymbol{U}(\phi_c \Delta\rho_c + \phi_s \Delta\rho_s) \tag{24.48}$$

From Eqs. (24.44) and (24.45) we immediately obtain (see Eq. (27) of Ref. 16)

$$\left(\frac{|\boldsymbol{U}|}{|\langle \boldsymbol{E} \rangle|} \right)_{\boldsymbol{g}=\boldsymbol{0}} = \left(\frac{|\langle \boldsymbol{i} \rangle|}{(\phi_c \Delta\rho_c + \phi_s \Delta\rho_s)|\boldsymbol{g}|} \right)_{\langle \boldsymbol{E} \rangle = \boldsymbol{0}} \tag{24.49}$$

On the left-hand side of Eq. (24.49), the particle velocity $\boldsymbol{U}$ is related to the electrophoretic mobility μ as $\boldsymbol{U} = \mu \boldsymbol{E}$. On the right-hand side of Eq. (24.49), the current density $\langle \boldsymbol{i} \rangle$ is the sedimentation current $\boldsymbol{i}_{\text{SED}}$, which is given by Eq. (24.34). Equation (24.49) thus yields

$$\boldsymbol{E}_{\text{SED}} = -\frac{(\phi_c \Delta\rho_c + \phi_s \Delta\rho_s)}{K^\infty} \left(\frac{|\boldsymbol{E}|}{|\langle \boldsymbol{E} \rangle|} \right) \mu \boldsymbol{g} \tag{24.50}$$

Since the two fields $\boldsymbol{E}$ and $\langle \boldsymbol{E} \rangle$ are related to each other by Eq. (24.31), Eq. (24.50) can further be written as

$$\boldsymbol{E}_{\text{SED}} = -\frac{(\phi_c \Delta\rho_c + \phi_s \Delta\rho_s)}{K^\infty} \left(\frac{K^*}{K^\infty} \right) \mu \boldsymbol{g} \tag{24.51}$$

For low potentials, by substituting Eq. (7.11) into Eq. (24.51), we obtain Eq. (24.31). We thus see that the general Onsager relation (24.51) reproduces Eq. (24.37) at low potentials.

There is an alternative definition of electrophoretic mobility for concentrated suspensions: $\boldsymbol{U}_{\text{E}} = \mu^* \langle \boldsymbol{E} \rangle$, where $\langle \boldsymbol{E} \rangle$ is the average electric field in the suspension. It follows from the continuity condition (24.31) of electric current, that is, $K^\infty \boldsymbol{E} = K^* \langle \boldsymbol{E} \rangle$, that $\mu^*/\mu = K^*/K^\infty$ (in the dilute case $\mu^* = \mu$). In terms of μ^*, Eq. (24.47) is simply written as

$$\boldsymbol{E}_{\mathrm{SED}} = -\frac{(\phi_{\mathrm{c}}\Delta\rho_{\mathrm{c}} + \phi_{\mathrm{s}}\Delta\rho_{\mathrm{s}})}{K^{\infty}}\mu^{*}\boldsymbol{g} \tag{24.52}$$

24.7 DIFFUSION COEFFICIENT OF A SOFT PARTICLE

Although Eq. (24.40) has been derived for soft particles in a field of gravity $\boldsymbol{g}$, these relations generally hold for the ratio of the velocity of the steady motion of charged particles under any nonelectrostatic field to that when uncharged. Since the velocity $\boldsymbol{U}$ of a particle in a field $\boldsymbol{F}$ is expressed as $\boldsymbol{U} = \boldsymbol{F}/f$, f being the drag coefficient, $\boldsymbol{U}^0_{\mathrm{SED}}/\boldsymbol{U}_{\mathrm{SED}}$ equals the reciprocal of the ratio of the drag coefficient of a charged soft particle, f, to that when uncharged, f_{o}, namely,

$$\frac{|\boldsymbol{U}_{\mathrm{SED}}|}{|\boldsymbol{U}^0_{\mathrm{SED}}|} = \frac{f_{\mathrm{o}}}{f} \tag{24.53}$$

where $\boldsymbol{U}_{\mathrm{SED}}$ and $\boldsymbol{U}^0_{\mathrm{SED}}$ are, respectively, given by Eqs. (24.25) and (24.26). Schumacher and van de Ven [17] noticed that the well-known Einstein equation between the diffusion coefficient D_{o} and the drag coefficient f_{o} for an uncharged particle, namely,

$$\boldsymbol{D}_{\mathrm{o}} = \frac{kT}{f_{\mathrm{o}}} \tag{24.54}$$

needs modification for a charged soft particle and the diffusion coefficient of a charged soft particle, D, is related to that when uncharged, D_{o}, via

$$\frac{D}{D_{\mathrm{o}}} = \frac{f_{\mathrm{o}}}{f} = \frac{|\boldsymbol{U}_{\mathrm{SED}}|}{|\boldsymbol{U}^0_{\mathrm{SED}}|} \tag{24.55}$$

REFERENCES

1. F. Booth, *J. Chem. Phys.* 22 (1956) 1954.
2. D. A. Saville, *Adv. Colloid Interface Sci.* 16 (1982) 267.
3. H. Ohshima, T. W. Healy, L. R. White, and R. W. O'Brien, *J. Chem. Soc., Faraday Trans.* 2 80 (1984) 1299.
4. H. Ohshima, in: H. Ohshima and K. Furusawa (Eds.), *Electrical Phenomena at Interfaces, 2nd edition*, Marcel Dekker, New York, 1998, Chapter 2.
5. H. Ohshima, *Theory of Colloid and Interfacial Electric Phenomena*, Elsecier/Academic Press, Amsterdam, 2006.
6. H. J. Keh and Y. C. Liu, *J. Colloid Interface Sci.* 195 (1997) 169.
7. S. Kuwabara, *J. Phys. Soc. Jpn.* 14 (1959) 527.
8. S. Levine, G. H. Neale, and N. Epstein, *J. Colloid Interface Sci.* 57 (1976) 424.
9. H. Ohshima, *J. Colloid Interface Sci.* 208 (1998) 295.

10. H. Ohshima, *J. Colloid Interface Sci.* 229 (2000) 140.
11. H. Ohshima, *Colloids Surf. A: Physicochem. Eng. Aspects* 195 (2001) 129.
12. J. H. Masliyah, G. Neale, K. Malysa, and T. G. M. van de Ven, *Chem. Eng. Sci.* 42 (1987) 245.
13. J. C. Maxwell, *A Treatise on Electricity and Magnetism*, Vol. 1, Dover, New York, 1954, Art. 313.
14. S. R. de Groot, P. Mazur, and J. Th. G. Overbeek, *J. Chem. Phys.* 20 (1952) 1825.
15. J. Th. G. Overbeek, *J. Colloid Sci.* 8 (1953) 420.
16. A. S. Dukhin, V. N. Shilov, and Yu. B. Borkovskaya, *Langmuir* 15 (1999) 3452.
17. G. A. Schumacher and T. G. M. van de Ven, *Faraday Discuss. Chem. Soc.* 83 (1987) 75.

25 Dynamic Electrophoretic Mobility of a Soft Particle

25.1 INTRODUCTION

The dynamic electrophoretic mobility of colloidal particles in an applied oscillating electric field plays an essential role in analyzing the results of electroacoustic measurements of colloidal dispersions, that is, colloid vibration potential (CVP) and electrokinetic sonic amplitude (ESA) measurements [1–20]. This is because CVP and ESA are proportional to the dynamic electrophoretic mobility of colloidal particles. In this chapter, we develop a theory of the dynamic electrophoretic mobility of soft particles in dilute suspensions [21].

25.2 BASIC EQUATIONS

Consider a spherical soft particle moving with a velocity $\boldsymbol{U}\exp(-i\omega t)$ in a liquid containing a general electrolyte in an applied oscillating electric field $\boldsymbol{E}\exp(-i\omega t)$, where ω is the angular frequency (2π times frequency) and t is time (Fig. 25.1). The dynamic electrophoretic mobility $\mu(\omega)$, which is a function of ω, of the particle is defined by

$$U = \mu(\omega)E \tag{25.1}$$

where $U = |\boldsymbol{U}|$ and $E = |\boldsymbol{E}|$. We assume that the particle core of radius a is coated with an ion-penetrable layer of polyelectrolytes with a thickness d. The polyelectrolyte-coated particle has thus an inner radius a and an outer radius $b \equiv a + d$ (Fig. 25.1). The origin of the spherical polar coordinate system (r, θ, φ) is held fixed at the instantaneous center of the particle and the polar axis ($\theta = 0$) is set parallel to $\boldsymbol{E}$. Let the electrolyte be composed of M ionic mobile species of valence z_j and drag coefficient λ_j ($j = 1, 2, \ldots, M$), and n_j^∞ be the concentration (number density) of the jth ionic species in the electroneutral solution.

As in the case of static electrophoresis, we adopt the model of Debye–Bueche [22, 23] where the polymer segments are regarded as resistance centers distributed in the polyelectrolyte layer, exerting a frictional force on the liquid flowing in the

Biophysical Chemistry of Biointerfaces By Hiroyuki Ohshima

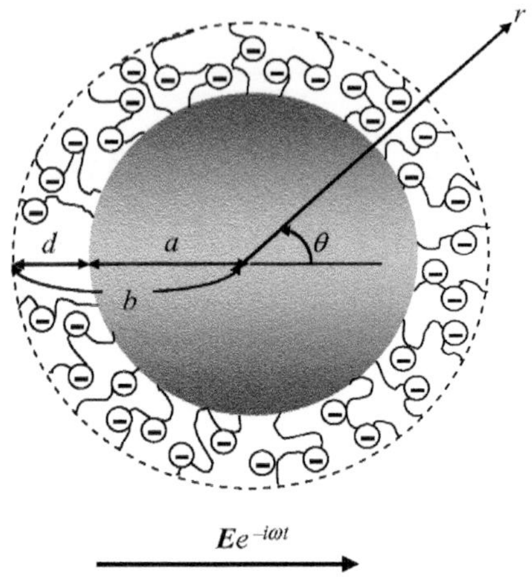

FIGURE 25.1 A soft particle in an oscillating electric field $\boldsymbol{E}\,e^{-i\omega t}$. a = radius of the particle core and d = thickness of the polyelectrolyte layer coating the particle core. $b = a + d$.

polyelectrolyte layer. Let the polymer segments be spherical particles of radius a_s and distributed at a number density of N_s, then the frictional force is given by [22]

$$\boldsymbol{F}_s = -\nu \boldsymbol{u} \tag{25.2}$$

Here, the frictional coefficient ν is given by [24]

$$\nu = 6\pi\eta a_s N_s \left\{ 1 - i\gamma a_s - \frac{1}{9}(\gamma a_s)^2 \right\} \tag{25.3}$$

with

$$\gamma = \sqrt{\frac{i\omega\rho_o}{\eta}} = (i+1)\sqrt{\frac{\omega\rho_o}{2\eta}} = (i+1)\frac{1}{\delta} \tag{25.4}$$

where ρ_o is the mass density of the liquid, η is the viscosity, and $\delta = (2\eta/\omega\rho_o)^{1/2}$ is the hydrodynamic penetration depth in the liquid. For a typical case where $a_s = 0.7$ nm, $\eta = 0.89 \times 10^{-3}$ Pa s, $\rho_o = 1 \times 10^3$ kg/m^3, and $\omega/2\pi = 1$ MHz, we have $|\gamma| a_s = 2.6 \times 10^{-3}$. We may thus assume that

$$|\gamma| a_s \ll 1 \tag{25.5}$$

so that Eq. (25.3) can be approximated by

$$\nu = 6\pi\eta a_s N_s \tag{25.6}$$

which is the same as the Stokes formula in the static case ($\omega = 0$). We also assume that fixed charges are distributed with a density of ρ_{fix}. If dissociated groups of valence Z is distributed with a constant density (number density) N within the polyelectrolyte layer, then we have $\rho_{\text{fix}} = ZeN$. The fundamental electrokinetic equations as well as the boundary conditions are essentially the same as those used in static electrophoresis, except that the flow velocity of the liquid $\boldsymbol{u}(\boldsymbol{r}, t)$ at position r and that of the jth mobile ionic species $\boldsymbol{v}_j(\boldsymbol{r}, t)$ are now functions of time t and may be written as

$$\boldsymbol{u}(\boldsymbol{r}, t) = \boldsymbol{u}(\boldsymbol{r}) \exp(-iwt) \tag{25.7}$$

$$\boldsymbol{v}_j(\boldsymbol{r}, t) = \boldsymbol{v}_j(\boldsymbol{r}) \exp(-i\omega t), \quad (j = 1, 2, \ldots, M) \tag{25.8}$$

so that the Navier–Stokes equations become

$$\begin{aligned} \rho_o \frac{\partial}{\partial t}[\boldsymbol{u}(\boldsymbol{r}, t) + \boldsymbol{U} \exp(-i\omega t)] \\ = -\eta \nabla \times \nabla \times \boldsymbol{u}(\boldsymbol{r}, t) - \nabla p(\boldsymbol{r}, t) - \rho_{\text{el}}(\boldsymbol{r}, t) \nabla \psi(\boldsymbol{r}, t) - \nu \boldsymbol{u}(\boldsymbol{r}, t), \quad a < r < b \end{aligned} \tag{25.9}$$

$$\begin{aligned} \rho_o \frac{\partial}{\partial t}[\boldsymbol{u}(\boldsymbol{r}, t) + \boldsymbol{U} \exp(-i\omega t)] \\ = -\eta \nabla \times \nabla \times \boldsymbol{u}(\boldsymbol{r}, t) - \nabla p(\boldsymbol{r}, t) - \rho_{\text{el}}(\boldsymbol{r}, t) \nabla \psi(\boldsymbol{r}, t), \quad r > b \end{aligned} \tag{25.10}$$

and the continuity equation for the ionic flow becomes

$$\frac{\partial n_j(\boldsymbol{r}, t)}{\partial t} + \nabla \cdot (n_j(\boldsymbol{r}, t) \boldsymbol{v}_j(\boldsymbol{r}, t)) = \boldsymbol{0} \tag{25.11}$$

where $\rho_{\text{el}}(\boldsymbol{r}, t)$ is the charge density resulting from the mobile charged ionic species, $\psi(\boldsymbol{r}, t)$ is the electric potential, and $n_j(\boldsymbol{r}, t)$ is the concentration (number density) of the jth ionic species at $\boldsymbol{r}$ and t.

In Eqs. (25.9) and (25.10), the term $\rho_o(\boldsymbol{u} \cdot \nabla)\boldsymbol{u}$ has been omitted, since we assume that the liquid flow is slow as in the case of static electrophoresis. The term involving the particle velocity $\boldsymbol{U} \exp(-i\omega t)$ in Eqs. (25.9) and (25.10) arises from the fact that the particle has been chosen as the frame of reference for the coordinate system.

25.3 LINEARIZED EQUATIONS

As in the case of static electrophoresis, for a weak field E, the electrical double layer around the particle is only slightly distorted. Then, we may write

$$n_j(\boldsymbol{r}, t) = n_j^{(0)}(r) + \delta n_j(\boldsymbol{r}) \exp(-i\omega t) \tag{25.12}$$

$$\psi(\boldsymbol{r},t) = \psi^{(0)}(r) + \delta\psi(\boldsymbol{r})\exp(-i\omega t) \tag{25.13}$$

$$\mu_j(\boldsymbol{r},t) = \mu^{(0)} + \delta\mu_j(\boldsymbol{r})\exp(-i\omega t) \tag{25.14}$$

$$\rho_{\text{el}}(\boldsymbol{r},t) = \rho_{\text{el}}^{(0)}(r) + \delta\rho_{\text{el}}(\boldsymbol{r})\exp(-i\omega t) \tag{25.15}$$

where the quantities with superscript (0) refer to those at equilibrium, that is, in the absence of E, and $\mu_j^{(0)}$ is a constant independent of r.

Further, symmetry considerations permit us to write

$$\boldsymbol{u}(\boldsymbol{r}) = (u_r, u_\theta, u_\varphi) = \left(-\frac{2}{r}h(r)E\cos\theta,\ \frac{1}{r}\frac{d}{dr}(rh(r))E\sin\theta,\ 0\right) \tag{25.16}$$

$$\delta\mu_j(\boldsymbol{r}) = -z_j e\phi_j(r)E\cos\theta \tag{25.17}$$

$$\delta\psi(\boldsymbol{r}) = -Y(r)E\cos\theta \tag{25.18}$$

Here, $\phi_j(r)$, $Y(r)$, and $h(r)$ are functions of r only. In terms of $\phi_j(r)$, $Y(r)$, and $h(r)$, the fundamental electrokinetic equations including Eqs. (25.9)–(25.11) can be rewritten as

$$L(Lh - \lambda^2 h + \gamma^2 h) = G(r), \quad a < r < b \tag{25.19}$$

$$L(Lh + \gamma^2 h) = G(r), \quad r > b \tag{25.20}$$

$$L\phi_j - \kappa^2\gamma_j(\phi_j - Y) = \frac{dy}{dr}\left(z_j\frac{d\phi_j}{dr} - \frac{2\lambda_j}{e}\frac{h}{r}\right) \tag{25.21}$$

$$\delta\rho_{\text{el}}(\boldsymbol{r}) = \sum_{j=1}^{M} z_j e\delta n_j(\boldsymbol{r}) = -\varepsilon_r\varepsilon_o\Delta\delta\psi(\boldsymbol{r}) = \varepsilon_r\varepsilon_o E\cos\theta\cdot LY \tag{25.22}$$

with

$$\lambda = (\nu/\eta)^{1/2} \tag{25.23}$$

$$\gamma_j = -\frac{i\omega\lambda_j}{\kappa^2 kT} \tag{25.24}$$

where κ is the Debye–Hückel parameter, L is a differential operator (Eq. (21.24)), and $G(r)$ is defined by Eq. (21.25).

25.4 EQUATION OF MOTION OF A SOFT PARTICLE

Consider the forces acting on an arbitrary sphere S enclosing the soft sphere at its center. We take the radius of S tends to infinity. Since the net electric charge within S is zero, there is no net electric force acting on S and one needs consider only the hydrodynamic force F_h. The equation of motion for S is thus given by

$$\rho_o \int_0^{\pi}\int_a^r \frac{d}{dt}\left[(u_r \cos\theta - u_\theta \cos\theta + U)e^{-i\omega t}\right]2\pi r^2 \sin\theta\, dr\, d\theta + \left(\rho_c \frac{4}{3}\pi a^3 + \rho_s V_s\right)\frac{d}{dt}\left[Ue^{-i\omega t}\right] = F_h \tag{25.25}$$

where ρ_c and ρ_s are, respectively, the mass densities of the particle core and the polymer segments, and V_s is the total volume of the polymer segments per particle. The hydrodynamic force F_h is given by

$$F_h = \int_0^{\pi} (\sigma_{rr}\cos\theta - \sigma_{r\theta}\sin\theta)2\pi r^2 \sin\theta\, d\theta \tag{25.26}$$

where σ_{rr} and σ_{rs} are, respectively, the normal and tangential components of the hydrodynamic stress.

25.5 GENERAL MOBILITY EXPRESSION

The dynamic electrophoretic mobility $\mu(\omega)$ of a soft particle can be obtained by solving Eqs. (25.19) and (25.20) with the result that

$$\begin{aligned}\mu(\omega) = {} & \frac{2b^2M_2}{9M_1R_1}\left[\frac{\gamma^2}{\beta^2}\int_a^b \left(1-\frac{r^3}{b^3}\right)G(r)dr - \int_b^\infty \left(1-\frac{r^3}{b^3}\right)G(r)dr\right] \\ & - \frac{2}{3\beta^2R_1}\int_a^b \left[R_2 - \frac{3a}{2\beta^2b^3M_1}\left(R_3 - \frac{2\gamma^2b^2M_2}{3M_3}\right)\right. \\ & \times\{(M_3 + M_4\beta r)\cosh[\beta(r-a)] - (M_4 + M_3\beta r)\sinh[\beta(r-a)]\} \\ & \left.- \frac{\gamma^2bM_2}{\beta M_1M_3}\{\beta r\cosh[\beta(r-a)] - \sinh[\beta(r-a)]\}\right]G(r)dr \\ & + \frac{2M_3R_3}{3\beta^2M_1R_1}\int_a^\infty \left(1+\frac{r^3}{2b^3}\right)G(r)dr \\ & - \frac{2R_2}{3\gamma^2R_1}\int_b^\infty \left\{\frac{1-i\gamma r}{1-i\gamma b}e^{i\gamma(r-b)} - 1\right\}G(r)dr\end{aligned} \tag{25.27}$$

where

$$R_1 = \left(1 - \frac{M_1}{M_3} + \frac{3a}{2M_3 b} - \frac{3\Gamma}{2\gamma^2 b^2}\right)\frac{2\gamma^2 b^2 M_2}{3M_1} + \left\{1 + \left(1 - \frac{3\Gamma}{2\gamma^2 b^2}\right)\frac{\gamma^2 M_3}{\beta^2 M_1}\right\}R_3 \tag{25.28}$$

$$R_2 = \frac{M_2(1 - i\gamma b)}{i\gamma b M_1 + M_2(1 - i\gamma b)}\left(1 + \frac{\gamma^2}{\beta^2} - \frac{3\gamma^2 a}{2\beta^2 b M_1}\right) \tag{25.29}$$

$$R_3 = R_2 - \frac{\gamma^2 b^2 M_2}{M_1}\left(1 - \frac{2M_1}{3M_3} + \frac{a}{bM_3}\right) \tag{25.30}$$

$$\begin{aligned} M_1 = {} & \left(1 + \frac{a^3}{2b^3} + \frac{3a}{2\beta^2 b^3} - \frac{3a^2}{2\beta^2 b^4}\right)\cosh[\beta(b-a)] \\ & - \left(1 - \frac{3a^2}{2b^2} + \frac{a^3}{2b^3} + \frac{3a}{2\beta^2 b^3}\right)\frac{\sinh[\beta(b-a)]}{\beta b} \end{aligned} \tag{25.31}$$

$$M_2 = \left(1 + \frac{a^3}{2b^3} + \frac{3a}{2\beta^2 b^3}\right)\cosh[\beta(b-a)] + \frac{3a^2}{2b^2}\frac{\sinh[\beta(b-a)]}{\beta b} - \frac{3a}{2\beta^2 b^3} \tag{25.32}$$

$$M_3 = \cosh[\beta(b-a)] - \frac{\sinh[\beta(b-a)]}{\beta b} - \frac{a}{b} \tag{25.33}$$

$$M_4 = \sinh[\beta(b-a)] - \frac{\cosh[\beta(b-a)]}{\beta b} + \frac{\beta a^2}{3b} + \frac{2\beta b^2}{3a} + \frac{1}{\beta b} \tag{25.34}$$

$$\Gamma = \frac{\gamma^2[V_c(\rho_c - \rho_o) + V_s(\rho_s - \rho_o)]}{6\pi b \rho_o} = \frac{2(\gamma b)^2\{f_c(\rho_c - \rho_o) + f_s(\rho_s - \rho_o)\}}{9\rho_o} \tag{25.35}$$

$$f_c = \left(\frac{a}{b}\right)^3 \tag{25.36}$$

$$f_s = \frac{V_s}{4\pi b^3/3} \tag{25.37}$$

$$V_c = \frac{4\pi}{3}a^3 \tag{25.38}$$

For the limiting case of $a \to 0$, the particle core vanishes, so a spherical soft particle becomes a spherical polyelectrolyte. In this case, Eq. (25.27) tends to

$$\begin{aligned}
\mu(\omega) = & \frac{2b^2}{9R_1^0\{1-\tanh(\beta b)/\beta b\}}\left[\frac{\gamma^2}{\beta^2}\int_0^b\left(1-\frac{r^3}{b^3}\right)G(r)dr - \int_b^\infty\left(1-\frac{r^3}{b^3}\right)G(r)dr\right] \\
& -\frac{2(1-i\gamma b)\{1+(\gamma/\beta)^2\}}{3\beta^2R_1^0\{1-i(\gamma/\beta)\tanh(\beta b)\}}\int_0^b\left[1-\frac{\cosh(\beta r)-\sinh(\beta r)/\beta r}{\cosh(\beta b)-\sinh(\beta b)/\beta b}\right]G(r)dr \\
& +\frac{2}{3\beta^2R_1^0}\left[\frac{(1-i\gamma b)\{1+(\gamma/\beta)^2\}}{\{1-i(\gamma/\beta)\tanh(\beta b)\}}-\frac{(\gamma b)^2}{3\{1-\tanh(\beta b)/\beta b\}}\right]\int_0^\infty\left(1+\frac{r^3}{2b^3}\right)G(r)dr \\
& -\frac{2(1-i\gamma b)\{1+(\gamma/\beta)^2\}}{3\gamma^2R_1^0\{1-i(\gamma/\beta)\tanh(\beta b)\}}\int_b^\infty\left\{\frac{1-i\gamma r}{1-i\gamma b}e^{i\gamma(r-b)}-1\right\}G(r)dr
\end{aligned} \tag{25.39}$$

with

$$\begin{aligned}
R_1^0 = & \left[1+\left\{1-\frac{3\Gamma^0}{2(\gamma b)^2}\right\}\left(\frac{\gamma}{\beta}\right)^2\right]\left[\frac{(1-i\gamma b)\{1+(\gamma/\beta)^2\}}{1-i(\gamma/\beta)\tanh(\beta b)}-\frac{(\gamma b)^2}{3\{1-\tanh(\beta b)/(\beta b)\}}\right] \\
& -\frac{\Gamma^0}{\{1-\tanh(\beta b)/(\beta b)\}}
\end{aligned} \tag{25.40}$$

$$\Gamma^0 = \frac{\gamma^2 V_s(\rho_s-\rho_o)}{6\pi b\rho_o} = \frac{2(\gamma b)^2 f_s(\rho_s-\rho_o)}{9\rho_o} \tag{25.41}$$

25.6 APPROXIMATE MOBILITY FORMULA

In this section, we treat the practically important case where potential is not very high so that dynamic relaxation effect is negligible. In this case, we have $\delta\rho_{el}(\boldsymbol{r}) = 0$ or $\delta n_j(\boldsymbol{r}) = 0$. Consider first the case where potential is low and $a \to 0$ and where $\rho_{fix} = $ constant, which corresponds a uniformly charged spherical polyelectrolyte of radius b.

We thus obtain the following expression for the dynamic mobility of a spherical polyelectrolyte:

$$\mu(\omega) = \frac{\rho_{\text{fix}}}{\eta\beta^2 R_1^0}\left[\frac{(1-i\gamma b)\{1+(\gamma/\beta)^2\}}{1-i(\gamma/\beta)\tanh(\beta b)} - \frac{(\gamma b)^2}{3\{1-\tanh(\beta b)/(\beta b)\}}\right]$$
$$+\frac{\rho_{\text{fix}}}{3\eta\kappa^2 R_1^0}\frac{\{1+(\gamma/\beta)^2\}(1-e^{-2\kappa b})}{\{1-i(\gamma/\beta)\tanh(\beta b)\}}\left[\left(\frac{1-i\gamma b-i\gamma/\kappa}{1-i\gamma/\kappa}\right)(\coth(\kappa b)-1/\kappa b)\right.$$
$$\left.+\left\{\frac{(1-i\gamma b)(1+1/\kappa b)}{(\beta/\kappa)^2-1}\right\}\left\{\frac{\beta}{\kappa}\left(\frac{\coth(\kappa b)-1/\kappa b}{\coth(\beta b)-1/\beta b}\right)-1\right\}\right] \tag{25.42}$$

When $|\beta|b \gg 1$, $\kappa b \gg 1$, $|\beta| \gg |\gamma|$, and $\kappa \gg |\gamma|$, Eq. (25.42) becomes

$$\mu(\omega) = \frac{\rho_{\text{fix}}}{\eta\beta^2\left\{1-i\gamma b-\frac{(\gamma b)^2}{3}-\Gamma^0\right\}}\left[1-i\gamma b-\frac{(\gamma b)^2}{3}+\frac{2}{3}(1-i\gamma b)\left(\frac{\beta}{\kappa}\right)^2\left(\frac{1+\beta/2\kappa}{1+\beta/\kappa}\right)\right] \tag{25.43}$$

which, for $\omega \to 0$ ($\beta \to \lambda$, $\gamma \to 0$, and $\Gamma^{\text{o}} \to 0$), further becomes Herman and Fujita's formula (21.54) for the static electrophoretic mobility of a spherical polyelectrolyte. That is, Eq. (25.43) is the generalization of Herman and Fujita's formula to the dynamic case.

Consider next the case where the electrolyte is symmetrical with a valence z and bulk concentration n (that is, we set $M = 2$, $z_1 = -z_2 = z$, and $n_1^\infty = n_2^\infty = n$), and $\rho_{\text{fix}}(r) = \rho_{\text{fix}}$ (= constant), and where

$$|\beta|b \gg 1,\ \kappa b \gg 1,\ |\beta|d = |\beta|(b-a) \gg 1,\ \kappa d = \kappa(b-a) \gg 1,\ |\beta| \gg |\gamma|,\ \kappa \gg |\gamma| \tag{25.44}$$

In this case, the potential inside the polyelectrolyte layer can be approximated by Eq. (21.47).

Equation (25.39) in this case gives

$$\mu(\omega) = \frac{2\varepsilon_r\varepsilon_o}{3\eta}\left[\frac{1-i\gamma b}{1-i\gamma b-(\gamma^2b^2/3)-\Gamma}\right]\left(1+\frac{a^3}{2b^3}\right)\left[\frac{\psi_o/\kappa_m+\psi_{\text{DON}}/\beta}{1/\kappa_m+1/\beta}\right]$$
$$+\left[\frac{1-i\gamma b-(\gamma^2b^2/3)(1-a^3/b^3)}{1-i\gamma b-(\gamma^2b^2/3)-\Gamma}\right]\frac{\rho_{\text{fix}}}{\eta\beta^2} \tag{25.45}$$

Equation (25.45) is the required expression for the dynamic mobility of a soft particle, applicable for most practical cases. When $\omega \to 0$ ($\beta \to \lambda$, $\gamma \to 0$, and $\Gamma \to 0$), Eq. (25.45) tends to Eq. (21.51) for the static case. When the polyelectrolyte layer is very thin, that is, for $a \approx b$ (or $d \ll a$), Eq. (25.45) reduces to

$$\mu(\omega) = \frac{\varepsilon_r \varepsilon_o}{\eta} \left[\frac{1 - i\gamma b}{1 - i\gamma b - (\gamma^2 b^2/3) - \Gamma} \right] \left[\frac{\psi_o/\kappa_m + \psi_{DON}/\beta}{1/\kappa_m + 1/\beta} + \frac{\rho_{fix}}{\eta \beta^2} \right] \tag{25.46}$$

In the opposite limiting case of very thick polyelectrolyte layer, that is, for $b \gg a$, Eq. (25.45) becomes

$$\begin{aligned} \mu(\omega) = & \frac{2\varepsilon_r \varepsilon_o}{3\eta} \left[\frac{1 - i\gamma b}{1 - i\gamma b - (\gamma^2 b^2/3) - \Gamma} \right] \left[\frac{\psi_o/\kappa_m + \psi_{DON}/\beta}{1/\kappa_m + 1/\beta} \right] \\ & + \left[\frac{1 - i\gamma b - (\gamma^2 b^2/3)}{1 - i\gamma b - (\gamma^2 b^2/3) - \Gamma} \right] \frac{\rho_{fix}}{\eta \beta^2} \end{aligned} \tag{25.47}$$

Some results of the calculation of the magnitude and phase of the dynamic mobility via Eq. (25.45) are given in Figs 25.2 and 25.3. All the calculations are performed at $\rho_{fix} = ZeN = 4.8 \times 10^6\ \mathrm{C/m^3}$, $1/\lambda = 1$ nm, $d = 10$ nm, $\rho_o = 1 \times 10^3\ \mathrm{kg/m^3}$, $\rho_c = \rho_s = 1.1 \times 10^3\ \mathrm{kg/m^3}$, $f_s = 0.1 \times (1 - f_c)$, $\eta = 0.89$ mPa s, and $T = 298$K. Figures 25.2 and 25.3 show the dependence of the magnitude (Fig. 25.2) and phase (Fig. 25.3) of the dynamic mobility on the frequency ω of the applied electric field for the case where $a = 1$ μm for a 1:1 electrolyte of concentration of $n = 0.01$ and 0.1 M. It is seen that the ω-dependence is negligible $\omega/2\pi < 10^4$ Hz and becomes appreciable for $\omega/2\pi > 10^4$ Hz. That is, $|\mu|$ is essentially equal to its static value ($\mu(\omega)$ at $\omega = 0$)

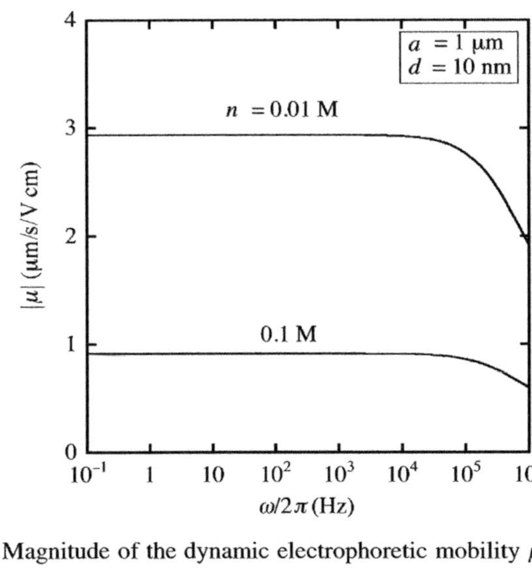

FIGURE 25.2 Magnitude of the dynamic electrophoretic mobility μ as a function of the frequency $\omega/2\pi$ of the applied oscillating electric field for a 1:1 electrolyte of concentration $n = 0.01$ and 0.1 M, calculated via Eq. (25.45) for $a = 1$ μm, $d = 10$ nm, $\rho_{fix} = ZeN = 4.8 \times 10^6\ \mathrm{C/m^3}$ (which corresponds to $N = 0.05$ M and $Z = 1$), $1/\lambda = 1$ nm, $\rho_o = 1 \times 10^3\ \mathrm{kg/m^3}$, $\rho_c = \rho_s = 1.1 \times 10^3\ \mathrm{kg/m^3}$, $f_s = 0.1 \times (1 - f_c)$, $\eta = 0.89$ mPa s, and $T = 298$K.

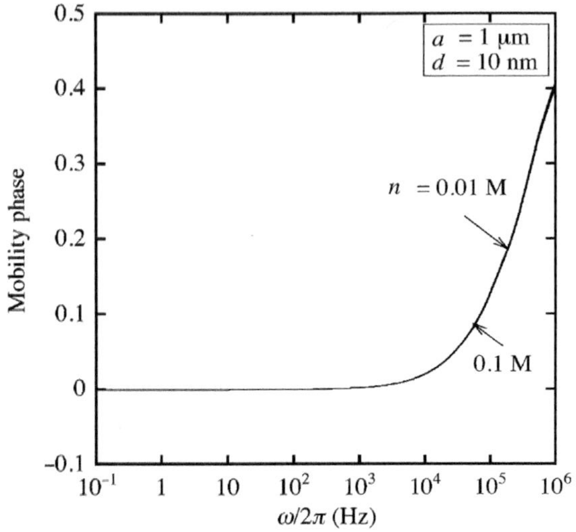

FIGURE 25.3 Phase of the reduced dynamic electrophoretic mobility μ as a function of the frequency $\omega/2\pi$ of the applied oscillating electric field calculated via Eq. (25.45). Numerical values used are the same as in Fig. 25.2.

for $\omega/2\pi < 10^4$ Hz and drops sharply to zero for $\omega/2\pi > 10^4$ Hz, while the phase of $\mu(\omega)$ is zero for $\omega/2\pi < 10^4$ Hz and increases sharply for the frequency range $\omega/2\pi > 10^4$ Hz.

It is to be noted that in the limit of very high electrolyte concentrations, Eq. (25.45) tends to a nonzero limiting value given by

$$\mu(\omega) \rightarrow \mu^{\infty}(\omega) = \left[\frac{1 - i\gamma b - (\gamma^2 b^2/3)(1 - a^3/b^3)}{1 - i\gamma b - (\gamma^2 b^2/3) - \Gamma}\right] \frac{\rho_{\text{fix}}}{\eta \beta^2} \tag{25.48}$$

which is not subject to the ionic shielding effect. This is a characteristic of electrophoresis of soft particles (see Eq. (21.62) for the static electrophoresis problem).

REFERENCES

1. R. W. O'Brien, *J. Fluid Mech.* 190 (1988) 71.
2. A. J. Babchin, R. S. Chow, and R. P. Sawatzky, *Adv. Colloid Interface Sci.* 30 (1989) 111.
3. A. J. Babchin, R. P. Sawatzky, R. S. Chow, E. E. Isaacs, and H. Huang, in: K. Oka (Ed.), *International Symposium on Surface Charge Characterization: 21st Annual Meeting of Fine Particle Society*, San Diego, CA, August 1990, p. 49.
4. R. P. Sawatzky and A. J. Babchin, *J. Fluid Mech.* 246 (1993) 321.
5. M. Fixman, *J. Chem. Phys.* 78 (1983) 1483.
6. R. O. James, J. Texter, and P. J. Scales, *Langmuir* 7 (1991) 1993.

7. C. S. Mangelsdorf and L. R. White, *J. Chem. Soc., Faraday Trans.* 88 (1992) 3567.
8. C. S. Mangelsdorf and L. R. White, *J. Colloid Interface Sci.* 160 (1993) 275.
9. H. Ohshima, *J. Colloid Interface Sci.* 179 (1996) 431.
10. M. Lowenberg and R. W. O'Brien, *J. Colloid Interface Sci.* 150 (1992) 158.
11. R. W. O'Brien, *J. Colloid Interface Sci.* 171 (1995) 495.
12. H. Ohshima, *J. Colloid Interface Sci.* 185 (1997) 131.
13. H. Ohshima, in: H. Ohshima and K. Furusawa (Eds.), *Electrical Phenomena at Interfaces, 2nd edition*, Dekker, New York, 1998, Chapter 2.
14. H. Ohshima and A. S. Dukhin, *J. Colloid Interface Sci.* 212 (1999) 449.
15. A. S. Dukhin, H. Ohshima, V. N. Shilov, and P. J. Goetz, *Langmuir* 15 (1999) 3445.
16. A. S. Dukhin, V. N. Shilov, H. Ohshima, and P. J. Goetz, *Langmuir* 15 (1999) 6692.
17. A. S. Dukhin, V. N. Shilov, H. Ohshima, and P. J. Goetz, *Langmuir* 16 (2000) 2615.
18. J. A. Ennis, A. A. Shugai, and S. L. Carnie, *J. Colloid Interface Sci.* 223 (2000) 21.
19. J. A. Ennis, A. A. Shugai, and S. L. Carnie, *J. Colloid Interface Sci.* 223 (2000) 37.
20. H. Ohshima, *Theory of Colloid and Interfacial Electric Phenomena*, Elsevier/Academic Press, 2006.
21. H. Ohshima, *J. Colloid Interface Sci.* 233 (2001) 142.
22. P. Debye and A. Bueche, *J. Chem. Phys.* 16 (1948) 573.
23. J. J. Hermans and H. Fujita, *Koninkl. Ned. Akad. Wetenschap. Proc. Ser. B.* 58 (1955) 182.
24. L. D. Landau and E. M. Lifshitz, *Fluid Mechanics*, Pergamon, Oxford, 1966, Sec. 24

26 Colloid Vibration Potential in a Suspension of Soft Particles

26.1 INTRODUCTION

When a sound wave is propagated in an electrolyte solution, the motion of cations and that of anions may differ from each other because of their different masses so that periodic excesses of either cations or anions should be produced at a given point in the solution, generating vibration potentials (Fig. 26.1). This potential is called ion vibration potential (IVP) [1–6]. A similar electroacoustic phenomenon occurs in a suspension of colloidal particles. Since colloidal particles are much larger and carry a much greater charge than electrolyte ions, the potential difference in the suspension is caused by the asymmetry of the electrical double layer around each particle rather than the relative motion of cations and anions (Fig. 26.2) [7–15]. It has been shown that the colloid vibration potential (CVP) of a suspension of colloidal particles is proportional to the dynamic electrophoretic mobility of the particles. Approximate expressions for the dynamic electrophoretic mobility of soft particles are given in Chapter 25. It must also be noted that in a colloidal suspension in an electrolyte solution, IVP and CVP are both generated simultaneously. Recently, we have developed a general acoustic theory for a suspension of spherical rigid particles, which accounts for both of CVP and IVP [16–18]. In the present paper, we apply this theory to the suspension of spherical soft particles with the help of an approximate expression for the dynamic electrophoretic mobility of soft particles [19].

26.2 COLLOID VIBRATION POTENTIAL AND ION VIBRATION POTENTIAL

Consider a dilute suspension of N_p spherical soft particles moving with a velocity $\boldsymbol{U}\exp(-i\omega t)$ in a symmetrical electrolyte solution of viscosity η and relative permittivity ε_r in an applied oscillating pressure gradient field $\nabla p\exp(-i\omega t)$ due to a sound wave propagating in the suspension, where ω is the angular frequency (2π times frequency) and t is time. We treat the case in which ω is low such that the dispersion of ε_r can be neglected. We assume that the particle core of radius a is coated

Biophysical Chemistry of Biointerfaces By Hiroyuki Ohshima

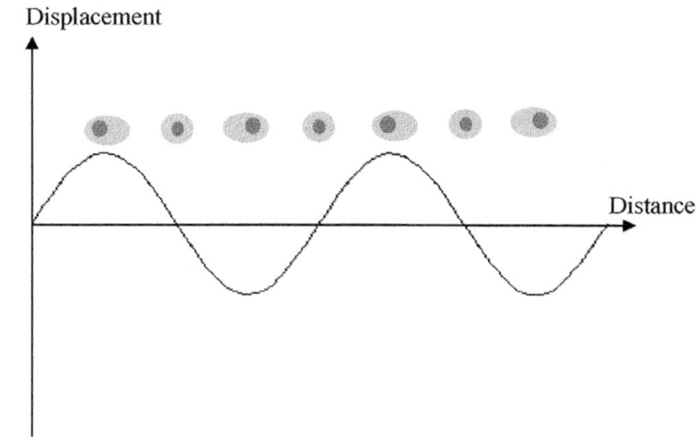

FIGURE 26.1 Ion vibration potential or periodic excesses of either cations or anions caused by the relative motion of cations and anions.

with an ion-penetrable layer of polyelectrolytes with a thickness d. The polyelectrolyte-coated particle has thus an inner radius a and an outer radius $b = a + d$ (Fig. 26.3). Let the valence and bulk concentration (number density) of the electrolyte be z and n, respectively. We also denote the mass, valence, and drag coefficient of cations by m_+, V_+, and λ_+, respectively, and those for anions by m_-, V_-, and λ_-. The drag coefficients λ_+ and λ_- are related to the corresponding limiting conductance of cations, Λ_+^0, and that of anions, Λ_-^0, by

$$\lambda_\pm = \frac{N_A e^2 z}{\Lambda_\pm^0} \tag{26.1}$$

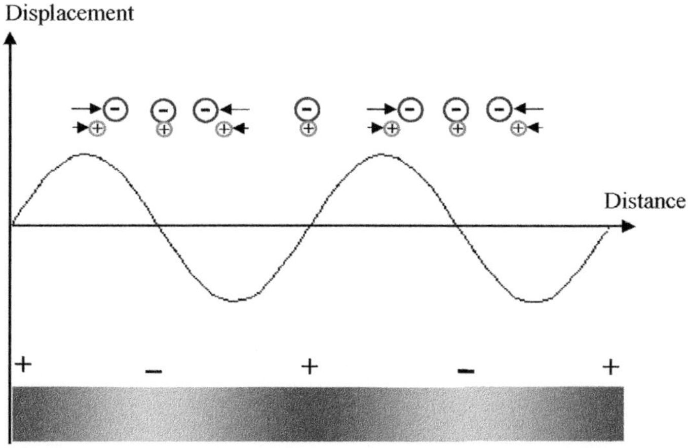

FIGURE 26.2 Colloid vibration potential caused by the asymmetry of the electrical double layer around the particles.

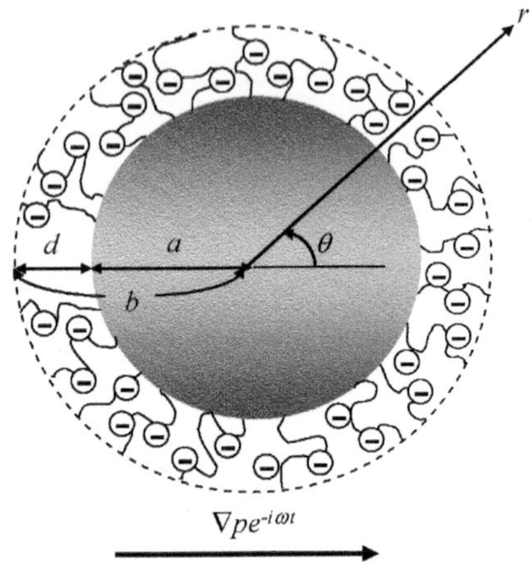

FIGURE 26.3 A spherical soft particle in an applied pressure gradient field. a = radius of the particle core. d = thickness of the polyelectrolyte layer covering the particle core. $b = a + d$.

where N_A is Avogadro's number. We adopt the model of Debye–Bueche (Chapter 21) that the polymer segments are regarded as resistance centers distributed in the polyelectrolyte layer, exerting a frictional force on the liquid flowing in the polyelectrolyte layer, where the frictional coefficient is ν. We also assume that fixed--charge groups of valence Z are distributed at a uniform density of N in the polyelectrolyte layer.

We have recently proposed a general acoustic theory for a dilute suspension of particles, which accounts for both of CVP and IVP [16–18]. Experimentally, the total vibration potential (TVP) between two points in the suspension, which is given by the sum of IVP and CVP, is observed. That is,

$$\mathrm{TVP} = \mathrm{IVP} + \mathrm{CVP} \tag{26.2}$$

where IVP and CVP are given by [19]

$$\mathrm{IVP} = \frac{zen}{\rho_o K^*}\left(\frac{m_+ - \rho_o V_+}{\lambda_+} - \frac{m_- - \rho_o V_-}{\lambda_-}\right)\Delta P \tag{26.3}$$

$$\mathrm{CVP} = \frac{[\phi_c(\rho_c - \rho_o) + \phi_s(\rho_s - \rho_o)]}{\rho_o K^*}\mu(\omega)\Delta P \tag{26.4}$$

with

$$K^* = K^{\infty} - i\omega\varepsilon_r\varepsilon_o \tag{26.5}$$

$$K^{\infty} = z^2 e^2 n \left(\frac{1}{\lambda_+} + \frac{1}{\lambda_-} \right) \tag{26.6}$$

$$\phi = \frac{(4\pi/3)a^3 N_p}{V} \tag{26.7}$$

$$\phi_c = \frac{V_c N_p}{V} \tag{26.8}$$

$$V_c = \frac{4}{3}\pi a^3 \tag{26.9}$$

$$\phi_s = \frac{V_s N_p}{V} \tag{26.10}$$

Here ∇p has been replaced with the pressure difference between the two points is ΔP, K^{∞}, and K^* are, respectively, the usual conductivity and the complex conductivity of the electrolyte solution in the absence of the particles, ϕ is the particle volume fraction, ϕ_c is the volume fraction of the particle core, V_c is the volume of the particle core, ϕ_s is the volume fraction of the polyelectrolyte segments, V_s is the total volume of the polyelectrolyte segments coating one particle, ρ_c and ρ_o, are respectively, the mass density of the particle core and that of the electrolyte solution, and ρ_s is the mass density of the polyelectrolyte segment, V is the suspension volume, and $\mu(\omega)$ is the dynamic electrophoretic mobility of the particles. Equation (26.4) is an Onsager relation between CVP and $\mu(\omega)$, which takes a similar form for an Onsager relation between sedimentation potential and static electrophoretic mobility (Chapter 24).

For a spherical soft particle, an approximate expression for $\mu(\omega)$ for the dynamic electrophoretic mobility is given by Eq. (25.45), which is a good approximation when the following conditions are satisfied:

$$|\beta|b \gg 1,\ \kappa b \gg 1,\ |\beta|d = |\beta|(b - a) \gg 1,\ \kappa d = \kappa(b - a) \gg 1,\ |\beta| \gg |\gamma|,\ \kappa \gg |\gamma| \tag{26.11}$$

which hold for most practical cases.

Some results of the calculation of the magnitude and phase of the dynamic mobility via Eqs. (26.2)–(26.4) are given in Figs 26.4 and 26.5, in which we have used the following values: $\rho_o = 1 \times 10^3$ kg/m^3, $\varepsilon_r = 78.5$ (water at 25°C), and $\rho_p = \rho_s = 1.1 \times 10^3$ kg/m^3 in an aqueous KCl solution at 25°C ($\Lambda_+^o = 73.5 \times 10^{-4}$ m$^2/\Omega$/mol and $m_+ = 39.1 \times 10^{-3}$ kg/mol for K^+, and $L_-^o = 76.3 \times 10^{-4}m^2/\Omega$/mol and $m_- = 35.5 \times 10^{-3}$ kg/mol for Cl^-). For the ionic volumes for K^+ and Cl^- ions, we have

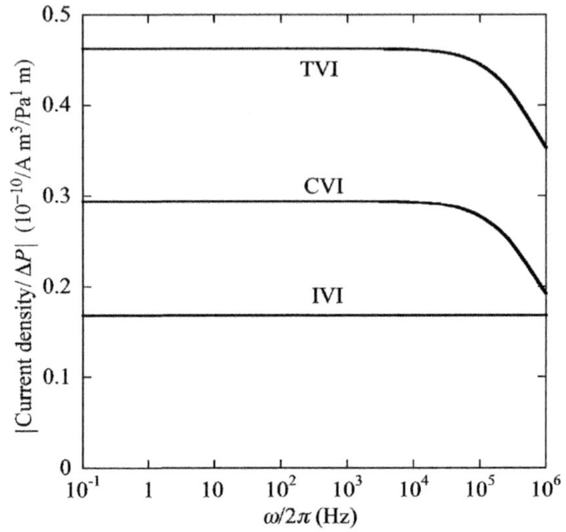

FIGURE 26.4 Magnitudes of CVI, IVI, and TVI divided by ΔP as a function of the frequency $\omega/2\pi$ of the applied pressure gradient field for a suspension of soft particles in a KCl solution of concentration $n = 0.01$ M. Calculated via Eqs. (26.2)–(26.4) as combined with Eq. (9.45) for $a = 1\ \mu$m, $d = 10$ nm, $N = 0.05$ M, $Z = 1$, $1/\lambda = 1$ nm, $\rho_o = 1 \times 10^3$ kg/m^3, $\rho_c = \rho_s = 1.1 \times 10^3$ kg/m^3, $\phi_s = 0$, $f_s = 0$, $\eta = 0.89$ mPa s, $T = 298$K. $\varepsilon_r = 78.5$, $\Lambda_+^o = 73.5 \times 10^{-4}$ m^2/Ω/mol, $V_+ = 3.7 \times 10^{-6}$ m^3/mol, $m_+ = 39.1 \times 10^{-3}$ kg/mol, $\Lambda_-^o = 76.3 \times 10^{-4}$ m^2/Ω/mol, $V_- = 22.8 \times 10^{-6}$ m^3/mol, and $m_- = 35.5 \times 10^{-3}$ kg/mol. From Ref. 19.

used the values of their partial molar volumes reported by Zana and Yeager [5], that is, $V_+ = 3.7 \times 10^{-6}$ m^3/mol (for K$^+$) and $V_- = 22.8 \times 10^{-6}$ m^3/mol (for Cl$^-$). The values of ϕ_s and f_s have approximately been set equal to zero. It is to be noted that the phase of CVI agrees with that of dynamic electrophoretic mobility $\mu(\omega)$ (Eq. (25.45)) and the phase of IVI is zero (in the present approximation).

Figures 26.4 and 26.5 show the dependence of the magnitude (Fig. 26.4) and phase (Fig. 26.5) of each of CVI, IVI, and TVI on the frequency ω of the pressure gradient field due to the applied sound wave for the case where $a = 1\ \mu$m for an aqueous KCl solution of concentration $n = 0.01$ M. It is seen that the ω-dependence is negligibly small $\omega/2\pi < 10^4$ Hz and becomes appreciable for $\omega/2\pi > 10^4$ Hz. That is, CVI is essentially equal to its static value at $\omega = 0$ for $\omega/2\pi < 10^4$ Hz and drops sharply to zero for $\omega/2\pi > 10^4$ Hz, while the phase of CVI is zero for $\omega/2\pi < 10^4$ Hz and increases sharply for the frequency range $\omega/2\pi > 10^4$ Hz. The magnitude of IVI, on the other hand, is constant independent of ω, while the phase of IVI is always zero, since in the present approximation IVI is a real quantity.

It can be shown that the CVI tends to a nonzero limiting value at very high electrolyte concentrations, as in the case of other electrokinetics of soft particles (Chapters 21 and 24). This is a characteristic of the electrokinetic behavior of soft particles, which comes from the second term of the right-hand side of Eq. (25.45).

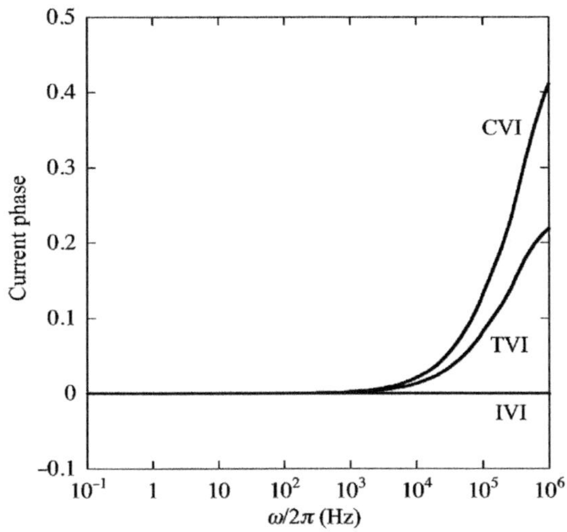

FIGURE 26.5 Phases of CVI, IVI, and TVI as a function of the frequency $\omega/2\pi$ of the applied pressure gradient field for a suspension of spherical soft particles in a KCl solution. Calculated via Eqs. (26.2)–(26.4) as combined with Eq. (9.45). Numerical values used are the same as in Fig. 26.4. From Ref. 19.

The limiting CVI value is obtained from Eq. (26.4) (as combined with Eq. (25.45)) with the result that

$$\mathrm{CVI} = \frac{[\phi_c(\rho_p - \rho_o) + \phi_s(\rho_s - \rho_o)]}{\rho_o}\left[\frac{1 - i\gamma b - (\gamma^2 b^2/3)(1 - a^3/b^3)}{1 - i\gamma b - (\gamma^2 b^2/3) - \Gamma}\right]\frac{ZeN}{\eta\beta^2}\Delta P \tag{26.12}$$

REFERENCES

1. P. Debye, *J. Chem. Phys.* 1 (1993) 13.
2. S. Oka, *Proc. Phys.-Math. Soc. Jpn.* 13 (1933) 413.
3. J. J. Hermans, *Phil. Mag.* 25 (1938) 426.
4. J. Bugosh, E. Yeager, and F. Hovorka, *J. Chem. Phys.* 15 (1947) 592.
5. R. Zana and E. Yeager, *Mod. Aspects Electrochem.* 14 (1982) 3.
6. A. S. Dukhin and P. J. Goetz, *Ultrasound for Characterizing Colloids, Particle Sizing, Zeta Potential, Rheology*, Elsevier, Amsterdam, 2002.
7. J. A. Enderby, *Proc. Phys. Soc.* 207A (1951) 329.
8. F. Booth and J. A. Enderby, *Proc. Phys. Soc.* 208A (1952) 321.
9. R. W. O'Brien, *J. Fluid Mech.* 190 (1988) 71.
10. H. Ohshima and A. S. Dukhin, *J. Colloid Interface Sci.* 212 (1999) 449.
11. A. S. Dukhin, H. Ohshima, V. N. Shilov, and P. J. Goetz, *Langmuir* 15 (1999) 3445.

12. A. S. Dukhin, V. N. Shilov, H. Ohshima, and P. J. Goetz, *Langmuir* 15 (1999) 6692.
13. A. S. Dukhin, V. N. Shilov, H. Ohshima, and P. J. Goetz, *Langmuir* 16 (2000) 2615.
14. V. N. Shilov, Y. B. Borkovskaja, and A. S. Dukhin, *J. Colloid Interface Sci.* 277 (2004) 347.
15. B. J. Marlow and R. L. Rowell, *J. Energy Fuels* 2 (1988) 125.
16. H. Ohshima, *Langmuir* 21 (2005) 12000.
17. H. Ohshima, *Colloids Surf. B; Biointerfaces* 56 (2006) 16.
18. H. Ohshima, *Theory of Colloid and Interfacial Electric Phenomena*, Academic Press, 2006.
19. H. Ohshima, *Colloids Surf. A, Physicochem. Eng. Aspects* 306 (2007) 83.

27 Effective Viscosity of a Suspension of Soft Particles

27.1 INTRODUCTION

The effective viscosity η_s of a dilute suspension of uncharged colloidal particles in a liquid is greater than the viscosity η of the original liquid. Einstein [1] derived the following expression for η_s:

$$\eta_s = \eta\left(1 + \frac{5}{2}\phi\right) \tag{27.1}$$

where ϕ is the particle volume fraction.

For concentrated suspensions, hydrodynamic interactions among particles must be considered. The hydrodynamic interactions between spherical particles can be taken into account by means of a cell model, which assumes that each sphere of radius a is surrounded by a virtual shell of outer radius b and the particle volume fraction ϕ is given by

$$\phi = (a/b)^3 \tag{27.2}$$

Simha [2] derived the following equation for the effective viscosity η_s of a concentrated suspension of uncharged particles of volume fraction ϕ:

$$\eta_s = \eta\left\{1 + \frac{5}{2}\phi S(\phi)\right\} \tag{27.3}$$

with

$$S(\phi) = \frac{4(1-\phi^{7/3})}{4(1+\phi^{10/3}) - 25\phi(1+\phi^{4/3}) + 42\phi^{5/3}} = \frac{4(1-\phi^{7/3})}{4(1-\phi^{5/3})^2 - 25\phi(1-\phi^{2/3})^2} \tag{27.4}$$

Biophysical Chemistry of Biointerfaces By Hiroyuki Ohshima

where $S(\phi)$ is called Simha's function. As $\phi \to 0$, $S(\phi)$ tends to 1 so that Eq. (27.3) reduces back to Einstein's equation (27.1).

If the liquid contains an electrolyte and the particles are charged, then the effective viscosity η_s is further increased. This phenomenon is called the primary electroviscous effect [3–18] and the effective viscosity η_s can be expressed as

$$\eta_s = \eta\left\{1 + \frac{5}{2}(1+p)\phi\right\} \tag{27.5}$$

where p is the primary electroviscous coefficient. The standard theory for this effect and the governing electrokinetic equations as well as their numerical solutions were given by Watterson and White [9]. Ohshima [17,18] derived an approximate analytic expression for p in a dilute suspension of spherical particles with arbitrary zeta potentials and large κa (where κ is the Debye–Hückel parameter and a is the particle radius). The standard theory of the primary electroviscous effect has been extended to cover the case of concentrated suspensions of charged spherical particles with thin electrical double layers under the condition of nonoverlapping electrical double layers of adjacent particles by Ruiz-Reina et al. [19] and Rubio-Hernández et al. [20]. Ohshima [21] has recently derived an approximate analytic expression for p in a moderately concentrated suspension of spherical particles with arbitrary zeta potential and large κa.

The effective viscosity of a suspension of particles of types other than rigid particles has also been theoretically investigated. Taylor [22] proposed a theory of the electroviscous effect in a suspension of uncharged liquid drops. This theory has been extended to the case of charged liquid drops by Ohshima [17]. Natraj and Chen [23] developed a theory for charged porous spheres, and Allison et al. [24] and Allison and Xin [25] discussed the case of polyelectrolyte-coated particles.

In this chapter, we first present a theory of the primary electroviscous effect in a dilute suspension of soft particles, that is, particles covered with an ion-penetrable surface layer of charged or uncharged polymers. We derive expressions for the effective viscosity and the primary electroviscous coefficient of a dilute suspension of soft particles [26]. We then derive an expression for the effective viscosity of uncharged porous spheres (i.e., spherical soft particles with no particle core) [27].

27.2 BASIC EQUATIONS

Consider a dilute suspension of spherical soft particles in an electrolyte solution of volume V in an applied shear field. We assume that the uncharged particle core of radius a is coated with an ion-penetrable layer of polyelectrolytes of thickness d. The polymer-coated particle has thus an inner radius a and an outer radius $b = a + d$. The origin of the spherical polar coordinate system (r, θ, φ) is held fixed at the center of one particle. We consider the case where dissociated groups of valence Z are distributed with a uniform density N in the polyelectrolyte layer so that the density of the fixed charges ρ_{fix} in the surface layer is given by $\rho_{\text{fix}} = ZeN$, where e is the elementary electric charge. Let the electrolyte be composed of M ionic mobile species of valence z_i, bulk concentration (number density) n_i^∞, and drag coefficient λ_i $(i = 1, 2, \ldots, M)$. The drag coefficient λ_i of the ith ionic species is related to the limiting conductance Λ_i^0 of that ionic species by Eq. (21.9). We adopt the model of Debye–Bueche, in which the polymer segments are regarded as resistance centers distributed in the polyelectrolyte layer, exerting frictional forces $-\gamma \boldsymbol{u}$ on the liquid flowing in the polymer layer, where u is the liquid velocity relative to the particle and γ is the frictional coefficient (Chapter 21). The main assumptions in our analysis are as follows: (i) The Reynolds numbers of the liquid flows outside and inside the polyelectrolyte layer are small enough to ignore inertial terms in the Navier–Stokes equations and the liquid can be regarded as incompressible. (ii) The applied shear field is weak so that electrical double layer around the particle is only slightly distorted. (iii) The slipping plane (at which the liquid velocity relative to the particle becomes zero) is located on the particle core surface. (iv) No electrolyte ions can penetrate the particle core. (v) The polyelectrolyte layer is permeable to mobile charged species. (vi) The relative permittivity ε_r takes the same value both inside and outside the polyelectrolyte layer.

Imagine that a linear symmetric shear field $\boldsymbol{u}^{(0)}(\boldsymbol{r})$ is applied to the system so that the velocity $\boldsymbol{u}(\boldsymbol{r})$ at position $\boldsymbol{r}$ outside the particle core is given by the sum of $\boldsymbol{u}^{(0)}(\boldsymbol{r})$ and a perturbation velocity. Thus, $\boldsymbol{u}(\boldsymbol{r})$ obeys the boundary condition

$$\boldsymbol{u}(\boldsymbol{r}) \to \boldsymbol{u}^{(0)}(\boldsymbol{r}) \quad \text{as} \quad r \to \infty \tag{27.6}$$

where $f(r)$ is a function of r ($r = |r|$). We can express $\boldsymbol{u}^{(0)}(\boldsymbol{r})$ as

$$\boldsymbol{u}^{(0)}(\boldsymbol{r}) = \boldsymbol{\alpha} \cdot \boldsymbol{r} \tag{27.7}$$

where $\boldsymbol{\alpha}$ is a symmetric traceless tensor so that $\boldsymbol{u}^{(0)}(\boldsymbol{r})$ becomes irrotational, namely,

$$\nabla \times \boldsymbol{u}^{(0)} = \boldsymbol{0} \tag{27.8}$$

Also we have from assumption (i) the following continuity equations:

$$\nabla \cdot \boldsymbol{u}^{(0)} = 0 \tag{27.9}$$

$$\nabla \cdot \boldsymbol{u} = 0 \tag{27.10}$$

From symmetry considerations $\boldsymbol{u}(\boldsymbol{r})$ takes the form

$$\boldsymbol{u}(\boldsymbol{r}) = \boldsymbol{u}^{(0)}(\boldsymbol{r}) + \nabla \times \nabla \times [(\boldsymbol{\alpha} \cdot \nabla) f(r)] \tag{27.11}$$

where $f(r)$ is a function of r and the second term on the right-hand side corresponds to the perturbation velocity. We introduce a function $h(r)$, defined by

$$h(r) = \frac{d}{dr}\left(\frac{1}{r}\frac{df}{dr}\right) \tag{27.12}$$

then Eq. (27.11) can be rewritten as

$$\begin{aligned} \boldsymbol{u}(\boldsymbol{r}) &= \boldsymbol{u}^{(0)}(\boldsymbol{r}) + \nabla \times \left[\frac{h(r)}{r}\boldsymbol{r} \times \boldsymbol{u}^{(0)}(\boldsymbol{r})\right] \\ &= \left(1 - \frac{dh}{dr} - \frac{2h}{r}\right)\boldsymbol{u}^{(0)} + \frac{1}{r^2}\left(\frac{dh}{dr} - \frac{h}{r}\right)\boldsymbol{r}(\boldsymbol{r}\cdot\boldsymbol{u}^{(0)}) \end{aligned} \tag{27.13}$$

where $h(r)$ is a function of r only and we have used Eq. (27.8) (irrotational flow), Eq. (27.9) (incompressible flow), and $(r\cdot\nabla)u^{(0)} = u^{(0)}$ (linear flow). The basic equations for the liquid flow $u(r)$ and the velocity $v_i(r)$ of the ith ionic species are similar to those for the electrophoresis problem. The boundary conditions for u and $v_i(r)$ at $r = a$ and $r = b$ are the same as those for the electrophoresis problem.

27.3 LINEARIZED EQUATIONS

For a weak shear field (assumption (ii)), as in the case of the electrophoresis problem (Chapter 21), one may write

$$n_i(\boldsymbol{r}) = n_i^{(0)}(r) + \delta n_i(\boldsymbol{r}) \tag{27.14}$$

$$\psi(\boldsymbol{r}) = \psi^{(0)}(r) + \delta\psi(\boldsymbol{r}) \tag{27.15}$$

$$\mu_i(\boldsymbol{r}) = \mu^{(0)} + \delta\mu_i(\boldsymbol{r}) \tag{27.16}$$

$$\rho_{\mathrm{el}}(\boldsymbol{r}) = \rho_{\mathrm{el}}^{(0)}(r) + \delta\rho_{\mathrm{el}}(\boldsymbol{r}) \tag{27.17}$$

where the quantities with superscript (0) refer to those at equilibrium (i.e., in the absence of the shear field), and $\delta n_i(r)$, $\delta\psi(r)$, $\delta\mu_i(r)$, and $\delta\rho_{\mathrm{el}}(r)$ are perturbation quantities.

Symmetry considerations permit us to write

$$\delta\mu_i(\boldsymbol{r}) = -z_i e \frac{\phi_i(r)}{r^2}\boldsymbol{r}\cdot\boldsymbol{u}^{(0)}(\boldsymbol{r}) \tag{27.18}$$

Here $\phi_i(r)$ is a function of r only. In terms of $\phi_i(r)$ and $h(r)$, electrokinetic equations can be written as

$$L(Lh - \lambda^2 h) = G(r), \quad a < r < b \tag{27.19}$$

$$L(Lh) = G(r), \quad r > b \tag{27.20}$$

$$L\left(\frac{\phi_i}{r}\right) = \frac{dy}{dr}\left[\frac{z_i}{r}\frac{d\phi_i}{dr} + \frac{\lambda_i}{e}\left(1 - \frac{3h}{r}\right)\right] \tag{27.21}$$

with

$$\lambda = \sqrt{\frac{\gamma}{\eta}} \tag{27.22}$$

$$G(r) = -\frac{2e}{\eta r^2}\frac{dy}{dr}\sum_{i=1}^{M} z_i^2 n_i^{\infty} \exp(-z_i y)\phi_i \tag{27.23}$$

$$L \equiv \frac{d}{dr}\frac{1}{r^4}\frac{d}{dr}r^4 = \frac{d^2}{dr^2} + \frac{4}{r}\frac{d}{dr} - \frac{4}{r^2} \tag{27.24}$$

The boundary conditions for $u(r)$ can be expressed in terms of $h(r)$ as follows.

$$h(a) = \frac{a}{3} \tag{27.25}$$

$$\left.\frac{dh}{dr}\right|_{r=a^+} = \frac{1}{3} \tag{27.26}$$

$$h(b^+) = h(b^-) \tag{27.27}$$

$$\left.\frac{dh}{dr}\right|_{r=b^+} = \left.\frac{dh}{dr}\right|_{r=b^-} \tag{27.28}$$

$$\left.\frac{d^2h}{dr^2}\right|_{r=b^+} = \left.\frac{d^2h}{dr^2}\right|_{r=b^-} \tag{27.29}$$

$$\left.\frac{d^3h}{dr^3}\right|_{r=b^+} = \left.\frac{d^3h}{dr^3}\right|_{r=b^-} + \lambda^2\left\{1 - \left.\frac{dh}{dr}\right|_{r=b^-} - \frac{2h(b^-)}{b}\right\} \tag{27.30}$$

$$h(r) \rightarrow 0 \quad \text{as } r \rightarrow \infty \tag{27.31}$$

$$h\frac{dh}{dr} \rightarrow 0 \quad \text{as } r \rightarrow \infty \tag{27.32}$$

Equation (27.30) results from the continuity condition of pressure $p(r)$. Similarly, the boundary conditions for $\delta\mu_i(r)$ can be expressed in terms of $\phi_i(r)$ as

$$\left.\frac{d\phi_i}{dr}\right|_{r=a^+} = 0 \tag{27.33}$$

$$\phi_i(r) \rightarrow 0 \quad \text{as } r \rightarrow \infty \tag{27.34}$$

27.4 ELECTROVISCOUS COEFFICIENT

Following the theory of Watterson and White [9], we consider the volume averaged stress tensor $\langle\boldsymbol{\sigma}\rangle$ defined by

$$\langle\boldsymbol{\sigma}\rangle = \frac{1}{V}\int_V \boldsymbol{\sigma}\, \mathrm{d}V \tag{27.35}$$

where $\langle\boldsymbol{\sigma}\rangle$ stands for the average of the stress tensor σ over the suspension volume V. We use the following identity:

$$\begin{aligned}\langle\boldsymbol{\sigma}\rangle &= - < p(\boldsymbol{r}) > \boldsymbol{I} + \eta(\langle\nabla\boldsymbol{u}(\boldsymbol{r})\rangle + \langle\nabla\boldsymbol{u}(\boldsymbol{r})^{\mathrm{T}}\rangle) \\ &\quad + \frac{1}{V}\int_V [\boldsymbol{\sigma} + p(\boldsymbol{r})\boldsymbol{I} - \eta(\nabla\boldsymbol{u}(\boldsymbol{r}) + \nabla\boldsymbol{u}(\boldsymbol{r})^{\mathrm{T}})]\mathrm{d}V \\ &= \eta(\langle\nabla\boldsymbol{u}(\boldsymbol{r})\rangle + \langle\nabla\boldsymbol{u}(\boldsymbol{r})^{\mathrm{T}}\rangle) + \frac{1}{V}\int_V [\boldsymbol{\sigma} - \eta(\nabla\boldsymbol{u}(\boldsymbol{r}) + \nabla\boldsymbol{u}(\boldsymbol{r})^{\mathrm{T}})]\mathrm{d}V\end{aligned} \tag{27.36}$$

where we have omitted the term in $\langle p(r)\rangle$, since the average pressure is necessarily zero. The integral in the second term on the right-hand side of Eq. (27.36) may be calculated for a single sphere as if the other were absent and then multiplied by the particle number N_{p} in the volume V, since the integrand vanishes beyond the double layer around the particle and the suspension is assumed to be dilute. We transform the volume integral into a surface integral over an infinitely distant surface S containing a single isolated sphere at its center. Then, Eq. (27.33) becomes

$$\langle \boldsymbol{\sigma} \rangle = \eta(\langle \nabla \boldsymbol{u} \rangle + \langle (\nabla \boldsymbol{u})^T \rangle) + \tfrac{N_p}{V} \int_S \frac{1}{r} \{ \boldsymbol{\sigma} \cdot \boldsymbol{rr} - (\boldsymbol{ur} + \boldsymbol{ru}) \} \, dS$$

$$= \eta(\langle \nabla \boldsymbol{u} \rangle + \langle (\nabla \boldsymbol{u})^{\mathrm{T}} \rangle) + \frac{3\phi}{4\pi b^3} \int_S \frac{1}{r} \{ \boldsymbol{\sigma} \cdot \boldsymbol{rr} - (\boldsymbol{ur} + \boldsymbol{ru}) \} dS \tag{27.37}$$

where

$$\phi = \frac{(4\pi/3) b^3 N_{\mathrm{p}}}{V} \tag{27.38}$$

is the volume fraction of soft spheres of outer radius b. It can be shown that the second term on the right-hand side of Eq. (27.36) becomes

$$\frac{3\phi}{4\pi b^3} \int_S \frac{1}{r} \{ \boldsymbol{\sigma} \cdot \boldsymbol{rr} - (\boldsymbol{ur} + \boldsymbol{ru}) \} dS = \frac{6\phi}{b^3} \eta D_3 \tag{27.39}$$

where D_3 is defined by

$$D_3 = \lim_{r \to \infty} [r^2 h(r)] \tag{27.40}$$

Also, to lowest order in the particle concentration,

$$\langle \nabla \boldsymbol{u} \rangle + \langle (\nabla \boldsymbol{u})^T \rangle = 2\boldsymbol{\alpha} \tag{27.41}$$

Equation (27.33) thus becomes

$$\langle \boldsymbol{\sigma} \rangle = 2\eta \boldsymbol{\alpha} + \frac{6\phi}{a^3} \eta D_3 \boldsymbol{\alpha} = 2\eta \left(1 + \phi \frac{3D_3}{a^3} \right) \boldsymbol{\alpha} \tag{27.42}$$

so that the effective viscosity η_s of a concentrated suspension of porous spheres is given by

$$\eta_{\mathrm{s}} = \eta \left(1 + \phi \frac{3D_3}{b^3} \right) \tag{27.43}$$

By evaluating D_3 from $h(r)$, we obtain

$$D_3 = \frac{5b^3 L_2}{6L_1} + \frac{b^5}{20L_1} \int_b^{\infty} \left\{ L_3 - \frac{5L_2}{3} \left(\frac{r}{b} \right)^2 + \frac{2L_1}{3} \left(\frac{r}{b} \right)^5 \right\} G(r) dr$$
$$- \frac{a^5}{3\lambda^2 b^2 L_1} \int_a^b \left\{ L_4 - \frac{3b^3 L_5(r)}{2a^3} \left(\frac{r}{a} \right)^2 - L_6 \left(\frac{r}{a} \right)^5 + \frac{15}{\lambda^2 ab} \left(\frac{r}{a} \right)^2 L_7(r) \right\} G(r) dr \tag{27.44}$$

where the definitions of L_1–L_4, $L_5(r)$, L_6, and $L_7(r)$ are given in Appendix 27A.

We first consider the case of a suspension of uncharged soft particles, that is, $ZeN = 0$. In this case $D_3 = 5b^3L_2/L_1$ and Eq. (27.40) becomes

$$\eta_s = \eta\left(1 + \frac{5}{2}\Omega\phi\right) \tag{27.45}$$

where

$$\Omega = \frac{L_2}{L_1} \tag{27.46}$$

The coefficient $\Omega(\lambda a, a/b)$ in Eq. (27.45) expresses the effects of the presence of the uncharged polymer layer upon η_s. As $\lambda \to \infty$ or $a \to b$, soft particles tend to hard particles and $\Omega(\lambda a, a/b) \to 1$ so that Eq. (27.45) tends to Eq. (27.1). For $\lambda \to 0$, on the other hand, the polymer layer vanishes so that a suspension of soft particles of outer radius b becomes a suspension of hard particles of radius a and the particle volume fraction changes from ϕ to ϕ', which is given by

$$\phi' = \frac{(4/3)\pi a^3}{V} = \frac{a^3}{b^3}\phi \tag{27.47}$$

Indeed, in this case $\Omega(\lambda a, a/b)$ tends to a^3/b^3 so that η_s becomes

$$\eta_s = \eta\left(1 + \frac{5}{2}\phi'\right) \tag{27.48}$$

Figure 27.1 shows $\Omega(\lambda a, a/b)$ as a function of λa for several values of a/b. The case of $a/b = 1$ corresponds to a suspension of hard particles with no surface structures. It is seen that for $\lambda a \geq 100$, $\Omega(\lambda a, a/b)$ is almost equal to 1 so that a suspension of uncharged soft particles with $\lambda a \geq 100$ behaves like a suspension of uncharged hard particles. For the special case of $a = 0$, Eq. (27.46) becomes

$$\Omega(\lambda b) = \frac{\lambda^2 b^2\{(3 + \lambda^2 b^2)\sinh(\lambda b) - 3\,\lambda b\cosh(\lambda b)\}}{(30 + 10\lambda^2 b^2 + \lambda^4 b^4)\sinh(\lambda b) - 30\,\lambda b\cosh(\lambda b)} \tag{27.49}$$

which is applicable for a suspension of uncharged porous spheres of radius b and coincides with the result of Natraj and Chen [23].

For a suspension of charged soft particles, we write

$$\eta_s = \eta\left\{1 + \frac{5}{2}(1 + p)\phi\Omega\right\} \tag{27.50}$$

The primary electroviscous coefficient p is thus given by

$$p = \frac{6D_3}{5b^3\Omega} - 1 = \frac{6D_3L_1}{5b^3L_2} - 1 \tag{27.51}$$

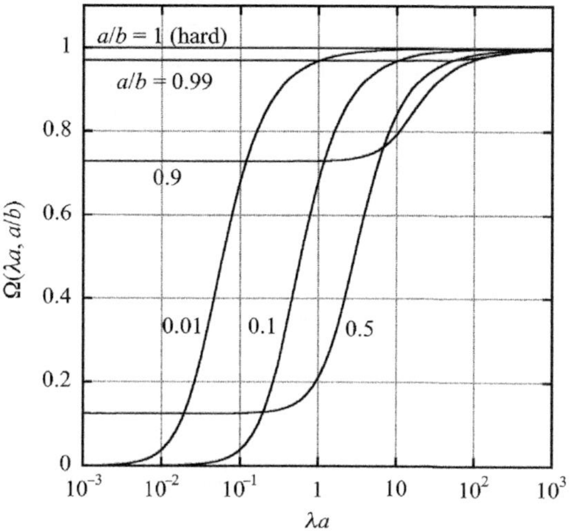

FIGURE 27.1 Function $\Omega(\lambda a, a/b)$ for a suspension of uncharged spherical soft particles as a function of λa for several values of a/b. The line for $a/b = 1$ corresponds to a suspension of spherical hard particles. From Ref. 26.

By substituting Eq. (27.44) into Eq. (27.51), we obtain

$$p = \frac{3b^2}{50L_2}\int_b^{\infty}\left\{L_3 - \frac{5L_2}{3}\left(\frac{r}{b}\right)^2 + \frac{2L_1}{3}\left(\frac{r}{b}\right)^5\right\}G(r)dr$$
$$-\frac{2}{5\lambda^2 L_2}\left(\frac{a}{b}\right)^5\int_a^b\left\{L_4 - \frac{3b^3L_5(r)}{2a^3}\left(\frac{r}{a}\right)^2 - L_6\left(\frac{r}{a}\right)^5 + \frac{15}{\lambda^2 ab}\left(\frac{r}{a}\right)^2 L_7(r)\right\}G(r)dr \tag{27.52}$$

27.5 APPROXIMATION FOR LOW FIXED-CHARGE DENSITIES

We derive an approximate formula for the effective viscosity η_s applicable for the case where the fixed-charge density ZeN is low. We use Eqs. (4.54)–(4.56) for the equilibrium potential $\psi^{(0)}(r)$. Then, from Eq. (27.52) we obtain

$$p = \left(\frac{6\varepsilon_r\varepsilon_o kT}{5\eta e^2}\right)\left(\frac{\sum_{i=1}^{M} z_i^2 n_i^{\infty}\lambda_i}{\sum_{i=1}^{M} z_i^2 n_i^{\infty}}\right)L(\kappa a, \lambda a, a/b)\left(\frac{e\psi_o}{kT}\right)^2 \tag{27.53}$$

with

$$L(\kappa a, \lambda a, a/b) = \frac{\kappa^2 b^2}{50L_2} \int_b^\infty \left\{ L_3 - \frac{5L_2}{3}\left(\frac{r}{b}\right)^2 + \frac{2L_1}{3}\left(\frac{r}{b}\right)^5 \right\} \left(\frac{1}{\psi_o} \frac{d\psi^{(0)}}{dr} \right) H(r)dr$$
$$- \frac{2\kappa^2}{15\lambda^2 L_2}\left(\frac{a}{b}\right)^5 \int_a^b \left\{ L_4 - \frac{3b^3 L_5(r)}{2a^3}\left(\frac{r}{a}\right)^2 - L_6\left(\frac{r}{a}\right)^5 \right.$$
$$\left. + \frac{15}{\lambda^2 ab}\left(\frac{r}{a}\right)^2 L_7(r) \right\} \left(\frac{1}{\psi_o} \frac{d\psi^{(0)}}{dr} \right) H(r)dr \tag{27.54}$$

where $H(r)$ is defined by

$$H(r) = \left(1 + \frac{2a^5}{3r^5}\right) \int_a^\infty \left(\frac{1}{\psi_o} \frac{d\psi^{(0)}}{dx} \right) \left(1 - \frac{3h}{x}\right) dx$$
$$- \int_a^r \left(\frac{1}{\psi_o} \frac{d\psi^{(0)}}{dx} \right) \left(1 - \frac{x^5}{r^5}\right) \left(1 - \frac{3h}{x}\right) dx \tag{27.55}$$

One can calculate the primary electroviscous coefficient p for a suspension of soft particles with low ZeN via Eq. (27.53).

For a suspension of charged spherical soft particles carrying low ZeN, the electroviscous coefficient p can be calculated via Eq. (27.53) as combined with Eq. (27.54) for $L(\kappa a, \lambda a, a/b)$. Figure 27.2 show some examples of the calculation of $L(\kappa a, \lambda a, a/b)$ as a function of κa obtained via Eq. (27.54) at $\lambda a = 50$. This figure shows that as κa increases, $L(\kappa a, \lambda a, a/b)$ first decreases, reaching a minimum around ca. $\kappa a = 10$, and then increases. Note that the primary electroviscous coefficient p for a suspension of hard particles (shown as a curve with $a/b = 1$ in Fig. 27.2) does not exhibit a minimum.

A simple approximate analytic expressions for $L(\kappa a, \lambda a, a/b)$ without involving numerical integrations can be derived for the case where

$$\kappa a \gg 1,\ \lambda a \gg 1,\ \kappa d = \kappa(b - a) \gg 1, \text{ and } \lambda d = \lambda(b - a) \gg 1 \tag{27.56}$$

which holds for most practical cases. In this case Eq. (27.54) becomes

$$L(\kappa a, \lambda a, a/b) = \left\{ \frac{3}{\kappa^2 b^2} + \frac{\kappa}{10(\kappa + \lambda)} \right\} \left[1 + \frac{2a^5}{3b^5} + \frac{\kappa^2}{\lambda^2} \left\{ \frac{2}{3}\left(1 - \frac{a^5}{b^5}\right) \right. \right.$$
$$\left. \left. + \left(1 + \frac{2a^5}{3b^5}\right)\left(\frac{\kappa}{\kappa + \lambda}\right) \right\} \right] \tag{27.57}$$

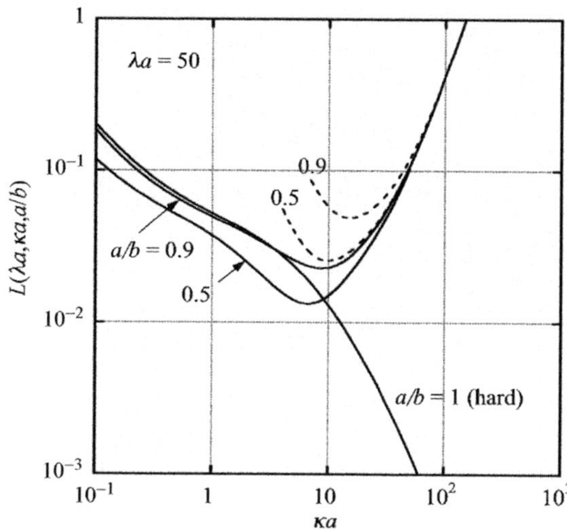

FIGURE 27.2 Function $L(\kappa a, \lambda a, a/b)$ for a suspension of soft particles as a function of κa for $a/b = 0.5$ and 0.9 at $\lambda a = 50$. The solid lines represent the results calculated via Eq. (27.54) and the dotted lines approximate results calculated via Eq. (27.57). The curve for $a/b = 1$ corresponds to a suspension of spherical hard particles. From Ref. 26.

Consider several limiting cases for Eq. (27.57).

1. For the case of particles covered with a very thin polyelectrolyte layer ($a \approx b$), Eq. (27.54) becomes

$$L(\kappa a, \lambda a, a/b) = \left\{\frac{5}{\kappa^2 b^2} + \frac{\kappa}{6(\kappa+\lambda)}\right\}\left[1 + \frac{\kappa^2}{\lambda^2}\left(\frac{\kappa}{\kappa+\lambda}\right)\right] \tag{27.58}$$

2. In the opposite limiting case of a very thick polyelectrolyte layer ($a \ll b$). Eq. (27.57) tends to

$$L(\kappa a, \lambda a, a/b) = \left\{\frac{3}{\kappa^2 b^2} + \frac{\kappa}{10(\kappa+\lambda)}\right\}\left[1 + \frac{\kappa^2}{\lambda^2}\left\{\frac{2}{3} + \left(\frac{\kappa}{\kappa+\lambda}\right)\right\}\right] \tag{27.59}$$

   which is also applicable for a suspension of spherical polyelectrolyte with no particle core ($a = 0$). The above two limiting cases, $L(\kappa a, \lambda a, a/b)$ does not depend on the value of the inner radius a.

3. In the limit of $\lambda \to \infty$ ($\lambda \gg \kappa$), Eq. (27.57) tends to

$$L(\kappa a, \lambda a, a/b) = \frac{3}{\kappa^2 b^2}\left(1 + \frac{2a^5}{3b^5}\right) \tag{27.60}$$

When $a = b$, Eq. (27.57) further becomes

$$L(\kappa a, \lambda a, a/b) = \frac{5}{\kappa^2 b^2} \tag{27.61}$$

which agrees with the result for a suspension of hard spheres of radius b[16].

4. For $\kappa \to \infty$ ($\kappa \gg \lambda$)

$$L(\kappa a, \lambda a, a/b) = \frac{\kappa^2}{6\lambda^2} \tag{27.62}$$

Approximate results calculated via Eq. (27.57) are also shown as dotted lines in Fig. 27.2. It is seen that $\kappa a \geq 100$, the agreement with the exact result is excellent. The presence of a minimum of $L(\kappa a, \lambda a, a/b)$ as a function of κa can be explained qualitatively with the help of Eq. (27.57) as follows. That is, $L(\kappa a, \lambda a, a/b)$ is proportional to $1/\kappa^2$ at small κa and to κ^2 at large κa, leading to the presence of a minimum of $L(\kappa a, \lambda a, a/b)$. As is seen in Fig. 27.3, for the case of a suspension of hard particles, the function $L(\kappa a)$ decreases as κa increases, exhibiting no minimum. This is the most remarkable difference between the effective viscosity of a suspension of soft particles and that for hard particles. It is to be noted that although $L(\kappa a, \lambda a, a/b)$ increases with κa at large κa, the primary electroviscous coefficient p itself decreases with increasing electrolyte concentration. The reason is that the

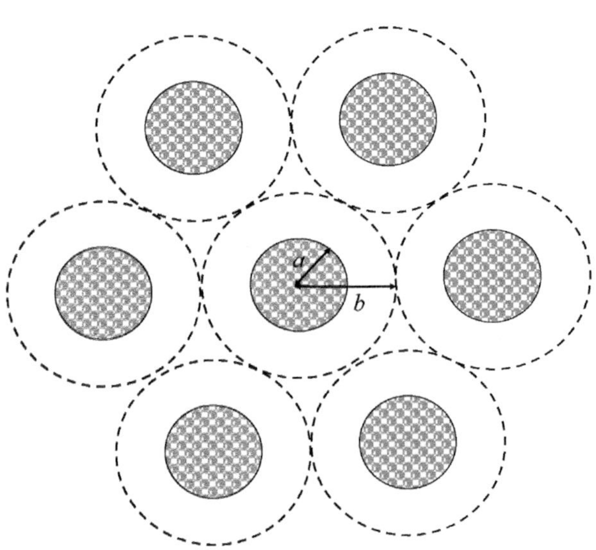

FIGURE 27.3 A cell model for a concentrated suspension of porous spheres. Each porous sphere of radius a is surrounded by a virtual shell of outer radius b. The particle volume fraction ϕ is given by $\phi = (a/b)^3$.

surface potential ψ_o, which becomes for large κa (Eq. (4.29))

$$\psi_o = \frac{ZeN}{2\varepsilon_r \varepsilon_o \kappa^2} \tag{27.63}$$

is proportional to $1/\kappa^2$ so that for large κa, Eq. (27.54) becomes

$$p = \frac{(ZN)^2}{20\eta\lambda^2} \frac{\left(\sum_{i=1}^{M} z_i^2 n_i^\infty \lambda_i \right)}{\left(\sum_{i=1}^{M} z_i^2 n_i^\infty \right)^2} \tag{27.64}$$

Equation (27.64) shows that the electroviscous coefficient p for large κa decreases with increasing κa.

27.6 EFFECTIVE VISCOSITY OF A CONCENTRATED SUSPENSION OF UNCHARGED POROUS SPHERES

Consider a concentrated suspension of porous spheres of radius a in a liquid of viscosity η [27]. We adopt a cell model that assumes that each sphere of radius a is surrounded by a virtual shell of outer radius b and the particle volume fraction ϕ is given by Eq. (27.2) (Fig. 27.3). The origin of the spherical polar coordinate system (r, θ, φ) is held fixed at the center of one sphere. According to Simha [2], we the following additional boundary condition to be satisfied at the cell surface $r = b$:

$$\boldsymbol{u}(\boldsymbol{r}) = \boldsymbol{u}^{(0)}(\boldsymbol{r}) \quad \text{at } r = b \tag{27.65}$$

which means that the perturbation velocity field is zero at the outer cell surface. Equation (27.65) can be rewritten in terms of $h(r)$ as

$$h(b) = 0 \tag{27.66}$$

$$\left. \frac{dh}{dr} \right|_{r=b^-} = 0 \tag{27.67}$$

The effective viscosity η_s of the suspension can be expressed as [27]

$$\eta_s = \eta \left(1 + \frac{5}{2} \phi \Omega(\lambda a, \phi) \right) \tag{27.68}$$

with

$$\Omega(\lambda a, \phi) = \frac{6}{5a^3} D_3 \tag{27.69}$$

where the coefficient D_3 can be given by

$$D_3 = \frac{b^4}{10} \left(\left. \frac{d^2 h}{dr^2} \right|_{r=b^-} + \frac{b}{3} \left. \frac{d^3 h}{dr^3} \right|_{r=b^-} \right) \tag{27.70}$$

The value of D_3 is found to be

$$D_3 = \frac{5a^3 M_1}{6M_2} \tag{27.71}$$

where

$$\begin{aligned} M_1 = & \left\{ 1 - \phi^{7/3} + \frac{3}{(\lambda a)^2} - \frac{45\phi^{7/3}}{(\lambda a)^2} - \frac{105\phi^{7/3}}{(\lambda a)^4} \right\} \sinh(\lambda a) \\ & + \left\{ -\frac{3}{\lambda a} + \frac{10\phi^{7/3}}{\lambda a} + \frac{105\phi^{7/3}}{(\lambda a)^3} \right\} \cosh(\lambda a) \end{aligned} \tag{27.72}$$

$$\begin{aligned} M_2 = & \left\{ 1 - \frac{25\phi}{4} + \frac{21\phi^{5/3}}{2} - \frac{25\phi^{7/3}}{4} + \phi^{10/3} + \frac{10}{(\lambda a)^2} - \frac{75\phi}{4(\lambda a)^2} + \frac{315\phi^{5/3}}{2(\lambda a)^2} - \frac{775\phi^{7/3}}{4(\lambda a)^2} \right. \\ & \left. + \frac{45\phi^{10/3}}{(\lambda a)^2} + \frac{30}{(\lambda a)^4} + \frac{315\phi^{5/3}}{(\lambda a)^4} - \frac{1500\phi^{7/3}}{(\lambda a)^4} + \frac{105\phi^{10/3}}{(\lambda a)^4} - \frac{2625\phi^{7/3}}{(\lambda a)^6} \right\} \sinh(\lambda a) \\ & + \left\{ \frac{75\phi}{4(\lambda a)} - \frac{105\phi^{5/3}}{2(\lambda a)} + \frac{175\phi^{7/3}}{4(\lambda a)} - \frac{10\phi^{10/3}}{\lambda a} - \frac{30}{(\lambda a)^3} - \frac{315\phi^{5/3}}{(\lambda a)^3} \right. \\ & \left. + \frac{625\phi^{7/3}}{(\lambda a)^3} - \frac{105\phi^{10/3}}{(\lambda a)^3} + \frac{2625\phi^{7/3}}{(\lambda a)^5} \right\} \cosh(\lambda a) \end{aligned} \tag{27.73}$$

Then we obtain

$$\Omega(\lambda a, \phi) = \frac{M_1}{M_2} \tag{27.74}$$

We have shown that the effective viscosity η_s of a concentrated suspension of uncharged porous spheres of radius a and volume fraction ϕ in a liquid is given by Eq. (27.68) (as combined with Eq. (27.74)). The coefficient $\Omega(\lambda a, \phi)$ expresses how the presence of porous spheres affects the viscosity η_s of the original liquid.

In the limit of a dilute suspension of porous spheres $\phi \to 0$, $\Omega(\lambda a, \phi)$ tends to

$$\Omega(\lambda a, 0) = \frac{\lambda^2 a^2 \{(3 + \lambda^2 a^2)\sinh(\lambda a) - 3\lambda a \cosh(\lambda a)\}}{(30 + 10\lambda^2 a^2 + \lambda^4 a^4)\sinh(\lambda a) - 30\lambda a \cosh(\lambda a)} \qquad (27.75)$$

which agrees with the result of Natraj and Chen [23] for the effective viscosity of a dilute suspension of uncharged porous spheres of radius a. In the limit of $\lambda a \to \infty$, on the other hand, $\Omega(\lambda a, \phi)$ becomes Simha's function $S(\phi)$ given by Eq. (27.4).

Figure 27.4 shows $\Omega(\lambda a, \phi)$ as a function of λa for several values of the particle volume fraction ϕ. We see that $\Omega(\lambda a, \phi) \approx 0$ for $\lambda a \leq 0.1$. That is, the viscosity η of the original liquid is not altered by the presence of porous spheres. The curve for $\phi = 0$ corresponds to a dilute suspension of porous spheres, in which case $\Omega(\lambda a, \phi)$ agrees with $\Omega(\lambda a, 0)$ derived by Natraj and Chen (Eq. (27.75)) [23]. It is also seen that for $\lambda a \geq 10^3$, $\Omega(\lambda a, \phi)$ is almost equal to $S(\phi)$ so that a suspension of uncharged porous spheres with $\lambda a \geq 100$ behaves like a suspension of uncharged hard spheres.

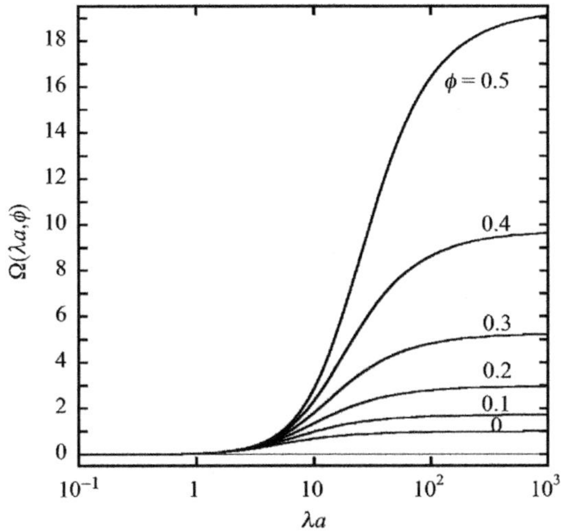

FIGURE 27.4 Function $\Omega(\lambda a, \phi)$ for a suspension of uncharged porous spheres of radius a as a function of λa for several values of the particle volume fraction ϕ. The curve for $\phi 0$ corresponds to a dilute suspension of porous spheres. As $\lambda a \to \infty$, $\Omega(\lambda a, \phi)$ becomes Simha's function $S(\phi)$ given by Eq. (27.4). From Ref. 27.

APPENDIX 27A

Expressions for L_1–L_7 are given below.

$$L_1 = \left(1 + \frac{2a^5}{3b^5} + \frac{10}{\lambda^2 b^2} + \frac{10a^3}{\lambda^2 b^5} - \frac{30}{\lambda^4 ab^3} + \frac{30}{\lambda^4 b^4}\right)\cosh[\lambda(b-a)]$$
$$+\left(1 + \frac{4a^5}{b^5} + \frac{10}{\lambda^2 b^2} + \frac{10a^3}{\lambda^2 b^5} - \frac{30}{\lambda^2 ab^3} - \frac{30a}{\lambda^2 b^3} + \frac{30}{\lambda^4 b^4}\right)\frac{\sinh[\lambda(b-a)]}{\lambda a} - \frac{20a^2}{\lambda^2 b^4} \tag{27A.1}$$

$$L_2 = \left(1 + \frac{2a^5}{3b^5} - \frac{3}{\lambda^2 ab} + \frac{3}{\lambda^2 b^2} + \frac{10a^3}{\lambda^2 b^5} - \frac{12a^4}{\lambda^2 b^6} + \frac{2a^5}{\lambda^2 b^7} - \frac{30a^2}{\lambda^4 b^6} + \frac{30a^3}{\lambda^4 b^7}\right)\cosh[\lambda(b-a)]$$
$$+\left(1 - \frac{3a}{b} + \frac{4a^5}{b^5} - \frac{2a^6}{b^6} + \frac{3}{\lambda^2 b^2} + \frac{10a^3}{\lambda^2 b^5} - \frac{30a^4}{\lambda^2 b^6} + \frac{12a^5}{\lambda^2 b^7} + \frac{30a^3}{\lambda^4 b^7}\right)\frac{\sinh[\lambda(b-a)]}{\lambda a} \tag{27A.2}$$

$$L_3 = \left(1 + \frac{2a^5}{3b^5} + \frac{10a^5}{3\lambda^2 b^7} + \frac{15}{\lambda^2 b^2} - \frac{20a^4}{\lambda^2 b^6} + \frac{10a^3}{\lambda^2 b^5} - \frac{5}{\lambda^2 ab}\right.$$
$$\left. - \frac{30}{\lambda^4 ab^3} + \frac{30}{\lambda^4 b^4} - \frac{50a^2}{\lambda^4 b^6} + \frac{50a^3}{\lambda^4 b^7}\right)\cosh[\lambda(b-a)]$$
$$+\left(1 - \frac{5a}{b} + \frac{4a^5}{b^5} - \frac{10a^6}{3b^6} + \frac{15}{\lambda^2 b^2} - \frac{30a}{\lambda^2 b^3} + \frac{10a^3}{\lambda^2 b^5}\right.$$
$$\left. - \frac{50a^4}{\lambda^2 b^6} + \frac{20a^5}{\lambda^2 b^7} + \frac{30}{\lambda^4 b^4} + \frac{50a^3}{\lambda^4 b^7}\right)\frac{\sinh[\lambda(b-a)]}{\lambda a} - \frac{10a^2}{3\lambda^2 b^4} \tag{27A.3}$$

$$L_4 = \left(1 + \frac{15}{\lambda^2 a^2} + \frac{3}{\lambda^2 b^2} - \frac{18}{\lambda^2 ab} + \frac{45}{\lambda^4 a^2 b^2} - \frac{45}{\lambda^4 a^3 b}\right)\cosh[\lambda(b-a)]$$
$$+6\left(1 - \frac{a}{2b} + \frac{5}{2\lambda^2 a^2} - \frac{15}{2\lambda^2 ab} + \frac{3}{\lambda^2 b^2} + \frac{15}{2\lambda^4 a^2 b^2}\right)\frac{\sinh[\lambda(b-a)]}{\lambda a} + \frac{3b^3}{2a^3} \tag{27A.4}$$

$$L_5(r)=\left(1+\frac{2a^5}{3b^5}+\frac{3}{\lambda^2r^2}-\frac{3}{\lambda^2ra}+\frac{10a^3}{\lambda^2b^5}+\frac{2a^5}{\lambda^2r^2b^5}-\frac{12a^4}{\lambda^2rb^5}-\frac{30a^2}{\lambda^4rb^5}+\frac{30a^3}{\lambda^4r^2b^5}\right)\cosh[\lambda(r-a)]$$
$$+\left(1+\frac{4a^5}{b^5}-\frac{2a^6}{rb^5}-\frac{3a}{r}+\frac{10a^3}{\lambda^2b^5}+\frac{3}{\lambda^2r^2}+\frac{12a^5}{\lambda^2r^2b^5}-\frac{30a^4}{\lambda^2rb^5}+\frac{30a^3}{\lambda^4r^2b^5}\right)\frac{\sinh[\lambda(r-a)]}{\lambda a} \quad (27A.5)$$

$$L_6=\left(1-\frac{3}{\lambda^2ab}+\frac{3}{\lambda^2b^2}\right)\cosh[\lambda(b-a)]+\left(1-\frac{3a}{b}+\frac{3}{\lambda^2b^2}\right)\frac{\sinh[\lambda(b-a)]}{\lambda a}-\frac{a^2}{b^2} \quad (27A.6)$$

$$L_7(r)=\left(1-\frac{b}{r}-\frac{3}{\lambda^2rb}+\frac{3}{\lambda^2r^2}\right)\cosh[\lambda(r-b)]$$
$$+\left(1+\frac{\lambda^2b^2}{3}+\frac{3}{\lambda^2r^2}+\frac{b^2}{r^2}-\frac{3b}{r}\right)\frac{\sinh[\lambda(r-b)]}{\lambda b} \quad (27A.7)$$

REFERENCES

1. A. Einstein, *Ann. Phys.* 19 (1906) 289.
2. R. J. Simha, *Appl. Phys.* 23 (1952) 1020.
3. M. Smoluchowski, *Kolloid Z.* 18 (1916) 194.
4. W. Krasny-Ergen, *Kolloidzeitschrift* 74 (1936) 172.
5. F. Booth, *Proc. R. Soc. A* 203 (1950) 533.
6. W. B. Russel, *J. Fluid Mech.* 85 (1978) 673.
7. D. A. Lever, *J. Fluid Mech.* 92 (1979) 421.
8. D. J. Sherwood, *J. Fluid Mech.* 101 (1980) 609.
9. I. G. Watterson and L. R. White, *J. Chem. Soc., Faraday Trans.* 2 77 (1981) 1115.
10. E. J. Hinch and J. D. Sherwood, *J. Fluid Mech.* 132 (1983) 337.
11. A. S. Duhkin and T. G. M. van de Ven, *J. Colloid Interface Sci.* 158 (1993) 85.
12. F. J. Rubio-Hernández, E. Ruiz-Reina, and A. I. Gómez-Merino, *J. Colloid Interface Sci.* 206 (1998) 334.
13. F. J. Rubio-Hernández, E. Ruiz-Reina, and A. I. Gómez-Merino, *J. Colloid Interface Sci.* 226 (2000) 180.
14. J. D. Sherwood, F. J. Rubio-Hernández, and E. Ruiz-Reina, *J. Colloid Interface Sci.* 228 (2000) 7.
15. F. J. Rubio-Hernández, E. Ruiz-Reina, and A. I. Gómez-Merino, *Colloids Surf. A: Physicochem. Eng. Aspects* 192 (2001) 349.
16. F. J. Rubio-Hernández, E. Ruiz-Reina, A. I. Gómez-Merino, and J. D. Sherwood, *Rheol. Acta* 40 (2001) 230.

17. H. Ohshima, *Langmuir* 22 (2006) 2863.
18. H. Ohshima, *Theory of Colloid and Interfacial Electric Phenomena*, Academic Press, Amsterdam, 2006.
19. E. Ruiz-Reina, F. Carrique, F. J. Rubio-Hernández, A. I. Gómez-Merino, and P. García-Sánchez, *J. Phys. Chem. B* 107 (2003) 9528.
20. F. J. Rubio-Hernández, F. Carrique, and E. Ruiz-Reina, *Adv. Colloid Interface Sci.* 107 (2004) 51.
21. H. Ohshima, *Langmuir* 23 (2007) 12061.
22. G. I. Taylor, *Proc. R. Soc. London, Ser. A* 138 (1932) 41.
23. V. Natraj and S. B. Chen, *J. Colloid Interface Sci.* 251 (2002) 200.
24. S. Allison, S. Wall, and M. Rasmusson, *J. Colloid Interface Sci.* 263 (2003) 84.
25. S. Allison and Y. Xin, *J. Colloid. Interface. Sci.* 299 (2006) 977.
26. H. Ohshima, *Langmuir* 24 (2008) 6453.
27. H. Ohshima, *Colloids Surf. A: Physicochem. Eng. Aspects* 347 (2009) 33.

PART IV
Other Topics

28 Membrane Potential and Donnan Potential

28.1 INTRODUCTION

When two electrolyte solutions at different concentrations are separated by an ion--permeable membrane, a potential difference is generally established between the two solutions. This potential difference, known as membrane potential, plays an important role in electrochemical phenomena observed in various biomembrane systems. In the stationary state, the membrane potential arises from both the diffusion potential [1,2] and the membrane boundary potential [3–6]. To calculate the membrane potential, one must simultaneously solve the Nernst–Planck equation and the Poisson equation. Analytic formulas for the membrane potential can be derived only if the electric field within the membrane is assumed to be constant [1,2]. In this chapter, we remove this constant field assumption and numerically solve the above-mentioned nonlinear equations to calculate the membrane potential [7].

28.2 MEMBRANE POTENTIAL AND DONNAN POTENTIAL

Imagine the stationary flow of electrolyte ions across a membrane separating solutions I and II that contain uni–uni symmetrical electrolyte at different concentrations n_{I} and n_{II}, respectively. We employ a membrane model [7] in which each side of the membrane core of thickness d_{c} is covered by a surface charge layer of thickness d_{s}. In this layer, membrane-fixed negative charges are assumed to be distributed at a uniform density N. We take the x-axis perpendicular to the membrane with its origin at the boundary between the left surface charge layer and the membrane core (Fig. 28.1). The potential in the bulk phase of solution II is set equal to zero.

The potential in bulk solution I is thus equal to the membrane potential, which we denote by E_{m}. First, consider the regions $x < 0$ and $x > d_{\mathrm{c}}$. We assume that the distribution of electrolyte ions in the surface charge layer and in solutions I and II is scarcely affected by ionic flows so that it is in practice at thermodynamic equili-

Biophysical Chemistry of Biointerfaces By Hiroyuki Ohshima

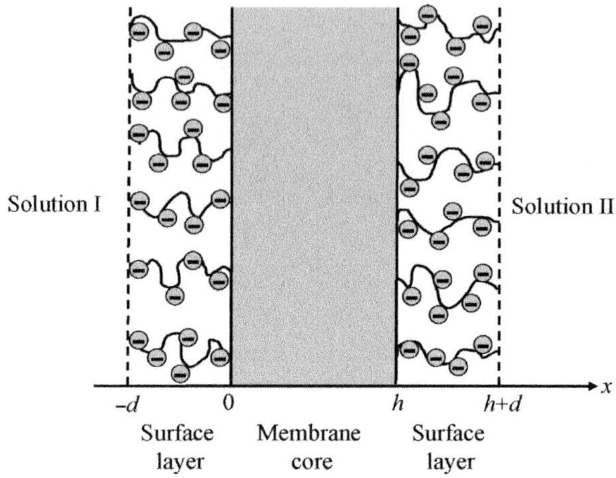

FIGURE 28.1 Model for a charged membrane. The membrane core (of thickness d_c) is covered by a surface charge layer (of thickness d_s). The membrane-fixed charges are assumed to be negative and are encircled.

brium. Accordingly, the potential $\psi(x)$ in these regions is given by the Poisson–Boltzmann equations, namely,

$$\frac{d^2\psi}{dx^2} = \frac{2en_{\rm I}}{\varepsilon_r\varepsilon_o}\sinh\left[\frac{e(\psi - E_{\rm m})}{kT}\right], \quad x < -d_s \tag{28.1}$$

$$\frac{d^2\psi}{dx^2} = \frac{2en_{\rm I}}{\varepsilon_r\varepsilon_o}\sinh\left[\frac{e(\psi - E_{\rm m})}{kT}\right] + \frac{eN}{\varepsilon_r\varepsilon_o}, \quad -d_s < x < 0 \tag{28.2}$$

$$\frac{d^2\psi}{dx^2} = \frac{2en_{\rm II}}{\varepsilon_r\varepsilon_o}\sinh\left(\frac{e\psi}{kT}\right) + \frac{eN}{\varepsilon_r\varepsilon_o}, \quad d_c + d_s < x < d_c \tag{28.3}$$

$$\frac{d^2\psi}{dx^2} = \frac{2en_{\rm II}}{\varepsilon_r\varepsilon_o}\sinh\left(\frac{e\psi}{kT}\right), \quad x > d_c + d_s \tag{28.4}$$

where ε_r is the relative permittivity of solutions I and II. Next, consider the region $0 < x < d_c$ (i.e., the membrane core). We regard this region as a different phase from the surrounding solution phase and introduce the ionic partition coefficients between the membrane core and the solution phase. We denote by b_+ and b_-, respectively, the partition coefficients for cations and anions. The electric potential $\psi(x)$ in this region is given by the following Poisson equation:

$$\frac{d^2\psi}{dx^2} = \frac{e}{\varepsilon_r'\varepsilon_o}\{n_+(x) - n_-(x)\}, \quad 0 < x < d_c \tag{28.5}$$

where $n_+(x)$ and $n_-(x)$ are the respective concentrations of cations and anions at position x ($0 < x < d_c$) and ε_r' is the relative permittivity of the membrane core. The flow of cations, denoted by J_+, can be expressed as the product of the number density $n_+(x)$ and velocity v_+ of cations. Similarly, the flow J_- of anions can be expressed as the product of the number density $n_-(x)$ and velocity v_- of anions. That is,

$$J_\pm = n_\pm v_\pm \tag{28.6}$$

The ionic velocities $v_\pm$ are given by

$$v_\pm = -\frac{1}{\lambda_\pm}\frac{d\mu_\pm}{dx} \tag{28.7}$$

Here, $\lambda_\pm$ are the ionic drag coefficients of cations and anions, respectively, and $\mu_\pm$ are the electrochemical potentials of cations and anions, respectively, which are

$$\mu_\pm = \mu_\pm^{\mathrm{o}} + kT\ln n_\pm(x) \pm ze\psi(x) \tag{28.8}$$

From Eqs. (28.6)–(28.8), we obtain

$$J_\pm = -D_\pm\left(\frac{dn_\pm}{dx} \pm \frac{n_\pm e}{kT}\frac{d\psi}{dx}\right) \tag{28.9}$$

with

$$D_\pm = \frac{kT}{\lambda_\pm} \tag{28.10}$$

where D_+ and D_- are the diffusion coefficients of cations and anions, respectively, in the membrane core. In the stationary state, div $J_\pm = dJ_\pm/dx = 0$ and the net electric current must be zero in the absence of external field so that

$$J_+ = J_- = \text{constant (independent of } x) \tag{28.11}$$

The boundary conditions for $\psi(x)$ are

$$\psi(x) \to 0 \quad \text{as } x \to +\infty \tag{28.12}$$

$$\psi(x) \to E_{\mathrm{m}} \quad \text{as } x \to -\infty \tag{28.13}$$

The other boundary conditions for $\psi(x)$ are continuity conditions of $\psi(x)$ and the electric displacement at $x = -d_s$, 0, d_c, and $d_c + d_s$. The solution of coupled equations, Eqs. (28.1)–(28.5), (28.9) and (28.11) determines the whole potential profile across a membrane.

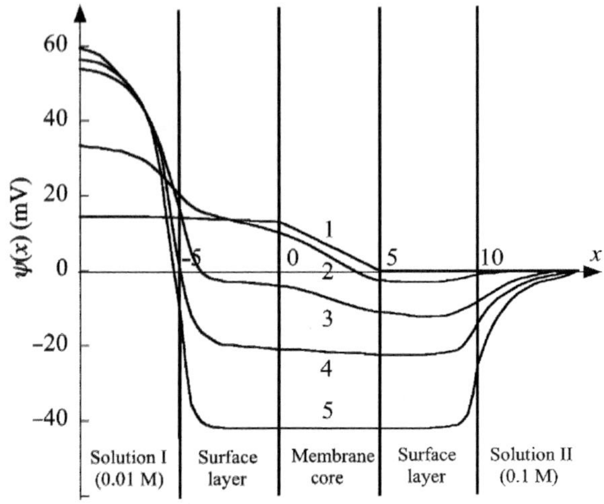

FIGURE 28.2 Potential distribution $\psi(x)$ across a membrane in the stationary state for several values of the density of the membrane-fixed charges N. Curves: (1) $N = 0.005$ M, (2) $N = 0.02$ M, (3) $N = 0.1$ M, (4) $N = 0.2$ M, (5) $N = 0.5$ M, calculated when $n_{\rm I} = 0.01$ M, $n_{\rm II} = 0.1$ M, $d_{\rm s} = d_{\rm c} = 50$ A, $b_{\pm} = 0.001$, $\varepsilon_{\rm r} = 78.5$, $\varepsilon'_{\rm r} = 2$, and T=298K. From Ref. [7].

An example of the numerical calculation for several values of N (expressed in M) is given in Figs 28.2 and 28.3. We have used the following typical numerical values: $n_{\rm I} = 0.01$ M, $n_{\rm II} = 0.1$ M, $\varepsilon_{\rm r} = 78.5$, $\varepsilon'_r = 2$, $d_{\rm c} = d_{\rm s} = 5$ nm, $D_+/D_- = 2$, $b_{\pm} = 10^{-3}$, and $T = 298$K.

As Fig. 28.2 shows, the potential diffuses at the boundaries $x = d_{\rm s}$, $x = 0$, $x = d_{\rm c}$, and $x = d_{\rm c} + d_{\rm s}$. Note that the potential at the boundary between the surface charge layer and the solution (i.e., $\psi(-d_{\rm s})$ and $\psi(d_{\rm c} + d_{\rm s})$) is different from that far inside the surface charge layers. It follows from Eqs. (28.2) and (28.3) that when $d_{\rm s} \geq 1/\kappa_{\rm I}$, $1/\kappa_{\rm II}$, where $\kappa_{\rm I} = (2n_{\rm I}e^2/\varepsilon_{\rm r}\varepsilon_{\rm o}kT)^{1/2}$ and $\kappa_{\rm II} = (2n_{\rm II}e^2/\varepsilon_{\rm r}\varepsilon_{\rm o}kT)^{1/2}$ are the Debye–Hückel parameters of solutions I and II, respectively ($1/\kappa_{\rm I} = 30$ nm and $1/\kappa_{\rm II} = 1$ nm for $n_{\rm I} = 0.01$ M and $n_{\rm II} = 0.1$ M, respectively), the potential at points far away from the boundaries in the surface charge layer (where $d^2\psi/dx^2 \approx 0$) relative to the bulk solution phase is effectively equal to the Donnan potential, which is, for the region $d_{\rm s} < x < 0$,

$$\psi^{\rm I}_{\rm DON} = -\frac{kT}{e}\,{\rm arcsinh}\left(\frac{N}{2n_{\rm I}}\right) \tag{28.14}$$

and for the region $-d_{\rm c} < x < d_{\rm c} + d_{\rm s}$,

$$\psi^{\rm II}_{\rm DON} = -\frac{kT}{e}\,{\rm arcsinh}\left(\frac{N}{2n_{\rm II}}\right) \tag{28.15}$$

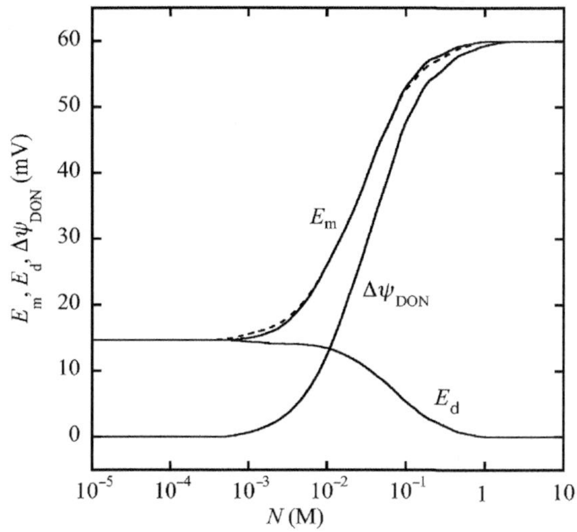

FIGURE 28.3 Membrane potential E_m, contributions from the diffusion potential E_d and the Donnan potential difference $\Delta\psi_{DON} = \psi^{II}_{DON} - \psi^{I}_{DON}$ as functions of the density of membrane-fixed charges N. The values of the parameters used in the calculation are the same as those in Fig. 28.2. The dashed line is the approximate result for E_m (Eq. (28.21)). As $N \to \infty$, E_m tends to the Nernst potential for cations (59 mV in the present case). From Ref. [7].

In this case, by integrating Eqs. (28.1)–(28.4) once and using the appropriate boundary conditions, we have

$$\psi(-d_s) = \psi^{I}_{DON} - \frac{kT}{e}\tanh\left(\frac{e\psi^{I}_{DON}}{2kT}\right) + E_m \tag{28.16}$$

$$\psi(d_c + d_s) = \psi^{II}_{DON} - \frac{kT}{e}\tanh\left(\frac{e\psi^{II}_{DON}}{2kT}\right) \tag{28.17}$$

which relate the membrane surface potentials $\psi(-d_s)$ and $\psi(d_c + d_s)$ to the Donnan potentials ψ^{I}_{DON} and ψ^{I}_{DON}. When the potential far inside the surface charge layer is effectively equal to the Donnan potential, the membrane potential can be expressed as

$$E_m = \{\psi(0) - \psi(d_c)\} + \{\psi^{II}_{DON} - \psi^{I}_{DON}\} + [\{\psi^{I}_{DON} + E_m - \psi(0)\} - \{\psi^{II}_{DON} - \psi(d_c)\}] \tag{28.18}$$

where the first term enclosed by brackets on the right-hand side corresponds to the diffusion potential, the second to the contribution from the Donnan potentials in the surface charge layers, and the third to that from deviation in the boundary potentials at $x = 0$ and $x = d_c$ due to the ionic flows. The contribution from the third term is normally very small so that the diffusion potential E can practically be regarded as arising from the diffusion potential and the Donnan potential. Note that the

Donnan potentials (Eqs. (28.14) and (28.15)) are independent of the surface layer thickness d_s and therefore the membrane potential depends little on d_s, provided that $d_s \geq 1/\kappa_I$, $1/\kappa_{II}$.

Figure 28.2 (in which the potential far inside the surface layer is almost equal to the Donnan potential) shows how the contributions from the diffusion potential and from the Donnan potential change with density N of the membrane-fixed charges. For very low N, the potential in the surface charge layer differs little from that in the bulk solution phase (the Donnan potential itself is very small) and the membrane potential is mostly due to the potential gradient in the membrane core. In other words, the membrane potential is almost equal to the diffusion potential. As N is increased, however, the potential gradient in the membrane core becomes small, which means that the contribution from the diffusion potential decreases. Instead, the potential difference between the surface charge layer and the surrounding solution becomes appreciable and the membrane potential for very large N is determined almost solely by the potential difference in each surface charge layer. As stated before, the potential difference is in practice equal to the Donnan potential in each surface charge layer if $d_s \geq 1/\kappa_I$, $1/\kappa_{II}$ so that the membrane potential E_m for large N is expressed as

$$
\begin{aligned}
E_m &= \psi_{DON}^{II} - \psi_{DON}^{I} \\
&= \frac{kT}{e} \ln \left[\frac{(N/2n_I) + \{(N/2n_I)^2 + 1\}^{1/2}}{(N/2n_{II}) + \{(N/2n_{II})^2 + 1\}^{1/2}} \right] + \psi_s^{II} - \psi_s^{I}
\end{aligned}
\tag{28.19}
$$

To illustrate more clearly the above-mentioned N dependence, we have displayed in Fig. 28.3 the contributions from the diffusion and Donnan potentials to the membrane potential as a function of N.

The decrease of the contribution from the diffusion potential with increasing N can be explained as follows. For large N, that is, for the case where the surface layers are highly negatively charged, coions (i.e., anions) are repelled by the membrane so that the membrane is in practice impermeable to coions and their flow within the membrane, J_-, decreases. Since the net electric current must be zero ($J_+ - J_- = 0$), the flow of counterions (i.e., cations), J_+, must decrease to the same extent as J_-. That is, all ionic flows decrease, resulting in a smaller diffusion potential. In the limit of $N \to \infty$, the membrane becomes fully impermeable to anions and an equilibrium is reached with respect to cations so that the membrane potential tends to the Nernst potential for cations, namely,

$$
E_m \to \frac{kT}{e} \ln \left(\frac{n_{II}}{n_I} \right) \tag{28.20}
$$

which is obtained by taking the limit $N \to \infty$ in Eq. (28.19). In this limit, the membrane behaves like a semipermeable membrane (permeable to cations only). Note that Eq. (28.20) also holds when $b_- \to 0$, irrespective of the value of N.

As shown in Fig. 28.2, the potential gradient in the membrane core is slightly concave (see curve 3, in particular). However, deviation from linearity is seen to be small in Fig. 28.2. This means that the usual assumption of constant field within the membrane is not a bad approximation. If we employ this approximation, Eq. (28.9) can easily be integrated in the following two limiting cases. When $d_s \geq 1/\kappa_I$, $1/\kappa_{II}$, in which case the potential far inside the membrane surface layer is in practice the Donnan potential, the constant field assumption yields [3]

$$E_m = \frac{kT}{e} \ln \left[\frac{P_+ n_{II} \exp(-e\psi_{DON}^{II}/kT) + P_- n_I \exp(e\psi_{DON}^{I}/kT)}{P_+ n_I \exp(-e\psi_{DON}^{I}/kT) + P_- n_{II} \exp(e\psi_{DON}^{II}/kT)} \right] + \psi_{DON}^{II} - \psi_{DON}^{I} \tag{28.21}$$

with

$$P_\pm = b_\pm D_\pm / d_c \tag{28.22}$$

where P_+ and P_- are the permeabilities of cations and anions, respectively. The approximate result calculated from Eq. (28.21) is plotted in Fig. 28.3 in comparison with the exact numerical results. A good agreement is obtained: the relative error is less than a few percent. On the other hand, in the limit $d_s \to 0$ with Nd_s kept constant, we can then obtain the following formula [5,6]:

$$E_m = \frac{kT}{e} \ln \left[\frac{P_+ n_{II} \exp(-e\psi_s^{II}/kT) + P_- n_I \exp(e\psi_s^{I}/kT)}{P_+ n_I \exp(-e\psi_s^{I}/kT) + P_- n_{II} \exp(e\psi_s^{II}/kT)} \right] + \psi_s^{II} - \psi_s^{I} \tag{28.23}$$

where

$$\psi_s^{I} = \frac{2kT}{e} \operatorname{arcsinh} \left[\frac{\sigma}{(8n_I \varepsilon_r \varepsilon_o kT)^{1/2}} \right] \tag{28.24}$$

$$\psi_s^{II} = \frac{2kT}{e} \operatorname{arcsinh} \left[\frac{\sigma}{(8n_{II} \varepsilon_r \varepsilon_o kT)^{1/2}} \right] \tag{28.25}$$

Equations (28.24) and (28.25) are the Gouy–Chapman double-layer potential of the membrane surface with a surface charge density $\sigma = -eNd_s$.

REFERENCES

1. D. E. Goldman, *J. Gen. Physiol.* 27 (1943) 37.
2. A. L. Hodgkin and B. G. Katz, *J. Physiol.* 108 (1949) 37.

3. M. J. Polissar, in: F. H. Johnson, H. Eyring, and M. J. Polissar (Eds.), *Kinetic Basis of Molecular Biology*, Wiley, New York, 1954, p. 515.
4. S. Ohki, in: J. F. Danielli (Ed.), *Progress in Surface and Membrane Science*, Vol. 10, Academic Press, New York, 1976, p. 117.
5. S. Ohki, *Phys. Lett.* 75A (1979) 149.
6. S. Ohki, *Physiol. Chem. Phys.* 13 (1981) 195.
7. H. Ohshima and T. Kondo, *Biophys. Chem.* 29 (1988) 277.

INDEX

Adsorption
ion 111, 129, 200
constant 115, 202, 422
of cations 421
of potential-determining ion 200

Bessel function 33, 59, 81, 168, 181, 280
modified 100, 181, 198, 372, 448, 476
Bilayer membrane 420
Boltzmann distribution 64, 72, 134, 143, 147, 389, 487
Boltzmann's law 4, 6, 84

Cell model 468, 475, 480, 485, 515, 526
Charge regulation model 201
Colloid vibration current 512
Colloid vibration potential 497
Complex conductivity 511
Concentrated suspension of soft particles
dynamic electrophoretic mobility 508
electrical conductivity 480
electrophoretic mobility 468
sedimentation potential in 485
sedimentation velocity in 485
Conductivity (see Electrical conductivity)
Constant surface charge density model (see also Double layer interaction at constant surface charge density) 201, 205, 223, 241, 252, 273, 352
Constant surface potential model (see also Double layer interaction at constant surface potential) 198, 223, 225
Counterion 3, 11, 28
Co-ion 3, 11, 28
CVP (see Colloid Vibration Potential)
CVI (see Colloid Vibration Current)
Cylindrical polyelectrolyte (see also porous cylinder) 372, 448, 477
Cylindrical soft particle (see Soft cylinder)

Debye-Bueche model 435, 468, 487, 497, 510, 517
Debye length 5, 11, 62, 87, 90, 100, 265, 270, 288, 298, 303, 367, 372, 433, 458
Debye-Hückel approximation 5, 10, 39
Debye-Hückel equation 5
Debye-Hückel linearization 18, 23, 290
Debye-Hückel linearized solution 10, 12, 21, 27
Debye-Hückel parameter 5, 9, 14, 20, 22, 38, 41, 47, 56, 59, 62, 74, 192
effective 67, 76
in a salt-free medium 134, 143, 390
in the surface charge layer 441
local 80
Debye-Hückel theory 78
Derjaguin-Landau-Verwey-Overbeek theory (see DLVO theory)
Derjaguin's approximation 203, 298, 323, 353
curvature correction to 290, 341
for interaction between two spheres 283, 363, 406
for interaction between two crossed cylinders 294, 370, 412
for interaction between two parallel cylinders 292, 369, 411
for interaction between two parallel torus particles 416
validity of 310

Derjaguin's formula (see Derjaguin's approximation)
Diffusion potential 535
Dissociable group 116, 129, 202
Dissociation 199
  constant 202
DLVO theory 186, 226, 420
Donnan potential 83, 87, 90, 92, 95, 101, 128, 357
  linearized 87
Donnan-potential-regulated interaction 298, 368, 375
  between
    porous media 298
    two parallel membranes 311
    two porous cylinders 310
    two porous plates 298
    two porous spheres 306
Donnan potential regulation model 265, 298, 319
Double layer (see Electrical double layer)
Double layer interaction
  at constant surface charge density 208, 214, 234, 244, 252, 285, 294, 329, 334, 350, 364, 371
  at constant surface potential 211, 224, 231, 251, 285, 290, 293, 328, 334, 348
  between
    a plate and a cylinder 348
    a plate and a sphere 342
    soft particles 357
    two crossed scylinders 294
    two cylinders 279
    two dissimilar soft cylinders 369
    two dissimilar soft spheres 363
    two parallel cylinders 292, 348
    two parallel dissimilar plates 241, 270
    two parallel dissimilar soft plates 357
    two parallel membranes 311
    two parallel porous cylinders 310
    two parallel porous cylinders 298
    two parallel similar plates 203, 65
    two spheres 278, 283, 327
    two porous spheres 306
  series expansion representation for 323
Drag coefficient
  of ionic species 435, 537
  of an uncharged soft particle 446, 495
Dynamic electrophoresis (see dynamic electrophoretic mobility)
Dynamic electrophoretic mobility
  of a spherical soft particle 433
  of concentrated soft particles 468
Dynamic mobility (see Dynamic electrophoretic mobility)

Effective surface potential
  of a cylinder 42
  of a plate 38
  of a soft plate 102
  of a soft cylinder 104
  of a soft sphere 103
  of a sphere 41
Electrical diffuse double layer (see double layer)
Electrical double layer
  thickness of 4
  free energy of 118
Electrical conductivity
  complex 511
  of a suspension of soft particles 480
  of electrolyte solution 470
  of concentrated suspension 470
Electroacoustic
  measurement 497
  phenomenon 508
Electrokinetic sonic amplitude 497
Electrokinetic equation 436, 480, 485, 499, 516
Electrochemical potential
  of ions 53, 112, 437, 537
Electroosmosis 451
Electroosmotic velocity 451, 457, 475
Electrophoresis (see Electrophoretic mobility)
Electrophoretic mobility
  dynamic 497
  in concentrated suspension 468
  limiting value of 443
  of hard particles 433
  of a soft sphere 440
  of a soft cylinder 447
  of latex particles 456
  of RAW 117-P cells 455

of soft particles 433
of cylindrical polyelectrolyte 448
of spherical particles 434
of spherical polyelectrolytes 441
Electrophoretic softness 438
Electrophoretic velocity 433, 469, 480, 489
Electrostatic interaction (see also Double layer interaction)
between
nonuniformly charged membranes 375
two parallel soft plates after their contact 381
ion-penetrable membranes in a salt-free medium 391
of point charges 165
Electroviscous effect (see Primary Electrovisocus effect)
Equation of motion of a soft particle 501
ESA (see Electrokinetic sonic amplitude)
Exponential integral 434

Fixed charge 84
Free energy
of a charged surface 111
of double layer interaction 186
surface 114, 118

Gravitational force 488
Gravitational field 485

Hamaker constant 403, 417, 423
Henry's function
for a spherical particle 434
approximate expression for 434
for a cylindrical particle 434
approximate expression for 435
Hogg-Healy-Fuestenau formula
for plates 252
for spheres 285
HHF formula (see Hogg-Healy-Fuestenau formula)
Hydrodynamic force 439, 488, 501
Hydrodynamic stress 501

Image 81, 174
method of 325
Image interaction 82
Image potential 325
Interaction
between soft spheres 425
Ion adsorption 111
Langmuir-type 113, 422
Ion vibration current 512
Ion vibration potential 508
Isodynamic curve 252, 311
IVP (see Ion Vibration Potential)
IVI (see Ion Vibration Current)

Kuwabara's cell model (see Cell model)

Lamella phase 420
Linear superposition approximation 265
Linearization approximation 234, 239, 270
Debye-Hückel 231, 290
London-van der Waals constant 400
LSA (see Linear Superposition Approximation)

Maxwell's method 43
Maxwell's relation 483, 492
Maxwell's stress 43, 186, 206, 243, 270, 314, 438
Membrane
Bilayer 420
Nonunformly charged 375
Potential 535
Method of images 325

Navier-Stokes equation 436, 487, 499, 517
one-dimensional 442, 449
Nearly spherical spheroidal particle

Oblate spheroid 43
Onsager relation 452, 492, 494, 511
Osmotic pressure 186, 206, 243, 270, 314, 382

Partial molar volume 512
Penetration depth 498
Planar soft surface (see Soft plate)
Poisson-Boltzmann equation
cylindrical 31
in a salt-free medium 134

Poisson-Boltzmann equation (*Continued*)
in the surface charge layer
of a cylinder 100
of a plate 84
of a sphere 93
modified 63
planar 7
spherical 16, 25
two-dimentional 47
Poisson equation 4, 165
Polyelectrolyte (see Cylindrical polyelectrolyte and Spherical polyelectrolyte)
Polyelectrolye layer (see also Surface charge layer) 83
Polyelectrolye-coated particle (see soft particle)
Polymer segments 435
Porosity 475
Porous cylinder (see also cylindrical polyelectrolyte) 310, 448
Porous sphere (see also spherical polyelectrolyte) 306, 367, 480
Porous plate 298
Porous particle 298
Potential determining ion 201
Potential distribution
across a surface charge layer 83, 87
around
a cylinder 31, 42
a plate 8, 38
a nearly spherical particle 43
a sphere 16, 41
a soft cylinder 100
a soft plate 83
a soft sphere 93
in a salt-free medium 132
Pressure gradient 449
field 508
Primary electroviscous effect
of a suspension of soft particles 516
coefficient 516
Prolate spheroid 43

Quasilinearization approximation 106

Retardation effect 400
Resistance center 435
Relaxation effect 435, 440

Salt-free medium 132
Sedimentation 485
Sedimentation field 485
Sedimentation potential 485
Sedimentation velocity 485
Smoluchowski's formula 433
Soft cylinder
double layer interaction between 369
electrophoretic mobility of 447
potential distribution around 100
surface potential-Donnan potential relationship for 101
Soft particle 83
diffuse 464
electrophoresis of 433
double layer interaction between 357
Soft plate
potential distribution around 87
double layer free enegy of 127
surface potential-Donnan potential relationship for a thick 90
Soft sphere
double layer interaction between two dissimilar 363
electrophoretic mobility of 440
potential distribution around 93
surface potential-Donnan potential relationship for 95
Soft surface 83, 111, 458
free energy of 123
with dissociated groups 128
Softness parameter 452, 457
Spherical polyelectrolyte (see also porous sphere) 161, 357, 358, 367, 444, 472, 480, 493, 503
Spherical soft particle (see Soft sphere)
Stokes drag 446
Stokes formula 499
Streaming potential 452
Surface charge layer 83, 357, 381, 425, 441, 535
Surface charge density 5, 47, 198, 205
Surface free energy 114, 118
Surface layer (see Surface charge layer)

Surface potential (see also Effective surface potential)
  of a hard particle 3
  of a soft particle 83
  unperturbed (see Unperturbed surface potential)
Surface potential-Donnan potential relationship
  of a soft sphere 94
  of a soft cylinder 101
  of a thick soft plate 90
Surface potential-surface charge density relationship
  of a cylinder 33
  of a plate 6
  of a sphere 18

Total vibration current 512
TVI (see Total vibration current)

Unit cell 486
Unperturbed potential 193, 210, 242, 265, 298, 324, 362, 380
Unperturbed surface potential 195, 208, 242, 265, 285, 301, 330, 362, 426

Van der Waals interaction
  between
    a molecule and a cylinder 408
    a molecule and a rod 407
    a molecule and a sphere 404
    two spheres 405
    two crossed cylinders 412
    two molecules 399
    two parallel cylinders 410
    two parallel plates 402
      covered with surface layers 418
    two parallel rings 412
    two parallel rods 408
    two parallel torus-shaped particles 413
    two particles immersed in a medium 417
Viscosity
  effective 515
Vorticity 469, 489

Zeta potential 433, 444, 493, 516

Printed and bound by CPI Group (UK) Ltd, Croydon, CR0 4YY
16/09/2026
14963465-0001